Automatic Control with Interactive Tools

José Luis Guzmán · Ramon Costa-Castelló ·
Manuel Berenguel · Sebastián Dormido

Automatic Control
with Interactive Tools

 Springer

José Luis Guzmán (iD)
Systems Engineering and Automatic
Control Division, Department of Informatics
School of Engineering
CIESOL-ceiA3
University of Almería
Almería, Spain

Manuel Berenguel (iD)
Systems Engineering and Automatic Control
Division, Department of Informatics, School of
Engineering, CIESOL-ceiA3
University of Almería
Almería, Spain

Ramon Costa-Castelló (iD)
Departament d'Enginyeria de Sistemes,
Automàtica, i Informàtica Industrial (ESAII),
Escola Tècnica Superior d'Enginyeria Industrial de
Barcelona (ETSEIB)
Universitat Politècnica de Catalunya (UPC)
Barcelona, Spain

Sebastián Dormido (iD)
Department of Computer Science and Automatic
Control, School of Computer Engineering
National University of Distance Education
Madrid, Spain

ISBN 978-3-031-09922-9   ISBN 978-3-031-09920-5   (eBook)
https://doi.org/10.1007/978-3-031-09920-5

Previously published by Pearson as *Control automático con herramientas interactivas*, 978-8-483-22750-3, 2010
Translation from the Spanish language Edition: *Control automático con herramientas interactivas* by José Luis
Guzmán Sánchez, Ramón Costa Castelló, Manuel Berenguel Soria, Sebastián Dormido Bencomo. Copyright © 2012
Pearson Educación, S.A. All Rights Reserved.
© Springer Nature Switzerland AG 2023

This Springer imprint is published by the registered company Springer Nature Switzerland AG
The registered company address is: Gewerbestrasse 11, 6330 Cham, Switzerland

To Jimena, Julieta, and Aurelia (J. L. Guzmán); to Marta and Belén (R. Costa); to Mari Paz, Laura, and Ana (M. Berenguel); and to María Antonia, Sebastián, Raquel, David, Enrique, and Javier (S. Dormido)

Foreword

The engineers, when designing or projecting a plant, make use of a representation of that system in their own conception process. This fact allows them to have a "dialogue" with what they are imagining. For this reason, geometry in general, and drawing in particular, have been among the classical tools of every engineer, allowing them to have approximate images of what does not yet exist, but which, thanks to their work, will eventually come into existence. Both buildings and public works, as well as machines, have benefited from graphic representations for their design. In this way, images have been a basic tool in the engineer's activity. However, recourse to geometry is restricted to static conceptions, and not to dynamic ones in which time becomes an essential dimension and which, as will be seen below, are those of the control engineer. Students starting out in automatic control must become familiar with the time evolution of systems.

If we look at feedback systems, the problem with feedback is that, despite the simplicity of the principle on which it is based, its implementation can produce instability problems, since it is a circular causal chain, which, together with the unavoidable delays and inertias along the chain, tends to produce pernicious oscillatory effects, in the form of undesirable transients. It is precisely one of the genuine tasks of the control engineer to make these transients fulfill the specifications. However, it is obvious that these processes propagate over time, so that a proper study of feedback systems inevitably requires the availability of tools that allow for a comfortable and efficient handling of this domain. Initially, what could be called "indirect techniques" were used, in the sense that what was studied were mathematical characteristics of the transients, such as the eigenvalues of the linearized form around the operating point, with which the relative stability of the system was determined. This fact began to be systematically investigated in the nineteenth century, using algebraic methods.

The methods of analysis of feedback systems reached their maturity in the twentieth century, especially with the work at Bell Laboratories, which list of employees included the telecommunications engineer Harold Black, who conceived the negative feedback amplifier to compensate for the loss of quality in the long-distance transmission of electrical signals over telephone lines. This electronic device was systematically studied using frequency-domain methods with which telecommunication engineers were familiar. These methods provided a general solution to the problem of the design of feedback systems, making it possible to calculate the compensation elements necessary for the system to meet certain specifications. But these methods were developed in the frequency domain, so specifications in the time domain had to be interpreted in that domain in a way that was not intuitive. However, the flexibility and possibilities offered by the frequency domain meant that they eventually prevailed even in areas where the frequency-domain interpretation was somewhat forced. Moreover, they were limited to the study of linear systems, so their scope of application was restricted. Nevertheless, their success was crucial for the consolidation of feedback systems design methods and for the genesis of control engineering itself.

In any case, the appearance of oscillations, attenuated or not, inherent to the introduction of feedback, makes transients a distorting factor of the expected behavior. These transients evolve in the time domain, so it is to this domain that the analysis and design of control systems must ultimately be resorted to.

The advent of the computer led to the idea that numerical procedures, considered to be more accurate, would replace graphical ones. This even influenced theoretical approaches. From the frequency-domain methods, a change to state variables, imbued with methods based on matrix algebra, was carried out, which implementation implied the use of numerical calculations. It seemed that graphical methods were destined to disappear, replaced by the more precise numerical ones. However, those image-based methods were reluctant to extinction because they possessed, by their very graphic nature, an intuitive character that facilitated the design process.

Fortunately, progress in computer science has allowed the traditional, and sometimes cumbersome, graphical methods to acquire possibilities that engineers could not have dreamed of in the 1930s, when control system design methods were in their infancy, and which they would have welcomed with delight. In this way, computer graphics methods have allowed the classical feedback systems design methods to regain their relevance and strength.

But perhaps the greatest influence of computational methods in the conception and design of control systems has been provided by the possibility of simulating the system's behavior in the time domain, in addition to allowing animations that complement the mere evolution of the variables over time. This opens a splendid path for control systems students. This is the path that the authors of this book are brilliantly following, who also benefit from "friendly" (comfortable and flexible) interactive simulation methods that open up unusual opportunities for the student of feedback systems.

The result of the insights afforded by these new computing methods is evident in the book that the readers hold in their hands, which presents an exuberant exploration of these opportunities. Its contributions are quite remarkable and constitute a valuable tool for those new to control engineering. The world of possibilities that unfold in the pages of this book is certainly fascinating, so it will be very well received by young people entering the study of feedback control, who will find in the pages that follow a list of attractive tools to familiarize themselves with traditional control methods in an easy and elegant way.

Thus, the readers will find in this book a set of computer-based interactive tools that will help them to understand the basic concepts related to the study of feedback systems. This will be done through a series of examples of progressive difficulty which illustrate these concepts through learning based on those examples.

Therefore, the student who is trained in automatic control through feedback systems learns and consolidates the concepts that progressively appear, interacting with the help of powerful computer tools, with which, by simply using the mouse, it is possible to visualize the different representations that are associated with the problem under study. This solves something that was both the dream and the nightmare of those who used frequency-domain methods: the comfortable and efficient transition from the frequency domain to the time domain, and vice versa; and at the same time, the behavior associated with the positions of poles and zeros in the complex plane can be easily verified. There are also animations that give a glimpse of the time response of the systems studied. These interactive tools are accompanied by cards explaining the learning objectives, a summary of the theory associated with these objectives, the main bibliographical references, a detailed description of the interactive tool, and a list of exercises designed to solve them with the help of those interactive tools. In this way, the work is adapted to pedagogical models that encourage autonomous learning by students, with the active assistance of lecturers, for which the potential of e-learning has enormous possibilities.

Seville, Spain
Javier Aracil Santonja
Emeritus Professor of Systems Engineering
and Automatic Control at the University of Seville
Numerary Academician of the Royal Spanish
Academy of Engineering

Preface

The culmination of a new book always provokes in its authors a certain feeling of contained satisfaction when seeing that the work so many times planned finally comes to an end. In a certain sense, it is like the runners who see the finish line after overcoming the many obstacles that have been placed in their way. No matter what has happened, the target is already in their hands. This text was not going to be an exception to this rule, and after the natural ups and downs, it has reached its final port. These lines that appear in the first pages of the book are in fact the last ones that have been written. For this reason, we thought it appropriate to share with our readers, with a certain degree of complicity, the way it was conceived and why we became interested in this subject of the study and learning of automatic control through the use of *interactive tools*. The authors come from the academic world and, as educators, are interested in the challenges posed by control education.

The goal of any engineering course is for students to acquire the knowledge and skills that make them valid for the industry. The labor market looks for engineers with a solid background in the theory behind engineering concepts, but who at the same time have the ability to apply these concepts with ingenuity and dexterity. The second of these skills must be learned through repeated application of the concepts in a wide variety of situations. In doing so, the student acquires an insight into the problem under study. For many years, the basic method of teaching engineering concepts has been classroom interaction and the exercises performed by students. These methods are considered adequate to convey the mathematics behind the scenes, but they often leave students orphaned of the intuition that is necessary to apply the concepts.

Students learn in several ways and in a variety of settings. They learn through lectures, in informal study groups, or alone at their desks or in front of a computer. Wherever the location, students learn most efficiently by solving problems, with frequent feedback from an instructor, following a worked-out problem as a model. Worked-out problems have several positive aspects. They can capture the essence of a key concept (often better than paragraphs of explanation), they provide methods for acquiring new knowledge and for evaluating its use, they provide a taste of real-life issues and demonstrate techniques for solving real problems, and, most importantly, they encourage active participation in learning. The main obstacle in acquiring intuition about a concept is the student's inability to visualize the ideas to be learned. Many concepts have complex visual representations, which cannot be sufficiently explored through normal classroom experience. Moreover, when we look at an element or a certain graphic, we generally not only think about what that element represents, but also about what kind of activities or actions can be carried out using it. That is, we try to look for some interaction with the elements, where not only visual information is required, but also a cause–effect relationship. This feature is identified as *interactivity*. For these reasons, the need arises to build a set of applications that provide students with the opportunity to, visually and interactively, explore the concepts without having to use "pencil and paper". In recent years, there has been a growing demand within the educational community for the development of such modules, which provide the students with more complete control of their own learning process. With such tools, one is not constrained to the examples given in a textbook. Moreover, doing it this way improves their understanding and intuition on the subject. Thus, important for both educators and students is the premise on which these interactive tools are based: students learn best when they are actively involved in their own learning. Many instructors now ask, "Are we simply teaching students the latest technology or are we helping them to reason?" We believe that these two alternatives need not be mutually exclusive. In fact, the interactive tools used in this book were created on the belief that computer solutions and theory can be mutually reinforcing. Properly applied, computing can illuminate theory and help students to think, analyze, and reason in meaningful ways. Rather than attempting to teach students all the latest knowledge, educators are now striving to teach them to reason: to understand the relationships and connections between new information and existing knowledge and to cultivate problem-solving skills, intuition, and critical thinking.

Automatic control is a relatively recent contribution to engineering that emerged as a discipline in the 1950s. A very attractive feature of feedback control is its generality, being part of the curriculum of aerospace, industrial, mechanical, electrical, and chemical engineering students, among others. Moreover, automatic control is spreading its influence in biotechnology, economics, agriculture, ecology, medicine, etc., existing "a clear need to make the basic principles of feedback and control known to a wider community". This feature does, however, also create difficulties in teaching the engineering aspects of the field.

In the Preface to the book "Geometry and the Imagination" [1], Hilbert and Cohn-Vossen wrote: "In mathematics, as in any scientific research, we find two tendencies present. On the one hand, the tendency toward *abstraction* seeks to crystallize the logical relations inherent in the maze of material that is being studied, and to correlate the material in a systematic and orderly manner. On the other hand, the tendency toward *intuitive understanding* fosters a more immediate grasp of the objects one studies, a live rapport with them, so to speak, which stresses the concrete meaning of their relations". This passage effectively captures the spirit and aim of this book. Paraphrasing Hilbert and Cohn-Vossen, it is our goal to explore how, with the help of visual imagination, one can illuminate the many facts and problems that arise in the study of automatic control and in many cases, how it is also possible to represent the underlying geometric structure in most of the concepts that are introduced in a basic automatic control course.

Computer-aided control systems design (CACSD) tools are an important aspect of teaching since they make it possible to significantly increase the personal efficiency in problem-solving. The availability of the classic MATLAB® has had a profound effect on the development of computer-aids for control system analysis and design. With the availability of computer-aids, instructors have now the opportunity to address more realistic problems making teaching more interesting. The use of interactive computer-aids in instruction is particularly relevant because it provides practical insight into control system fundamentals. They make possible an early introduction of concepts such as linearization, effects of time delay, and integrator windup. At the same time, they allow for the demonstration of design iterations interactively. For example, a pole on a lead controller may be dragged on a root locus plot using the mouse and the effect on the transient response displayed instantly. Hence, design iterations using different control design methods (root locus, frequency-domain design using Bode diagrams, pole placement, LQG, H_∞, among others) may be carried out and demonstrated in an instructive way. Since the phenomena dealt with in automatic control are mostly of dynamic nature, it is difficult to describe them by traditional pedagogical means. Consequently, there was, and still is, a need for developing new teaching and learning tools, which can help students understand these phenomena and facilitate their analysis. Teachers in the field of automatic control have been active in pioneering new teaching and learning approaches. The use of interactive graphical-oriented tools may reinforce the active participation of students.

A significant impact on teaching control theory is also provided by the use of computer animation for displaying a physical system's response, pushing visualization to a new level so that intuition can be developed from visual feedback. This is in contrast with parameterized plots used most frequently. For example, different control structures may be related to physical behavior in an effective manner. Another significant benefit is derived from the extensibility of these tools. An instructor can quickly develop new plants, including the dynamics in symbolic form and accompanying animation. These plants can then be used in control design assignments wherein the student demonstrates the tradeoffs involving competing control strategies. In the same way, advanced topics may be demonstrated at a much earlier stage in a student's course with such interactive tools.

It is our firm conviction that the automatic control community should "invest in new approaches to education and outreach for the dissemination of control concepts and tools to nontraditional audiences" [2]. An important element of automatic control education is that it focuses on the continuous use of experiments and the development of new computer-based laboratories and applications. The idea of *interactive tools* has given an effective means for our students to reinforce the learning of new concepts related to automatic control, to understand the relationships between different representations of a dynamical system, and to acquire certain skills and abilities that we consider essential for obtaining a good training in this matter.

This type of technological development should be integrated as part of the educational content. We have addressed these needs by developing a set of cards to facilitate the learning of the relevant concepts of a basic course in automatic control. Each card gives the user complete control over the learning process of a certain concept or control technique, by providing all the necessary elements for its study in a fully interactive way: Theory brief, interactive tool, and exercises. Interactive tools are useful for both teachers and students. For the former because they allow them to present the concepts in a visual and intuitive way and thus supporting the underlying theory. For the latter because they facilitate the understanding and autonomous learning of basic concepts of automatic control and the construction of numerous examples and scenarios to test their skills and abilities. They can also be a useful element in assessing such learning. In a way, it follows the idea of R. Bellman [3] of seeking a good compromise between theory and practice in learning.[1]

The interest in interactive tools came from the "Interactive Tools for Education in Automatic Control" project, which was developed at the Department of Automatic Control at the University of Lund (Sweden) in the mid-nineties of the last century, led by Prof. K. J. Åström. It has served as a reference to many of the ideas and concepts later produced in this field. *ICTools*, as this set of tools was known, were implemented in MATLAB® 5.3 and were freely distributed.

In April 2000, on the occasion of his investiture as Doctor Honoris Causa by the UNED (Spanish National University of Distance Education), Prof. K. J. Åström visited Madrid and we had the opportunity to have an in-depth conversation about the role this type of interactive tools might have in improving the learning of the basic concepts of an automatic control course. He encouraged us to work on the topic and showed us some interactive applications developed in a new language called *Sysquake*, which had been developed by Yves Piguet, a Ph.D. student at the Ecole Polytechnique Fédérale de Lausanne (EPFL). The base language was compatible with MATLAB®, but incorporated a set of primitives and functionalities that facilitated the writing of interactive applications. It was, in his view, the right tool to start working on and developing a new family of interactive modules dedicated to the study of automatic control.

In 2001, we started developing our first interactive applications using Sysquake, and the results encouraged us to continue exploring new forms of interactivity and proposing applications and control techniques in different domains. At the same time, we suggested successive improvements in the Sysquake environment to Yves Piguet, who incorporated them while maintaining full backward compatibility with previous works. This close collaboration

---

[1]"Theory without application is like the smile of the Cheshire cat; application without theory is blind man's buff".

with Yves has been maintained over time and it is, in our opinion, one of the key factors that has allowed us to realize many ideas that would not have been possible without his help.

Along this path, a crucial event took place in order to continue generating new interactive tools that was the incorporation of José Luis Guzmán at the end of 2002 as a Ph.D. student to carry out his Doctoral Thesis on the subject of "Interactive design of control systems". Ramon Costa also began to develop a set of interactive tools that he used in his lectures at the Polytechnic University of Catalonia and which represent the first germ on which this text is based. As part of his training plan in the development of his thesis, José Luis made several stays in Lund, where we started a fruitful collaboration with K. J. Åström and T. Hägglund for the development of a series of modules known as "Interactive Learning Modules" (ILM) adapted to the "Advanced PID Control" book [4] they had published at that time. The ILM project (https://arm.ual.es/ilm/), [5, 6], which is still active, was intended to support the teaching and learning of basic automatic control concepts. To date, the following modules have been implemented:

1. Modeling.
2. Proportional–Integral–Derivative (PID) control fundamentals.
3. PID control design methods (including loop shaping).
4. Interaction of integral actions and saturations (*windup*) in PID controllers.
5. Dead-time compensators.
6. Introduction to the problem of the interaction between control loops.
7. Fundamentals of feedforward controllers design.

In fact, all the works by K. J. Åström, T. Hägglund, and co-workers have been an inspiration for the writing of this book, as the underlying theory can be found in their works.

Other interactive tools that we have developed (in collaboration with other colleagues) during these years and that the interested reader can download are the following:

1. SISO-GPCIT: Interactive tool for the study of *Generalized Predictive Control* in single-input single-output systems (J. L. Guzmán, M. Berenguel, F. Rodríguez, S. Dormido, [7]).
   https://arm.ual.es/siso-gpcit/.
2. MIMO-GPCIT: Interactive tool for the study of *Generalized Predictive Control* in multiple-input multiple-output systems (J. L. Guzmán, M. Berenguel, S. Dormido, [8]).
   https://arm.ual.es/mimo-gpcit/.
3. MRIT: Interactive tool to explain fundamentals of *mobile robotics* (J. L. Guzmán, O. López, M. Berenguel, F. Rodríguez, S. Dormido, [9]).
   https://arm.ual.es/mrit/.
4. *Filtered Smith Predictor* interactive tool: An Unified approach for dead-time process control (J. E. Normey-Rico, J. L. Guzmán, S. Dormido, M. Berenguel, E. F. Camacho, [10]).
   https://arm.ual.es/interactiveFSP/.
5. ITCRI: Interactive tool for *control relevant identification* (J. D. Álvarez, J. L. Guzmán, D. E. Rivera, M. Berenguel, S. Dormido, [11, 12]).
   https://arm.ual.es/ITCRI/.
6. ITSIE: Interactive tool for *system identification education* (J. L. Guzmán, D. E. Rivera, S. Dormido, M. Berenguel, [13, 14]).
   https://arm.ual.es/ITSIE/.
7. ITCLI: Interactive tool for *closed-loop identification* (J. L. Guzmán, D. E. Rivera, M. Berenguel, S. Dormido, [15]).
   https://arm.ual.es/ITCLI/.

8. SISO-QFTIT: Interactive tool for the study of *Quantitative Feedback Theory* (QFT) in single-input single-output systems (J. M. Díaz, S. Dormido, J. Aranda, [16]).
`https://www2.uned.es/itfe/QFTIT/`.

9. i-pIDtune: Interactive tool for *integrated system identification and PID control* (J. L. Guzmán, D. E. Rivera, M. Berenguel, S. Dormido, [17]).
`https://arm.ual.es/i-pidtune/`.

10. ITTSAE: Interactive tool for *time series analysis education* (J. M. Díaz, S. Dormido, D. E. Rivera, [18]).
`https://www2.uned.es/itfe/ITTSAE/ITTSAE.html`.

11. ITTSAE-TSG: Interactive tool for *time series analysis education - Time series generator* (J. M. Díaz, R. Costa-Castelló, S. Dormido, [18]).
`https://www2.uned.es/itfe/ITTSAE/ITTSAE_TSG.html`.

12. ITTSAE-TSA: Interactive tool for *time series analysis education - Time series analyser* (J. M. Díaz, R. Costa-Castelló, S. Dormido, [18]).
`https://www2.uned.es/itfe/ITTSAE/ITTSAE_TSA.html`.

13. LCSD: Interactive tool for *linear control system design* (J. M. Díaz, R. Costa-Castelló, S. Dormido, [19, 20]).
`https://www2.uned.es/itfe/LCSD/LCSD.html`.

14. ITCLSD: Interactive tool for *closed loop shaping linear control design* (J. M. Díaz, R. Costa-Castelló, S. Dormido, [21]).
`https://www2.uned.es/itfe/ITCLSD/ITCLSD.html`.

15. RCLSD: Interactive tool for *robust closed-loop shaping linear control systems design* (J. M. Díaz, R. Costa-Castelló, S. Dormido, [22]).
`https://www2.uned.es/itfe/RCLSD/RCLSD.html`.

16. SMCITOOL: Interactive tool for *teaching, analyzing, designing and interactively simulating sliding mode control* (R. Costa-Castelló, N. Carrero, S. Dormido, E. Fossas, [23]).
`https://sites.google.com/site/ramoncostacastello/smcitool`.

A feature that all these tools often share is that, despite having a user manual and associated publications explaining their characteristics, they contain a lot of information and are difficult to use for inexperienced users without the direct help of a teacher to guide them in the learning process. From these considerations and taking as a starting point the experience gained during many years developing interactive tools, we decided to start the project of writing a new text. The objective we set ourselves was that it could be used as a vehicle through which to explain basic concepts of an introductory course in automatic control and facilitate learning for the newcomer. This generic objective was specified in a series of premises that we considered essential from the outset:

1. *Simplicity.* The tools should be simple and focus on the specific concept to be transmitted, without overloading their content. The virtue of simplicity is therefore a priority objective.
2. *Uniformity.* The tools must share a structure in their visual aspect that identifies them as belonging to the same family. The reusability of components facilitates their use and learning.
3. *Self-contained.* Each card can be accessed as an independent study unit that provides all that is necessary for its learning.

These premises also led us to take the decision to concentrate on single-input single-output (SISO) systems including disturbances acting on the control loop and not to include system analysis and design tools in the state-space, since, except in dimension 2, the sought-after characteristics of understanding concepts through visualization are lost. The interested reader can access a simplified version of a tool dealing with analysis and design in the state-space for two-dimensional systems at http://www2.ual.es/icontrol/supplementary-material/, together with an explanatory text. The same applies to some analysis methods like the Routh–Hurwitz stability test.

The readers are expected to have a basic background in derivatives and integrals and differential equations. Basic knowledge on signals and systems should be desirable.

The focus of the book and all these considerations are aligned with the recent study led by the IEEE Technical Committee on Control Education [24, 25], where some important reflections are made based on a survey made along the world both for academia and industry with 495 participants:

- "It is more important that students understand why feedback is important and understand its impact, rather than become fully mathematically literate with a range of analysis and design tools".
- "The assessment of a first course is that it should not include too much algebra and proofs. Instead, it should focus on understanding concepts, perhaps supported by software for number crunching and experiments".
- "There was also consensus about the importance of first principles modeling, dynamics, and quantification of behaviors. However, only some disciplines were keen on including state-space approaches, with most believing these models could come in a later course. Although not an overwhelming consensus, there was still a majority view that Laplace transform tools were appropriate for a first course".
- "There was a fairly universal desire for some exposure to proportional–integral–derivative tuning to be in a first course".
- "A core skill for graduate engineers is the ability to learn independently and be confident in applying that learning to unseen scenarios (for example, through problem-solving)".
- Staff has to "provide students with guidance and support on how to develop their independent learning and self-assessment skills".
- Regarding analysis and design, there is no consensus in [25] if including frequency response in a first course. We believe it is important that students learn both in time and frequency domains and in fact, one of the advantages of the used tools is that they can easily be used to simultaneously analyze what happens in both domains under different modeling and control scenarios.
- Regarding control design, both academy and industry agree on six topics, most of which are covered in this book (not those related to hardware): Feedback loop concepts, definitions, and hardware components, PID, control loop requirements, control performance, disturbances, and design with frequency response.
- The top five priority areas are as follows: Modeling of simple systems, Laplace and transfer functions, stability, feedback (concept, hardware required for the implementation of feedback loops), and PID.

In developing the theory and examples in this book, we have also taken into account the fundamental books on classical control listed in the survey [26].

The interactive tools and the Spanish preliminary version of the book [27] have been used by our students for 10 years, and much of the modifications made have been the result of feedback received from them. We hope that readers will enjoy the set of interactive tools that have been developed following this philosophy.

Almería, Spain   José Luis Guzmán
Barcelona, Spain   Ramon Costa-Castelló
Almería, Spain   Manuel Berenguel
Madrid, Spain   Sebastián Dormido

Website of the Book

Automatic control with interactive tools includes related information and documentation in the website: http://www2.ual.es/icontrol/.

The interactive tools are optimized for Mac OS and Windows and will be continuously updated to include new features, account for new versions of operating systems, and correct errors. The changes performed on the tools or the associated information will be updated on the website. Supplementary material will also be included, as solved exercises and slides for instructors. Notice that this is a living project and the authors welcome suggestions for improving the tools and supplementary material.

References

1. Hilbert, D., & Cohn-Vossen, S. (1983). *Geometry and the imagination.* Chelsea Publication Company.
2. Murray, R. M., Åström, K. J., Boyd, S. P., Brockett, R. W., & Stein, G. (2003). Future directions in control in an information-rich world. *IEEE Control System Magazine, 9*(2), 20–33.
3. Bellman, R. (1971). *Introduction to the mathematical theory of control processes: Vol. II, nonlinear systems.* Academic Press.
4. Åström, K. J., & Hägglund, T. (2006). *Advanced PID control.* ISA - The Instrumentation, Systems and Automation Society.
5. Guzmán, J. L., Åström, K. J., Dormido, S., Hägglund, T., Berenguel, M., & Piguet, Y. (2008). Interactive learning modules for PID control. *IEEE Control Systems Magazine, 28*(5), 118–134.
6. Guzmán, J. L., Åström, K. J., Dormido, S., Hägglund, T., & Piguet, Y. (2006). Interactive learning modules for PID control. *IFAC Proceedings Volumes, 39*(6), 7–12.
7. Guzmán, J. L., Berenguel, M., & Dormido, S. (2005). Interactive teaching of constrained generalized predictive control. *IEEE Control Systems Magazine, 25*(2), 52–66.
8. Guzmán, J. L., Berenguel, M., & Dormido, S. (2004). MIMO-GPCIT: Interactive generalized predictive control tool for multivariable constrained systems. *Revista Iberoamericana de Automática e Informática Industrial, 1*(1), 57–68.
9. Guzmán, J. L., Rodríguez, F., Berenguel, M., & Dormido, S. (2008). An interactive tool for mobile robot motion planning. *Robotics and Autonomous Systems, 56*(5), 396–409.
10. Normey-Rico, J. E., Guzmán, J. L., Dormido, S., Berenguel, M., & Camacho, E. F. (2009). An unified approach for DTC design using interactive tools. *Control Engineering Practice, 17*, 1234–1244.
11. Álvarez, J. D., Guzmán, J. L., Rivera, D. E., Berenguel, M., & Dormido, S. (2011). ITCRI: An interactive software tool for control-relevant identification education. *IFAC Proceedings Volumes, 44*(1), 6367–6372.
12. Álvarez, J. D., Guzmán, J. L., Rivera, D. E., Dormido, S., & Berenguel, M. (2012). ITCRI: An interactive software tool for evaluating control-relevant identification. *IFAC Proceedings Volumes, 45*(16), 1529–1534.
13. Guzmán, J. L., Rivera, D. E., Dormido, S., & Berenguel, M. (2009). ITSIE: An interactive software tool for system identification education. *IFAC Proceedings Volumes, 42*(10), 752–757.
14. Guzmán, J. L., Rivera, D. E., Dormido, S., & Berenguel, M. (2012). An interactive software tool for system identification. *Advances in Engineering Software, 45*(1), 115–123.
15. Guzmán, J. L., Rivera, D. E., Berenguel, M., & Dormido, S. (2014). ITCLI: An interactive tool for closed-loop identification. *IFAC Proceedings Volumes, 47*(3), 12249–12254.
16. Díaz, J. M., Dormido, S., & Aranda, J. (2007). An interactive software tool for learning robust control design using Quantitative Feedback Theory methodology. *International Journal of Engineering Education, 23*(5), 1011–1023.
17. Guzmán, J. L., Rivera, D. E., Berenguel, M., & Dormido, S. (2012). I-PIDtune: An interactive tool for integrated system identification and PID control. *IFAC Proceedings Volumes, 45*(3), 146–151.
18. Díaz, J. M., Dormido, S., & Rivera, D. E. (2016). ITTSAE: A set of interactive software tools for time series analysis education. *IEEE Control Systems, 36*(3), 112–210.
19. Díaz, J. M., Costa-Castelló, R., Muñoz, R., & Dormido, S. (2017). An interactive and comprehensive software tool to promote active learning in the loop shaping control system design. *IEEE Access, 5*, 10533–10546.
20. Vargas, H., Marín, L., de la Torre, L., Heradio, R., Díaz, J. M., & Dormido, S. (2020). Evidence-based control engineering education: Evaluating the LCSD simulation tool. *IEEE Access, 8*, 170183–170194.
21. Díaz, J. M., Costa-Castelló, R., & Dormido, S. (2019). Closed loop shaping linear control system design: An interactive teaching/learning approach. *IEEE Control Systems, 39*(5), 58–74.
22. Díaz, J. M., Costa-Castelló, R., & Dormido, S. (2021). An interactive software tool to learn/teach robust closed-loop shaping control systems design. *IEEE Access, 9*, 125805–125821.

23. Costa-Castelló, R., Carrero, N., Dormido, S., & Fossas, E. (2018). Teaching, analyzing, designing and interactively simulating sliding mode control. *IEEE Access, 6*, 16783–16794.
24. Rossiter, J. A., Hedegren, J., & Serbezov, A. (2021). Technical committee on control education: A first course in systems and control engineering [technical activities]. *IEEE Control Systems Magazine, 41*(1), 20–23.
25. Rossiter, J. A., Serbezov, A., Visioli, A., Zakova, K., & Huba, M. (2020). A survey of international views on a first course in systems and control for engineering undergraduates. *IFAC Journal of Systems and Control, 13*, 100092.
26. Davison, D. E., Chen, J., Ploen, O. R., & Bernstein, D. S. (2007). What is your favorite book on classical control? responses to an informal survey. *IEEE Control Systems Magazine, 27*(3), 89–99.
27. Guzmán, J. L., Costa-Castelló, R., Berenguel, M., & Dormido, S. (2012). *Control automático con herra – mientas interactivas (Automatic control with interactive tools)*. Pearson.

Acknowledgements

First of all, we would like to express our deepest gratitude to Professor Javier Aracil, who kindly wrote the book's foreword. A special thanks go to our colleague and friend José Manuel Díaz, who has helped us a lot in the improvement of previous versions of the tools, in aspects such as rescaling, zoom in the graphics, and the development of the loop shaping interactive tool. Thank you very much José Manuel for your great work and support.

We would like also to express our gratitude to our families for the patience they have shown during the development of this book, as well as to all the people who have helped and inspired us in tackling this work and sent us suggestions for improvement: Yves Piguet, Denis Gillet, Karl J. Åström, Tore Hägglund, Daniel Rivera, David Muñoz de la Peña, Fernando Morilla, José Sánchez, Manuel R. Arahal, Manuel G. Ortega, Marga Marcos, Pedro García, Pedro Albertos, Sebastián Dormido Canto, Joaquín Aranda, Francisco Gordillo, Alfonso Baños, Antonio Barreiro, Julio E. Normey, Antonio Visioli, José D. Álvarez, Francisco Rodríguez, José C. Moreno, Agustín Pérez, Ignacio Fernández, Carlos Rodríguez, Eduardo F. Camacho, Luis Basañez, and many others with whom we have collaborated in the development of interactive tools. We also gratefully acknowledge all our colleagues from the Automatic Control, Robotics, and Mechatronics research group for being beta-testers of the interactive tools and the many students who have provided careful reviews and suggestions for improvement, specially Malena Caparroz for helping with the solved exercises.

Finally, we would like to thank the Spanish Ministry of Science and Innovation, the Spanish Automatic Control Committee (Control Education Group), the Degrees Commissioner of the University of Almería, and the Social Councils of the UNED and the University of Almería for the funding and the awards granted to various initiatives related to the development of virtual and remote laboratories for teaching and learning automatic control concepts.

Almería, Spain — José Luis Guzmán
joseluis.guzman@ual.es

Barcelona, Spain — Ramon Costa-Castelló
ramon.costa@upc.edu

Almería, Spain — Manuel Berenguel
beren@ual.es

Madrid, Spain — Sebastián Dormido
sdormido@dia.uned.es

Contents

Terminology and Abbreviations

Terminology

Automatic control is a branch of engineering (also called *control engineering*) aimed at behavioral self-government of *systems* or *processes*, these being understood as a set of parts or elements organized and related which interact to achieve an objective. It is therefore a discipline that applies the *automatic control theory* to design systems with predictable behavior. It is said that a system is automated when working alone, without human intervention. In practice, *sensors*[2] are used to measure the evolution of the output of the process to be *controlled* and that measurement is used within a feedback loop, so that actions can be performed on the process through the *actuators*,[3] in order to achieve the desired behavior.

Feedback, which is the process by which a certain proportion of the output signal[4] of a system is redirected to the input, is the key concept in automatic control.

As pointed out by K. J. Åström and R. M. Murray [1], "Feedback has many interesting and useful properties. It makes it possible to design precise systems from imprecise components and to make relevant quantities in a system change in a prescribed fashion. An unstable system can be stabilized using feedback, and the effects of external disturbances can be reduced. Feedback also offers new degrees of freedom to a designer by exploiting sensing, actuation, and computation".

An *open-loop system* uses an actuator to control the process directly without using feedback, and thus the output has no effect on the input signal. A *closed-loop system* uses a measurement of the output signal and a comparison with the desired output (*reference, setpoint*, or *command signal*) to generate an error signal which is used to compute the control signal to be sent to the actuator [2]. The main related definitions are as follows:

- *Output* or controlled variable/s (signal/s): Information of the process that is to be kept in a certain range or prescribed value or which behavior is to be modified (examples: temperature of a room, speed of a vehicle, etc.).
- *Manipulated*, control variable/s (signal/s) or actuating signal: Signals that allow acting on the system to modify its behavior (examples: state of a valve, ventilation opening, etc.).
- *Reference* (*setpoint, command signal*): The desired value (or profile) controlled variables must track (examples: desired temperature, velocity, etc.).
- *Error*: Difference between the reference and the controlled variable. Based on the error produced, the control system will generate the appropriate control signal.
- *Disturbances*: External or exogenous variables that affect (generally negatively) the behavior of the process to be controlled (example: external temperature, slope of a road, etc.). When disturbances are measurable, they can be used within *feedforward* control

---

[2] A device capable of measuring physical or chemical magnitudes and transforms them into electrical ones.

[3] Device that provides the motive power to the process.

[4] A signal, in a broad sense, means any physical quantity that evolves over time. It is also required to have a certain informational content. The types of signals generally used in control systems are electrical voltages or currents, mechanical displacements, and pneumatic or hydraulic pressures.

schemes, not treated in this book (although adequate references to learn the basic concepts will be introduced).

Following [3, 4], a feedback control system is thus "a control system that operates to achieve prescribed relationships between selected system variables by comparing functions of these variables and using the comparison to effect control". The following definitions are also used:

- *Servomechanism* (servo): A mechanical system that ensures zero steady-state error for a constant reference (although the concept has been extended to the tracking of any reference). The feedback synthesis problem is usually posed as the design of a dynamic feedback controller such that the output of the resulting closed-loop system tracks (i.e. converges to), some a priori given reference signal. This problem is known as the *servo* problem [5].
- *Regulator*: Systems with constant steady-state output for a constant reference. For a certain control system that is subjected to external disturbances, the problem is here to design a dynamic feedback controller such that the output of the closed-loop system converges to the constant reference as time tends to infinity, regardless of the disturbance and the initial state [5].

In automatic control, a graphical representation (called *block diagram*) such as the one shown in the following section (basic block diagram and associated signals) is used. It is the most widespread graphical representation language used to represent dynamical systems in engineering. The purpose of a block diagram is to emphasize the flow of information and to abstract from the details of the system [1]. The authors suggest the reader to take a look at the examples shown in [6]. In a block diagram, the different elements of the process are described using boxes or blocks, each of which has inputs represented by an incoming arrow and outputs symbolized by an arrow exiting the block. The inputs denote the variables that influence the process, the outputs denote the signals of interest or that influence other subsystems [1], and the signals may be added or subtracted. The representation using block diagrams allows us not only to derive models from the constitutive and structural relationships, but also to use certain analysis techniques directly on the diagrams, known as *block algebra* (Chap. 7), as a consequence of the application of the properties of the *Laplace transform*. In this text, the blocks represent a *linear time-invariant* (LTI) *dynamical system* (Chap. 2) and may therefore contain different descriptions of it, such as an equation or a system of *linear constant-coefficient differential equations* (Chap. 2), a *transfer function* (Chap. 3), or a *state-space description* (not treated in this book, but included in the supplementary material, as pointed out in the Preface). Notice that, as done in many books, the term *system* is used to refer both to the process under study itself and to its representative model.

Abbreviations

2-DoF	Two-Degree-of-Freedom
AS	Asymptotically Stable
BIBO	Bounded-Input Bounded-Output
BW	BandWidth
CACSD	Computer-Aided Control Systems Design
DC	Direct Current
EMF	ElectroMotive Force
EPFL	Ecole Polytechnique Fédérale de Lausanne
FOTD	First-Order system with Time Delay
GA	Gain Amplification
GBW	Gain–BandWidth product
GM	Gain Margin
GR	Gain Reduction
IAE	Integral Absolute Error
ILM	Interactive Learning Modules
ISA	Instrument Society of America
ISE	Integral Square Error
ITAE	Integral of Time multiplied by Absolute Error
LCSD	Linear Control System Design
LHP	Left Half s-Plane
LTI	Linear Time Invariant
MATLAB®	MATrix LABoratory[5]
MIMO	Multiple-Input Multiple-Output
MP	Minimum Phase
NMP	NonMinimum Phase
ODE	Ordinary Differential Equation
OS	Maximum OverShoot
PID	Proportional–Integral–Derivative
PM	Phase Margin
QFT	Quantitative Feedback Theory
RHP	Right Half s-Plane
SISO	Single-Input Single-Output
SM	Security Margin
SOTD	Second-Order system with Time Delay
ZIR	Zero-Input Response

Basic Block Diagram and Associated Signals

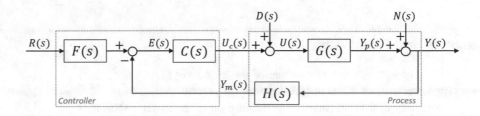

---

[5]MATLAB and Simulink are registered trademarks of The Mathworks, Inc. See https://www.mathworks.com/trademarks for a list of additional trademarks.

Standard block diagram of a two-degree-of-freedom (2-DoF) closed-loop system including representative transfer functions (Chap. 3) of the main components. Throughout the text the case of unit feedback ($H(s) = 1$) will be used. Notice that different textbooks also add disturbance/noise signals in other parts of the loop, and equivalences through block algebra can be obtained.

Basic Signals and Transfer Functions

$d(t), D(s)$	Disturbance at the process input (load disturbance)
$e(t), E(s)$	Reference tracking error signal
$n(t), N(s)$	Measurement noise
$r(t), R(s)$	Reference, setpoint, or command signal to the feedback loop
$u(t), U(s)$	Process input (input signal)
$u_c(t), U_c(s)$	Controller output (control signal)
$y(t), Y(s)$	Measured process output (output signal)
$y_m(t), Y_m(s)$	Measured process output if $H(s) \neq 1$
$y_p(t), Y_p(s)$	Process output
$C(s)$	Transfer function of the controller
$F(s)$	Transfer function of the reference filter
$G(s)$	Transfer function of the process
$H(s)$	Transfer function of the feedback loop (sensor dynamics)

Combined Transfer Functions

$G_{cl}(s)$	Transfer function of the closed loop: $G_{cl}(s) = \frac{F(s)C(s)G(s)}{1+C(s)G(s)H(s)}$ (from reference to measured process output)
$J(s)$	Closed-loop characteristic polynomial: $J(s) = 1 + C(s)G(s)H(s)$, (return difference)
$J(s) = 0$	Closed-loop characteristic equation: $J(s) = 1 + C(s)G(s)H(s) = 0$
$L(s)$	Loop transfer function: $L(s) = C(s)G(s)H(s)$
$P(s)$	Transfer function of a process taken as an example
$Q(s)$	Direct transmission transfer function (forward transfer function), also known as direct term: $Q(s) = C(s)G(s)$, ($Q(s) = L(s)$ when $H(s) = 1$)
$S(s)$	Sensitivity function: $S(s) = \frac{1}{1+C(s)G(s)}$
$T(s)$	Complementary sensitivity function: $T(s) = \frac{C(s)G(s)}{1+C(s)G(s)}$, ($T(s) = G_{cl}(s)$ when $F(s) = 1$, $H(s) = 1$)

Variables and Parameters

a	Area (section) of the outlet orifice of the tank level system (m^2, cm^2)
A	Area (section) of the tank level system (m^2, cm^2)
b	Coefficient of friction (inverted pendulum on a cart system) (N/(m/s))
b_l	Viscous-friction coefficient of the combination of the motor and load referred to as the motor shaft (DC motor system) (Nm/rad/s)
BW	Bandwidth (rad/s)

c	Residue in partial fraction decomposition for inverse Laplace transform
e_a	Armature voltage (DC motor system) (V)
e_b	Back EMF (DC motor system) (V)
e_f	Field voltage (DC motor system) (V)
e_{ss}	Steady-state error (%)
e_{ss_a}	Steady-state error related to acceleration (%)
e_{ss_p}	Steady-state error related to position (%)
e_{ss_v}	Steady-state error related to velocity (%)
g	Acceleration of gravity at sea level (m/s^2, cm/s^2)
GM	Gain margin (–, dB)
h	Fluid tank height (tank level systems) (cm)
i_a	Armature current (DC motor system) (A)
i_f	Field current (DC motor system) (A)
I	Moment of inertia of the pendulum with respect to its center of gravity (inverted pendulum on a cart system) (kg m^2)
I_a	Moment of inertia of the beam (ball & beam system) (kg cm^2)
I_b	Moment of inertia of the ball (ball & beam system) (kg cm^2)
j	Unit imaginary number
J_l	Moment of inertia of the combination of the motor and load referred to as the motor shaft (DC motor system) (kg m^2)
k	Static gain (DC gain, canonical gain, and zero frequency gain)
K	Proportional gain of the controller (in time constants format)
K_a	Acceleration error constant
K_b	Back EMF constant of the DC motor (V/rad/s)
K_c	Proportional gain of the controller (in pole–zero format)
K_d	Derivative gain of the PID controller
K_f	Proportionality constant of the magnetic flux generated by the field (DC motor system) (Wb/A)
K_i	Integral gain of the PID controller
K_m	Torque proportionality constant (DC motor system) (Nm/(Wb A))
K_p	Position error constant
K_t	Motor torque constant (DC motor system) (Nm/A)
K_v	Velocity error constant
l	Pendulum length (inverted pendulum on a cart system) (m)
L_a	Armature inductance (DC motor system) (H)
L_f	Field inductance (DC motor system) (H)
m	Ball mass (ball & beam system) (kg)
	Pendulum mass (inverted pendulum on a cart system) (kg)
M	Cart mass (inverted pendulum on a cart system) (kg)
M_r	Peak magnitude or resonant peak value (–, dB)
M_s	Maximum sensitivity (–, dB)
$\aleph$	System type (number of poles at the origin of the s-plane)
N	Number of clockwise encirclements of the origin of plane $J(s)$ or around the critical point $(-1, 0)$ in the $L(s)$-plane
n_d	Number of experimental data used for model fitting
N_d	Parameter of the filter of the derivative action in PID control
OS	Maximum or peak overshoot, percent overshoot (–, %)
P	Number of poles of the characteristic polynomial
p	Pole of a system
p_c	Pole of a phase-lead or phase-lag controller
PM	Phase margin (°, rad)
q	Fluid inflow into the tank (tank level systems) (cm^3/s)

r	Rolling of the ball (ball & beam system) (cm)
	Radius of the spherical tank (variable section tank level system) (cm)
R_0	Reference amplitude
R_a	Armature resistance (DC motor system) (Ω)
R_f	Field resistance (DC motor system) (Ω)
s	Complex variable of the Laplace transform $s = \sigma + j\omega$ (rad/s)
SM	Security margin in the frequency-domain design (°, rad)
s_m	Stability margin
T_d	Derivative time in a PID controller (s)
t_d	Time delay (s)
T_i	Integral time in a PID controller (s)
T_l	Load torque disturbance (DC motor system) (Nm)
T_m	Motor torque (DC motor system) (Nm)
t_p	Peak time (s)
t_r	Rise time (s)
t_s	Settling time (s)
U_0	Control input amplitude
u_d	Derivative action
u_p	Unit parabola function
u_r	Unit ramp function
u_s	Unit step function
u_{sin}	Sinusoidal input
$\bar{u}$	Input defining the operating point
u_i	Integral action
u_P	Feedforward term in a P controller or proportional action
v	Translational velocity of the ball (ball & beam system) (cm/s)
V	Volume (variable section tank level system) (cm^3)
x	State variable
	Translational position (ball & beam system) (cm)
	Cart position coordinate (inverted pendulum on a cart system) (m)
$\bar{y}$	Output defining the operating point
z_c	Zero of a phase-lead or phase-lag controller
Z	Number of zeros of the closed-loop characteristic polynomial

Greek Letters

α	Attenuation factor in phase-lead controllers
α_i	Coefficients of the closed-loop characteristic polynomial
β	Time constant associated with a zero (s)
γ	Angles of the asymptotes of root locus (°, rad)
Γ	Contour in the complex plane
δ	Unit impulse function
ε	Very small number ($\varepsilon \to 0$)
ζ	Relative damping factor
η	Centroid of the root locus
θ	Vector of model parameters
	Angle of the beam shaft (ball & beam system) (°, rad)
	Angle the inverted pendulum on a cart makes to the vertical (°, rad)
	Angular displacement of the motor shaft (DC motor system) (°, rad)

κ	Gain term (transfer constant, gain–bandwidth product) of the transfer function in the pole–zero format
λ	Lambda-tuning method closed-loop time constant (τ_{cl}) (s)
v	Normalized frequency ω/ω_n
ρ	Density of fluid (tank level systems) (kg/m^3, kg/cm^3)
	Magnitude of $G(j\omega)$ ($|G(j\omega)|$) (–, dB)
$\rho_{\omega_{\downarrow\downarrow}}$	Low-frequency magnitude asymptote (–, dB)
$\rho_{\omega_{\uparrow\uparrow}}$	High-frequency magnitude asymptote (–, dB)
σ	Real part of complex poles (rad/s)
	Damping factor (rad/s)
	Variance of the noise signal
τ	Time constant associated with a pole (s)
τ_{cl}	Specification of the closed-loop time constant (s)
ϕ	Phase angle ($\lfloor G(j\omega))$ ($°$, rad)
	Magnetic flux inside the motor (DC motor system) (Wb)
ϕ_{p_i}	Angle formed by the real axis with the vector joining the i-th pole with a complex pole or zero of the root locus ($°$, rad)
ϕ_{z_ℓ}	Angle formed by the real axis with the vector joining the ℓ-th zero with a complex pole or zero of the root locus ($°$, rad)
ϕ_m	Maximum/minimum phase angle in phase-lead/lag control ($°$, rad)
$\phi_{\omega_{\downarrow\downarrow}}$	Low-frequency phase asymptote ($°$, rad)
$\phi_{\omega_{\uparrow\uparrow}}$	High-frequency phase asymptote ($°$, rad)
φ	Angle formed by the roots (poles and zeros) with the x-axis $\varphi = 180° - \phi_{p_i}$ ($°$, rad)
	Rotational angle of the ball with respect to the shaft (ball & beam) ($°$, rad)
ω	Frequency and imaginary part of the complex poles (rad/s)
	Angular velocity of the ball (ball & beam system) (rad/s)
ω_B	Frequency defining the control band in loop shaping (rad/s)
ω_C	Frequency defining the cutoff band in loop shaping (rad/s)
ω_c	Cutoff frequency (rad/s)
ω_{cf}	Corner frequency (rad/s)
ω_{gc}	Gain crossover frequency (rad/s)
ω_d	Damped natural frequency(rad/s)
ω_m	Frequency of maximum (minimum) phase contribution in phase-lead (lag) controllers (rad/s)
ω_{ms}	Maximum sensitivity frequency (rad/s)
ω_n	Natural frequency (undamped natural frequency) (rad/s)
ω_{pc}	Phase crossover frequency (rad/s)
ω_r	Resonant frequency(rad/s)
ω_{sc}	Sensitivity crossover frequency (rad/s)
ς	Relationship between open- and closed-loop time constant

Interactive Symbols

o	Characteristic point of the response
o	System zero
×	System pole
o	Differentiator
×	Integrator
△	Closed-loop poles

◇   Centroid of root locus
□   Breakaway points (saddle points, bifurcation points) of the root locus
□   Poles representing closed-loop specifications
□   Pole of the filter of the reference
□   Points of maximum/minimum magnitude and phase in phase-lead and phase-lag controllers
◇   Determines the frequency in frequency diagrams
◇   Security margin (*SM*) in the design of phase-lead and phase-lag compensators
○···   Interactive circle to modify the time and amplitude of a reference step
○   Interactive circle to modify the time and amplitude of a load disturbance step
○   Interactive circle to change the time and variance of the output noise
Sincro   Activation button
□☒   Checkbox
◉○   Radio button
▭   Slider
$k[-]$ 1.000   Textbox
⚙   Changing the full scale in the x and y-axes in the graphs
🖑   Cursor changes to indicate an interactive element

Functions

$\arccos(\cdot)$ Inverse cosine
$\arcsin(\cdot)$ Inverse sine
$\arctan(\cdot)$ Inverse tangent
$\arg(\cdot)$ Argument
$\arg\min(\cdot)$ Minimum argument
$\cos(\cdot)$ Cosine
$\det(\cdot), |\cdot|$ Determinant
$\exp(\cdot), e^{(\cdot)}$ Exponential
$\mathrm{Im}(\cdot)$ Imaginary part
$\lim(\cdot)$ Limit
$\ln(\cdot)$ Natural logarithm
$\log(\cdot)$ Decimal logarithm
$\mathrm{Re}(\cdot)$ Real part
$\sin(\cdot)$ Sine
$\tan(\cdot)$ Tangent

Other Symbols

$|\cdot|$ Module of a complex number
$\|\cdot\|$ Euclidean norm
$\angle\cdot$ Phase of a complex number
$\{\cdot\}_0$ Initial value
$\{\bar{\cdot}\}, \{\cdot\}_{ss}$ Operating point, steady state
$\{\cdot\}_{cl}$ Closed loop
$\{\tilde{\cdot}\}$ Deviation variable

$\{\hat{\cdot}\}$ Estimated variable
$\{\dot{\cdot}\}$ Time derivative $(d\{\cdot\}/dt)$
Δ Increment

Complex Number Theory in Relation with the Laplace Transform

The following notation regarding complex numbers theory and the Laplace transform will be used throughout the text.

Denoting by $\mathbb{C}$ the body of complex numbers with the addition and product operations, and $\mathbb{R}$ the subset of real numbers, if $\sigma \in \mathbb{R}$ and $\omega \in \mathbb{R}$, a complex variable s has two components: a real component $\sigma = \mathrm{Re}(s)$ and an imaginary component $\omega = \mathrm{Im}(s)$. Any symbol s that represents any element of a set $D \subseteq \mathbb{C}$ is called a complex variable. Graphically the real component of s is plotted on the abscissa axis (σ-axis), while the imaginary component is measured along the ordinate axis ($j\omega$-axis) in the complex plane (s-plane). The next figure illustrates the complex s-plane, in which any arbitrary point $s = s_1$ is defined by coordinates $\sigma = \sigma_1$ and $\omega = \omega_1$, or simply $s_1 = \sigma_1 + j\omega_1$, with $j = \sqrt{-1}$; $s_1^* = \sigma_1 - j\omega_1$ represents the complex conjugate (same real part, opposite sign imaginary parts).

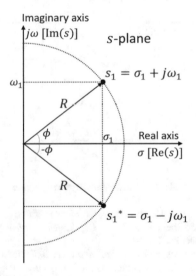

This representation is called binomic form, while the representation $s = (\sigma_1, \omega_1)$ is called the cartesian form. The polar form is given by the radius R and angle ϕ in the figure:

$$\sigma_1 = R\cos\phi, \quad \omega_1 = R\sin\phi, \quad R = |s_1| = \sqrt{\sigma_1^2 + \omega_1^2}, \quad \phi = \lfloor s_1 = \arctan\left(\frac{\omega_1}{\sigma_1}\right), \quad \text{so that}$$

$s_1 = R\cos\phi + jR\sin\phi$. Moreover, as $e^{j\phi} = \cos\phi + j\sin\phi$, s_1 can be represented in polar form as $s_1 = Re^{j\phi} = R\lfloor\phi$. As the conjugate can be represented as $s_1^* = \sigma_1 - j\omega_1 = R\cos\phi - jR\sin\phi = Re^{-j\phi}$, therefore $s_1 s_1^* = R^2 = \sigma_1^2 + \omega_1^2$. Notice that ϕ is the angle formed by the real axis and the vector defined by the origin of coordinates and the affix of s_1.

The complex exponential function is defined as $e^{\sigma + j\omega} = e^{\sigma}(\cos\omega + j\sin\omega)$.

The following table provides the main mathematical operations with complex numbers using the different representations used in the text.

	Binomic form	Polar form
	$\begin{cases} s_1 = \sigma_1 + j\omega_1; s_1^* = \sigma_1 - j\omega_1 \\ s_2 = \sigma_2 + j\omega_2; s_2^* = \sigma_2 - j\omega_2 \end{cases}$	$\begin{cases} s_1 = R_1 e^{j\phi_1}; s_1^* = R_1 e^{-j\phi_1} \\ s_2 = R_2 e^{j\phi_2}; s_2^* = R_2 e^{-j\phi_2} \end{cases}$
Addition	$s = s_1 + s_2 = (\sigma_1 + \sigma_2) + j(\omega_1 + \omega_2)$	$s = s_1 + s_2 = R_1 e^{j\phi_1} + R_2 e^{j\phi_2}$
Subtraction	$s = s_1 - s_2 = (\sigma_1 - \sigma_2) + j(\omega_1 - \omega_2)$	$s = s_1 - s_2 = R_1 e^{j\phi_1} - R_2 e^{j\phi_2}$
Multiplication	$s = s_1 s_2 = (\sigma_1 \sigma_2 - \omega_1 \omega_2) + j(\sigma_1 \omega_2 + \sigma_2 \omega_1)$	$s = s_1 s_2 = R_1 R_2 e^{j(\phi_1 + \phi_2)}$
Division	$s = \frac{s_1}{s_2} = \frac{s_1}{s_2} \frac{s_2^*}{s_2^*} = \frac{(\sigma_1 \sigma_2 + \omega_1 \omega_2) + j(\sigma_2 \omega_1 - \sigma_1 \omega_2)}{\sigma_2^2 + \omega_2^2}$	$s = \frac{s_1}{s_2} = \left(\frac{R_1}{R_2}\right) e^{j(\phi_1 - \phi_2)}$

As shown in the next figure, $G(s)$ is a complex variable function if for each complex value of s there is another complex value $G(s) = \text{Re}[G(s)] + j\text{Im}[G(s)]$. A function $G(s)$ of complex variable s is called an analytic function in a region of the s-plane if both the function and its derivatives exist in that region.

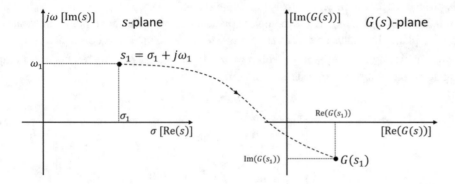

References

1. Astrom, K. J. & Murray, R. M. (2014). *Feedback systems: An introduction for scientists and engineers* (2nd ed.). Princeton University Press.
2. Dorf, R. C. & Bishop, R. H. (2011). *Modern control systems* (12th ed.). Prentice Hall.
3. D'Azzo, J. J., Houpis, C. H., & Sheldon, S. N. (2003). *Linear control system analysis and design with MATLAB®* (5th ed.). Marcel Dekker Inc.
4. IEEE (1984). *IEEE standard dictionary of electrical and electronics terms* (3rd ed.). Wiley-Interscience.
5. Trentelman, H. L., Stoorvogel, A. A., & Hautus, M. (2001). *Control theory for linear systems*. Springer.
6. Albertos, P., & Mareels, I. (2010). *Feedback and control for everyone*. Springer Verlag.

List of Figures

List of Tables

Part I
About Interactivity and Main Objectives of the Book

Many engineering curricula have only an introductory course in automatic control, which usually takes one semester. It contains a large number of new concepts to be assimilated by students, who generally find it very difficult to a thorough understanding of the subject. This also poses a challenge for teachers.

One objective of this text is to promote the continued study and to encourage individual student work through a set of interactive materials, which are useful for both teachers and students. Rather than the concepts to be introduced, the book has focused on how to present each of them. Once the ideas that are relevant have been identified, consideration has been given to how to visualize and formulate them through example-based learning.

The methodology of study proposed in this book also allows students to solve exercises and provide the solution in the form of short videos, so that the teacher can check whether the students are simply "playing" with the tool looking for a solution, or whether they are understanding the concept through the use of the tool, as the main aim is to promote "what if" analysis. The authors agree with the recent survey [1] in the fact that "while heuristics are useful and an easy way in, we must encourage students to learn the value of systematic and rigorous analysis which is more generalizable".

The role of this new interactive learning experience in automatic control curricula is two fold [2]:

- To provide a new method for generating learning material that allows introducing the concepts of control systems engineering through interactive applications.
- To provide a new opportunity to innovate in laboratory work where students can analyze, design and modify control engineering systems through interactive tools.

In Chapter 1 a conceptual diagram of the book is introduced (Figure 1.1), helping as guide on how-to-use the book by the readers.

References

1. Rossiter, J. A., Serbezov, A., Visioli, A., Zakova, K., & Huba, M. (2020). A survey of international views on a first course in systems and control for engineering undergraduates. *IFAC Journal of Systems and Control, 13*, 100092.
2. Dormido, S. (2004). Control learning: Present and future. *Annual Reviews in Control, 28*(1), 115–136.

Introduction

1.1 Introduction: Visualization and Interactivity

In recent years, active participation of students in their own learning process has been sought, changing the culture from that of "information receivers" to that of "knowledge seekers", an ideal scenario in which teachers and students share the exciting adventure of discovery. From this point of view, teaching can mean two completely different things. Firstly, it can simply mean presenting information, so that when a class is taught on a certain topic, it can be said that it has been taught, regardless of whether anyone has learned. The second meaning of teaching is to help someone to learn. According to this meaning, which the authors of this text personally accept, when a concept is taught and the students have not learned, we can conclude that the students have not been taught. The traditional approach of "teaching a course" implicitly uses the first meaning. A syllabus is prepared, making explicit the topics that are planned and are to be covered; the topics are presented in class, and for doing this task, teachers earn their salaries. It does not matter how many students learn; if the syllabus has been explained, the work has been done. The alternative approach is sometimes called *outcomes-based education*. Instead of defining a course by simply writing a syllabus, it attempts to specify in as much detail as possible the knowledge, skills, attitudes, and competencies that students should acquire by the end of the course.

In the *teaching–learning* process, teachers should present and explain the specified knowledge, provide practical exercises, give timely feedback throughout the course, and offer aids and models for the attitudes that are considered important for students to adopt. Students must use all means at their disposal to meet the learning goals and acquire the competencies and skills related to the subject under study. Perhaps, the most direct way to analyze whether the learning process has been successful is to observe how students do something that demonstrates their knowledge or understanding (give a few "reference points" or "benchmarks" against which they can verify their understanding). The more explicit these objectives are, especially those requiring high levels of critical and creative thinking, the more likely it is that students will achieve them.

Learning styles are the different ways that students characteristically adopt to process new information. Students operate with a panoply of strategies in learning situations. Some prefer to deal with specific information (facts, observations, or experimental data), and others are more comfortable with abstract concepts and mathematical models. Some retain more visual information (figures, diagrams, or graphs) than strictly verbal information (spoken or written words), and some progress better in the opposite way. Most science and engineering students learn in an active way and gain their highest degree of assimilation when they do something (solve problems or exchange ideas with others), preferring courses based on exercises, case studies, and laboratories (which are notable exceptions today), to those that follow a master class format. They are uncomfortable if they cannot see connections between what they are learning and the "real world". They can be considered "visual learners".

1.2 What Is Meant by Visualization?

The term *visualization* is employed to describe the process of producing or using geometric or graphical representations of control concepts, principles, or problems, whether represented by hand or generated by computer. Science, engineering, and, in a somewhat more limited sense, mathematics are enjoying a renaissance of interest in visualization. To some extent, this revival is being driven by technological developments. Computer graphics are expanding the scope and potential of visualization in all fields. Visualization transforms the symbolic into the geometric, allowing simulations and calculations to be observed in a much clearer way and without requiring highly specialized computer skills. Visualization offers a method of making visible the invisible.

© Springer Nature Switzerland AG 2023
J. L. Guzmán et al., *Automatic Control with Interactive Tools*,
https://doi.org/10.1007/978-3-031-09920-5_1

One of the reasons why the teaching of automatic control has a certain technical complication is because the concepts that control experts consider "intuitive" are not so for students. The reason is quite simple. Intuition is a global resonance in the brain, which is a function of each individual's cognitive structure, which in turn is dependent on one's prior experience. There is no reason at all to assume that the newcomer to a field will have the same intuitions as the expert, even when considering seemingly simple visual situations. Automatic control has elements that are spatial, elements that are arithmetic or algebraic, elements that are verbal or programmatic, elements that are logical or didactic, and elements that are intuitive or even counter-intuitive.

Our use of the term *visualization* differs somewhat from that used in everyday language and psychology, where the meaning of visualization is closer to its fundamental meaning of forming a *mental image*. Vision is not visualization; seeing is not necessarily understanding. The kind of visualization in which we are interested in automatic control is not an end in itself, but a means to an end which is its learning. Visualization in control is the process of forming images and diagrams (mentally, with pencil and paper or with the aid of technology) and using them effectively for the discovery and understanding of the concept to be conveyed and learned. It provides depth and meaning to understanding, and it serves as a reliable guide to problem-solving and inspires creative discovery, but without isolating itself from other forms of numerical or symbolic representation.

Visualization also has its distinctive disadvantages. The problem is that images can often suggest theorems or conjectures that are false, so the strengths and weaknesses of visual imagery need to be considered and weighed against the notions of intuition and rigor, which should never be opposed but rather complementary. From a formal and theoretical point of view, in control theory, the "proof of a theorem" is the last stage of the abstraction process and synthesizes knowledge in a particular field. However, before a proof exists, there must be an idea of what the theorem is trying to prove or what theorems might be true. This exploratory phase of abstract thinking about the objects over which control operates can benefit from correct visualization. It is not accidental that when we think we understand something, we say "oh, I see!" The following anecdote attributed to Norbert Wiener, the father of cybernetics, illustrates better than anything this fact. Wiener was teaching at Massachusetts Institute of Technology developing a complicated mathematical proof, with the blackboard full of formulas, when he suddenly stopped and stared, for a long time, at the last mathematical expression he had written down. All his students, in stony silence, thought that their teacher was at a dead end. Wiener, however, without saying anything, went to a corner of the blackboard that was still free of his formulae and drew a few figures that his broad shoulders hid from his students. He suddenly changed his face and with a sense of relief, erased those enigmatic figures, returned to the point of the proof where he had stopped and continued flawlessly and smoothly to the end.

A good example of the value of visualization is the famous Cauchy theorem in complex analysis, which is essential in the proof of the Nyquist stability criterion. This theorem states that the integral of an analytical function around a closed curve with no singularities inside is zero. When Cauchy stated the first version of this theorem, he thought of a complex number $z = x + jy$ analytically in terms of its real and imaginary parts. By analogy with the real case, he defined the contour integral $\int_{z_1}^{z_2} f(z)dz$ between two points z_1 and z_2 along a curve whose real and imaginary parts are both either monotonic increasing or decreasing.

As a formal generalization of the real case, this restriction of the type of curve is natural. However, if we open our eyes and look at a picture, we see that such graphs (for increasing x and y) are a restricted set of curves that are contained in a rectangle with opposite vertices in z_1 and z_2.

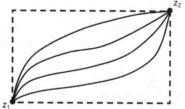

Cauchy had to visualize the situation for a more general curve in the complex plane in order to formulate his theorem in the form which is known today. If great mathematicians need to think "visually", why keep such a thought process away from students?

1.3   The Role of Visualization in Automatic Control Teaching and Learning

Visualization, understood as the way of paying explicit attention to graphic representations that allow explaining abstract contents, is highly developed in the field of computer science, where it is possible to find objects in three dimensions with excellent visual quality.

The fundamentals, ideas, concepts, and procedures of automatic control present a great wealth of visual content that supports an intuitive geometric representation. Clear examples are the time response, the frequency response diagrams (Bode, Nyquist, and Nichols diagrams), the poles and zeros plane (s-plane), the root locus, the phase plane, etc. Their use is very useful in the presentation and handling of the related concepts and procedures, as well as their manipulation in a graphical and interactive way for solving problems. Experts have visual images that act as intuitive mechanisms to understand the fundamentals of their discipline and that at the same time are of great value and effectiveness in the development of their creative work. From these images, they are able to relate and project in a very versatile way a huge amount of facts that allow them to select, in a natural way and without apparent effort, which is the most effective line of attack to solve the complex problems they face. This way of acting, with explicit attention to the possible concrete representations that reveal the abstract relationships of interest to the control engineering specialist, constitutes what we call *visualization in control*. The fact that visualization is a particularly important aspect of the control expert's activity comes naturally when one takes into account the applied mathematics characteristics of control theory. Visualization thus appears as something deeply inherent in the transmission and communication of our knowledge. One of the important tasks that we teachers have is to transmit to our students not only the logical and formal structure of our own discipline, but also, and with much more emphasis, the most intuitive and motivating facets of the subject we are addressing, so facilitating their learning. These are precisely the aspects that are much more difficult to explain and to be assimilated because very often they are in the less conscious substratum of our activity as teachers.

Given the nature of visualization, there will be many highly subjective elements. The ways to visualize and make the ideas of automatic control more approachable and intuitive, in order to implement them in certain situations and apply them to specific problems, largely depend on the mental structure of each person. The degree of visual support certainly varies considerably from one analysis to another, and what may be useful for one person may be a complication for another. However, these differences should not be an impediment to offering those instruments that have been useful in the work on visualization and without which this task would have been much more difficult, obscure, and tedious.

The technical language used by automatic control specialists is a mixture of natural language and language formalized in mathematical terms, a strange jargon consisting of elements of natural language, more or less esoteric words, and mathematical and logical symbols. In this context, explicit or not so explicit references are made to scientific conventions that have been established over time and that are interspersed with visual and intuitive connotations.

1.4   What Is Meant by Interactivity?

There are two essential aspects in the study of dynamical systems that today's computers allow to address and to which much attention has not been paid. One is the visualization referred to in the previous sections and the other is the interactivity of the user with the dynamical system, in its different forms of representation, through the visualization itself. Interactivity is a concept based on the well-known cause–effect relationship. It is about providing the dynamical system with powerful mechanisms so that the user can answer any imaginable question of the types "what would happen if ...?" or "could you do that ...?" simply by acting on the visual elements that constitute the graphical expression that represents it. More often than one might think, ideas in engineering arise from specific and visual situations. The perception of reality is essentially visual and it is not surprising that this support is so deeply rooted in our everyday way of working.

Symbolic treatments, visual diagrams, and other forms of imaginative processes are often used to acquire an intuition of what, in its most formal aspect, has a much more abstract content. It is precisely here where it makes sense to have powerful and user-friendly environments that, using the current state-of-the-art computer technology, allow the creation of interactive dynamic simulations with high visual content. The idea of changing properties and immediately being able to see the effects of the changes is very helpful for both analysis and design. The dynamics of the changes provide additional information that is not available in a static graph.

Bricklin and Frankston made a pioneering effort in the use of interactivity in the late 1970s when they developed VisiCalc,[1] which was based on the metaphor of an electronic sheet. VisiCalc contained a grid of rows and columns of data accessible for financial calculations. Its implementation in an Apple II was one of the reasons why personal computers started to be used in companies. VisiCalc transformed the spreadsheet from a calculation application to a modeling and optimization tool. Excel (which is derived from these ideas) is now standard software in all offices and companies. The spreadsheets were easy to implement because they only operated on numbers; however, in automatic control, much richer graphical objects are present. The possibility of having them all interconnected so that they can be directly manipulated and updated interactively and concurrently has great potential to enhance learning. The VisiDyn program was a pioneering implementation in this direction [1]. Another attempt was made by Blomdell [2], who developed a program for an Apple Macintosh computer. His system had some novel features such as a graphical diagram representing the pole–zero structure of a transfer function. It also had an original way of interacting with Bode diagrams. These early tools were very useful, but their implementation required a great deal of effort, which limited their use as educational tools.

Therefore, the goal of interactivity as discussed in this text is to enhance learning, by exploiting the advantages of immediately seeing the effects of changes that can never be shown by images of a static nature, e.g. on a blackboard. The use of interactivity in education provides a wide range of possibilities to both teachers and students. Teachers can make use of interactive presentations where not only the meaning of a certain concept is shown, but also how that concept may be related to others or how it may be influenced by certain external factors (e.g. some input modifications). For the teaching staff, interactivity provides a way to assess the main ideas on a given topic and the level of difficulty involved in assimilating these ideas by the students. On the other hand, students can make use of interactive websites or computer-based interactive tools to study theoretical concepts abstracted through interactive objects.

From a software implementation point of view, interactivity can be defined as a *fourth dimension*. In this way, the saying "a picture is worth a thousand words" could be modified to "interactivity is worth a thousand images" [3].

Interactive tools can be used to investigate precisely what needs to be done to make students aware of theoretical concepts. In this way, the virtue of simplicity becomes a fact in learning research on the design and use of this type of tool. The main reasons why interactive tools are useful for teaching are as follows [4]:

- They facilitate student-focused learning, allowing choice in the pathways for learning and the rate at which new material is introduced.
- They can address several learning styles and modalities, providing a rich variety of instructional approaches which can teach in most of the ways what students learn best.
- They motivate students' interaction, experimentation, and cooperative learning.
- Students often work together on computer projects as they never did on paper-and-pencil projects.
- They facilitate storylines or thematic learning, where a pathway for exploration can easily be woven around a particular concept dynamics.
- They promote the constructivist view of learning.

Interactivity also has drawbacks. Students can try to learn and solve problems simply by manipulating the interactive applications, but without understanding what they are doing. Interactive tools should challenge students by making them relate theory to practice, as interactive objects are abstractions from theoretical concepts. Neither should teachers develop modules of this type without correctly managing the different levels of abstraction (from the idea to be shown using interactivity and taking into account the student's perspective to the creation of the final product). It can be said that interactivity can be very useful in the teaching–learning process, but it must be used with control.

1.5   The Role of Interactivity in Automatic Control Teaching and Learning

The concepts of visualization and interactivity are becoming increasingly relevant in the teaching of automatic control, where the underlying ideas, concepts, and methods are very rich in visual content, which can be represented intuitively and geometrically. Calculations and representations include responses in the time and frequency domains, poles and zeros in the complex plane, Bode/Nyquist/Nichols diagrams, phase plane, etc. Often, such magnitudes are strongly related and represent different views of the same reality. The understanding of these relationships is one of the keys to achieving good learning of the basic concepts and allows the student to be able to accurately design automatic control systems [5].

---

[1] http://bricklin.com/visicalc.htm.

Traditionally, systems design is carried out following an iterative process. Specifications of the problem are not usually used to calculate the value of the system parameters, because there is no explicit formula that relates them directly. This is the reason for dividing each iteration into two phases. The first, known as *synthesis*, consists of calculating the unknown parameters of the system based on a set of design variables, which are related to the specifications. During the second phase, called *analysis*, the performance of the system is evaluated and compared to the specifications. If they do not match, the design variables are modified and a new iteration is performed. It is possible, however, to merge both phases into one, so that when a parameter affecting the design is modified, the analysis results are updated and observed immediately, or vice versa, so that it is possible to visually modify representative elements of the analysis phase and obtain the design parameters needed to achieve these objectives. In this way, the design process becomes truly dynamic and the students perceive the gradient of change in the performance criteria for the elements they are manipulating. This interactive capability makes it much easier to identify compromises that can be achieved.

The computing resources and programming environments available today have enabled the emergence of a new generation of environments for interactive learning of automatic control. These tools are based on objects that support direct graphic manipulation. During these manipulations, objects are updated immediately, so that the relationships between the objects are maintained at all times. Ictools and CCSdemo developed at the Department of Automatic Control at the University of Lund [6, 7] and Sysquake at the Institute of Automatic Control of the Federal Institute of Technology in Lausanne [8] are, among others [9–11], excellent examples of this new educational philosophy of automatic control.

For newcomers in the field of automatic control, many of the concepts are not very intuitive when first approached, since their properties are expressed in two different domains: The time domain and the frequency domain. Transient behavior (e.g. settling time and peak overshoot) and saturation risk are typically analyzed in the time domain; while concepts such as stability, noise rejection, and robustness are more easily expressed in the frequency domain. The basic mechanisms that relate them and other phenomena, such as the inclusion of nonlinearities, can be illustrated very effectively through interactivity, as is the case with the applications included in this text.

1.6   Objectives and Organization of the Contents

As pointed out by [12], the process of control system design often follows a step-by-step procedure:

1. Study the system (process or plant) to be controlled to get initial information about control objectives.
2. Model the system with a degree of detail depending on the final use (simulation, control, or optimization). When necessary, simplify the model.
3. Analyze and determine the properties of the resulting model.
4. Decide which variables are to be controlled (controlled outputs).
5. Decide what sensors are required to measure the controlled outputs and disturbances and what actuators have to be used to implement the manipulated variables (and where sensors and actuators will be placed).
6. Select the control configuration.
7. Take decisions on the controller structure to be used.
8. Set performance specifications based on control objectives.
9. Design a controller using the most adequate method according to the selected model, control objectives, and control architecture.
10. Analyze the obtained closed-loop system to verify that specifications are fulfilled. If they are not satisfied, modify the type of controller or the specifications.
11. Simulate the resulting controlled closed-loop system.
12. Repeat from step 2, if necessary.
13. Choose hardware and software to implement the controller.
14. Test and validate the control system on the real plant and re-tune the controller online if necessary.

From all these steps there are several that are related to implementation issues in the real plant (5 in part, 13 and 14). With the aim of facilitating the understanding of basic concepts related to the steps in the control system design that do not require the real plant, a set of *interactive tools* has been designed as a support to this book, which is made up of the following:

- A graphical application, called *interactive tool*, that visualizes each concept and includes numerous interactive objects in order to enhance interaction with the application.
- The application incorporates predefined examples that illustrate a specific phenomenon or particularity of the concept introduced.
- Each application is accompanied by a theoretical summary that presents the concept formally and analytically. It is not intended to replace a textbook. Therefore, the theoretical contents are reduced to the minimum necessary to understand the specific application. It contains references to specific books for users to broaden their knowledge. The package consisting of the interactive tool and the related documentation is what we have called a *learning card* (other authors refer to learning pills).
- The learning cards have been designed to be self-contained, so the summary includes a small manual of the interactive tool describing all its functionalities. Interactive symbols are explained in the Terminology introduction of the book, while common aspects of all interactive tools are explained at the end of this chapter to avoid the text being somewhat repetitive for the reader. We have tried to find a tradeoff so that the understanding of each interactive tool does not require previous analysis of others.
- The documentation or each learning card contains different exercises to be solved using the interactive tools. The purpose of these examples is to initially guide the student's steps to reduce the time required to learn each of the concepts and to emphasize the important points. The objectives follow a *learning-by-discovering* approach and aim to combine *qualitative knowledge* obtained through interactivity with *quantitative knowledge* obtained through the different examples. Once the exercises have been completed, the users will be able to create their own scenarios and self-assess the knowledge acquired. Each part of the book (open-loop analysis and closed-loop analysis and design) includes solved problems that make use of the different tools analyzed. Moreover, the website of the book will be updated with new exercises and solutions.

As a complement to the previous information, some comments related to the way in which the tools have been developed are briefly included as follows:

- Sysquake [8] has been used as a development tool due to its programming language, very close to the *de facto standard*, MATLAB®. Interactivity is fully integrated into the language. This provides a high-level development framework that is very convenient for non-professional programmers.
- The applications contain a large number of common components, such as the input signal or the pole–zero editors, which have been included in a library used by all applications. This simplifies the development and maintenance of the toolkit. These components are described at the end of this chapter. We encourage users to read it.
- The applications are lightweight and distributed via executables that do not require pre-installation for Mac OS and Windows.

The contents have been designed to cover a basic course on classical control techniques based on the external description (transfer function), with a strong focus on those having graphical visualization features that can be interacted with. In the preface, we have provided the arguments to focus the text on the use of transfer functions which, on the other hand, are the natural description in process control [13]. A brief introduction to state-space analysis and design (constrained to two-dimensional systems) can be found on the website of the book. Figure 1.1 shows a conceptual diagram of the book, showing all the topics covered, which are those most important for any engineer to know about dynamical systems, such as dynamic behavior, physical and empirical modeling, computer simulation, feedback and stability, basic control approaches, and analysis and design tools. The applications cover the following fields:

- **From nonlinear physical models to linear models:** This chapter explains the concept of dynamical systems, the linearization of the mathematical representations of nonlinear systems around an operating point, and the analysis of their behavior. Five processes are described that will be used as examples throughout the text: A tank with water and discharge to the atmosphere (two versions are included: constant area and variable area), a ball and beam system, an inverted pendulum on a cart system (simplified version), and a direct current (DC) motor. As only the tank level systems are open-loop stable, they will have three related interactive tools each one: one devoted to linearizing the nonlinear representative model, another one to relate the model parameters to the representative transfer function, and the last one to feedback control of the fluid level. The other processes have one interactive tool described in Chap. 9 dedicated to closed-loop control, although the summary of theory and exercises will be included in the appropriate chapters.

- **Time response:** This group contains seven interactive tools designed to illustrate the time response (step response) of continuous linear time-invariant (LTI) systems. An interactive tool has also been incorporated to explain the fitting of linear models from experimental data.
- **Frequency response:** This group includes eight interactive tools aimed at illustrating the concept of frequency response and its analysis for the same systems treated in the previous group. The main graphical representations associated with this concept are also introduced, as well as an interactive tool for fitting models from Bode diagrams obtained experimentally.
- **Relationship of model parameters to physical models:** This chapter illustrates the relationship between physical models and the main elements of models used in controller design (transfer functions), such as poles, zeros, and time response. The parameters of the nonlinear physical models discussed in Chap. 2 are related to those of the linearized models.
- **Problems on open-loop analysis:** This chapter includes four solved exercises where all the interactive tools introduced in the previous chapters are used as support to the theory.
- **Closed-loop systems and stability:** The main techniques for analyzing the stability of closed-loop linear systems are introduced: Root locus, Nyquist criterion, and phase and gain margins. An interactive tool is also included to illustrate the limitations imposed by the time delay in a feedback loop.
- **Control system design:** This chapter brings together the analysis of steady-state errors in systems with unit feedback and the main controller design techniques used in introductory courses, both in the time and frequency domains, covered in eleven tools. It covers aspects of the analytical design of simple PID-type controllers for systems without delay (basically pole cancellation and pole placement), heuristic or approximate designs for systems with delay (Ziegler–Nichols and lambda method), the inclusion of filters in the reference to decouple the reference tracking problem and disturbance rejection, the classical frequency design of phase-lead and phase-lag controllers in the Bode diagram, and also the analytical design of these kinds of compensators based on typical frequency response specifications. Finally, two more complex tools aimed at explaining how to design controllers by the *loop shaping* technique are included. Although to a certain extent these tools visually break with the simplicity of the previous ones, being the last controller design tools, they encompass practically all the concepts analyzed in the previous ones. We believe that the reader will find these last tools very interesting for combining all the concepts acquired.
- **Control of physical systems:** The design of controllers and the simulation of the closed-loop system are studied for the processes explained in Chap. 2. Although the interactive tools include selected control techniques, the reader can solve exercises on these processes using the different interactive tools included in the text.
- **Problems on closed-loop analysis and design:** Three solved exercises are included, where control system design is performed using different interactive tools as support.

It should again be noted that the text is not restricted and oriented to sequential reading by chapters, as is usual in a conventional book. The users can locate the concept they want to analyze in the table of contents and go to the self-contained card where the concept is treated. In fact, the book can be used as a dictionary of basic control concepts, under the support of interactive tools. The conceptual map shown in Fig. 1.1 also helps to locate the main concepts (state-space modeling and design are not treated but can be found on the web page of the book).

Chapters 2 and 5 cover the main concepts related to modeling approaches indicated in the conceptual diagram, including linearization and the explanation of the use of steady-state models for evaluating the operating point. Linear models are dealt with in detail in Chap. 3 (time domain) and Chap. 4 (frequency domain). The main tools for representing transfer functions are also treated in these chapters (time response plots, pole–zero maps, and frequency-domain diagrams). Some hints about experimental identification are included in the last interactive tool of both chapters. Solved problems on open-loop analysis are included in Chap. 6. Basic closed-loop stability analysis tools are introduced in Chap. 7, mainly root locus, Nyquist criterion, and stability margins, also considering the negative effects of time delays in the feedback loop. Chapter 8 deals with steady-state performance with unit feedback and closed-loop transient specifications, both in time and frequency domains. The control design is based on one- and two-degree-of-freedom (2-DoF) control schema for PID controllers and lead–lag controllers using both classical methods and loop shaping, as already explained. Chapter 9 relates to closed-loop simulation and analysis using the five representative processes introduced in Chap. 2. Finally, Chap. 10 summarizes all concepts through solved problems on closed-loop analysis and design.

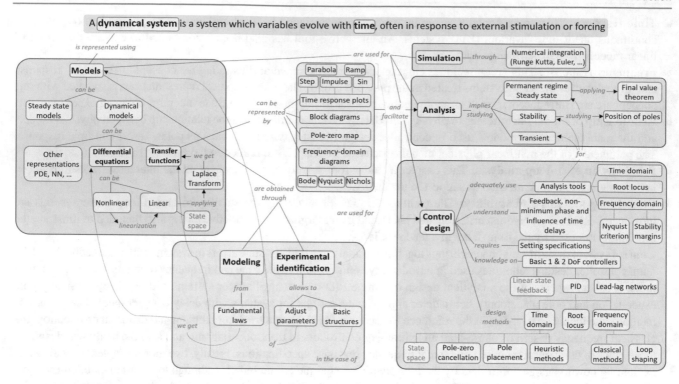

Fig. 1.1 Conceptual map of the book

1.7 Components and Structures of the Interactive Tools

1.7.1 Distribution of Elements and Interactive Objects in the Tools

This section briefly explains the types of interactive objects, graphic and text elements that can be found in the interactive tools, as well as their spatial distribution. Notice that the size of the tools is scalable to a maximum. Interactive objects are characterized by the fact that when the mouse pointer is placed over them, their form of representation changes to a hand ✋, indicating that this element is interactive. For instance, placing the mouse pointer over a certain point of the time or frequency response curves activates a label that shows the properties associated with that point.

In addition to the interactive elements, other non-interactive elements will be used, such as symbolic representations of formulas or transfer functions, legends on graphs, etc., which help to understand the concepts that are intended to be analyzed with the tools. In any case, the values associated with these elements can be modified using the available interactive elements.

1.7.1.1 Distribution of Graphic Windows in the Tools

Most interactive tools used in this text have been designed following certain patterns for the distribution of objects on the screen (patterns which in turn have many similarities between them). Table 1.1 shows the main patterns used for the different interactive tools, which are described in the following:

Pattern 1. Interactive tools related to physical systems: The organization in this pattern by quadrants is as follows:

Interactive parameters and equations	Time response of the system
Process representation	Input signal to the system

Table 1.1 includes as example of this pattern the tool tank_level_lin.

Pattern 2. Interactive tools related to the analysis of the time response, design in the time domain, and root locus: The organization of the interactive tools associated with this pattern follows the same structure (with small differences in the size of the objects), being shown as an example in Table 1.1 those corresponding to the tools t_second_order (Pattern 2.a) and PI_lambda (Pattern 2.b):

Interactive parameters, characteristics, and equations	Time response of the system
Pole–zero diagram	Input signal to the system

Pattern 3. Interactive tools related to the analysis of the frequency response (without including the open-loop time response): The layout of the interactive tools associated with this pattern (the one corresponding to the interactive tool f_second_order is included in Table 1.1) follows the same structure:

Interactive parameters, characteristics, and equations	Bode magnitude plot
Pole–zero diagram	Bode phase plot

This group has a uniform distribution in which the main working graph (Bode plots) occupies the entire right side of the tool, so its distribution is optimal for the user to analyze the associated concepts.

Pattern 4. Interactive tools for analysis of the frequency response (including the open-loop time response), stability analysis, and design in the frequency domain: Table 1.1 presents as examples of this group the tools f_concept and f_design_lead. The layout of all these tools is the same:

Interactive parameters, characteristics, and equations	Selectable frequency diagrams
Pole–zero diagram	Time response (hideable)

The fundamental elements in this group are the representations in the frequency domain, but it may be of interest for the user to also analyze the open-loop (Pattern 4.a) or closed-loop (Pattern 4.b) system response, these representations occupying a different area on the screen (smaller in the case of open-loop ones). By default, when the tool is started, the time response curve will appear in the lower-right quadrant in all displays, but there will be an entry in the Options menu (explained later) that allows to hide them to increase the size of the frequency plot (Bode, Nyquist, Nichols, or all four at once). In pattern 4.b., the option in the Options menu to show the time response enables (on user demand) the evolution of the system output or input.

Pattern 5. Interactive tools devoted to explaining the concepts associated with PID controllers and the influence of time delays: The layout of all the interactive tools associated with this group (Table 1.1 includes as example the tool PID_concept) is the same, the main difference being the substitution of the pole–zero diagram by the Nyquist plot.

Interactive parameters, characteristics, and equations	Time response of the system
Nyquist plot	Input signal to the system

As commented before, the interactive tool loop_shaping used to explain the loop shaping design approach is more complex, as this method is difficult to explain with a basic layout. It will be explained in more detail in the corresponding section.

1.7.1.2 Color Codes

The predefined order of the color code in the library is red, blue, green, magenta, and cyan.

In the time and frequency response analysis interactive tools, by default, the color used for the representation of text and graphics associated with a given system is black. When the representation of two systems is carried out simultaneously, the representative colors are red color and blue color. When multiple systems are displayed, the predefined order is used and black color is used to highlight the active system, i.e. the one that represents the current parameter values. In the case of the interactive tools treating the controllers' design, the color codes used are as follows:

- **Red** to represent the transfer function of the open-loop system, its associated time response (dashed line), and its frequency response (solid line).
- **Blue** to represent the transfer function of the controller, the closed-loop time response (solid line), and the frequency response of the loop transfer function in frequency-domain plots (thin line).

Table 1.1 Typical examples of object layout in interactive tools

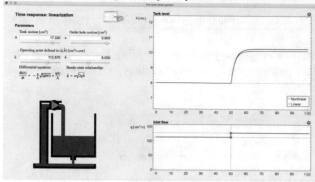

Pattern 1. Cards related to physical systems

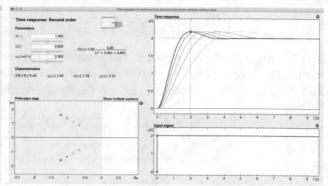

Pattern 2.a. Open-loop analysis

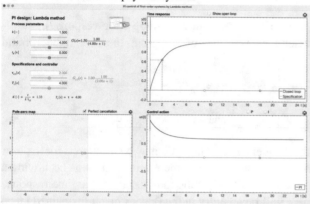

Pattern 2.b. Closed-loop analysis

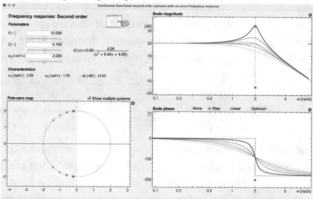

Pattern 3. Cards related to frequency response analysis

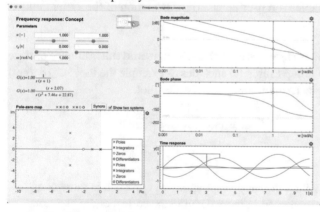

Pattern 4.a. Open-loop analysis

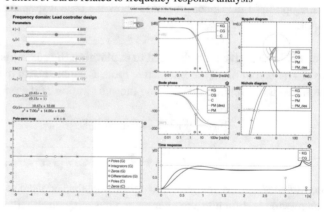

Pattern 4.b. Closed-loop responses

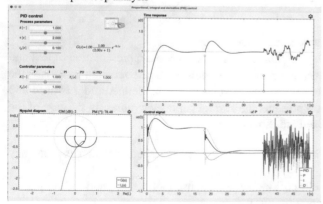

Pattern 5. Cards related to PID controllers

- **Green** to represent the frequency response of the controller, the closed-loop transfer function defining the specifications, and its associated time response (solid line). The reference (setpoint) is drawn in a dashed line.
- **Magenta** is employed to represent other lines, for instance, the root locus.

These colors are also used to represent some performance and stability metrics such as phase lags and relative stability margins (details are provided in each card).

1.7.1.3 Menu Bar

The menu bar contains three menus in all tools:

- Session: This menu includes the so-called Reset option, allowing to restart the tool to its initial settings. Moreover, there are two interesting features: Save session and Restore session, providing the possibility of saving and restoring scenarios.
- Options: A drop-down menu with different options appears on the top toolbar. An attempt has been made to standardize as much as possible the information represented in this menu, the most common options being those presented in Table 1.2, which are explained in detail in each card.
- Info: Includes information about the tools in two sub-menus: About and support.

The last two interactive tools in Chap. 8 (loop shaping design) do not follow this basic layout, as they practically summarize the contents of the book, requiring more specific options which are explained in the corresponding cards.

1.7.1.4 Interactive Elements Associated with the Selection of Values and Options
- **Textboxes** k [-] 1.000 : Allow changing the value of a parameter by entering it directly via the keyboard. By clicking the box with the mouse, the text is highlighted. After introducing the new text, the *enter* key must be pressed and the new value will be refreshed throughout the tool. The point (not the comma) is used as a decimal separator.
- **Sliders or scrollbars** : Are used to change the value of a certain parameter. Moving to the right increases the value of the parameter and moving to the left decreases its value.
- **Radio buttons** ◉○ : Are used to activate a certain option (alternative or exclusive to the one being used) by clicking on them.
- **Check boxes** □☒ : Are used to activate a certain option by clicking on them.
- **Activation buttons** Sincro : Serve to perform a certain action.

1.7.1.5 Interactive Elements Associated with Graphical Representations

Common to all representations

- **Scaling in graphs** ⚙ : By clicking the gearwheel symbol, two possibilities arise: Locked scale, allowing the user to modify the scale settings on the abscissa and ordinate axes of the corresponding plot (x_{min}, x_{max}, y_{min}, y_{max}), both introducing the values or moving the slides; Variable scale, where the scale is automatically updated.
- **Interactive lines:** In the plots of physical systems, the boundary lines that compose the objects have a greater thickness (–), which usually indicates that they are interactive objects (in addition to the change in cursor representation) and that therefore the dimensions of the graphic object can be modified directly by clicking and dragging the mouse over them[2] .

Associated with the representation in the s-plane and the geometrical root locus

- **Poles:** The poles of a dynamical system are described using a cross: ×.
- **Zeros:** The zeros of a dynamical system are described using a circle: ○.
- **Integrators:** They are symbolized by a bold cross: **×**.

---

[2] *Clicking and dragging* is a way to move certain objects on the screen. To move an object, place the mouse cursor over it, press and hold down the left mouse button, then move the mouse while still holding down the left mouse button. When you have *dragged* the object to the location you want, let go of the mouse button.

Table 1.2 Options menu

Options		Tools including the option
Gain effect		t_first_order
Time constant effect		f_first_order
Damping factor effect	Constant imaginary part	t_second_order
Natural frequency effect	Constant real part	f_second_order
P1(s)=k/s∧n	$P_1(s) = \dfrac{k}{s^n}$	t_generic*
P2(s)=k/(Ts+1)	$P_2(s) = \dfrac{k}{(\tau s + 1)}$	f_concept*
P3(s)=k(bs+1)/s∧n	$P_3(s) = \dfrac{k(\beta s + 1)}{s^n}$	f_generic*
P4(s)=k(bs+1)/(Ts+1)	$P_4(s) = \dfrac{k(\beta s + 1)}{(\tau s + 1)}$	f_nonminimum_phase*
P5(s)=k/(s(Ts+1))	$P_5(s) = \dfrac{k}{s(\tau s + 1)}$	root_locus*
P6(s)=k(bs+1)/(s(Ts+1))	$P_6(s) = \dfrac{k(\beta s + 1)}{s(\tau s + 1)}$	Nyquist_criterion**
P7(s)=k(bs+1)/((T1s+1)(T2s+1))	$P_7(s) = \dfrac{k(\beta s + 1)}{(\tau_1 s + 1)(\tau_2 s + 1)}$	stability_margins
P8(s)=k/(s(T1s+1)(T2s+1))	$P_8(s) = \dfrac{k}{s(\tau_1 s + 1)(\tau_2 s + 1)}$	steady_state
P9(s)=k/((T1s+1)(T2s+1)(T3s+1))	$P_9(s) = \dfrac{k}{(\tau_1 s + 1)(\tau_2 s + 1)(\tau_3 s + 1)}$	PID_concept*
P10(s)=k wn∧2/(s(s∧2+2 z wn s+wn∧2))	$P_{10}(s) = \dfrac{k\omega_n^2}{s(s^2 + 2\zeta\omega_n s + \omega_n^2)}$	f_design_lag*
P11(s)=k wn∧2/((Ts+1)(s∧2+2 z wn s+wn∧2))	$P_{11}(s) = \dfrac{k\omega_n^2}{(\tau s + 1)(s^2 + 2\zeta\omega_n s + \omega_n^2)}$	f_design_lead*
P12(s)=k wn∧2 (bs+1)/(s∧2+2 z wn s+wn∧2)	$P_{12}(s) = \dfrac{k\omega_n^2(\beta s + 1)}{(s^2 + 2\zeta\omega_n s + \omega_n^2)}$	f_design_lead_lag*

Note: The tools with (*) only incorporate as examples some of the $P_i(s)$. Those with (**) include more examples

Introduce plant (NUM,DEN)		t_generic
Introduce plant (ZPK)		t_dominance
Allow defining examples of plants in MATLAB format		f_concept
(NUM,DEN): Numerator and denominator in polynomial format		f_generic
$G(s) = \frac{b_m s^m + \cdots + b_1 s + b_0}{a_n s^n + \cdots + a_1 s + a_0} \rightarrow$ [bm,...,b1,b0],[an,...,a1,a0]		f_nonminimum_phase root_locus
ZPK: Zero, pole, gain formats		stability_margins
$G(s) = \kappa \dfrac{(s - z_1)(s - z_2)}{(s - p_1)(s - p_2)(s - p_3)} \rightarrow$ [z1;z2],[p1;p2;p3],k.		steady_state
Complex poles or zeros are introduced in the format [x+yj;x-yj]		

Bode diagram		f_concept
Nyquist diagram		f_nonminimum_phase
Nichols diagram		lead_lag_concept
Complete view		f_design_lag
Graphical representations of the frequency response		f_design_lead
		f_design_lead_lag
Experiment $i, i = 1, \ldots, 4$		t_model_fitting
Analyze dominance	Dominance example 1	t_dominance
Dominance ratio	Dominance example 2	
Lag compensator	Several lead compensators	lead_lag_concept
Lead compensator	Lag vs lead compensators	f_design_lead_lag
Several lag compensators		

Note: lead_lag_concept also includes an option to switch the Nichols diagram to a time response plot

- **Differentiator:** They are represented using a bold circle o.
- **Closed-loop poles:** They are represented by a blue triangle △.
- **Desired closed-loop poles (obtained from the specifications):** They are represented using a green square (□).
- **Centroid of root locus:** The intersection of the linear asymptotes is represented by the cyan diamond symbol ◇.
- **Entry or exit points of the real axis of the root locus:** Where the root locus leaves or reaches the real axis (breakaway point, saddle points, or bifurcation points) are represented by a magenta square symbol □.

Associated with time representation

- **Characteristic point of the response:** There may exist characteristic points of the response to which an interactive behavior can be associated, such as the time constant in first-order systems and overshoot in second-order systems. These points are usually represented by a circle (o) that allows to modify the shape of the response dragging it.
- **Exogenous input (⊥):** Dashed vertical and horizontal lines with circles may be displayed in the time response of systems (output and input plots) that allow modifying the amplitude of the step used as input in open loop or that of the closed-loop reference (o · · ·), as well as the step-shaped load disturbance (o). These signals are measured on the axes of the plots on which they are represented, and their associated values (activation instant and amplitude) are displayed in the lower-left corner of the tool. The variance of the noise signal is also represented with the same symbol (o). When dragged vertically, the value of said variance is shown in the lower-left corner of the tool, the axes of the graphs not being valid in this case because they would provide a noise value without physical sense. The formula implemented in the tool for the noise generator in the form of a difference equation is given by [14]

$$x(t_i) = 0.7x(t_{i-1}) + \sqrt{0.5\sigma^2}\ \text{randn},$$

where subscript i represents discrete-time instants, σ is the noise variance, and "randn" is a random number generator normally distributed.

When the circles determining the amplitude of the signal are separated from the time axis, the circles which remain on the said axis are used to change the time at which the disturbance or noise is introduced.

Associated with frequency representation

- **Characteristic point of the response:** There may exist characteristic points of the response to which an interactive behavior can be associated.
  - In the case of low-frequency gain, the interactive element that appears is a dash line (–) associated with the log-magnitude axis.
  - The frequency linked to the position of a pole in the Bode plots (corner or break frequency of the representative asymptotes) is usually represented by a cross (×) placed on the frequency axis, while that associated with a zero uses a circle (o).
  - Some characteristic points of the response, such as a resonant peak or a relative stability margin, can also be represented. To make them interactive, a black circle is used (o), linked to that point. These characteristic points are those that facilitate the modification of the shape of the frequency response in the corresponding graphs.
  - The maximum or minimum magnitude and phase design points of phase-lead or phase-lag controllers are symbolized by a green square (□). The design frequency and the safety margin are drawn using a green and white diamond, respectively (◇, ◇).

1.7.2   A Remark on Numerical Implementation

The time responses in a closed loop have been implemented using a discrete numerical approach, as it is the way to consider the presence of time delays and to allow introducing load disturbances and noise at any time. A tradeoff solution between numerical performance and avoiding loss of interactivity has been found using a sampling time of 0.05 s in the discrete-time approximation, as a lower value has a negative impact on interactivity. This means that in systems with very fast responses (e.g. using derivative action in closed-loop control), numerical errors may appear in the time responses in special cases. Where this may occur, a clarification has been added in the associated text.

References

1. Granbom, E., & Olsson, T. (1987). VISIDYN: Ett program för interaktiv analys av reglersystem (VISIDYN: An interactive program for design of linear dynamic systems). Master thesis TFRT-5375. Technical report, Department of Automatic Control. Lund Institute of Technology, Lund, Sweden.
2. Blomdell, A. (1989). Spread-sheet for dynamic systems: A graphic teaching tool for automatic control. *Wheels for the Mind Europe, 2*, 46–47.
3. Guzmán, J. L., Dormido, S., & Berenguel, M. (2013). Interactivity in education: An experience in the automatic control field. *Computer Applications in Engineering Education, 21*(2), 360–371.
4. Gilliver, R. S., Randall, B., & Pok, Y. M. (1998). Learning in cyberspace: Shaping the future. *Journal of Computer Assisted Learning, 14*(14), 212–222.
5. Dormido, S. (2004). Control learning: Present and future. *Annual Reviews in Control, 28*(1), 115–136.
6. Åström, K. J., & Wittenmark, B. (1997). CCSDEMO. Technical report, Department of Automatic Control. Lund Institute of Technology.
7. Johansson, M., Gäfvert, M., & Åström, K. J. (1998). Interactive tools for education in automatic control. *IEEE Control Systems Magazine, 18*(3), 33–40.
8. Piguet, Y. (2009). *Sysquake 5: User manual*. Laussane, Switzerland: Calerga Sarl.
9. Guzmán, J. L., Åström, K. J., Dormido, S., Hägglund, T., Berenguel, M., & Piguet, Y. (2008). Interactive learning modules for PID control. *IEEE Control Systems Magazine, 28*(5), 118–134.
10. Johansson, M., & Gäfvert, M. (2003). *ICTools—Interactive learning tools for control*. Department of automatic control. Technical report, Lund Institute of Technology, Lund, Sweden.
11. Longchamp, R. (2006). *Comande numériques de systémes dynamiques*. Laussane, Switzerland: Cours d'Automatique. Presses Polytechniques et Universitaires Romandes.
12. Skogestad, S., & Postlethwaite, I. (2005). *Multivariable feedback control, analysis and design* (2nd ed.). Wiley-Interscience.
13. Seborg, D. E., Edgar, T. F., Mellichamp, D. A., & Doyle III, F. J. (2011). *Process dynamics and control. International student version* (3rd ed.). Wiley.
14. Guzmán, J. L., Berenguel, M., & Dormido, S. (2005). Interactive teaching of constrained generalized predictive control. *IEEE Control Systems Magazine, 25*(2), 52–66.

Part II
Open-Loop Analysis

The first part of the book is devoted to analyzing open-loop dynamical systems.

The analytical models for most engineering systems are nonlinear, so that Chap. 2 focuses on how to obtain dynamic linear models from nonlinear physical models through linearization using a Taylor series expansion about a nominal operating point, to make clear that the linearized models are only valid in the vicinity of the selected operating point. Related definitions are also introduced. Several examples of systems with different characteristics are included: The tank level system, variable section tank level system, ball and beam system, inverted pendulum on a cart system, and DC motor position system. Notice that only two interactive tools are included in this chapter and Chap. 5, as those systems that are open-loop unstable are treated in the interactive tools introduced in Chap. 9, corresponding to closed-loop control (these systems are made stable through feedback).

Then, it is justified how a linear system has outputs linearly dependent on its inputs through proportionality constants that depend on the selected operating point. Superposition and homogeneity properties of linear systems are explained and the rest of the chapters are dedicated to classical time response and frequency response of linear time invariant systems, starting from first and second-order ones and analyzing the influence of zeroes and also higher-order dynamics.

Chapter 3 deals with the time response of continuous time linear time invariant dynamical systems, represented using transfer functions. Seven interactive tools are included, to understand the time response of linear systems of different orders and complexities and how to adjust linear models from experimental data.

Chapter 4 deals with the frequency response of the same systems (and corresponding models) studied in Chap. 3. An interactive tool to explain nonminimum phase systems is included, as well as another one devoted to fitting models in the frequency domain.

Chapter 5 is devoted to exploring the relationships between the parameters of the physical models of the systems studied in Chap. 2 with their representative transfer functions (in nonlinear processes obtained by linearizing the system around an operating point).

Most of the tools use as input for analysis a step input in the time domain. Notice that the properties of linear systems allow the user to analyze impulse, ramp or parabola responses just by adding differentiators or integrators to the open-loop system under study (obviously this is not possible in closed-loop analysis). This is allowed in those interactive tools that include a pole-zero editor and associated repository, facilitating also the analysis of systems with more poles and zeros. In fact, students are encouraged to analyze, on the basis of the step response, what the output would be compared to other types of inputs (impulse, ramp, parabola, etc.). Sinusoidal inputs are restricted to their natural domain, frequency response, very rich in visual content through the Bode, Nyquist and Nichols diagrams. Notice that although Nichols diagram is explained, is not extensively used as most of the analysis and design is based on reasonings over Bode and Nyquist diagrams. Moreover, regarding closed-loop control, it is not used in this book for design purposes (as does, for instance, Quantitative Feedback Theory (QFT) technique). In the authors' experience, both Bode and Nyquist diagrams are better accepted by students.

Chapter 6 includes solved exercises using different interactive tools covering the most important concepts analyzed in the previous chapters.

From Nonlinear Physical Models to Linear Models 2

2.1 Introduction

Automatic control is the branch of engineering that aims to design devices that work autonomously. In this field, a *system* is understood as a set of elements, components or entities that interact with each other to achieve a common goal, not possible with any of the individual parts. Although processes can be understood as all the related activities (parts) inside the system that work together to make it function, in this text, a system or process is understood as a device or plant which is under control (including physical, biological, organizational, and other entities, and combinations thereof, which can be represented through a common mathematical symbolism [1]).

All industrial processes involve information and power. Their operation is determined by the actions (command or control signals) that allow the energy to be dosed at different points in the process, so that the operation of the whole is as desired. Thus, systems have *inputs*, which can be data, energy or matter that they receive from outside; and *outputs*, providing information, energy or matter from the system. In addition, they are subject to disturbances, a class of inputs that affect the outputs of the system but are not manipulable, that is, their value cannot be determined at any time. They are generally unwanted signals that corrupt the input or output of a plant or process.

As shown in Fig. 2.1, it is natural to represent a system by means of a *block*, with inputs (control, command or input signals, $u_i, i = 1 \ldots q$), disturbances (uncontrollable inputs, $d_i, i = 1 \ldots p$), and outputs (products produced by the process, $y_i, i = 1 \ldots r$). As pointed out in Chap. 1, this book only deals with single-input single-output (SISO) systems (input u, output y), including disturbances acting on it (d). Blocks are also used to represent other elements like controllers or filters, not only the process or plant under control.

2.1.1 Static Versus Dynamical Systems

Static systems constitute the simplest relationship between the signals $u(t)$ and $y(t)$: an algebraic equation. From an elementary consideration of physical feasibility, it is clear that in such a case the relationship can be expressed as $y(t) = f(u(t))$, where, for the cases of practical interest, $f(\cdot)$ is a uniform function (see Fig. 2.2). The systems that admit this form of representation are called *static systems*, and are those in which the value taken by the output signal $y(t)$ at a certain time t depends exclusively on the value taken by the input signal $u(t)$ at that instant of time t, and not on the values taken by $u(t)$ and $y(t)$ in the past. Combinational logic systems are an example of static systems. In fact, no physical system can react instantaneously to a change in input, although some electrical systems can react very quickly [2].

A *dynamical system*, on the other hand, is one that evolves over time and in which the effects of a given action on it do not occur instantaneously. They are, therefore, systems whose response evolves over time with memory of the past response. That is, for a SISO system, its output $y(t)$ depends on actual value of the input $u(t)$ and past values of y and u. The external description of a dynamical system is defined by an input–output function $f(\cdot)$, which makes the set of values taken by the input signal u in a certain time interval $[t_0, t]$ correspond to the value taken by the output $y(t)$ at time t, so that $y(t) = f(u[t_0, t])$. Since a signal is a physical quantity that evolves over time and a system is a signal processor, consideration of the dynamic behavior of a system plays a major role.

Supplementary Information The online version contains supplementary material available at https://doi.org/10.1007/978-3-031-09920-5_2.

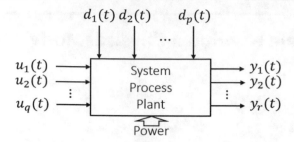

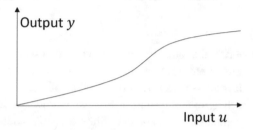

Fig. 2.1   Block representing a system or process

Fig. 2.2   Representation of a static system

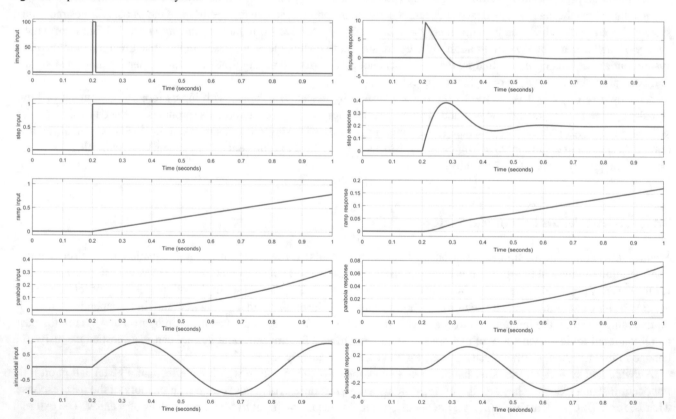

Fig. 2.3   Examples of responses of a dynamical system to typical inputs

Figure 2.3 shows an example of the response of a dynamical system to the different types of inputs used in automatic control, both to characterize the dynamics of a process and as references to a closed-loop system. Obviously, any input profile can be used to excite dynamical systems, but the selected inputs allow to determine interesting features of the time response. Another characteristic that can be analyzed in the responses in the figure is the cause-effect relationship, since in causal systems, there can be no output without an excitation at the input. In mathematical terms, the inputs in Fig. 2.3 are all 0 for

$t < t_0$, t_0 being the time instant in which the input is introduced ($t_0 = 0.2$ s in the plots in Fig. 2.3, although $t_0 = 0$ is the usual value). These input signals can be expressed as:

- **Unit impulse input:** $u(t) = \delta(t) = \infty$, for $t = t_0$ and 0 for $t \neq t_0$. This signal fulfills the property: $\int_{-\infty}^{\infty} \delta(t)dt = 1$.
- **Unit step input:** $u(t) = u_s(t) = 1$, $t \geq t_0$.
- **Unit ramp input:** $u(t) = u_r(t) = t$, $t \geq t_0$.
- **Unit parabola input:** $u(t) = u_p(t) = \frac{t^2}{2}$, $t \geq t_0$.
- **Unit sinusoidal input:** $u(t) = u_{sin}(t) = \sin(\omega t)$, $t \geq t_0$.

The mathematical definition of the impulse makes it impossible to be applied in real life, as it is not feasible to create a signal of amplitude infinite during an infinitesimal time. Using its properties, in practice, it is implemented as a pulse of area 1 with a very small time duration, as shown in Fig. 2.3.

Notice that there is a clear relationship between the first four inputs (also called *singularity functions*, [1]), each one being the integral/derivative of the previous/next one (the singularity functions can be obtained from one another by successive differentiation or integration)

$$\delta(t) = \frac{du_s(t)}{dt}, \ u_s(t) = \int_{t_0}^{t} \delta(\xi)d\xi, \ u_s(t) = \frac{du_r(t)}{dt}, \ u_r(t) = \int_{t_0}^{t} u_s(\xi)d\xi, \dots.$$

In linear systems, these properties can be applied to the corresponding outputs (as superposition principle holds, as will be explained in the following paragraphs), so, knowing the response to an input signal (the step one is the most used and that included in all interactive tools in this book), the response to the others can be obtained by derivation or integration. The response to a sinusoidal input in steady state will be used to define the frequency response of an LTI system in Chap. 4.

2.1.2 Mathematical Models

To understand and be able to make predictions about the behavior of a dynamical system, mathematical representations called *models* are used. A *model* is a mathematical representation of a physical, biological or information system. Mathematical models are fundamental in automatic control and that is the reason why special attention is devoted to them in automatic control courses. Mathematics has a dual use in engineering [3]:

- Mathematics can be used as a *language* when problems are to be formulated with the help of mathematical concepts in order to achieve precision and clarity.
- Mathematics can be used as a *tool* when, once the problem has been formulated in mathematical terms, the resulting equations are solved (analytically or by simulation).

A dynamical system can be modeled in multiple ways. In general, a first classification distinguishes between *models based on first principles* (mass, energy, and momentum balances) and *models based on experimental data*. In both cases, it is possible to obtain *nonlinear* or *linear* models, which will be discussed below. The complexity and precision of the models will depend on their purpose (simulation, controller design, and optimization) and on the characteristics of the dynamical system to be modeled. These possibilities are included in the conceptual diagram of the book in Fig. 1.1.

When using models, it is important to keep in mind that they are an approximation of the underlying system. Analysis and design using models must always be done carefully, to ensure that, the limits of the model are respected [4]. As pointed out in [2], any mathematical representation is, at best, an approximation of the behavior of the real physical system. The most important limitations of mathematical models are:

- **Parameter inaccuracies:** In many cases, the parameters of physical models cannot be determined with absolute precision. These inaccuracies can have many origins, e.g. concerning the data acquisition of experiments, experimental conditions do not exactly reproduce the operating conditions, parameter values may change over time due to usage or may even differ from those in the data sheets, due to deviations in the component manufacturing process itself, etc.
- **Unmodeled dynamics:** During the creation of the mathematical model of a physical system, it is often desirable to keep the model as simple as possible, as long as it can provide enough information to design an effective controller. In this

way, higher-order effects that make modeling difficult are often neglected (as happens when linearizing a model about a particular operating point). One must always remember that such higher-order effects are always present, even when they do not appear explicitly in the model. These unmodeled dynamics are generally of high frequency, so if simplifications of the model are being carried out, it should be taken into account in the design of the control algorithms that fast responses should not be demanded, since at high frequency, the modeling uncertainty is greater. The same applies to the rejection of possible disturbances affecting the process.

- **Presence of nonlinearities:** In physical systems, there are usually static nonlinearities, such as control signal saturation, valve opening rate limits (slew rate), dead zones in power electronics components or mechanical devices, etc. These static nonlinearities cannot be treated with linear systems control theory, and require more advanced techniques for their treatment (e.g. the use of descriptive functions) or, alternatively, their effect is analyzed or handled at the simulation level or at the design stage of the control system (an example is the anti-windup mechanism in PID controllers, [5]). There are also nonlinearities associated with the intrinsically nonlinear behavior of systems, which are described by nonlinear differential equations, from which, linear approximations can be obtained that are valid for describing the dynamics of a system around a given operating point (linearization process).

Examples of the previous limitations will be commented on through the text.

This chapter introduces the concepts associated with obtaining linear models of a dynamical system from modeling based on first principles. Following the philosophy of example-based learning, several highly didactic SISO systems (tank with liquid with discharge to the atmosphere—both constant and variable section cases are treated, ball & beam system, inverted pendulum on a cart system, and DC motor position system) have been chosen to introduce different concepts, such as modeling based on nonlinear differential equations, steady-state operating point, linearization using a Taylor series expansion, linear modeling and simulation of dynamic behavior, open-loop stable and unstable systems, and others.

2.1.3   Linear Systems

The models of dynamical systems used in most textbooks are linear, time invariant (LTI), and causal. The reason is that there exists a well-established theory both for analysis and design purposes. Notice that, as treated in this chapter, linear systems may come from the approximation of the dynamics of a nonlinear system around an operating point, defined by its boundary conditions in steady state.

- A *linear system* is one that satisfies the properties of scaling (homogeneity[1]) and superposition (additive), while a *nonlinear system* is any system that does not meet at least one of these properties. Linear systems comply with the superposition principle: If the system has an input that can be expressed as a sum of signals, then the response of the system can be expressed as the sum of the individual responses to the respective signals. Let's consider a SISO system with input u and output y in steady state. Suppose that when it is subjected to an excitation $u_1(t)$, it provides a response $y_1(t)$. After restoring the system to a steady state, it is subjected to an excitation $u_2(t)$, providing a response $y_2(t)$. For a linear system, it is necessary that the excitation $u_1(t) + u_2(t)$ results in a response $y_1(t) + y_2(t)$. This is usually called the principle of superposition [6]. Other authors [7] include a combination with the homogeneity property, so that if the composite input $u(t) = \upsilon_1 u_1(t) + \upsilon_2 u_2(t)$ is applied, if superposition applies,[2] then the response will be $y(t) = \upsilon_1 y_1(t) + \upsilon_2 y_2(t)$ ($\upsilon_1, \upsilon_2 \in \mathbb{R}$).

For instance, although equation $\dot{y}(t) + 3u(t) - 6 = 0$ is linear, the system it represents is nonlinear, because it does not comply with the superposition principle, as if the input $u_1(t) + u_2(t)$ is applied, the response should be $\dot{y}_1(t) + \dot{y}_2(t)$:

$$\dot{y}_1 = -3u_1 + 6; \quad \dot{y}_2 = -3u_2 + 6; \text{ applying } u_1 + u_2,$$

results in

$$\dot{y}_{12} = -3(u_1 + u_2) + 6 = -3u_1 - 3u_2 + 6 \neq \dot{y}_1 + \dot{y}_2 = -3u_1 - 3u_2 + 12.$$

---

[1] It is the property of a system whereby a scaled input results in an equally scaled output, that is, the response of a system to a signal $bu(t)$ is equal to b times the response to $u(t)$.

[2] Superposition will apply if and only if the system is linear.

The linear combination of inputs does not produce the linear combination of outputs due to the existence of the independent term. This simple example shows an interesting issue discussed in this chapter, and related to the deviation variables used in the linearization process and the fact that Laplace transform can only be applied to linear systems. Independent terms are usually associated with a determined operating point and they disappear from the equation once the linearization steps are applied, so that only the linear system parameters will depend on the operating point, reflecting that the linearized system is only a valid approximation of the behavior about the operating point selected in the linearization process. An example of a linear system could be one described by $\dot{y}(t) - y(t) + 3u(t) = 0$ (notice that in this case the superposition principle holds).

Linear time-invariant (LTI) systems are characterized by the fact that the relationship between the system input and output is independent of the time at which the input is applied (and therefore, the coefficients of the representing differential equations are constant) [1].

- A causal system is one that is non anticipatory, that is, the outputs depend on present and past inputs, but not on future inputs (the output at time t depends only on the input up to, and including, time t). *Strictly causal* systems depend only on past inputs[3] (the output depends only on the input preceding t).

2.1.4 Linearization

As pointed out before, Fig. 2.2 represents a nonlinear relationship (as it is not a straight line) between the input and the output of a static system, given by $y = f(u)$. The idea of linearization in this example is quite simple, but it helps to understand the underlying concepts. First, notice that obtaining a linear approximation to the curve $y = f(u)$ in this case, is just to obtain the tangent line to the curve at a given point. The election of such point is very important, as can be seen in Fig. 2.4, because different slopes will be obtained at different points. The selected point $\overline{y} = f(\overline{u})$ defines a value at which the linear approximation can be performed. It can easily be understood that such approximation is only valid for values of the input close to $\overline{u}$. As explained in the next paragraphs, in dynamical systems, this point will correspond to an operating point of the process (often associated with a steady state of the system), so that linear approximations that can be obtained through a linearization process are only valid for small deviations around that particular point. As pointed out in [8], if a nonlinear system operating at point A is considered, described by $(\overline{u}, \overline{y} = f(\overline{u}))$ in Fig. 2.4, small changes in the input can be related to changes in the output about the point by way of the slope of the curve at point A, s_A. Thus, a small change of the input about point A, $\tilde{u}$, produces a small change in the output, $\tilde{y}$, related by the slope at point A. Thus, $[f(u) - f(\overline{u})] \approx s_A(u - \overline{u})$, from which $\tilde{y} \approx s_A \tilde{u}$ and $y = f(u) \approx f(\overline{u}) + s_A(u - \overline{u}) \approx \overline{y} + s_A \tilde{u}$.

In the same way, a linear approximation of a nonlinear steady-state model is most accurate near the point of linearization. The same is true for dynamic process models [9]. Linearization is essentially an approximation of the nonlinear dynamics around the desired operating point [4]. With this idea in mind, when dealing with dynamical systems, the first step is to recognize the nonlinear components and write the nonlinear differential equation. A dynamical system with a single input $u(t)$ and a single output $y(t)$ can be represented by a general differential equation of the form[4] [8]

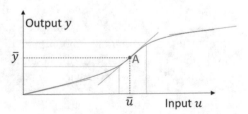

Fig. 2.4 Linearization of a static model

---

[3] In the scope of *transfer functions* modeling the system dynamics, this concept will be related to the difference or relative degree between the number of poles and finite zeros. If the number of poles and zeros is the same, the system is called causal and there is an instantaneous transmission from the input signal to the output, while if the transfer function is strictly proper (more poles than finite zeros), the system is called strictly causal.

[4] Notice that as usual, $\dot{y}$ represents the time derivative $\frac{dy(t)}{dt}$, $\ddot{y} = \frac{d^2 y(t)}{dt^2}$ and so on, where t is the independent variable (time), $y(t)$ is the dependent variable (output), and $u(t)$ is the input. Notation $d^n y(t)/dt^n$ is used to indicate the n-th derivative of y with respect to time. The dependence on time is often obviated to facilitate understanding.

$$f(u, \dot{u}, \ddot{u}, \ldots, y, \dot{y}, \ddot{y}, \ldots) = 0. \tag{2.1}$$

Throughout the book, it is assumed that f, u, and y are continuous and smooth enough so that they are differentiable to any required order.

An *equilibrium operating point* is obtained making all input and output derivatives equal to zero ($\dot{u} = \ddot{u} = \cdots = \dot{y} = \ddot{y} = \cdots = 0$), and the operating point $(\overline{u}, \overline{y})$ satisfies the original model so that $f(\overline{u}, 0, 0, \ldots, \overline{y}, 0, 0, \ldots) = 0$ at the input and output values defining the operating point $(u, y) = (\overline{u}, \overline{y})$. Thus, at equilibrium points there area no active dynamics of the system. There may be operating points around which the system is linearized that are not "natural" equilibrium points of the system, where, for instance, the time derivatives are not zero at the steady state. An example is given in Sect. 2.5, devoted to the inverted pendulum on a cart system, where the objective is to keep the pendulum upright and linearization is performed about the upright vertical position, which is not the natural steady state of the system, so that it is required that the cart is continuously moving to maintain the unstable vertical position. The situation is even more complicated if the operating point is defined out of the vertical position. So, although linearizing around an equilibrium point is the most frequent case, it is not the only possibility. An equilibrium point is a particular solution of a dynamical system, represented by a constant solution. It is possible to linearize a dynamical system around any particular solution, that should be a solution of interest, and often this is an equilibrium point. As indicated in the case of the inverted pendulum, there are situations where other non-constant solutions are of interest. Another example is the case of a harmonic oscillator, where a periodic solution, i.e. a cycle, is chosen for linearization (and in this case the steady state is often referred to as the steady-state regime). For this reason, this book mainly refers to operating points.

Remark

The model given by Eq. (2.1) can be expressed in terms of *state-space* description which, as indicated in the introduction, is not the one adopted in this text. The *states* of a dynamical system are defined as a collection of variables that summarize the dynamic past of the system and allow predicting its future behavior if future inputs are known. In physical systems, the state is composed of variables that represent the storage of mass, energy, and momentum. *State variables* are part of a vector $x \in \mathbb{R}^n$ called the *state vector* [4]. A dynamical system with one input $u(t)$ and one output $y(t)$ (SISO) can be represented by a set of ordinary differential equations (ODEs) of the form

$$\frac{dx(t)}{dt} = \dot{x}(t) = f(u(t), x(t)), \quad y(t) = g(u(t), x(t)),$$

where $f : \mathbb{R}^n \times \mathbb{R} \to \mathbb{R}^n$ and $g : \mathbb{R}^n \times \mathbb{R} \to \mathbb{R}$ are functions without discontinuities, and n is the dimension of the state vector (system *order*). A representation of this type is known as a *state-space model*. When the functions f and g are not explicitly time dependent, it is a time-invariant system. The function f provides the speed of change of the state vector as a function of the state $x(t)$ and the control signal $u(t)$, and the function g provides the measured values of the output as a function of the state $x(t)$ and the input $u(t)$ [4]. These functions determine the time response of the system [6] and $(\overline{u}, \overline{x})$ is said to be an *equilibrium state* (*point*) if $f(\overline{u}, \overline{x}) = 0$. If that condition is met, when all the constituents of the transient response have decayed (under the assumption that the system is stable), the system output reaches a *steady state*. An equilibrium state is a solution of the original differential equation in a steady state characterized by a constant position or oscillation (in the latter case the system state is called a steady-state regime). The equation $\dot{x} = 0$ may have several solutions. Which ones are acceptable depends on their stability, the stable ones are locally considered [10].

The transition from a general nonlinear model described by Eq. (2.1) to a linear one can be done through a linearization process, which allows a nonlinear differential equation to be approximated by a linear differential one that models the dynamics of the system around the operating point. This is obtained by removing the second and higher-order terms of the Taylor series expansion about the operating point (retaining only first derivative terms)

$$f(u, \dot{u}, \ddot{u}, \ldots, y, \dot{y}, \ddot{y}, \ldots) \approx f(\overline{u}, 0, 0, \ldots, \overline{y}, 0, 0, \ldots) +$$

$$+ \left[\left. \frac{\partial f}{\partial u} \right|_{(\overline{u},\overline{y})} \right] \underbrace{(u - \overline{u})}_{\tilde{u}} + \left[\left. \frac{\partial f}{\partial \dot{u}} \right|_{(\overline{u},\overline{y})} \right] \underbrace{(\dot{u} - \overline{\dot{u}})}_{\dot{\tilde{u}}} + \left[\left. \frac{\partial f}{\partial \ddot{u}} \right|_{(\overline{u},\overline{y})} \right] \underbrace{(\ddot{u} - \overline{\ddot{u}})}_{\ddot{\tilde{u}}} + \cdots +$$

$$+ \left[\left. \frac{\partial f}{\partial y} \right|_{(\overline{u},\overline{y})} \right] \underbrace{(y - \overline{y})}_{\tilde{y}} + \left[\left. \frac{\partial f}{\partial \dot{y}} \right|_{(\overline{u},\overline{y})} \right] \underbrace{(\dot{y} - \overline{\dot{y}})}_{\dot{\tilde{y}}} + \left[\left. \frac{\partial f}{\partial \ddot{y}} \right|_{(\overline{u},\overline{y})} \right] \underbrace{(\ddot{y} - \overline{\ddot{y}})}_{\ddot{\tilde{y}}} + \cdots$$

$$(2.2)$$

where in the case the operating point is not an equilibrium point:

- Subindex $(\overline{u}, \overline{y})$ referring to the variables defining the operating point should be written as $(\overline{u}, \overline{\dot{u}}, \overline{\ddot{u}}, \ldots, \overline{y}, \overline{\dot{y}}, \overline{\ddot{y}}, \ldots)$.
- The first term at the right of $\approx$ operator should be $f(\overline{u}, \overline{\dot{u}}, \overline{\ddot{u}}, \ldots, \overline{y}, \overline{\dot{y}}, \overline{\ddot{y}}, \ldots)$.

Expression (2.2) is the one usually found in textbooks because, as already mentioned, linearization is often performed about equilibrium operating points. *Deviation variables* are defined as the difference of the actual variable to the value defining the operating point (deviations about the defined operating point), such that the linearized function passes through the origin of the new coordinate system

$$\tilde{u} = (u - \overline{u}), \ \dot{\tilde{u}} = (\dot{u} - \overline{\dot{u}}), \ \ddot{\tilde{u}} = (\ddot{u} - \overline{\ddot{u}}), \ldots$$
$$\tilde{y} = (y - \overline{y}), \ \dot{\tilde{y}} = (\dot{y} - \overline{\dot{y}}), \ \ddot{\tilde{y}} = (\ddot{y} - \overline{\ddot{y}}), \ldots \quad (2.3)$$

These deviation variables have been identified in Eq. (2.2), where each of the terms in square brackets evaluates as a constant depending on the parameters of the system and the selected operating point. Notice that deviation variables represent small-signals. Taking into account (2.1) and that $f(\overline{u}, 0, 0, \ldots, \overline{y}, 0, 0, \ldots) = 0$, function f can be expressed in deviation variables as $f(\tilde{u}, \dot{\tilde{u}}, \ddot{\tilde{u}}, \ldots, \tilde{y}, \dot{\tilde{y}}, \ddot{\tilde{y}}, \ldots) = 0$, so that Eq. (2.2) can be written as a linear ordinary differential equation (ODE)

$$\cdots + \left[\left. \frac{\partial f}{\partial \ddot{y}} \right|_{(\overline{u},\overline{y})} \right] \ddot{\tilde{y}} + \left[\left. \frac{\partial f}{\partial \dot{y}} \right|_{(\overline{u},\overline{y})} \right] \dot{\tilde{y}} + \left[\left. \frac{\partial f}{\partial y} \right|_{(\overline{u},\overline{y})} \right] \tilde{y} = - \cdots - \left[\left. \frac{\partial f}{\partial \ddot{u}} \right|_{(\overline{u},\overline{y})} \right] \ddot{\tilde{u}} - \left[\left. \frac{\partial f}{\partial \dot{u}} \right|_{(\overline{u},\overline{y})} \right] \dot{\tilde{u}} - \left[\left. \frac{\partial f}{\partial u} \right|_{(\overline{u},\overline{y})} \right] \tilde{u}. \quad (2.4)$$

This is the general expression of a linearized dynamical model of whatever order of the derivatives and is also called the small-signal linear approximation of the original nonlinear equation. In classical textbooks, the ODEs are represented by the expression

$$a_n \frac{d^n y(t)}{dt^n} + a_{n-1} \frac{d^{n-1} y(t)}{dt^{n-1}} + \cdots + a_1 \frac{dy(t)}{dt} + a_0 y(t) = b_m \frac{d^m u(t)}{dt^m} + b_{m-1} \frac{d^{m-1} u(t)}{dt^{m-1}} + \cdots + b_1 \frac{du(t)}{dt} + b_0 u(t), \quad (2.5)$$

where $y(t)$ is the output of the system, $u(t)$ is the input and n is the order of the equation, holding $n \geq m$ in causal systems. Notice that, as pointed out in [11], the linearization is presented loosely in a lot of control theory textbooks. Moreover, the original coordinate system is shifted to the operating point, stopping to emphasize the incremental nature of the linearized model and deviation variables are not explicitly used for the sake of simplicity. For that reason, this text deals with nonlinear behavior and linearization in this chapter and Chaps. 5 and 9, and develops linear control theory in other chapters, keeping the notation simple (without using notation $(\tilde{\ })$ in the variables, as done in Eq. (2.5), where signals $u(t)$ and $y(t)$ have their origin in 0 as they represent linear dynamics). If the original model is linear, it makes no sense to apply Taylor series expansion, but the reader can check that substituting the original variables by the deviation variables following (2.3), the same equation in the deviation variables is obtained, as is logical when dealing with linear systems. Moreover, the election of the initial condition to correspond to an initial operating point ($y(0) = \overline{y}$) greatly simplifies the application of the Laplace transform, as will be analyzed in Chap. 3. Transforming an n-th order derivative to Laplace space requires the first $(n - 1)$ derivatives at $t = 0$. If deviation variables are used, these derivatives will all be zero.

Relating Eqs. (2.4) to (2.5), it is easy to see that

$$a_0 = \left[\left. \frac{\partial f}{\partial y} \right|_{(\overline{u},\overline{y})} \right]; \ a_1 = \left[\left. \frac{\partial f}{\partial \dot{y}} \right|_{(\overline{u},\overline{y})} \right]; \ \ldots; \ b_0 = \left[\left. \frac{\partial f}{\partial u} \right|_{(\overline{u},\overline{y})} \right]; \ b_1 = \left[\left. \frac{\partial f}{\partial \dot{u}} \right|_{(\overline{u},\overline{y})} \right]; \ \ldots,$$

and so on. So, in general, the coefficients of the linearized system depend both on the parameters of the nonlinear system and the chosen operating point. This will be analyzed in the examples shown in this chapter.

The previous is the general case, but very often, nonlinear dynamical models are derived from first principles (material, energy, or momentum balances) [9], providing first-order differential equations as

$$\frac{dy(t)}{dt} = f(u(t), y(t)),\tag{2.6}$$

where, as usual, $y(t)$ is the output and $u(t)$ is the input. First-order differential equations are also used in the framework of state-space representations of systems dynamics. In what follows, time dependency will not be included in order to simplify the explanation. It has been explained before that a linear approximation of this equation can be obtained by using a Taylor series expansion, truncating after the first-order terms (second-order and higher are removed). The system represented by (2.6) is linearized about the nominal steady-state operating point $(\overline{u}, \overline{y})$

$$f(u, y) \approx f(\overline{u}, \overline{y}) + \left(\frac{\partial f(u, y)}{\partial u} \bigg|_{(\overline{u}, \overline{y})} (u - \overline{u}) + \frac{\partial f(u, y)}{\partial y} \bigg|_{(\overline{u}, \overline{y})} (y - \overline{y}) \right) +$$

$$removed \nearrow \qquad + \frac{1}{2} \left(\frac{\partial^2 f(u, y)}{\partial u^2} \bigg|_{(\overline{u}, \overline{y})} (u - \overline{u})^2 + 2 \frac{\partial^2 f(u, y)}{\partial y \partial u} \bigg|_{(\overline{u}, \overline{y})} (u - \overline{u})(y - \overline{y}) \right.$$

$$\searrow \qquad \left. + \frac{\partial^2 f(u, y)}{\partial y^2} \bigg|_{(\overline{u}, \overline{y})} (y - \overline{y})^2 \right) + \text{higher-order terms.}\tag{2.7}$$

By definition, the steady-state condition (equilibrium) corresponds to $f(\overline{u}, \overline{y}) = 0$. Including deviation variables so that $u = \overline{u} + \tilde{u}$ and $y = \overline{y} + \tilde{y}$, the linearized differential equation in terms of $\tilde{u}$ and $\tilde{y}$ (after substituting $dy/dt = d\tilde{y}/dt$) is given by

$$\frac{d\tilde{y}}{dt} = \frac{\partial f(u, y)}{\partial u} \bigg|_{(\overline{u}, \overline{y})} \tilde{u} + \frac{\partial f(u, y)}{\partial y} \bigg|_{(\overline{u}, \overline{y})} \tilde{y},\tag{2.8}$$

This is a typical ODE defining a first-order system (the order of the ODE is one), of the form

$$\frac{dy(t)}{dt} + a_0 y(t) = b_0 u(t).\tag{2.9}$$

When Eq. (2.9) is obtained from the linearization of a nonlinear model, the output and input variables are replaced by the corresponding deviation variables about the operating point ($\tilde{u}(t) = u(t) - \overline{u}$, $\tilde{y}(t) = y(t) - \overline{y}$) and the coefficients depend both on the parameter of the nonlinear model and the selected operating point, so that $b_0 = \frac{\partial f}{\partial u} \big|_{(\overline{u}, \overline{y})}$ and $a_0 = -\frac{\partial f}{\partial y} \big|_{(\overline{u}, \overline{y})}$.

It must be taken into account that large changes in operating conditions for a nonlinear process cannot be approximated satisfactorily by linear expressions [9]. In many instances, however, nonlinear processes remain in the neighborhood of an operating point, around which a linearized model of the process may be sufficiently accurate.

One question that often arises when linearizing nonlinear systems is if the neighborhood of the operating point in which the linear approximation is considered valid can be quantified. Another question is related to the evaluation of the degree of nonlinearity of the model, that is still an open issue. These are not usually treated in detail in basic textbooks. In [10], some useful comments are included on pages 90 and 91 but using the state-space description. In [12], gap-metric-based[5] nonlinearity measures are used to help students correctly understand the significance of the neighborhood of linearization, using the inverted pendulum on a cart system as a case-driven pedagogical approach. Anyway, this analysis can be done by visualization and comparing the output of the original nonlinear model and that of the linearized model. It is not a mathematical way, but it serves to select the bounds on the input that makes the linear approximation valid from the point of view of the user of the model.

---

[5] The gap metric is an extension of the common measure of the ∞-norm of the difference between two systems and it is used to measure the gap between a linearization of a nonlinear system at its operating point and a fixed linear system [12].

To conclude this introduction, the difference between equilibrium and operating point is recalled. When linearizing around equilibrium points, conclusions can be drawn about the local stability of the system (neighborhood around the point where the nonlinear system is also stable, meaning that the system will evolve to the equilibrium point in some time). When dealing with linearization around an operating point, this is usually chosen to be an equilibrium, but linearization could also be performed around operating points that are not equilibrium points. In fact, this is sometimes what is achieved by feedback, so that the equilibrium point of the system is changed to the desired point through feedback in such a way that linear control theory can be applied to achieve stability and a degree of performance when operating about that particular operating point.

In what follows, several illustrative examples of how to obtain linear models from physical systems are explained.

2.2 The Tank Level System

2.2.1 Interactive Tool: tank_level_lin

2.2.1.1 Concepts Analyzed in the Card and Learning Outcomes
- Obtaining system models from first principles.
- Linearization of nonlinear models through Taylor series expansion around an operating point. Definition of the operating point and deviation variables.
- Effect of physical system parameters and choice of operating point on the dynamic response to a step input.
- Obtaining first-order linear models by linearizing smooth nonlinear first-order differential equations.

2.2.1.2 Summary of Fundamental Theory
The level control of fluid inside a system of tanks [13] or a single tank [6] is a popular modeling and control problem. This system consists of a constant-section tank that stores a fluid, with an outlet orifice or discharge section, and into which a variable flow of fluid enters by the action of a pump or a valve (actuator[6]).

As pointed out by [6] (Example 2.13, pp. 94–104), where a complete development of the contents summarized in this section can be found, the general equations of motion and energy describing fluid flow are quite complicated (coupled nonlinear partial differential equations). Therefore, simplifying assumptions have to be made to reduce the complexity of mathematical models. It is assumed that the fluid inside the tank is incompressible (therefore with constant density ρ) and the flow is inviscid (viscosity is neglected), irrotational[7] and steady[8] (low fluid speed, output hole small enough).

With these assumptions, a mathematical model can be obtained by using the principle of conservation of mass. The mass of water in the tank at any time is $m = \rho A h$, where h represents the level or height of water in the tank (distance between the outlet and the water level in the tank) and A the area or section of the constant-section tank (see Fig. 2.5). Taking the time derivative: $\dot{m} = \rho A \dot{h}$. The change in mass in the tank is equal to the mass that enters the tank ρq (q is the inlet flow rate) minus the mass that leaves the tank $\rho a v_a$, where a the area of the outlet hole and v_a the exit velocity (function of the water height).

From Bernoulli's law,[9] it is obtained

$$\frac{1}{2}\rho v^2 + P + \rho g h = \frac{1}{2}\rho v_a^2 + P_a,$$

---

[6] Actuators are the devices that provide the driving power to the process, and therefore, are in charge of converting the control signal into the manipulated variable.

[7] If each fluid element at each point in the flow has no net angular velocity about that point, the flow is termed irrotational.

[8] The velocity at a given point does not change with time.

[9] Bernoulli's principle can be derived from the principle of conservation of energy. It states that, in a steady flow, the sum of all forms of energy of a fluid along a streamline is the same at all points along the streamline. This requires that the sum of kinetic energy, potential energy, and internal energy remain constant. The hydrostatic pressure across the orifice is $\Delta P = \rho g h$. In applications using valves, the volumetric flow is multiplied by a coefficient of discharge.

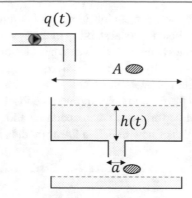

Fig. 2.5 Tank level system modeling

where g is the acceleration of gravity at sea level, v is the water velocity at the mouth of the tank and P and P_a are the atmospheric pressures at the input and output, respectively, which are equal to the atmospheric pressure, and a is sufficiently small so the water flows out slowly and v is negligible. Bernoulli's equation thus provides $v_a = \sqrt{2gh}$. Therefore

$$\dot{m} = \rho A \dot{h} = \rho q - \rho a v_a, \tag{2.10}$$

so that the system is described by the following nonlinear differential equation [13]:

$$A \frac{dh(t)}{dt} = q(t) - a\sqrt{2gh(t)}. \tag{2.11}$$

For a steady state ($dh(t)/dt = 0$) characterized by $\overline{h} = h(t)|_{t \to \infty}$ and $\overline{q} = q(t)|_{t \to \infty}$, from the model given by Eq. (2.11), it follows that the steady-state condition[10] is given by

$$\overline{q} = a\sqrt{2g\overline{h}}. \tag{2.12}$$

That is, if the tank is fed with a constant flow rate $\overline{q}$, the height of the fluid inside the tank in steady state will be $\overline{h} = \overline{q}^2/(2ga^2)$, in the corresponding units (see nomenclature).

Since the model (2.11) is nonlinear, to obtain a linear approximation, it has to be linearized around the operating point given by $(\overline{q}, \overline{h})$. The process input $q(t)$ and output $h(t)$ are then represented as the sum of the steady-state value that defines the operating point $(\overline{q}, \overline{h})$ plus the value of the deviation variable around that, given by $(\tilde{q}(t), \tilde{h}(t))$:

$$q(t) = \overline{q} + \tilde{q}(t), \tag{2.13}$$
$$h(t) = \overline{h} + \tilde{h}(t). \tag{2.14}$$

The deviation variables are small changes from the operating point $(\overline{q}, \overline{h})$ and represent the linear dynamics of the system. By applying Taylor series expansion of the nonlinear part of the model and by choosing only the terms up to the first-order one, it is obtained

$$\sqrt{h(t)}|_{\overline{h}} \approx \sqrt{\overline{h}} + \frac{1}{2\sqrt{\overline{h}}} \underbrace{(h(t) - \overline{h})}_{\tilde{h}(t)}, \tag{2.15}$$

so that taking into account that $dh(t)/dt = d\tilde{h}(t)/dt$, Eq. (2.11) can be simplified to

---

[10] Notice that $(\overline{q}, \overline{h})$ is used here to denote the actual operating point representing a steady state of the system. Other texts use different equivalent notations, as (q_0, h_0), (q_e, h_e), ...The notation (q_0, h_0) is sometimes preferred because it also represents the initial state from which the response to changes in the input is studied (initial condition).

$$A\frac{d\tilde{h}(t)}{dt} = \overline{q} + \tilde{q}(t) - \underbrace{a\sqrt{2g\overline{h}}}_{\overline{q}} - a\sqrt{\frac{g}{2\overline{h}}}\tilde{h}(t) \rightarrow \frac{d\tilde{h}(t)}{dt} = \frac{\tilde{q}(t)}{A} - \frac{a}{A}\sqrt{\frac{g}{2\overline{h}}}\tilde{h}(t). \tag{2.16}$$

The integration of the differential equation (2.16) starting from zero initial conditions, $\tilde{h}(0) = 0$, when considering a small step variation of input flow ($\tilde{q}(t) = \Delta q$) with respect to the actual operating point, provides the time evolution of the fluid level $\tilde{h}(t)$ from the point $\overline{h}$ to a new steady-state value $\overline{h}_1$ generally close to the previous one (as linear approximations are only valid in a neighborhood of the initial operating point):

$$\tilde{h}(t) = \frac{\Delta q}{a}\sqrt{\frac{2\overline{h}}{g}}\left(1 - e^{-\left(\frac{a}{A}\sqrt{\frac{g}{2\overline{h}}}\right)t}\right), \tag{2.17}$$

where the new steady-state height $\overline{h}_1 = \overline{h} + \Delta h = \overline{h} + \tilde{h}(t \rightarrow \infty) = \overline{h} + \frac{\Delta q}{a}\sqrt{\frac{2\overline{h}}{g}}$.

Usually, Eq. (2.16) can be written in a standardized way by making a variable change

$$\tau = \frac{A}{a}\sqrt{\frac{2\overline{h}}{g}} \quad \text{and} \quad k = \frac{\tau}{A},$$

so that the dynamics of system around the actual operating point can be roughly represented using a first-order linear dynamic model of the form:

$$\tau\frac{d\tilde{h}(t)}{dt} + \tilde{h}(t) = k\tilde{q}(t), \tag{2.18}$$

which main characteristics are analyzed in Sect. 3.2.

The above equations allow to simulate the behavior of the nonlinear system (2.11) and the linearized one (2.18), as analyzed in the interactive tool introduced in this section using step signals as input. Simulation consists of studying the time evolution of the system variables by solving the equations that determine its behavior [15]. Generally, as a test input for simulation, a *step input* is used, causing a *transient response* of the system until it reaches the new steady state. Steady-state response means the way the system output behaves when the time t tends to infinity. The transient response consists of the time evolution of the system output from the moment a change in the system input occurs until a state close to or equal to the steady state is reached. The addition of the steady-state component and the transient component gives the complete solution.

Note: $h(t)$ has been considered in this card as the output of the process, but following the same development, the outlet flow rate $q_a = a\sqrt{2gh}$ could be taken as process output, obtaining the following linearized equation relating the output flow rate to the input flow rate:

$$\frac{d\tilde{q}_a(t)}{dt} = \frac{a}{A}\sqrt{\frac{g}{2\overline{h}}}\left(\tilde{q}(t) - \tilde{q}_a(t)\right).$$

In steady state, $\overline{q} = \overline{q}_a = a\sqrt{2g\overline{h}}$.

Also notice that, in different textbooks, instead of a discharge hole, a valve is used to regulate the output flow rate $q_a = aC_d\sqrt{2gh}$ (a coefficient of discharge C_d is added). The valve resistance is computed (using an electrical analogy) through [16]

$$\left.\frac{dq_a}{dh}\right|_{\overline{h}} = aC_d\sqrt{\frac{g}{2\overline{h}}} = \frac{1}{R_a} \rightarrow \tilde{h}(t) = R_a(\overline{h})\tilde{q}_a(t).$$

2.2.1.3 References Related to this Concept

- [4] Åström, K. J., & Murray, R. M. (2014). *Feedback systems: An introduction for scientists and engineers* (2nd ed.). Princeton University Press. ISBN: 9780691193984. Exercise 4.2, pp. 4–33.
- [6] Dorf, R. C., & Bishop, R. H. (2011). *Modern control systems* (12th ed.). Prentice Hall. ISBN: 978-0-13-602458-3. Example 2.13, pp. 94–104.
- [13] Johansson, K. H. (2000). The quadruple-tank process: A multivariable laboratory process with an adjustable zero. *IEEE Transactions on Control Systems Technology, 8*(3), 456–465.
- [15] Bolzern, P., Scattolini, R., & Schiavoni, N. (2009). *Fundamentos de control automático (Fundamentals of automatic control)*. McGraw-Hill. ISBN: 978-84-481-6640-3. Example 1.13, p. 12.
- [17] Golnaraghi, F., & Kuo, B. C. (2017). *Automatic control systems* (10th ed.). McGraw Hill Education. ISBN: 978-1-25-964384-2.

- [18] Ogata, K. (2010). *Modern control engineering* (5th ed.). Prentice Hall. ISBN: 978-0-13-615673-4. Chapter 2, section 7, pp. 43–45; Chapter 4, section 2, pp. 101–106.

2.2.1.4 Further Reading

The tank level system is treated in many books, with different degrees of complexity. Table 2.1 provides information on where to find different approaches to the modeling and control of this plant (which is also treated in Chaps. 5 and 9).

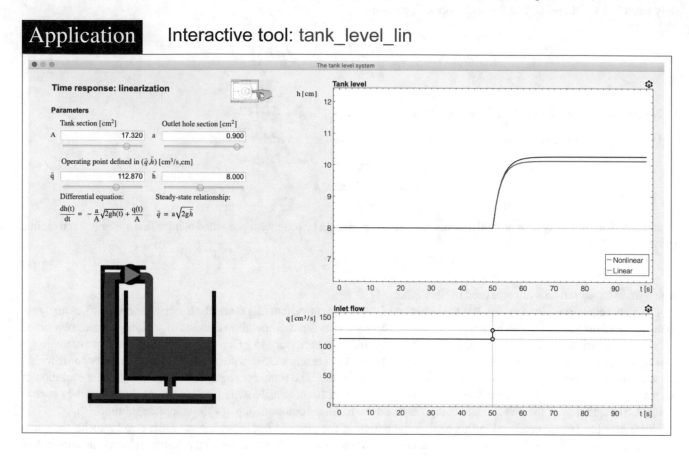

Time Response: Tank Level System Linearization

In this interactive tool, it is possible to compare, through simulation, the time responses (transient and steady-state responses) obtained using both the nonlinear and linear models of the tank level process. A step signal is used as a test input for the simulation. The tool also allows for analysis at different operating points and using different tank geometries (different areas of the tank and the discharge hole), providing quantitative and qualitative information on the validity range of the linear approximations. System parameters can be changed by interacting with the visual geometric representation of the tank.

The **Parameters** section is at the top left part of the tool, where it is possible to modify the values of the parameters and fundamental physical variables that define the response of the system, through textboxes and scroll bars: the Tank section (A), the Outlet hole section (a) and the variables that define the Operating point ($\bar{q}, \bar{h}$). The corresponding units are displayed.

It should be noted that to achieve a realistic effect, the centimeter (cm) has been used as the unit of length and the second (s) as the unit of time. If the scroll bars of $\bar{q}$ or $\bar{h}$ are accessed, the value of the linked variable is modified according to Eq. (2.12) and an automatic scaling of the valid intervals for these variables takes place. The maximum value of fluid level is 17.32 cm. There is also a linkage between the ranges of A and a so that the time evolutions shown in the upper-right image make physical sense. A symbolic representation of the nonlinear Differential equation used as a model of the system, as well as the relationship that determines the Steady-state relationship, are also included in this area.

Table 2.1 Tank level control references

Ref.	Section/page	Main contents
[1]	Section 2.11	Two tanks. Modeling and electrical analogy
[4]	Exercise 4.2, pp. 4–33	Simulation without input flow (problems of square root)
[6]	Example 2.13, pp. 94–104	Fluid flow modeling
	AP4.1, p. 293	Tank level regulator in response to disturbances
[8]	Exercise 2.66, p. 105	Linearization and basic control
[10]	Problem 1.16, p. 64	Explanation of the elements of a liquid tank system and control objectives
	Example 2.12, pp. 103–104	Three tanks in series
	Exercise 2.43, p. 186	Three tanks in series with discharge to the atmosphere
[15]	Example 1.13, p. 12	Interconnection of tanks and block diagrams
	Example 9.1, pp. 239–241	Modeling, linearization and block diagrams
[16]	Example 2.8, pp. 29–31	Modeling and solution of the linearized equation
[17]	Example 2-3-2	Modeling
	Example 2-3-4	Time response
	Example 2-3-5	Adding drainage pipe of length l
	Example 2-3-6	Double tank system
	Example 2-5-1	Obtaining k and τ from system equations
	Section 3-4-1	First-order prototype system
	Problem 7-42	Liquid level control
	Problem 7-59	PD control
	Problem 7-60	Ramp error, rise time and overshoot in closed loop
[18]	Section 4.2, pp. 101–105	Modeling
	Example A-4-1, pp. 140–141	Time response
[19]	Exercise 11.3, p. 482	Linearized control
[20]	Example 2.4, pp. 19–20	Level control
	Example 2.8, p. 29	Equilibrium conditions and linearization
	Example 4.4, pp. 136–137	Level control with feedforward action
	Example 4.5, pp. 145–156	Control design (pole placement)
	Example 4.8, pp. 151–152	Performance limitations due to actuator features
[21]	Example 2.10, pp. 25–26	Two tanks in series with outlet valves. Modeling, linearization, transfer functions and state equations
[22]	Section 7.2, pp. 200–203	Cascade tanks modeling, transfer function
[23]	Exercise 5.34 and 5.35	Modeling
	Toilet water tank, pp. 33–34, 95, 110	Differential equations, transfer functions, proportional control, disturbance rejection
[24]	Section 7.2, pp. 335–347	Liquid level systems: equations, conical tank
[25]	Example 2.16, pp. 54–55	Equations for describing water tank height
	Example 2.18, p. 56-57	Linearization of water tank height and outflow
	Example 9.3, pp. 607–608	Linearization of the water tank revisited
[26]	Section 1-1-3, 1-1-4, pp. 5–7	Open- vs closed-loop control
	Example 4-5-3, pp. 185–187	Modeling
	Example 4-10-4, pp. 215–216	Static gain, time constant and electrical analogies
[27]	Example 2.3, p. 33	Differential equation and conceptual level control
	Section 3.4, pp. 44–45	Cylindrical tank. Modeling from data
	Chapter 18 and the web site	Tank level estimation (two coupled tanks)
[28]	Example 3.2. pp. 94–95	Linear model with laminar flow
	Example 3.3. pp. 95–97	Time response
	Example 3.4. pp. 97–98	Tank as integrator
	Example 3.5. pp. 99–100	Time response using partial fraction expansion and inverse Laplace transform
	Section 3.12. Exercises 1, 2 and 3, pp. 184–185	Parameter estimation, proportional level control, PI level control
	Section 6.5.2, pp. 368–369	Adding time delay due to distance between the tank and the position of the input valve. Effect in closed-loop stability
	Example 7.2 and 7.3	State-space analysis

(continued)

Table 2.1 (continued)

Ref.	Section/page	Main contents
[29]	Section 2.6.2, pp. 116–119	Modeling of a simple liquid tank system
	pp. 119–121	A simple two tank system
	Example 1.5, p. 32	Linearization of the tank equation
[30]	Chapter 4, pp. 84–88	Modeling fluid systems assuming laminar and turbulent flows. Fluid capacitance
	Problem 4.2, p. 105	Modeling from data
	Example 5.10, pp. 130–132	Linearization
	Example 6.8., pp. 173–177	Simulation in Simulink®, both linear and nonlinear model
[31]	Exercise 1.12, p. 2	Differential equation of the linear behavior
	Exercises 3.1, 3.2, pp. 10–11	PID control of the tank using the transfer function
	Exercise 6.13, p. 48	Two interconnected tank systems
	Exercise 7.2, p. 51	P+feedforward control of the tank level
	Exercise 9.8, p. 62	Steady-state analysis considering valve dynamics
[32]	Example 5.10, pp. 174–176	Mathematical modeling and time response
	Example 11.14, pp. 11-37–11-41	Closed-loop transient response, and time-domain specifications
[33]	Exercise 1.5	Modeling and linearization
	Exercise 2.8.b	Determination of transfer function
	Exercise 7.2	Control level. PID and feedforward
[34]	Problem 1.3, pp. 17–18	Modeling
	Problem 1.4, pp. 18–19	Modeling two tanks with two restrictions
	Problem 1.28, pp. 59–62	Linearization of the models in Problem 1.4
[35]	Section 2.2.2, pp. 12–13	Flow control problem with time delay
	Exercise 2.6.1-2., pp. 31–32	Modeling and basic control
[36]	Section 2.1.4, pp. 36–38	Modeling of liquid-level systems
	Exercise 2.25, p. 60	Transfer function of interconnected tanks
[37]	Section 1.6, pp. 17–20	Modeling of a liquid level system
	Section 7.1, pp. 186–188	Two tanks. Block diagrams, transfer functions
[38]	Exercise 7.9, pp. 280–286	Level control, Bode plot and phase margin
[39]	Example pp. 39–40	Modeling of tank level and transfer function
	Question, pp. 41–42	About linear and nonlinear models
	Section 5.4, pp. 110–122	Level control, including sensors, actuators, time response
	Section 6.1, p. 150	First-order system model
	Section 10.4.1, pp. 253–257	Feeder tank, origin of zeros
	Section 16.2, pp. 471–477	Proportional control
	Problem, pp. 558–565	Practical aspects of PID control. Reverse acting controllers
	Section 21.7., pp. 673–675	Linearizing the model equations for a liquid level process

The graphs in the right part of the tool (**Tank level**) compare the time response of the linear model (red line) and the nonlinear model (black line), as a result of a step change in the fluid input flow rate. A legend displays the color code used. When the mouse pointer is placed over any point in the time responses, a label indicating the time associated with that point (t) and the value of the output (y) becomes visible. The black horizontal dotted line allows modifying the operating point defined by ($\bar{h}$), used as a base in the simulation, by vertically dragging it using the mouse. Logically, the value of the flow rate $\bar{q}$ will change according to the relationship (2.12) in the other areas of the tool. The same applies if changes are done on the **Inlet flow** plot, located in the lower-right part of the tool. Note that the base value of the change is defined by $\bar{q}$, which can be modified in this graph or in the scroll bars in the upper-left area (**Parameters**) and even in the process diagram, as will be explained later. On the step graph, there are two black circles (○) that determine the crossing point of two horizontal lines and a vertical dashed black line. By accessing the lower circle or any point on the vertical line, the time instant at which the step change in the pump flow rate occurs can be modified. Similarly, by accessing the lower horizontal dashed line, the operating point ($\bar{q}$) can be changed, while by moving the upper horizontal dashed line, the step amplitude (Δq in Eq. (2.17)) can be modified. It is important to note that when the mouse is placed over any of these lines, information regarding the time at which the step is entered (Step time) and the amplitude of the step (Step amplitude) is displayed in the lower-left corner of the tool and a label is also visible over the signal containing these values.

The default graph shows the entire possible range of inlet flow variation (red dotted horizontal lines), from the one that causes the level of the fluid inside the tank to overflow (in which case it remains at its maximum level producing a discontinuity in the level plot) to the one that corresponds to the emptying of the tank (notice that the computer code has been implemented so that to avoid difficulties in the simulation produced by a negative square root, the height being limited to be positive). All graphs allow locked and variable scaling through the settings associated with the gearwheel symbol.

The tool presents a schematic of the process to be modeled in the lower-left part. The figure is interactive, in the sense that its attributes can be modified:

- The area of the tank A by dragging the tank walls using the mouse (note that there is a limitation of minimum and maximum values so that the simulation makes physical sense and the graph can be represented in the space left for this purpose).
- The discharge outlet hole area a in the same way.
- The equilibrium level $\overline{h}$ can also be modified by dragging the liquid surface in the vertical direction.
- The constant flow rate $\overline{q}$ driven by the pump can be modified by dragging the liquid level in the vertical duct just before the elbow to the left of the pump.

In this tool, the Session menu includes the Reset option, which forces the tool to return to the default settings, as well as the possibility of saving and restoring sessions.

2.2.1.5 Homework

1. With the configuration that is shown when opening the tool, bring the system state to the operating point defined by $\overline{h} = 6$ cm and indicate the associated value of $\overline{q}$. Enter an open-loop step that increases the pump flow rate by approximately $\Delta q = 15$ cm^3/s and check that the value of the liquid level matches that expected from the theoretical calculations for the linear model. Justify the answer.
2. Repeat the previous step, but performing a change in the flow rate of approximately $\Delta q = 30$ cm^3/s. Comment on whether the linear model behaves better or worse than in the previous case and justify the answer.
3. Using Eqs. (2.12) and (2.17), compare the results obtained in the tool and with those equations in two different steady-state conditions. Is the same steady state obtained with the linear model and the nonlinear model? Justify the answer.
4. Choose a value of $A = 8$ cm^2 and $a = 0.3$ cm^2 and select an operating point defined by $\overline{h} = 5$ cm ($\overline{q} = 29.74$ cm^3/s). Indicate the maximum value of flow rate change Δq that can be applied so that the fluid in the tank does not overflow.
5. Select $A = 23$ cm^2, $a = 0.2$ cm^2 and $\overline{h} = 15$ cm. Indicate the time it would take to discharge the tank if the inlet flow rate is brought to 0 cm^3/s. Do the nonlinear and the linear model take the same time? Adjust the scales if necessary using the settings available in the gearwheel and justify the answer. This exercise is similar to Exercise 4.2 (pp. 4–33) in [4] and Example 2.4 (pp. 19–20) in [20].
6. Select $A = 15$ cm^2, $a = 0.3$ cm^2 and $\overline{q} = 25$ cm^3/s. Enter a step of amplitude $\Delta q = 10$ cm^3/s (up to 35 cm^3/s) at time $t = 20$ s. Indicate the steady-state value obtained with the nonlinear model and with its linear approximation. Repeat the procedure introducing a step of amplitude $\Delta q = -10$ cm^3/s (up to 15 cm^3/s). Are similar results obtained for positive and negative steps? Justify the answer.
7. Show that when there is a change in flow rate the new steady state reached by the model does not depend on the tank section (A). Does tank section A influence how fast the new operating point is reached?

2.3 Variable Section Tank Level System

2.3.1 Interactive Tool: spherical_tank_level_lin

2.3.1.1 Concepts Analyzed in the Card and Learning Outcomes

- Obtaining system models from first principles.
- Linearization of nonlinear models through Taylor series expansion around an operating point. Definition of the operating point and deviation variables.

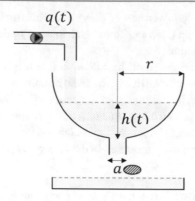

Fig. 2.6 Spherical tank level system modeling

- How reliable is a linearized model?
- Effect of physical system parameters and choice of the operating point on the dynamic response to a step input.

2.3.1.2 Summary of Fundamental Theory

As in the previous tool, the objective of this card is to obtain a linear approximation of a nonlinear model and compare the results and the validity of the linear model in the neighborhood of an operating point. In this case, the fluid level inside a spherical tank is studied, since the nonlinear behavior is more pronounced than in the previous example. The same nomenclature is used, where $q(t)$ is the inlet flow rate, $h(t)$ is the fluid level (height) inside the tank, a is the section of the outlet hole and r is the radius of the sphere (see Fig. 2.6).

The same simplifying assumptions that were in the previous section are used here. The reader is encouraged to analyze the mathematical development in that section. In this case, the variation of the volume of fluid inside the spherical tank is given by

$$\frac{dV(t)}{dt} = q(t) - a\sqrt{2gh(t)}, \tag{2.19}$$

where $V = \int_0^h A(\xi)d\xi$. As the fluid section is not constant, by considering a differential volume $dV = Adh$ (time dependence is omitted until reaching the final result for the sake of simplicity), it is obtained

$$\frac{dV}{dt} = \frac{dV}{dh}\frac{dh}{dt} = A(h)\frac{dh}{dt} = q - a\sqrt{2gh}. \tag{2.20}$$

The volume of a spherical cap is given by

$$V = \frac{\pi h^2}{3}(3r - h), \tag{2.21}$$

so that

$$A = \frac{dV}{dh} = \pi(2hr - h^2), \tag{2.22}$$

and Eq. (2.19) becomes

$$\pi(2rh(t) - h^2(t))\frac{dh(t)}{dt} = q(t) - a\sqrt{2gh(t)}$$
$$\frac{dh(t)}{dt} = \frac{1}{\pi(2rh(t) - h^2(t))}\left(q(t) - a\sqrt{2gh(t)}\right). \tag{2.23}$$

Equation (2.23) represents the nonlinear dynamics of the fluid level inside a spherical tank.

The steady-state condition is obtained by making the time derivative equal to zero, so that $\bar{q} = a\sqrt{2g\bar{h}}$. Linearizing Eq. (2.23) about the operating point defined by $(\bar{q}, \bar{h})$ a linear model can be obtained. In this case, the procedure is somewhat more complex than in the previous card, due to the fact that section A is not constant. If Eq. (2.23) is expressed as

Table 2.2 Variable section tank level control references

Ref.	Section/page	Main contents
[18]	Problem B-4-1, p. 152	Modeling and time response of a conical tank
[24]	Section 7.2, pp. 335–347	Liquid-level systems: equations, conical tank
[29]	Section 2.6.2, pp. 116–119	Examples of the description of continuous-time systems: Modeling of a simple liquid tank
[32]	Example 5.3, pp. 159–160	Capacitance of variable cross-section vessels
[34]	Problem 1.10, pp. 25–27	Modeling of a conical tank
[41]	Example 3.29	Main equations
	Example 3.32	Linearization

$$\frac{dh(t)}{dt} = f(q(t), h(t)) = \frac{1}{\pi(2rh(t) - h^2(t))}\left(q(t) - a\sqrt{2gh(t)}\right), \tag{2.24}$$

linearizing $f(q, h)$ (time dependence is omitted in what follows to obtain compact expressions) about the operating point $(\overline{q}, \overline{h})$), it results in

$$f(q, h) \approx f(\overline{q}, \overline{h}) + \left.\frac{\partial f}{\partial q}\right|_{(\overline{q},\overline{h})} (q - \overline{q}) + \left.\frac{\partial f}{\partial h}\right|_{(\overline{q},\overline{h})} (h - \overline{h}), \tag{2.25}$$

$$\left.\frac{\partial f}{\partial q}\right|_{(\overline{q},\overline{h})} = \frac{1}{\pi\overline{h}(2r - \overline{h})}, \tag{2.26}$$

$$\left.\frac{\partial f}{\partial h}\right|_{(\overline{q},\overline{h})} = \frac{1}{\pi\overline{h}^2(2r - \overline{h})^2}\left(-2\overline{q}(r - \overline{h}) - a\sqrt{\frac{g\overline{h}}{2}}(2r - \overline{h}) + 2a\sqrt{2g\overline{h}}(r - \overline{h})\right). \tag{2.27}$$

The steady-state condition implies $\overline{q} = a\sqrt{2g\overline{h}}$, so that $f(\overline{q}, \overline{h}) = 0$ (steady-state equilibrium condition) and first and third terms in Eq. (2.27) are equal but of different signs and cancel each other. Taking into account that $q(t) = \overline{q} + \tilde{q}(t)$, $h(t) = \overline{h} + \tilde{h}(t)$ and $dh(t)/dt = d\tilde{h}(t)/dt$, and simplifying the above expressions, the linear approximation to (2.23) is given by:

$$\frac{d\tilde{h}(t)}{dt} = \frac{1}{\pi\overline{h}(2r - \overline{h})}\left(\tilde{q}(t) - a\sqrt{\frac{g}{2\overline{h}}}\tilde{h}(t)\right). \tag{2.28}$$

The development and the obtained expression are similar to those performed in the previous card, where the tank has a constant area A.

Notice that the nearer the input signal $u(t)$ to its nominal value $\overline{u}$, the more valid the linearized model is. This is something the interactive tool introduced in this section helps to understand.

2.3.1.3 Reference Related to this Concept
- [40] Tavakolpour-Saleh, A., Setoodeh, A., & Ansari, E. (2016). Iterative learning control of two coupled nonlinear spherical tanks. *International Journal of Mechanical and Mechatronics Engineering, World Academy of Science, Engineering and Technology, 10*(11), 1862–1869.

2.3.1.4 Further Reading
The case of variable section tank level modeling and control is treated in a few books, often in the framework of nonlinear or gain-scheduling control techniques, see Table 2.2.

| Application | Interactive tool: spherical_tank_level_lin |

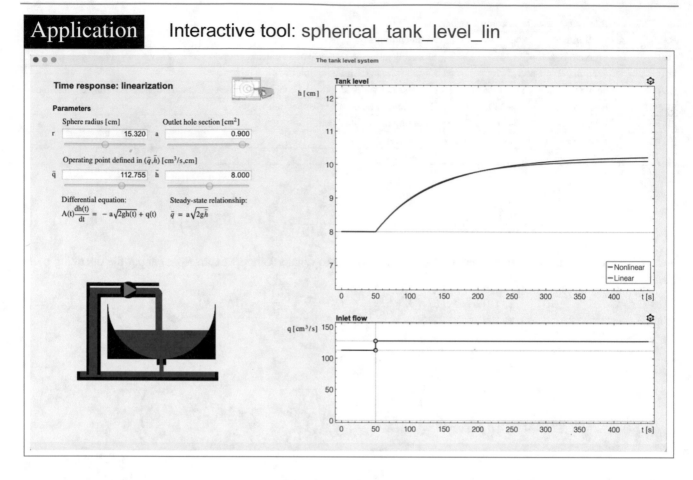

Time Response: Variable Section Tank Level System Linearization

In this interactive tool, it is possible to compare, through simulation, the step responses obtained using the nonlinear and linear models of a spherical tank level system. The tool is useful to visualize the degree of nonlinearity of the system and how it changes depending on the selected operating point, thus influencing the validity of the linearized model.

The explanation of this interactive tool is the same that can be found in Sect. 2.2, with the only difference that the radius of the sphere r is shown instead of the section of the tank A, as in this case, it depends on the level of fluid inside the tank.

2.3.1.5 Homework

1. Open the tool. Using the configuration of parameters that appears by default, select $\bar{h} = 6$ cm and indicate the associated value of $\bar{q}$. Enter an open-loop step that increases the pump flow rate by approximately $\Delta q = 15$ cm^3/s and check that the value of the liquid-level matches that expected from the theoretical calculations for the linear model. Justify the answer.
2. Repeat the previous step, but cause a change in the flow rate of approximately $\Delta q = 30$ cm^3/s. Comment on whether the linear model behaves better or worse than in the previous case and justify the answer.
3. Repeat the previous two exercises but with $\Delta q = -15$ and $\Delta q = -30$ cm^3/s respectively. Is there any symmetry in the response to what was observed in the previous exercises? Comment on the results.
4. Choose a value of $r = 10$ cm^2 and $a = 0.5$ cm^2 and select an operating point defined by $\bar{h} = 5$ cm ($\bar{q} = 49.573$ cm^3/s). Indicate the maximum value of flow rate change Δq that can be applied so that the fluid in the tank does not overflow.

2.4   Ball & Beam System

2.4.1   Interactive Tool: ball_and_beam_control

2.4.1.1 Concepts Analyzed in the Card and Learning Outcomes

- Obtaining system models from laws of mechanics.
- Linearization of nonlinear models through Taylor series expansion around an operating point. Definition of the operating point and deviation variables.
- Effect of physical system parameters and choice of the operating point on the dynamic response to a step input.

2.4.1.2 Summary of Fundamental Theory

The ball & beam system is described in [42, 43] and it is a well-known laboratory equipment for testing automatic control strategies. The main equations modeling its dynamics, that mix both linear and angular displacements, are summarized in this section. The ball & beam system is characterized as a beam coupled to a motor shaft which can be tilted to a desired angle. The ball rolls on the beam and the position of the ball is detected by different sensors [44]. As will be analyzed in Chap. 9, feedback control systems have to move the motor so as to regulate the position of the ball on the beam by changing the beam angle. Figure 2.7 shows a diagram of the system, where m is the ball mass, x is the translational position, θ is the angle of the beam shaft, R is the ball radius, r is the rolling or effective radius of the ball, and g is the acceleration of gravity at sea level.

The complete development of the model can be found in [42], as well as the modeling assumptions (dynamics, noise, and nonlinearities of motor and sensors are neglected, the system stores no potential energy and there is no slipping). The beam is considered a single lumped inertia and there are only two degrees of motion, the ball moving linearly on the beam and the beam rotating about a pivot point.

The system can be modeled either using a Lagrangian approach or Newtown's second law of mechanics [28, 42–46]. As pointed out by [46], the most straightforward means of modeling is by variational methods, where x and θ are complete independent set of generalized coordinates.

The potential energy of the system relates to the relative movement of the ball on the beam. Assuming that the location of the pivot corresponds to zero potential energy, the total potential energy is given by $U = -mgx \sin\theta$, while the kinetic energy is $E = \frac{1}{2}mv^2 + \frac{1}{2}I_b\omega^2 + \frac{1}{2}I_a\dot\theta^2$, where v is the linear velocity of the ball, ω is the angular velocity of the ball, I_b is the ball inertia,[11] and I_a is the beam inertia.

The translational position is given by $x = r\varphi$, that is, the product of the rolling or effective radius of the ball (see Fig. 2.7) times the rotational angle of the ball with respect to the shaft. Taking into account all involved angles, the angular distance rolled by the ball with respect to the fixed point is given by $\varphi + \theta = \frac{x}{r} + \theta$, so that the rotational velocity of the ball is given

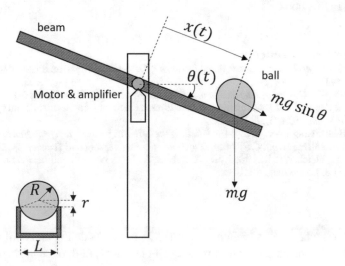

Fig. 2.7 Diagram of the ball & beam system

---

[11] $I_b = \frac{2}{5}mR^2$, R being the radius of the ball (see Fig. 2.7).

by $\omega = \frac{\dot{x}}{r} + \dot{\theta}$, while the translational one is obtained from kinematic relations

$$v = \sqrt{\dot{x}^2 + (x\dot{\theta})^2}. \tag{2.29}$$

Then, the kinetic energy equation can be written as

$$E = \frac{1}{2}\left(m\left(\dot{x}^2 + (x\dot{\theta})^2\right) + I_b \left(\frac{\dot{x}}{r} + \dot{\theta}\right)^2 + I_a\dot{\theta}^2 \right). \tag{2.30}$$

The Lagrangian is given by $L = E - U$ and the equations of motion

$$\frac{d}{dt}\left(\frac{\partial L}{\partial \dot{x}}\right) - \frac{\partial L}{\partial x} = 0, \tag{2.31}$$

then it is obtained

$$\left(\frac{I_b}{r^2} + m\right)\ddot{x} + \frac{I_b}{r}\ddot{\theta} - mx\dot{\theta}^2 - mg\sin\theta = 0, \tag{2.32}$$

by assuming that the shaft movements are small, the derivatives of the shaft angle can be disregarded.[12] Notice that the natural steady-state condition for this system is that the beam is in horizontal position $\overline{\theta} = 0$. Under these simplifications, the dynamics can be represented by:

$$\left(\frac{I_b}{r^2} + m\right)\ddot{x} = mg\sin\theta. \tag{2.33}$$

In fact, the above assumptions also impose that this representation is valid for small variations around an operating point, and therefore, the variables x and θ are deviation variables. Notice that the left hand of this equation is linear, while in the right part the nonlinear term is the sine function. This function can be easily linearized about the nominal operating point given by $\overline{\theta} = 0$, as $\sin\theta \approx \sin\overline{\theta} + \frac{d\sin\theta}{d\theta}\big|_{\overline{\theta}}(\theta - \overline{\theta}) = \theta = \tilde{\theta}$ (in this case the angle itself represents a deviation variable). In practice, it is always considered that for small angles, $\sin\theta \approx \theta$, so that the linear approximation of the dynamics is given by

$$\left(\frac{I_b}{r^2} + m\right)\ddot{x} = mg\theta \rightarrow \ddot{x} = k\theta. \tag{2.34}$$

As pointed out by [46], if the model is not simplified, the control problem dealt with in Sect. 9.4 should be very complex. In fact, a complete model of the system contains many modes which cannot be readily incorporated into a basic modeling and control study as that followed in this book.

2.4.1.3 References Related to this Concept

- [42] Hirsch, R. (2008). EDUMECH - Mechatronic Instructional Systems - Ball on Beam System. Shandor Motion Systems. Retrieved July 01, 2021, from https://cutt.ly/KUogfV1.
- [43] Shahian, B., & Hassul, M. (1993). *Control system design using MATLAB*®. Prentice Hall. ISBN: 0-13-174061-X.
- [44] Acharya, M., Bhattarai, M., & Poudel, B. (2014). Real time motion assessment for positioning in time and space critical systems. *International Journal of Applied Research and Studies (iJARS), 3*(7), 1–11.
- [45] Bolivar-Vincenty, C. G., & Beauchamp-Báez, G. (2014). *Modelling the ball-and-beam system from Newtonian mechanics and from Lagrange methods*. In *Twelfth LACCEI Latin American and Caribbean Conference for Engineering and Technology*, Guayaquil, Ecuador.
- [46] Wellstead, P. E., Chrimes, V., Fletcher, P. R., Moody, R., & Robins, A. J. (1989). Ball and beam control experiment. *The International Journal of Electrical Engineering & Education, 15*, 21–39.

2.4.1.4 Further Reading

Table 2.3 provides information on where to find different approaches to the modeling and control of this plant (which is also treated in Chaps. 5 and 9). Several books use a more complex plant called "Ball & plate", where the ball can freely move on a plane.

---

[12] As pointed out by [46], the validity of these assumptions is questionable, depending upon the relative physical features of the system. For many laboratory equipments, the approximations made are felt to be quite justifiable.

Table 2.3 Ball & beam system references

Ref.	Section/page	Main contents
[8]	VE 9.1, p. 466	PD control
[17]	Problem 2-11	Dynamic equation of motion (not the same layout used in this book)
	Problem 3-28	Dynamic equation of the motion, transfer function, step response using MATLAB®
[24]	Chapter 5, problem 14, pp. 260–261	Nonlinear equations of motion, state variable equations, linearized system, and Simulink® modeling
[28]	Chapter 14, pp. 825–868	Control of a ball and beam system. Design of a multi-loop control system using frequency response (Nyquist criterion and Bode diagrams) and the root locus method
	Chapter 4, pp. 226–228	Stability criteria and steady-state error, closed-loop control of a ball and beam system
	Section 5.2.8., pp. 291–306	Control of a ball and beam system using root locus and Routh stability criterion
	Section 6.7.5, pp. 403–414	Analysis and design examples. Stability analysis with Nyquist criterion, design using Bode diagrams
[31]	Exercise 11.5, p. 69	Modeling and digital control
[33]	Exercise 7.10	Model from Bode plot of the beam dynamics
[43]	Appendix A3, pp. 465–476	Ball-on-beam complete project
[47]	Design study E	Ball on beam homework problems
	E.2, p. 386	Kinetic energy
	E.3, p. 386	Equations of motion
	E.4, p. 386	Linearization of equations of motion
	E.5, pp. 386–387	Transfer function model
	E.6, p. 387	State space model
	E.8, pp. 387–388	Successive loop closure
	E.9, p. 388	Integrators and system type
	E.P.6, p. 388	Root locus PID
	E.10, pp. 388–389	Digital PID
	E.11, p. 389	Full state feedback
	E.12, p. 389	Full state with integrator
	E.13, p. 390	Observer based control
	E.14, p. 390	Disturbance observer
	E.15, pp. 390–391	Frequency Response
	E.16, p. 391	Loop gain
	E.17, pp. 391–392	Stability margins
	E.18, p. 392	Loop shaping design
[48]	Exercise 2.6.3	Dynamical equations of the rail+ball system from the Euler-Lagrange formalism

Time Response: Ball & Beam System Linearization

In this case, no explicit tool has been developed for the analysis of the open-loop time response of the nonlinear and linearized models because when excited with a step input, its output grows indefinitely. It represents an unstable system from the bounded-input bounded-output (BIBO) point of view (see Chapter 3), and therefore, its analysis only makes sense in Chap. 9 where it will be stabilized through feedback (see Sect. 9.4).

2.4.1.5 Homework

1. Integrate the ODE (2.34) considering a fixed value of $\theta(t) = \theta_0 < \pi/4$. A procedure to experimentally estimate the parameter k if the laboratory plant is available can be found in [28] (pp. 835–836).
2. If it is considered that a system is stable, if when entering a bounded input, the output is also bounded, why the ball & beam can be considered an unstable system?
3. What would happen with k if $r = 0$ and what does this mean? [28].

2.5   Inverted Pendulum on a Cart System

2.5.1   Interactive Tool: inverted_pendulum_control

2.5.1.1 Concepts Analyzed in the Card and Learning Outcomes

- Obtaining system models from laws of mechanics.
- Linearization of nonlinear models through Taylor series expansion around an operating point that is not the natural equilibrium of the system. Definition of the operating point and deviation variables around those, and its meaning in unstable systems.
- Obtaining a linear second-order unstable system by linearizing nonlinear second-order differential equations.

2.5.1.2 Summary of Fundamental Theory

The inverted pendulum on a cart problem is common in control theory. This device combines a simple structure which is very easy to model mathematically, with a remarkable complexity in control design [49]. It is also a remarkable example of the dual local and global issues in its behavior. The inverted pendulum is a rod connected via a hinge on its bottom end to a movable cart that moves under the action of a force, which is the control action intended to act on the position of the rod. As explained in [18, 43], the objective of this classical problem is to maintain a rod in the upright position as the cart to which it is hinged moves horizontally. An interesting analysis of how this system is used to help students understand the neighborhood of linearization can be found in [12]. In [47], a complete design study is presented. A simplified dynamical model can be obtained following the diagram presented in Fig. 2.8, where m is pendulum mass, l is pendulum length, M is the cart mass, I is the moment of inertia of the pendulum about its center of gravity, g is acceleration due to gravity, θ is the angle the pendulum makes to the vertical, x is the cart position coordinate, and u the control input force acting on the cart. The simplifying assumptions are that the rod has no mass (so the center of gravity of the pendulum is located at that of mass m as the mass is concentrated at the top of the rod) and the center of gravity is the center of the pendulum ball. Moreover, the gust of wind is disregarded. The cart is usually driven by an electric motor and cart position and rod angle can be measured with potentiometers or other sensors. Notice that, in this section, the complete model is developed, but its complexity for control analysis makes us focus only on the dynamics of the rod.

The inverted pendulum can be understood as a rigid body whose motion is limited to two dimensions (the cart and the pendulum are assumed to move in only one plane). The fundamental equations of plane motion of a rigid body are given by

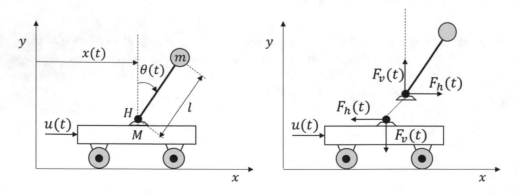

Fig. 2.8 Diagram of the inverted pendulum on a cart

Newton's second law for the horizontal and the vertical components of the force. To derive the equations of motion for the system, a free-body diagram can be used [18], so that to simplify the analysis of the system, the inverted pendulum can be divided into two bodies, the cart and the pendulum (Fig. 2.8). Two components appear, the reaction forces F_h (horizontal) and F_v (vertical) acting on the pivot point H (horizontal and vertical components of the force measured at point H). With the simplifying assumptions, the coordinates of the center of gravity of the pendulum coincide with the position of the ball $(x + l \sin \theta, l \cos \theta)$. The rotational, horizontal and vertical equations of motion of the ball, as well as the horizontal motion of the cart without considering friction, are given by

$$I \ddot{\theta} = F_v l \sin \theta - F_h l \cos \theta,$$

$$m \frac{d^2}{dt^2}(x + l \sin \theta) = m \left(\ddot{x} + l \frac{d^2}{dt^2}(\sin \theta) \right) = F_h,$$

$$m \frac{d^2}{dt^2}(l \cos \theta) = F_v - mg,$$

$$M \ddot{x} = u - F_h \tag{2.35}$$

and $Mg = F_v$ (all variables and parameters in the International System of Units). Because the objective is to keep the pendulum upright, linearization is performed about $\overline{\theta} = 0$, which is not the natural equilibrium of the system $\overline{\theta} = \pi$. At the upright position, it is required that the cart is moving to maintain the vertical position, so that the time derivatives at steady state $\overline{u}$ are not zero.[13]

If the angle of rotation is small, the previous four equations can be linearized so that $\sin \theta \approx \theta$ and $\cos \theta \approx 1$, and combining the expressions so that F_v and F_h do not appear in the final equations, it is obtained that

$$(m + M) \ddot{x} + ml \ddot{\theta} = u, \tag{2.36}$$

$$(I + ml^2) \ddot{\theta} + ml \ddot{x} = mgl\theta. \tag{2.37}$$

It can be considered that the moment of inertia of the pendulum about its center of gravity is negligible ($I = 0$), so that

$$(m + M) \ddot{x} + ml \ddot{\theta} = u, \tag{2.38}$$

$$ml^2 \ddot{\theta} + ml \ddot{x} = mgl\theta, \tag{2.39}$$

which can be modified to

$$M \ddot{x} = u - mg\theta, \tag{2.40}$$

$$Ml \ddot{\theta} = (M + m)g\theta - u. \tag{2.41}$$

Equations (2.40) and (2.41) describe the approximate linear dynamics of the plant about $\overline{\theta} = 0$. In fact, variables x and θ in those equations represent deviation variables $\tilde{x}$ and $\tilde{\theta}$ defining the new coordinates in that operating point, but are not written in that way for the sake of simplicity, as usual in all textbooks. Notice that equations taking into account the rod mass can be found in [18], Example 3–5, pp. 68–71.

The previous equations are a simplification of the real dynamics, that can be found in many textbooks (e.g. [43], p. 218) and web pages.[14]

Although in Sects. 5.5 and 9.5, Eq. (2.41) will be used to obtain the transfer function of the plant and for control purposes, respectively, in what follows, Eq. (2.35) will be developed without performing simplifications and including a friction term $b\dot{x}$ in the equation of motion in horizontal direction $M \ddot{x} + b\dot{x} = u - F_h$, b being the coefficient of friction for the cart. This is another example of linearization.

---

[13] As pointed out by [49], initially, in the 60s of the last century, this system was present in the control laboratories of the most prestigious universities. The demonstration consisted of initially placing the pendulum manually in the inverted vertical position, then releasing it and autonomously, by feeding back its position, the pendulum would continue in the inverted position by means of the appropriate action on the cart. The control problem, thus considered, is local and its interest lies in the fact that it involves stabilizing an open-loop unstable position. This problem, because of its local character, can be solved with linear methods, as discussed in this book. The problems arise when the pendulum has to be swung up from its natural equilibrium, and that the path of the cart is bounded, so that if one of the ends of the horizontal support is reached, the system stops working.

[14] https://cutt.ly/IUohO76; https://cutt.ly/SUohHyy.

$$F_h = m\ddot{x} + ml\ddot{\theta}\cos\theta - ml\dot{\theta}^2\sin\theta,$$
$$F_v = -ml\ddot{\theta}\sin\theta - ml\dot{\theta}^2\cos\theta + mg.$$

$$I\ddot{\theta} = mgl\sin\theta - ml^2\ddot{\theta}\sin^2\theta - ml^2\dot{\theta}^2\sin\theta\cos\theta - ml\ddot{x}\cos\theta - ml^2\ddot{\theta}\cos^2\theta + ml^2\dot{\theta}^2\sin\theta\cos\theta. \qquad (2.42)$$

Simplifying:

$$(I + ml^2)\ddot{\theta} = mgl\sin\theta - ml\ddot{x}\cos\theta. \qquad (2.43)$$

Now developing the equation of the horizontal motion

$$M\ddot{x} = u - b\dot{x} - m\ddot{x} - ml\ddot{\theta}\cos\theta + ml\dot{\theta}^2\sin\theta,$$
$$(M + m)\ddot{x} = u - b\dot{x} - ml\ddot{\theta}\cos\theta + ml\dot{\theta}^2\sin\theta. \qquad (2.44)$$

Equations (2.43) and (2.44) can be expressed, respectively, as

$$f_1(\ddot{x}, \ddot{\theta}, \theta) = 0,$$
$$f_2(\ddot{x}, \dot{x}, \ddot{\theta}, \dot{\theta}, \theta) = 0,$$

so that approximating the nonlinear terms by the Taylor series expansion about $(\overline{u}, \overline{\theta}) = (0, 0)$ and taking deviation variables:

$$\ddot{\theta}\cos\theta \approx \ddot{\overline{\theta}}\cos\overline{\theta} + (\cos\overline{\theta})\ddot{\tilde{\theta}} - (\ddot{\overline{\theta}}\sin\overline{\theta})\tilde{\theta} \approx \ddot{\tilde{\theta}},$$
$$\dot{\theta}^2\sin\theta \approx \dot{\overline{\theta}}^2\sin\overline{\theta} + (2\overline{\theta}\sin\overline{\theta})\dot{\tilde{\theta}} + (\dot{\overline{\theta}}^2\cos\overline{\theta})\tilde{\theta} \approx 0,$$
$$\ddot{x}\cos\theta \approx \ddot{\tilde{x}},$$
$$\sin\theta \approx \sin\overline{\theta} + (\cos\overline{\theta})\tilde{\theta} \approx \tilde{\theta}.$$

The linearized Eq. (2.43) and (2.44) are obtained as

$$(M + m)\ddot{\tilde{x}} = \tilde{u} - b\dot{\tilde{x}} - ml\ddot{\tilde{\theta}}, \qquad (2.45)$$
$$(I + ml^2)\ddot{\tilde{\theta}} = mgl\tilde{\theta} - ml\ddot{\tilde{x}}, \qquad (2.46)$$

which are the same as (2.36) and (2.37) but including the friction term.

2.5.1.3 References Related to this Concept

- [18] Ogata, K. (2010). *Modern control engineering* (5th ed.). Prentice Hall. ISBN: 978-0-13-615673-4. Chapter 3, example 3-5, pp. 68–72.
- [43] Shahian, B., & Hassul, M. (1993). *Control system design using MATLAB*®. Prentice Hall. ISBN: 0-13-174061-X. Chapter 7, section 9, pp. 217–219; Appendix A, section 4, pp. 476–488.
- [49] Aracil, J., & Gordillo, F. (2005). The inverted pendulum: A challenge for nonlinear control. *Revista Iberoamericana de Automática e Informática Industrial, 2*(2), 8–19.

2.5.1.4 Further Reading

Table 2.4 provides information on where to find different approaches to the modeling and control of this plant (also treated in Chaps. 5 and 9).

Table 2.4 Inverted pendulum on a cart system references

Ref.	Section/page	Main contents
[4]	Example 2.1-2, pp. 2.9–2.11	Modeling
	Example 2.8, pp. 2.24–2.25	Balance system
	Exercise 2.2, p. 2.37	Modeling
	Section 3.5, p. 3.18	Stabilization using PD
	Example 8.5, pp. 8.12–8.14	Transfer function and pole-zero diagram
	Exercise 8.3, p. 8.35	Transfer function obtaining
	Exercise 8.13, p. 8.37	PD control (tracking and disturbance rejection)
	Example 9.6, pp. 9.10–9.11	Application of Nyquist stability criterion
	Exercise 11.18, p. 11.36	Stabilization of an inverted pendulum with visual feedback
[6]	Example 3.4, pp. 186–187	Inverted pendulum control
	Example 6.9, p. 404	State-space representation
	CP8.8, p. 632	Block diagram and control
	Example 11.6, pp. 844–846	Control in state-space
	Example 11.10, pp. 853–856	Compensator design in state-space
[8]	Exercise 3.30, p. 151	Basic modeling and linearization
	Problem 9.33, pp. 511–512	Steady-state errors
	Problem 12.43, p. 701	PD control
	Problem 13.27, p. 758	PID control under time-domain specifications
[10]	Problem B.2, pp. 837–839	Modeling and linearization around the equilibrium point
	Problem 2.16, pp. 150–153	Double inverted pendulum (with two rods and two balls)
[17]	Problem 2-9	Free-body diagram and dynamic equation of the motion
	Problem 7-70	PD controller design (rise time and overshoot specifications)
	Problem 8-64	State-space model and state feedback control
[18]	Example 3-5, pp. 68–72	Modeling, transfer function and state-space description
	Example 10-5, pp. 746–749	Control in state-space
[19]	Example 2.1, pp. 29–30	Modeling
	Problem 2.1, p. 62	Modeling using a Lagrangian
[20]	Example 2.5, pp. 21–24	Modeling with Lagrange's equations and simulation
	Example 2.9, p. 30	Linearization about an operating point
	Exercise 2.19, p. 42	Two-pendula problem
	Example 3.3, pp. 57–59	Pole calculation (eigenvalues)
	Example 4.10, pp. 161–162	H_∞ design
	Example 5.7, p. 188	Root locus
[21]	Example 2.8, pp. 22–23	Modeling, linearization, transfer function
[23]	Section 5.6, pp. 198–201	Modeling
[24]	Example 5.14, pp. 219–222	Free-body and kinematic diagrams, equations of motion. Linearization and state-space
	Example 5.18, pp. 230–233	Energy method (modeling)
	Problem 10.10, pp. 576–577	Linearization, full-state feedback controller
[25]	Example 2.7, pp. 37–38	Equations of motion
	Exercise 6.48, p. 404	Frequency control design
	Example 7.23, pp. 460–462	Root locus of the linearized system
[27]	Chapters 3, 9, 24	and the web site
	Example 3.7, pp. 55–57	Nonlinear equations of motion. State space representation
	Example 4.3, p. 72	Transfer function
	Section 9.6, pp. 254–258	Example of design trade-offs
	Example 9.4, pp. 254–257	Inverted pendulum without angle measurement. Control design. Sensitivities
	Example 24.3, pp. 775–776	Example of non-square systems in MIMO control
[29]	Section 2.6.5, pp. 124–126	Description and modeling
	Section 3.4.3, pp. 152–154	State-space
[31]	Exercise 8.2, p. 53	Linearization of system dynamics

(continued)

Table 2.4 (continued)

Ref.	Section/page	Main contents
[35]	Example 2.2, pp. 18–19	Modeling
[39]	Problem, pp. 695–696	Modeling
[41]	Example 3.13, p. 112	Stabilization with PD control applying Nyquist theorem
	Example 3.27, pp. 129–130	Modeling
	Example 3.33, p. 136	Linearization
	Example 3.36, p. 142	Transfer function
	Example 5.13, p. 251	Fundamental limitations
[43]	Problem 7.6, pp. 217–218	Modeling
	Appendix 4, pp. 476–489	Inverted pendulum complete project
[47]	Design study B	Examples (with solutions)
	B.2, pp. 23–26	Kinetic energy
	B.3, pp. 39–43	Equations of motion
	B.4, pp. 56–58	Linearization of equations of motion
	B.5, pp. 67–69	Transfer function model
	B.6, pp. 81–83	State-space model
	B.8, pp. 116–121	Successive loop closure
	B.9, pp. 137–140	Integrators and system type
	B.P.6, pp. 473–474	Root locus
	B.10, pp. 153–155	Digital PID
	B.11, pp. 177–180	Full state feedback
	B.12, pp. 195–199	Full state with integrator
	B.13, pp. 222–227	Observer based control
	B.14, pp. 247–251	Disturbance observer
	B.15, pp. 273–277	Frequency Response
	B.16, pp. 295–296	Loop gain
	B.17, pp. 311–314	Stability margins
	B.18, pp. 345–358	Loop shaping design
[48]	Section 2.3.1, pp. 21–23	Modeling
[50]	Exercise 1.20, pp. 89–91	Equations of motion
[51]	Exercise 1.13, pp. 43–45	State-space oriented modeling
[52]	Example 7.6, pp. 139–140	Control limitations
[53]	Section 2.2.2.8, p. 91	Lagrangian mechanics
	Example 7.2, p. 496	Position control with unlimited motion
[54]	Section 14.7, pp. 240–247	Stabilizing an unstable system. Modeling, transfer function, feedback control, PI control, Routh array for stability analysis, root locus, PID control, phase-lead compensation (root locus design)
[55]	Example 25.3 pp. 25–12	Modeling and state-space control

Time Response: Inverted Pendulum on a Cart System Linearization

In this case, no explicit tool has been developed for the analysis of the time response of the nonlinear and linearized models because, when excited with a step input, its output grows indefinitely. It represents an unstable system from the BIBO point of view and therefore its analysis only makes sense in Chap. 9 where it will be stabilized through feedback (see Sect. 9.5).

2.5.1.5 Homework

1. Integrate the ODE (2.41) considering a fixed value of u. Analyze the obtained response.
2. If it is considered that a system is stable when entering a bounded input, the output is also bounded, why the inverted pendulum represented by (2.41) can be considered an unstable system?

2.6   DC Motor System

2.6.1   Interactive Tool: DC_motor_control

2.6.1.1 Concepts Analyzed in the Card and Learning Outcomes

- Modeling a linear electromechanical system.
- Meaning of deviation variables.
- Relationship between angular velocity and angular position dynamics.

2.6.1.2 Summary of Fundamental Theory

A DC motor is an actuating device that provides power to a load. It converts electrical energy in the form of direct current into rotational mechanical energy. In order to produce torque, a magnetic field containing electric conductors is required. When current flows through the electric conductor, it reacts with the magnetic field producing a force responsible for producing the torque and thus the movement of the motor. The principle of operation of DC generators is similar, inducing an electromotive force (EMF).

Magnetic poles (North, South) are used to produce the magnetic field, which are often created by a coil wound on a ferromagnetic core. The poles are usually located in the stator (static block of the machine), an area referred to as the field. The conductors referred to in the previous paragraph necessary for the production of torque are placed in a mobile ferromagnetic core, called rotor, to facilitate the rotation when the force is produced on them. This area of the motor is called the armature. Much of the torque generated in the rotor (armature) will be available to move an external load.

DC motors are used as servomotors in control systems, including drive amplifiers and gear boxes. Different configurations can be found, but basically an embedded tachometer is used to convert angular velocity to a proportional electric signal (voltage). By integrating this signal or by using position sensors (mainly potentiometers or different types of encoders), an electric signal is also obtained proportional to the angular position of the motor shaft. In many laboratory plants, the motor drives a steel disk acting as a load and an adjustable magnetic brake applies a viscous friction effect, thus allowing modification of the dynamics. Angular position and angular velocity are controlled by adjusting the voltage applied to the motor.

The most commonly used DC motor is that which has separately excited fields: the armature and the field. Figure 2.9 shows the electromechanical scheme of the motor. In this section, the model of an armature-controlled motor is developed, where the control signal is the voltage applied to the armature circuit of the DC motor and a fixed voltage is applied to the field winding.

The DC motor model is obtained for a linear approximation of a real motor and second-order effects are neglected, as well as hysteresis and voltage drop across the brushes (if they exist). The armature has been modeled as an electrical circuit with a resistance R_a in series with an inductance L_a, and a voltage source $e_b(t)$ representing the back electromotive force (back EMF) generated in the armature by the rotation of the rotor. A voltage $e_a(t)$ is applied to the armature circuit of the DC motor, so that current $i_a(t)$ circulates through the armature winding. The field winding has been represented as a resistance R_f in series with an inductance L_f. A voltage $e_f(t)$ is applied generating current $i_f(t)$.

According to the diagram in Fig. 2.9, there are two possibilities for motor control: Armature control through the voltage $e_a(t)$ and field control, using $e_f(t)$. It is assumed that

- The magnetic flux inside the motor $\phi(t)$ is proportional to the field current i_f

$$\phi(t) = K_f i_f(t), \tag{2.47}$$

where K_f is the proportionality constant of the magnetic flux generated by the field.
- The torque developed by the motor T_m is proportional to the magnetic flux $\phi(t)$ and the current in the armature $i_a(t)$, therefore

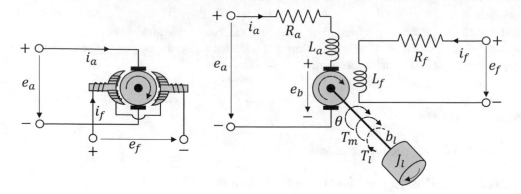

Fig. 2.9 Diagram of the DC motor

$$T_m(t) = K_m \phi(t) i_a(t), \tag{2.48}$$

where K_m is a torque proportionality constant.

In armature control, $e_a(t)$ is variable and $e_f(t) = e_f$ is held at a constant value[15], therefore $i_f(t) = i_f$ and $\phi(t) = \phi$ are also constants and from Eqs. (2.47) and (2.48)

$$T_m(t) = K_t i_a(t), \tag{2.49}$$

where K_t is the motor torque constant, usually provided by the manufacturer with units Nm/A.

For a constant flux $\phi(t) = \phi$, the voltage induced in the armature is directly proportional to the angular velocity. Thus

$$e_b(t) = K_b \frac{d\theta(t)}{dt}, \tag{2.50}$$

where K_b is the back EMF constant of the motor [18], also provided by the manufacturer with units V/rpm or V/rad/s and $\theta(t)$ is the angular displacement of the motor shaft.

The differential equation for the armature circuit from Kirchhoff's voltage law is

$$L_a \frac{di_a(t)}{dt} + R_a i_a(t) + e_b(t) = e_a(t). \tag{2.51}$$

As far as the conversion into mechanical energy is concerned, the torque developed by the motor shaft $T_m(t)$ is used to move a load subjected to friction in the mechanical elements (bearings). The equation for torque equilibrium is

$$J_l \frac{d^2\theta(t)}{dt^2} + b_l \frac{d\theta(t)}{dt} = T_m(t), \tag{2.52}$$

where J_l and b_l are, respectively, the inertia and the viscous friction coefficient of the combination of the motor and load referred to the motor shaft. Sometimes the left part of Eq. (2.52) includes a load torque disturbance T_l to model other effects besides inertia and friction.

Then, combining Eqs. (2.51) and (2.52), the fundamental equations that model the dynamics of the DC motor are obtained

$$L_a \frac{di_a(t)}{dt} + R_a i_a(t) + K_b \frac{d\theta(t)}{dt} = e_a(t), \tag{2.53}$$

$$J_l \frac{d^2\theta(t)}{dt^2} + b_l \frac{d\theta(t)}{dt} = K_t i_a(t). \tag{2.54}$$

---

[15] In [29] (Sect. 2.6.1), the model of the motor is obtained also for the case when the excitation voltage e_f is varying through a linearization process.

Notice that, under the assumptions and simplifications made, both differential equations are linear, that being the reason why the model of this plant is not usually formulated in terms of deviation variables, as it is supposed that the initial conditions defining the operating point are zero.

Regarding examples of mathematical models limitations explained at the introduction of the chapter and following [2], the parameters of the electrical part (R_f, L_f, R_a, L_a) are manufactured with a tolerance of 10%. The mechanical constants J_l and b_l can vary greatly as plant operating conditions change. Obviously, these parameters will be different if the motor is used to position an antenna platform or to drive a conveyor belt, for instance. There are therefore parameter inaccuracies. In the model developed, it is assumed that the motor torque is applied directly on the load, but there is actually a deformation at the junction of the motor with the load, so this junction could be modeled by two segments connected through a torsional spring. This is an example of unmodeled dynamics.

In the motor, it is also assumed that the torque is proportional to the current, but in reality, it is proportional to the magnetic flux created by the field current. This actually causes the relationship between current and torque to have a saturation and hysteresis characteristics. This is an example of the presence of nonlinearities in the modeling of the system.

Section 12.5 in Ref. [56] contains a deep explanation of the features and instrumentation of DC motors. Reference [28] also contains detailed information, summarized in Table 2.5.

2.6.1.3 References Related to this Concept

- [2] Rohrs, C. E., Melsa, J. L., & Schultz, D. G. (1994). *Sistemas de control lineal (Linear control systems)*. McGraw-Hill. ISBN: 978-0-07-041525-6. Example 2.2-4, pp. 26–28; Chapter 2, Section 2.7, pp. 77–88.
- [6] Dorf, R. C., & Bishop, R. H. (2011). *Modern control systems* (12th ed.). Prentice Hall. ISBN: 978-0-13-602458-3. Chapter 2, example 2.5, pp. 70–74.
- [18] Ogata, K. (2010). *Modern control engineering* (5th ed.). Prentice Hall. ISBN: 978-0-13-615673-4. Chapter 3, example A-3-9, pp. 95–97.

2.6.1.4 Further Reading

The DC motor system is used in many automatic control textbooks as a test-bed example. Table 2.5 provides information on where to find different approaches to the modeling and control of this plant (which is also treated in Chaps. 5 and 9).

Time Response: DC Motor System

In this case, no explicit tool has been developed for the analysis of the time response because when excited with a step input, its output grows indefinitely. It represents an unstable system from the BIBO point of view, and therefore, its analysis only makes sense in Chap. 9 where it will be stabilized through feedback (see Sect. 9.6).

2.6.1.5 Homework

1. Apply the steps followed for linearizing a model to Eqs. (2.53) and (2.54), writing the equations in deviation variables. Check that the same description as the original one is obtained in this case, as the model is linear. Can different operating points be defined in this system?
2. Integrate the ODEs (2.53) and (2.54) using a fixed value of e_a, both considering the angular velocity and the angular position as outputs of the plant. What is the steady-state value in both cases? Analyze the obtained response.
3. Add a torsional union between the motor and the load. Develop the model and the block diagram, comparing the result with the nominal one (Ref. [2], p. 79).
4. Obtain the linearized ODEs considering that the excitation voltage u_f is variable (Ref. [29], Sect. 2.6.1).

Table 2.5 DC motor system references

Ref.	Section/page	Main contents
[1]	Section 2.13	Modeling. Transfer function
	Section 5.5	Position control system - motor-generator control and associated block-diagram
[2]	Example 2.2.2	Field controlled DC motor moving a platform. Modeling. Transfer function
	Example 2.2.3	Block diagram
	Example 2.2.4	Armature controlled DC motor moving a load. Transfer function. Block diagram
	Section 2.5	State-space representation
[6]	Example 2.5, pp. 70–74	Transfer function
	Example 4.5, pp. 268–270	Speed control systems
	AP4.4. p. 295	Integral control
	AP5.6, p. 378	Steady-state errors
	DP5.6, p. 380	Motor position control
[8]	Case study, pp. 11–14, 92–95, 141–143, 202–205, 749–753	Antenna azimuth position control system
	Section 2.8, pp. 77–83	Electromechanical system transfer functions
	Exercise 3.18	State-space representation
	Section 4.9, pp. 192–195	Effect of nonlinearities upon time response
	Exercise 4.71	Time response
	Section 5.3, pp. 245–246	Analysis and design of feedback systems
	Exercise 5.72	Effect of feedback on transient response
	Exercise 9.43	Block diagrams and steady-state errors
	Exercise 9.44	Position control in time domain
	Example 11.1, pp. 616–617	Design in frequency domain
	Appendix I.	Derivation of a schematic for a DC motor (online)
[10]	Problem 2.14, pp. 146–147	Modeling (state-space)
	Exercise 2.21, p. 175	DC motor-load modeling
	Exercise 2.22, p. 175	Transfer function
	Exercise 2.23, p. 175	Model of a drum rotating by two motors
[15]	Example 2.6 p. 31	Modeling
[16]	Section 4.4.1, pp. 71–74	Modeling and transfer function
	Example 4.8, pp. 104–106	Steady-state analysis
	Example 4.9, pp. 106–107	Closed-loop transfer function for field controlled DC motor
[17]	Case study, Section 1-11	Intelligent vehicle obstacle avoidance. Uses a DC motor. Model and block diagram of PID control
	Section 2-6	Mechanical modeling
	Section 2-1-2	Modeling. Rigid and nonrigid coupling between motor and load
	Problem 3-31	Torque equations of the system. Transfer function. Characteristic equations. Steady-state analysis
	Problem 5-14	Asymptotic stability analysis
	Chapter 6	Modeling of active electrical elements. Sensors and encoders. Speed and position control of a DC motor. Position control. Practical examples. Motor modeling and characterization
	Section 7-10	Position control
	Appendix D-4	Simulation and experiments
[18]	Example A-3-9, pp. 95–97	Block diagrams. Modeling and transfer functions
[20]	Example 2.1, pp. 10–13	Model and simulation
	Example 2.6, p. 28	Equilibrium point and linearization
	Problem 2.4, p. 34	Simplified model with negligible inductance
	Problem 2.5, pp. 34–35	Servo with flexible shaft
	Example 3.6, pp. 63–64	Input–output and disturbance–output transfer functions
	Example 4.6, p. 146	DC motor control and sensitivity functions
	Example 5.3, pp. 179–180	Routh array DC servo (third-order)
	Example 5.6, pp. 186–188	Root locus DC servo (third-order)
	Example 6.2, pp. 235–236	Steady-state error for unit step reference input and unit step load torque

(continued)

Table 2.5 (continued)

Ref.	Section/page	Main contents
	Example 6.7, pp. 249–251	Proportional control: gain and phase margins specifications
	Example 6.8, pp. 254–255	Lag compensator: transient and steady-state specifications
	Example 6.10, pp. 263–264	Lead compensator classical design
	Example 6.11, pp. 265–268	Lead-lag compensator
	Example 6.12, pp. 268–269	PID design with phase margin and gain crossover frequency specifications
	Example 6.13, p. 270	Feedforward control
[22]	Section 7.2, pp. 202–204	Transfer function and state-space
[23]	Example 2.41, 2.46	Modeling
	Example 3.95, 3.99	Transfer function and time responses
	Example 4.29	Proportional feedback control
[24]	Section 6.4.2, pp. 295–299	Armature-controlled motors
	Section 6.6.3, pp. 316–323	Direct current motors (Simulink® and Simscape®)
[25]	Example 2.13, pp. 48–50	Modeling
	Example 3.17, pp. 98–99	Transfer function
	Example 4.4, p. 184	System type
	Example 4.7, pp. 190–191	PI control of position
	Example 6.15, pp. 351–356	Lead compensation
	Example 6.19, pp. 363–365	Lag compensation
	Example 7.5, pp. 420–421	Modeling in state-space
	Example 7.19, p. 454	Reference input
[27]	Example 3.4. pp. 47–49	Modeling: differential equations and state-space
	Example 4.2. pp. 67–70	Transfer function and time response
[28]	Section 1.3.1, pp. 7–9	A position control system
	Section 2.4.1. p. 59	Armature of a permanent magnet brushed DC motor
	Example 3.19, pp. 142–144	Model of permanent magnet brushed DC motor. Transfer function
	Section 3.8.1, pp. 158–163	Proportional control of velocity in a DC motor (controlling first- and second-order systems)
	Section 3.8.2, pp. 163–166	Proportional position control plus velocity feedback
	Section 3.8.3. pp. 166–170	PD position control of a DC motor
	Section 3.8.4. pp. 170–174	PI velocity control of a DC motor
	Exercise 3.11, p. 186	Proportional control with velocity feedback
	Exercise 3.18, p. 189	Natural and forced responses
	Example 4.17, pp. 221–224	Stability criteria and steady-state error. PI control of velocity
	Example 4.19, pp. 224–225	Steady-state errors in velocity with proportional control
	Example 4.20, pp. 225–227	PID control of DC motor position
	Example 4.21, pp. 228–229	Tracking a parabola
	Section 5.2, pp. 245–312	Root locus-based analysis and design
	Section 5.2.1, pp. 245–250	Proportional control of position
	Section 5.2.2, pp. 251–256	PD control of position
	Section 5.2.3. pp. 256–258	Position control using a lead compensator
	Section 5.2.4. pp. 258–266	PI control of velocity
	Section 5.2.5. pp. 266–277	PID control of position
	Section 5.3, pp. 306–312	Additional notes on the PID control of position
	Section 6.7, pp. 380–403	Frequency response-based design
	Section 6.7.1, pp. 380–388	PD position control of a DC motor
	Section 6.7.2., pp. 388–394	Redesign of the PD position control for a DC motor
	Section 6.7.3, pp. 394–400	PID position control of a DC motor
	Section 6.7.4, pp. 401–403	PI velocity control of a DC motor
	Example 7.7, pp. 475–477	State-space representation
	Chapter 10, pp. 605–655	Velocity control of a permanent magnet brushed DC motor
	Chapter 11, pp. 645–696	Position control of a permanent magnet brushed DC motor. Controller design using time response: P, PD, PI, PID, phase-lead controllers, 2-DoF controllers

(continued)

Table 2.5 (continued)

Ref.	Section/page	Main contents
[29]	Example 1.4, p. 31	Linearization of the moment equation of a DC motor
	Section 2.6.1, pp. 109–115	Description
[30]	DC Motor, pp. 59–62	DC motor modeling
	Example 5.17, pp. 142–143	Transfer function and block diagram
	pp. 124–125	State-space representation
	Example 10.1, pp. 326–327	Closed-loop transfer function
	Example 10.2. pp. 327–328	Poles of the closed-loop transfer function and step response
	Example 10.3, pp. 326–327	Repeat previous neglecting inductance
	Example 10.5, pp. 334–337	P and PI control of angular velocity
[31]	Exercise 2.1, p. 3	Modeling, transfer function
	Exercise 3.6, p. 12	Root locus
	Exercise 3.8, pp. 12–13	Cascade controlled DC-motor. Root locus with velocity and position feedback
	Exercise 3.17, p. 16	Oscillations in presence or time delay
	Exercise 3.25, p. 18	Block diagram and closed-loop transfer function under P and PI control. Steady-state errors
	Exercise 5.1, p. 29	Control moving a mechanical resonance
	Exercise 5.4, p. 30	Position servo including DC motor. Time-domain specifications and control
	Exercise 5.9, p. 33	Servo based on DC-motor. Models from bode plots and step responses. Control design
	Exercise 6.7, p. 46	Root locus with an uncertain parameter. Robustness criteria
[32]	Section 1.5.3, pp. 17–19	Modeling with load shaft
	Example 11.21	Feedback control in presence of disturbances
[36]	Section 2.1.3.3, pp. 35–36	Modeling of armature current controlled DC motor
	Exercise 2.10, p. 52	DC motor coupled to a disk by means of a shaft drive
	Problem 8.2, pp. 254–256	Root locus and proportional control
[37]	Example 2.1., pp. 28–32	Modeling and typical nonlinearities
	Examples 6.8, 6.9, pp. 168–171	Transfer function and time response through the inverse Laplace transform
	Example 7.6, pp. 207–210	Feedback control through an amplifier gain
[38]	Exercise 2.16, pp. 71–75	Block diagrams, transfer functions, steady state, time response
	Exercise 3.8, p. 118	Steady-state errors
	Exercises 4.6, 4.7, pp. 145–147	Position servomechanism. PD control
[39]	Section 5.6, pp. 135–144	Modeling of a DC motor in a conveying process
	Section 12.5, pp. 331–357	PD control
	Example, pp. 363–364	Root locus contour
	Problem, pp. 373–375	Root locus design
	Section 16.3, pp. 477–481	PI control
[41]	Example 2.7, pp. 62–63	Block diagram
	3.28, pp. 130–131	Modeling
[48]	Section 2.3.6, pp. 30–31	Modeling
[50]	Example 4.3, pp. 14–16	Modeling
	Exercise 2.71, pp. 206–207	State-space modeling
[51]	Example 1.3, pp. 14–15	Modeling
	Exercise 9.9, p. 408	State-space modeling
[53]	Section 11.2.4, p. 799	Models for mechanisms with geared actuators
[54]	Section 9.1 pp. 122–123	A head-positioning system
	Section 14.3.2, pp. 218–219	Closed-loop control with DC motors (hardware design)
[55]	Section 9.3.5.2., pp. 9–38	Type 0 feedback control system
	Section 9.4.5.1, pp. 9–65	Design of a proportional feedback gain
	Chapter 23	Modeling and control of brushless DC motors

(continued)

Table 2.5 (continued)

Ref.	Section/page	Main contents
[56]	Problem 7.10, 7.11, pp. 318–319	Motor modeling
	Problem 7.23, p. 325	Electrical and mechanical time constants
	Example 8.15, p. 374	Bode diagrams
	Example 9.1, p. 414	Position PD control
	Section 12.5, pp. 630–653	Instrumentation of DC motors
[57]	Example 11.1, pp. 384–385	Modeling. Block diagram of an electromechanical feedback control system
[58]	Example 3-3, pp. 3.3–3.7	Reaction wheel: a rotational first-order system
	Example 6-3, pp. 6.5–6.6	Response of velocity to a ramp input moment
	Chapter 14, pp. 14.1–14.18	Output operations for control of rotational position
	Problem 17.4, pp. 17.25–29	Estimation of the frequency response and closed-loop stability

References

1. D'Azzo, J. J., Houpis, C. H., & Sheldon, S. N. (2003). *Linear control system analysis and design with MATLAB*® (5th ed.). Marcel Dekker Inc.
2. Rohrs, C. E., Melsa, J. L., & Schultz, D. G. (1994). *Sistemas de control lineal (Linear control systems)*. McGraw-Hill.
3. Aracil, J., & Gómez-Estern, F. (2005). *Notes on automatic regulation*. School of Engineering, University of Seville.
4. Åström, K. J., & Murray, R. M. (2014). *Feedback systems: An introduction for scientists and engineers* (2nd ed.). Princeton University Press.
5. Åström, K. J., & Hägglund, T. (2006). *Advanced PID control*. ISA - The Instrumentation, Systems and Automation Society.
6. Dorf, R. C., & Bishop, R. H. (2011). *Modern control systems* (12th ed.). Prentice Hall.
7. Franklin, G. F., Powell, J. D., & Emani-Naeni, A. (2015). *Feedback control of dynamic systems* (7th ed.). Pearson.
8. Nise, N. S. (2015). *Control systems engineering* (7th ed.). Wiley.
9. Seborg, D. E., Edgar, T. F., Mellichamp, D. A., & Doyle, F. J., III. (2011). *Process dynamics and control* (3rd ed.). International Student Version. Wiley.
10. Bavafa-Toosi, Y. (2017). *Introduction to linear control systems*. Academic Press-Elsevier.
11. Roubal, J., Husek, P., & Stecha, J. (2010). Technical committee on control education: A first course in systems and control engineering [technical activities]. *IEEE Transactions on Education, 53*(3), 413–418.
12. Qian, D., Yi, J., & Tong, S. (2013). Understanding neighborhood of linearization in undergraduate control education. *IEEE Control Systems Magazine, 33*(4), 54–60.
13. Johansson, K. H. (2000). The quadruple-tank process: A multivariable laboratory process with an adjustable zero. *IEEE Transactions on Control Systems Technology, 8*(3), 456–465.
14. Guzmán, J. L., Vargas, H., Sánchez-Moreno, J., Rodríguez, F., Berenguel, M., & Dormido, S. (2007). Education research in engineering studies: Interactivity, virtual and remote labs. In *Distance Education Research Trends* (pp. 131–167). New York: Nova Science Publishers Inc.
15. Bolzern, P., Scattolini, R., & Schiavoni, N. (2009). *Fundamentos de control automático (Fundamentals of automatic control)*. McGraw-Hill.
16. Burns, R. S. (2006). *Advanced control engineering*. Butterworth Heinemann.
17. Golnaraghi, F., & Kuo, B. C. (2017). *Automatic control systems* (10th ed.). McGraw Hill Education.
18. Ogata, K. (2010). *Modern control engineering* (5th ed.). Prentice Hall.
19. Bechhoefer, J. (2021). *Control theory for physicists*. Cambridge University Press.
20. Bélanger, P. (1995). *Control engineering: A modern approach*. Saunders College Publishing.
21. Chen, C.-T. (1999). *Linear system theory and design* (3rd ed.). Oxford University Press.
22. de Larminat, P. (Ed.). (2007). *Analysis and control of linear systems*. ISTE.
23. de Oliveira, M. C. (2017). *Fundamentals of linear control. A concise approach*. Cambridge University Press.
24. Esfandiari, R. S., & Lu, B. (2018). *Modeling and analysis of dynamic systems* (3rd ed.). CRC Press, Taylor & Francis Group.
25. Franklin, G. F., Powell, J. D., & Emani-Naeni, A. (2010). *Feedback control of dynamic systems* (6th ed.). Pearson.
26. Golnaraghi, F., & Kuo, B. C. (2010). *Automatic control systems* (9th ed.). Wiley.
27. Goodwin, G. C., Graebe, S. F., & Salgado, M. E. (2001). *Control system design*. Prentice Hall.
28. Hernández-Guzmán, V. M., & Silva-Ortigoza, R. (2019). *Automatic control with experiments*. Springer.
29. Keviczky, L., Bars, R., Hetthéssy, J., & Bányász, C. (2019). *Control engineering*. Springer.
30. Kluever, C. A. (2015). *Dynamic systems: Modeling, simulation, and control*. Wiley.
31. KTH (2016). Royal Institute of Technology and Linköpings Universitet, Sweden. Reglerteknik ak med utvalda tentamenstal (automatic control exercises: Computer exercises, laboratory exercises). Retrieved July 01, 2011, from https://cutt.ly/jYkcFZV.
32. Lobontiu, N. (2010). *System dynamics for engineering students*. Academic Press.
33. Lund University, Department of Automatic Control, and Faculty of Engineering, Sweden. (2017). Automatic control exercises. Retrieved July 01, 2021, from https://cutt.ly/xUXjNue.
34. Najim, K. (2006). *Control of continuous linear systems*. ISTE.
35. Ozbay, H. (1999). *Introduction to feedback control theory*. CRC Press.
36. Tenreiro Machado, J., Lopes, A. M., Duarte, V., & Galhano, A. M. (2016). *Solved problems in dynamical systems and control*. The Institution of Engineering and Technology.

37. Tripathi, A. N. (1998). *Linear system analysis* (2nd ed.). New Age International Publishers.
38. Veloni, A., & Palamides, A. (2012). *Control system problems. Formulas, solutions and simulation tools*. CRC Press.
39. Wilkie, J., Johnson, M., & Katebi, R. (2002). *Control engineering. An introductory course*. Palgrave Macmillan.
40. Tavakolpour-Saleh, A., Setoodeh, A., & Ansari, E. (2016). Iterative learning control of two coupled nonlinear spherical tanks. *International Journal of Mechanical and Mechatronics Engineering, World Academy of Science, Engineering and Technology, 10*(11), 1862–1869.
41. Åström, K. J. (2004). *Introduction to control*. Department of Automatic Control, Lund Institute of Technology, Lund University.
42. Hirsch, R. (2008). EDUMECH - Mechatronic Instructional Systems - Ball on Beam System. Shandor Motion Systems. Retrieved July 01, 2021, from https://cutt.ly/KUogfV1.
43. Shahian, B., & Hassul, M. (1993). *Control system design using MATLAB®*. Prentice Hall.
44. Acharya, M., Bhattarai, M., & Poudel, B. (2014). Real time motion assessment for positioning in time and space critical systems. *International Journal of Applied Research and Studies (iJARS), 3*(7), 1–11.
45. Bolivar-Vincenty, C. G., & Beauchamp-Báez, G. (2014). *Modelling the ball-and-beam system from Newtonian mechanics and from Lagrange methods*. In *Twelfth LACCEI Latin American and Caribbean Conference for Enginering and Technology*, Guayaquil, Ecuador.
46. Wellstead, P. E., Chrimes, V., Fletcher, P. R., Moody, R., & Robins, A. J. (1989). Ball and beam control experiment. *The International Journal of Electrical Engineering & Education, 15*, 21–39.
47. Beard, R. W., McLain, T. W., Peterson, C., & Killpack, M. (2017). *Introduction to feedback control using design studies*. Chicago: Independently Published.
48. d'Andréa-Novel, B., & De Lara, M. (2013). *Control theory for engineers. A primer*. Springer.
49. Aracil, J., & Gordillo, F. (2005). The inverted pendulum: A challenge for nonlinear control. *Revista Iberoamericana de Automática e Informática Industrial, 2*(2), 8–19.
50. Antsaklis, P. J., & Michel, A. N. (2006). *Linear systems*. Birkhäuser.
51. Antsaklis, P. J., & Michel, A. N. (2007). *A linear systems primer*. Birkhäuser.
52. Braslavsky, J. H. (2001). Control automático 2: Notas de clase (Automatic control 2: Lecture notes). Technical report, Universidad Nacional de Quilmes.
53. Dodds, S. J. (2015). *Feedback control. Linear, nonlinear and robust techniques and design with industrial applications*. Springer.
54. Haidekker, M. A. (2013). *Linear feedback controls: The essentials*. Elsevier.
55. Levine, W. S. (Ed.). (2011). *The control handbook*. CRC Press.
56. de Silva, C. W. (2009). *Modeling and control of enginnering systems*. CRC Press, Taylor & Francis Group.
57. Boulet, B. (2006). *Fundamentals of signals & systems*. Charles River Media, An imprint of Thomson Learning Inc.
58. Hallauer, W. L. Jr. (2016). *Introduction to linear, time-invariant, dynamic systems for students of engineering*. Creative Commons.

Time Response

3.1 Introduction

As indicated in Chap. 2, the time response of a dynamical system has two components, the transient response and the steady-state one. The transient response consists of the time evolution of the system output from the time when the system input is modified until the time when it reaches a state near or equal to the steady state. The transient response occurs in the first period of time after the input is modified and reflects the evolution between the initial condition and the steady-state solution [1]. The portion of the response after the transient is called the steady-state response, that is how the output of the system behaves when time t tends to infinity. It reflects the long-term behavior of the system under the given inputs. For stable systems and inputs that are periodic, the steady-state response will be periodic, and for constant inputs, the response will often be constant [1].

This chapter deals with single-input single-output (SISO) linear time-invariant (LTI) dynamical systems, described by linear ordinary differential equations[1] of the form:

$$a_n \frac{d^n y(t)}{dt^n} + a_{n-1} \frac{d^{n-1} y(t)}{dt^{n-1}} + \cdots + a_1 \frac{dy(t)}{dt} + a_0 y(t) = \tag{3.1}$$

$$= b_m \frac{d^m u(t)}{dt^m} + b_{m-1} \frac{d^{m-1} u(t)}{dt^{m-1}} + \cdots + b_1 \frac{du(t)}{dt} + b_0 u(t),$$

where $y(t)$ is the system output, $u(t)$ the input, a_i $(i = 0 \ldots n)$ and b_ℓ $(\ell = 0 \ldots m)$ are real numbers, and n is the order of the equation.

To carry out the analysis of the time response, two descriptions are typically used as follows:

- **Internal description (state-space description):** As discussed in the introduction to Chap. 2, the internal description of a dynamical system is based on the concept of *state*. The state of a dynamical system is the smallest collection of variables whose value, in a certain instant, summarizes the dynamic past evolution of the system and it is sufficient to predict its future evolution. The variables that represent the state do not necessarily have to correspond to measurable physical quantities. Therefore, the internal description establishes an indirect relationship between the input and output signals of the system [2, 3].

- **External description (input–output description):** It provides an explicit direct relationship between the input and output signals, usually in the form of a *transfer function*, defined as the Laplace transform of the impulse response of the system [4]. From this definition, it also represents the quotient of the Laplace transform of the system output and the Laplace transform of the input with zero initial conditions.[2] Notice that the solution of a differential equation as (3.1) is given by the convolution integral[3]:

Supplementary Information The online version contains supplementary material available at https://doi.org/10.1007/978-3-031-09920-5_3.

[1] The input of the systems and their representative models will be represented by $u(t)$ and the corresponding outputs by $y(t)$ in all cases.

[2] This is a logical condition in linear causal systems and in linearized systems where the variables represent deviations from equilibrium or operating point. It is assumed that the reader has prior knowledge of the Laplace transform and its properties, which are usually included in classical automatic control texts. An excellent summary of the most common techniques for analysis in the complex variable domain (Laplace transform, Fourier series expansion, and Fourier transform) can be found in Appendix B of Ref. [5].

[3] The convolution integral can also be used in time-varying systems, while Laplace transforms cannot be used for such systems.

© Springer Nature Switzerland AG 2023
J. L. Guzmán et al., *Automatic Control with Interactive Tools*,
https://doi.org/10.1007/978-3-031-09920-5_3

Table 3.1 Main properties of the Laplace transform

Linearity	$\mathscr{L}\left(af(t) + bg(t)\right) = aF(s) + bG(s)$
Derivation in t	$\mathscr{L}\left[\dfrac{d^n f(t)}{dt^n}\right] = s^n F(s) - s^{n-1}f(0) - s^{n-2}f'(0) - \cdots - f^{n-1}(0)$
Integration in t	$\mathscr{L}\left[\int_0^t f(\xi)d\xi\right] = \dfrac{F(s)}{s} + \dfrac{\left[\int_0^t f(\xi)d\xi\right]_{0+}}{s}$
Time shifting	$\mathscr{L}\left(f(t - t_d)u_s(t - t_d)\right) = e^{-t_d s}F(s),\, t_d \geq 0$,
	u_s is the unit step (Heaviside) function.
Derivation in s	$\mathscr{L}^{-1}\left[\dfrac{dF(s)}{ds}\right] = -tf(t)$
Convolution	$f(t) * g(t) \xrightarrow{\mathscr{L}} F(s)G(s)$
Initial value theorem	$\lim\limits_{t\to 0^+} f(t) = \lim\limits_{s\to\infty} sF(s)$, if the time limit exists.
Final value theorem	$\lim\limits_{t\to\infty} f(t) = \lim\limits_{s\to 0} sF(s)$,
	if $sF(s)$ is analytic in the closed right half s-plane (RHP).

$$y(t) = \int_{-\infty}^{t} h(t - \xi)u(\xi)d\xi, \tag{3.2}$$

where $h(t)$ is the impulse response of the system (see Chap. 2). This impulse response has the causality or realizability property ($h(t) = 0,\quad \forall t < 0$) and system's stability requires convergence of the system, so that $\lim_{t\to\infty} h(t) = 0$ (asymptotic stability). A system is said to be *asymptotically stable* (AS) if all roots of the characteristic equation are in the open left side of the complex plane, open LHP (it will be analyzed in the given examples). In this text, the bounded-input bounded-output (BIBO) notion of stability is also used, so that if a system is stable, its output is bounded for every bounded input. For a continuous-time LTI system, the condition for BIBO stability is equivalent to the impulse response $h(t)$ being absolutely integrable:

$$\int_{-\infty}^{\infty} |h(t)|dt < \infty.$$

Notice that if an LTI system is AS, then it is also BIBO stable.[4] The transfer function will be the representation used in Chaps. 3–9 of this text.

3.1.1 Laplace Transform and Transfer Function

The Laplace transform is a very useful mathematical tool that has traditionally been used to obtain the response of linear systems to different inputs. In the automatic control field, its use is widespread through the treatment of the dynamic behavior of systems using the complex variable theory and has numerous properties of interest [1].

If a function $f(t)$ is considered, such that $f : \mathbb{R}^+ \to \mathbb{R}$ is integrable and that does not grow faster than $e^{s_0 t}$ for finite $s_0 \in \mathbb{R}$ and a time t sufficiently large, the Laplace transform of $f(t)$ is a function of the complex variable defined by [1]

$$\mathscr{L}(f(t)) = F(s) = \int_0^{\infty} e^{-st} f(t)dt, \quad \text{Re}(s) > s_0, \tag{3.3}$$

with $s = \sigma + j\omega$. The existence of the transform is conditioned by the convergence of the integral. This transformation has a number of very useful properties for linear systems, some of which will be used throughout this text. Table 3.1 summarizes the main properties and Table 3.2 contains some of the more common transforms.

As illustrative examples of the use of these properties, the *derivation in* t is the property used to obtain the transfer function from a linear differential equation considering zero initial conditions; the *final value theorem* allows an analysis of the steady state when working with transfer functions; the *time shifting* property facilitates the study of systems with time delay, etc. Moreover, the definition of transfer function states that it is the Laplace transform of the impulse response of the system:

---

[4] Always that there is no pole–zero cancellation (proof omitted).

Table 3.2 Table of most representative Laplace transforms

	Time domain	s domain
1. Impulse function	$f(t) = \delta(t) = \begin{cases} \infty, & t = 0 \\ 0, & t \neq 0 \end{cases} \quad (\int_{-\infty}^{\infty} \delta(t)dt = 1)$	$F(s) = \delta(s) = 1$
2. Unit step function	$f(t) = u_s(t) = \begin{cases} 0, & t < 0 \\ 1, & t \geq 0 \end{cases}$	$F(s) = U_s(s) = \dfrac{1}{s}$
3. Unit ramp function	$f(t) = u_r(t) = \begin{cases} 0, & t < 0 \\ t, & t \geq 0 \end{cases}$	$F(s) = U_r(s) = \dfrac{1}{s^2}$
4. Power	$f(t) = t^n \quad (t \geq 0)$	$F(s) = \dfrac{n!}{s^{n+1}}$
5. Exponential	$f(t) = e^{-at} \quad (t \geq 0)$	$F(s) = \dfrac{1}{s+a}$
6. Cosine	$f(t) = \cos(\omega t) \quad (t \geq 0)$	$F(s) = \dfrac{s}{s^2 + \omega^2}$
7. Sine	$f(t) = \sin(\omega t) \quad (t \geq 0)$	$F(s) = \dfrac{\omega}{s^2 + \omega^2}$
8. Damped cosine	$f(t) = e^{-at} \cos(\omega t) \quad (t \geq 0)$	$F(s) = \dfrac{s+a}{(s+a)^2 + \omega^2}$
9. Damped sine	$f(t) = e^{-at} \sin(\omega t) \quad (t \geq 0)$	$F(s) = \dfrac{\omega}{(s+a)^2 + \omega^2}$
10. Others	$f(t) = te^{-at} \quad (t \geq 0)$	$F(s) = \dfrac{1}{(s+a)^2}$
11.	$f(t) = t^n e^{-at} \quad (t \geq 0)$	$F(s) = \dfrac{n!}{(s+a)^{n+1}}$
12.	$f(t) = \dfrac{1}{a}(1 - e^{-at}) \quad (t \geq 0)$	$F(s) = \dfrac{1}{s(s+a)}$
13.	$f(t) = \dfrac{1}{(b-a)}(e^{-at} - e^{-bt}) \quad (t \geq 0)$	$F(s) = \dfrac{1}{(s+a)(s+b)}$
14.	$f(t) = \dfrac{\omega_n}{\sqrt{1-\zeta^2}} e^{-\zeta\omega_n t} \sin(\omega_n\sqrt{1-\zeta^2}\,t) \quad (0 < \zeta < 1)$	$F(s) = \dfrac{\omega_n^2}{s^2 + 2\zeta\omega_n s + \omega_n^2}$
15.	$f(t) = \sin(\omega t) - \omega t \cos(\omega t), \quad (t \geq 0)$	$F(s) = \dfrac{2\omega^3}{(s^2 + \omega^2)^2}$

$$H(s) = \int_0^{\infty} h(t)e^{-st}dt. \tag{3.4}$$

If an LTI system is considered and the Laplace transform is applied to the solution of equation (3.1) given by (3.2), taking into account that the Laplace transform of a convolution of two signals is the product of their transforms, it is obtained that $Y(s) = H(s)U(s)$, where it can be seen that $H(s)$ represents the transfer function of the system, that is usually represented as $G(s)$. Anyway, as explained before, in practice the transfer function is determined directly from the differential equation applying the property of derivation in t and directly obtaining $G(s) = Y(s)/U(s)$, always assuming zero initial conditions for the signals $y(t)$ and $u(t)$.

There also exists the *inverse Laplace transform* (denoted by $\mathscr{L}^{-1}$), which is an integral transformation of the function $F(s)$ from the s domain to the time domain $f(t)$ [4]. The inverse Laplace transform helps to provide the time evolution of the system output $y(t)$ based on its transfer function and depending on the kind of input. As $Y(s) = G(s)U(s)$, knowing the Laplace transform of the input, the product $G(s)U(s)$ can be decomposed into simple fractions (computation of residues) so that $y(t)$ can be computed through the corresponding $Y(s)$ using the tables of the most common transforms.[5] In this text, the transfer function will be primarily used as a mathematical tool for systems modeling, analysis, and design.

A generic linear dynamical system can be represented through the differential equation (3.1). Applying the derivation in t property of the Laplace transform with zero initial conditions, the following transfer function is obtained:

$$G(s) = \frac{Y(s)}{U(s)} = \frac{b_m s^m + b_{m-1} s^{m-1} + \cdots + b_1 s + b_0}{a_n s^n + a_{n-1} s^{n-1} + \cdots + a_1 s + a_0} = \frac{b(s)}{a(s)}, \quad (m \leq n), \tag{3.5}$$

---

[5] Some examples will be introduced along the text: $Y(s)$ has only real and simple poles (Sect. 3.2), complex poles (Sect. 3.3), or multiple poles (Sects. 3.3, 5.4, and 5.6).

whose order n (degree of the polynomial of the denominator $a(s)$, known as *characteristic polynomial* of the differential equation (3.5)) matches that of the original differential equation. As can be seen, the transfer function is a rational function that can be expressed as a quotient between two polynomials $b(s)/a(s)$, where s is a complex variable. As long as there are no common factors among the numerator and denominator, the values of s making zero the numerator of the transfer function are the *zeros* and those making zero the denominator (solutions of the *characteristic equation*) are the *poles*. As will be analyzed in Chap. 4, the interpretation in the frequency domain is that those frequencies at which a transfer function "blows up" (approaches infinity) are the transfer function poles, while those at which approaches zero are the zeros [6].

The condition $m \leq n$ is necessary to guarantee the causality of the system and that the model makes physical sense (otherwise the output of the model at the current instant would depend on future inputs). When the degree of $b(s)$ is less than or equal to the degree of $a(s)$, it is called a proper rational function (and thus the system it represents is causal), when the degree of $b(s)$ is strictly less than that of $a(s)$, the function is strictly proper (strictly causal system). Mathematically, if $\lim_{s \to \infty} G(s) < \infty$, $G(s)$ is said to be proper, otherwise is improper. If $\lim_{s \to \infty} G(s) = 0$, $G(s)$ is strictly proper. Strictly proper transfer functions have $(n - m)$ zeros at infinity, and improper transfer functions have $(m - n)$ poles at infinity. If there are no common roots between $b(s)$ and $a(s)$ (there are no pole–zero cancellations), $G(s)$ is *coprime* (or *irreducible*) [6]. An irrational function is one that cannot be expressed as the quotient between two polynomials in s.

By factoring the polynomials of the numerator and the denominator, the following expression can be obtained, which is called *zero–pole-gain factorization* (ZPK):

$$G(s) = \kappa \frac{(s - z_1)(s - z_2) \cdots (s - z_m)}{(s - p_1)(s - p_2) \cdots (s - p_n)}, \tag{3.6}$$

where κ is called the gain term or transfer constant of the transfer function and the coefficient of s in each factor is 1. The poles p_i $(i = 1 \ldots n)$ and the zeros z_ℓ $(\ell = 1 \ldots m)$ of a transfer function are represented in the complex plane called *s-plane*, which contains in the x-axis (abscissa axis) the real part of the poles or zeros (σ) and in the y-axis (ordinate axis) the imaginary part $(j\omega)$. Notice that the number of poles at the origin of the s-plane $(s = 0)$ is called the *system type* $\aleph$, which will be determinant in the analysis of the steady-state behavior of closed-loop systems (Chap. 8). In Eq. (3.6), poles and zeros can be real or complex conjugates. For systems with complex conjugate poles and/or zeros, the ZPK representation can also be written in general form as follows:

$$G(s) = \frac{\kappa \prod_{\ell=1}^{q} (s - z_\ell) \prod_{\ell=1}^{r} (s^2 + 2\zeta_{z_\ell} \omega_{n_{z_\ell}} s + \omega_{n_{z_\ell}}^2)}{s^{\aleph} \prod_{i=1}^{p} (s - p_i) \prod_{i=1}^{h} (s^2 + 2\zeta_i \omega_{n_i} s + \omega_{n_i}^2)} \tag{3.7}$$

with $q + 2r = m$, $\aleph + p + 2h = n$ and as will be analyzed when dealing with second-order systems, ζ is a parameter called relative damping factor and ω_n the undamped natural frequency of the complex conjugate poles or zeros.

The transfer function can also be represented in time constants format,[6] where $\tau_i = -1/p_i$, $(i = 1 \ldots n)$, $\beta_\ell = -1/z_\ell$, $(\ell = 1 \ldots m)$ in (3.6), so that the independent term in each factor is 1 and the static gain k (DC gain) appears explicitly in the transfer function:

$$G(s) = k \frac{(\beta_1 s + 1)(\beta_2 s + 1) \cdots (\beta_m s + 1)}{(\tau_1 s + 1)(\tau_2 s + 1) \cdots (\tau_n s + 1)}. \tag{3.8}$$

The static gain is the ratio of the steady-state output of a system to its constant input, i.e. the steady state of the unit step response. If the transfer function includes complex conjugate poles and/or zeros, (3.8) can be written in general form as follows[7]:

$$G(s) = \frac{k \prod_{\ell=1}^{q} (\beta_\ell s + 1) \prod_{\ell=1}^{r} \left(\left(\frac{s}{\omega_{n_{z_\ell}}} \right)^2 + \left(\frac{2\zeta_{z_\ell}}{\omega_{n_{z_\ell}}} \right) s + 1 \right)}{(s)^{\aleph} \prod_{i=1}^{p} (\tau_i s + 1) \prod_{i=1}^{h} \left(\left(\frac{s}{\omega_{n_i}} \right)^2 + \left(\frac{2\zeta_i}{\omega_{n_i}} \right) s + 1 \right)}, \tag{3.9}$$

---

[6] Notice that β_ℓ is used here to represent the time constant associated with each of the m zeros of the transfer function to simplify the writing. The usual notation uses τ_z to represent the time constant associated to a zero and τ_p in the case of a pole.

[7] Notice that the interactive tools using this format introduce the terms including complex conjugate poles/zeros in the form $\frac{\omega_n^2}{s^2 + 2\zeta \omega_n s + \omega_n^2}$, also having unit static gain, as they are the same as those included in Eq. (3.9).

with $q + 2r = m$, $\aleph + p + 2h = n$.

From the application of the final value theorem[8] of the Laplace transform (last line in Table 3.1) to the unit step response of the system, the value of k can be directly obtained by making $s = 0$ in the transfer function $G(0) = k$:

$$\lim_{t \to \infty} y(t) = \lim_{s \to 0} sY(s) = \lim_{s \to 0} sG(s)U_s(s) = \lim_{s \to 0} \cancel{s} G(s)\frac{1}{\cancel{s}} = G(0) = k. \tag{3.10}$$

Notice that Eqs. (3.6) and (3.8) provide the following relationship between k and κ:

$$G(0) = k = \kappa \frac{\prod_{\ell=1}^{m}(-z_\ell)}{\prod_{i=1}^{n}(-p_i)}.$$

Although this issue is treated in Chap. 8, if the system has a pole at the origin ($s = 0$), the steady-state output to a unit step input is a ramp, there is no steady-state value. In that case, k can be found as the ratio of the steady-state derivative of the output to its constant input. The same reasoning applies if the plant has more than one integrator. Although the conditions for applying the final value theorem do not hold in these cases, k can be obtained from (3.10) by removing the poles at the origin or either multiplying $G(s)$ by $s^\aleph$, $\aleph$ being the number of poles at the origin (notice that multiplying by s in Laplace domain is equivalent to perform a time derivative).

If the transfer function $G(s)$ has complex poles or zeros, for k to be explicit in (3.8), the independent term of the second-order polynomial must be one.

In this text, the most appropriate format will be used to explain each concept in the most intelligible way in the corresponding interactive tool.

As pointed out before, initial conditions are zero by default or when inputs and outputs represent deviations about an operating point. If the initial conditions would not be zero, then $Y(s) = \frac{b(s)}{a(s)}U(s) + \frac{ic(s)}{a(s)} = G(s)U(s) + \frac{ic(s)}{a(s)}$, where $ic(s)$ is a polynomial in s containing the terms arising from the initial conditions, in which case the total response is the sum of the contributions from the system input (*forced response*) and the initial conditions (*free response*). If the input is zero (unforced system), the *zero-input response* (ZIR) is obtained [6]. Both terms in the output equation have the same denominator polynomial (*characteristic polynomial*), in which roots (solution of the *characteristic equation*) are the poles, as mentioned above, also called system *modes*. Another way of analyzing this fact is through the solution to (3.1), which is the sum of two terms [1]: the general solution to the homogeneous equation, which does not depend on the input (ZIR), and a particular solution, which depends on the input. The homogeneous equation is obtained by equating the left-hand side term of (3.1) to zero, and its general solution (if there are no multiple roots[9]) is given by

$$y_h(t) = \sum_{i=1}^{n} c_i e^{p_i t},$$

c_i being arbitrary constants and p_i the roots of the characteristic equation. Since the coefficients in (3.1) are real, $e^{p_i t}$ decreases over time if $p_i < 0$, it is constant if $p_i = 0$, and increases if $p_i > 0$. For real roots, $\tau_i = -1/p_i$ is the time constant, while a complex root $p_i = -\sigma \pm j\omega$ corresponds to the time functions $e^{-\sigma t}\sin(\omega t)$ and $e^{-\sigma t}\cos(\omega t)$, providing oscillatory behavior. Notice that if $\mathbb{R}(p_i) < 0$ for all i, the system is AS; if $\mathbb{R}(p_i) > 0$ for any i, the system is unstable; and if $\mathbb{R}(p_i) = 0$ for any i, the output either remains constant or is sinusoidal (marginal stability). If the roots are repeated and have zero real parts, the system is unstable. A system with a simple mode at the origin or single pairs of complex modes on the $j\omega$-axis is marginally stable. Multiple modes at the origin or on the $j\omega$-axis indicate an unstable system [6]. Moreover, it is stated, without formal proof, that a system is BIBO stable if and only if all its poles lie in the open LHP. If any pole lies in the RHP, the system is unstable. A marginally stable system is not BIBO stable. AS implies BIBO stability but not vice versa. AS is determined by the modes, which are the roots of the characteristic equation. BIBO stability depends on the poles. The definitions are equivalent only in the case of coprime transfer functions [6] (no pole–zero cancellations).

---

[8] It should only be applied to stable systems [7], as in other cases the output is unbounded.

[9] With multiple roots $y_h(t) = \sum_{i=1}^{n} c_i(t)e^{p_i t}$, where $c_i(t)$ is a polynomial with degree less than the multiplicity of the root p_i [1].

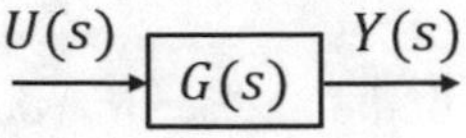

Fig. 3.1 Block diagram associated with a transfer function

Regarding the particular solution of (3.1), if $u(t) = e^{st}$ is selected [1], and assuming a particular solution $y(t) = G(s)e^{st}$ exists, then

$$\frac{du}{dt} = se^{st}, \ \frac{d^2u}{dt^2} = s^2e^{st}, \ \ldots, \ \frac{d^mu}{dt^m} = s^me^{st},$$

$$\frac{dy}{dt} = sG(s)e^{st}, \ \frac{d^2y}{dt^2} = s^2G(s)e^{st}, \ \ldots, \ \frac{d^ny}{dt^n} = s^nG(s)e^{st},$$

which inserted in (3.1) provides $G(s) = b(s)/a(s)$ already given in (3.5). So, the transfer function of the system describes a particular solution to the differential equation for the input e^{st}.

Due to the analytical complexity, it is common to study the time evolution of first- and second-order systems, extending the results to higher-order systems. Following this approach, this chapter introduces interactive tools dealing with the time response of first- and second-order systems and the associated concepts. The effect of adding a zero to the transfer function of these systems is also studied, and the concepts are generalized to the case of generic systems of any order.

Notice that the definition of transfer function allows representing it through *block diagrams*. The block diagram of a system is a graphical representation of the functions performed by each component and the signal flow between them. Unlike a purely abstract mathematical representation, a block diagram has the advantage of indicating in a more realistic way the signal flow of the real system. In a block diagram, the different variables of the system are linked by means of functional blocks. The functional block or simply block is a symbol to represent the mathematical operation that a block performs on the input signal to produce an output. Figure 3.1 represents the block diagram associated with the concept of a transfer function.

Structures and operations in block diagram algebra are analyzed in relationship with closed-loop analysis and design, and a summary can be found in Fig. 7.1, where the concept of transfer function will be extended to configurations of several connected linear systems, including cascade and parallel connection, feedback configurations, and others. Block diagrams are a representation of the interconnection of subsystems that form a system. In a linear system, the block diagram is made up of blocks (representing subsystems), arrows (representing signals), sum and union points (representing the algebraic sum of two or more signals), and distribution points (distributing a signal to several subsystems). A block diagram is composed of blocks that link system variables. Each block symbolizes a mathematical operation with a causal sense (it identifies which variables are *causes* and which are *effects*). It is the most widespread graphical representation language for dynamical systems in engineering. This representation allows not only to obtain models from the constitutive and structural relationships (generally ordinary differential equations), but also to use certain analysis techniques directly on the diagrams, known as *block algebra*.

3.2   Time Response of Continuous-Time First-Order Linear Systems without Zeros

3.2.1   Interactive Tool: t_first_order

3.2.1.1 Concepts Analyzed in the Card and Learning Outcomes
- Modeling of LTI dynamical systems from a first-order linear differential equation.
- Obtaining the transfer function of a first-order system from a linear differential equation.
- Time response of a first-order linear system for a step input (step response[10]).
- Concept of *static gain* and its effect on the system time response for a step input.
- Concept of *time constant* and its effect on the step response.
- Stability analysis of first-order linear systems.

---

[10] The step response describes the relationship between an input that changes from zero to a constant value abruptly (a step input) and the corresponding output [1].

3.2.1.2 Summary of Fundamental Theory

This card analyzes the continuous-time response of LTI first-order systems without zeros. First-order models are the simplest and most reasonable approximation of a physical system if the storage of mass, momentum, or energy can be captured by a single variable [8]. Typical examples are the velocity of a car on the road, the velocity of a rotating system, electric systems where energy is essentially stored in one component, incompressible fluid flow in a pipe, the level of a tank, pressure in a gas tank, the temperature in a body with essentially uniform temperature distribution, and so on.

The model of a first-order system is derived from a first-order differential equation of the form:

$$\tau \frac{dy(t)}{dt} + y(t) = ku(t). \tag{3.11}$$

In expression (3.11), $y(t)$ and $u(t)$ are the system output and input, respectively. The representative transfer function (in time constants format) can be written as follows:

$$G(s) = \frac{k}{\tau s + 1}, \tag{3.12}$$

wherein the denominator is called *characteristic polynomial*, whose only root (solution of the characteristic equation $a(s) = \tau s + 1 = 0$) is called *pole* of the transfer function, placed at $s = -1/\tau$. The two parameters that characterize the transfer function of a first-order system are as follows:

- k: *Static gain*, DC gain, canonical gain, or zero frequency gain of the system.
- τ: *Time constant*.

Notice that the system can also be expressed in ZPK factorization as follows:

$$G(s) = \kappa \frac{1}{(s - p)}, \quad \text{with } p = -\frac{1}{\tau} \text{ and } \kappa = \frac{k}{\tau}.$$

In what follows, the time constant format will be used as its parameters have a clear physical meaning.

The time response is linked to a certain excitation signal, traditionally the step ($U(s) = U_0/s$, with $U_0 = 1$ in the case of unit step). Therefore, using the inverse Laplace transform ($\mathscr{L}^{-1}$), one can obtain the time response of a first-order system when the input signal is a step. Using the expression of the output in the Laplace domain and a partial-fraction expansion:

$$Y(s) = G(s)U(s) = \frac{k}{\tau s + 1} \frac{U_0}{s} = \frac{c_1}{\tau s + 1} + \frac{c_2}{s},$$

where c_i is the residue of the pole at $s = p_i$, it is easy to compute the residues in this case:

$$c_1 = \left.\frac{kU_0(\tau s + 1)}{s(\tau s + 1)}\right|_{s=-1/\tau} = -kU_0\tau; \quad c_2 = \left.\frac{kU_0 s}{(\tau s + 1)s}\right|_{s=0} = kU_0,$$

so that

$$Y(s) = \frac{kU_0}{s} - \frac{\tau kU_0}{\tau s + 1}.$$

The inverse Laplace transform of each one of the terms can be found in Table 3.2 (lines 2 and 5), providing the time response when the input signal is a step:

$$y(t) = kU_0 \left(\mathscr{L}^{-1}\left\{\frac{1}{s}\right\} - \mathscr{L}^{-1}\left\{\frac{\tau}{\tau s + 1}\right\} \right) = kU_0(1 - e^{-\frac{1}{\tau}t}), \quad t \geq 0. \tag{3.13}$$

Notice that this time response could also be obtained by solving the differential equation with a step input:

$$u(t) = \begin{cases} 0, & t < 0, \\ U_0, & t \geq 0. \end{cases}$$

The behavior of the system output depends on the sign of τ:

- If $\tau > 0$, the output is *bounded*, and therefore, the system is stable (the exponential has negative coefficient).
- If $\tau < 0$, the output is *unbounded*, and therefore, the system is unstable.

As a generalization of this result, it is stated that a necessary and sufficient condition for a system to be stable is that all poles of its transfer function have negative real parts. In this case, the system is AS, since the output tends to a steady-state value asymptotically.

Note that if $\tau = 0$, the system is not a dynamic one and the relationship between the output and the input is determined by the static gain k, which corresponds to the ratio between the value taken by the system output in steady state[11] ($t \to \infty$) and the value of the input also in steady state, as zero initial conditions are assumed.[12] Applying the final value theorem of the Laplace transform to Eq. (3.13), it is obtained that

$$\lim_{t \to \infty} y(t) = kU_0 = \lim_{s \to 0} sY(s) = \lim_{s \to 0} \cancel{s} \frac{k}{\tau s + 1} \frac{U_0}{\cancel{s}} = kU_0.$$

As shown in the above equation, the static gain can be obtained by selecting $s = 0$ in the transfer function, $G(0) = k$. As indicated in [7], it is important to note that the final value theorem should only be applied to stable systems. For example, consider the steady-state value of

$$Y(s) = \frac{\kappa}{s(s - p)} \to y(\infty) = \lim_{s \to 0} sY(s) = -\frac{\kappa}{p},$$

that is a finite value. However, if $p > 0$, the output is unbounded, contrary to that predicted by the final value theorem (the conditions for applying this theorem do not hold):

$$y(t) = -\frac{\kappa}{p}\left(1 - e^{pt}\right).$$

Following Eq. (3.13), τ is called *time constant* as this parameter is indicative of the speed of the transient response of the system. The higher τ is, the slower the transient response of the system, which will take more time to reach its final value. Thus, according to Eq. (3.13), the following hold:

- For $t = \tau$, the output of the system corresponds to $kU_0(1 - e^{-1})$, which is approximately 63.2% of the final value, $y(\tau) \approx 0.632kU_0$.
- For $t = 3\tau$, the system output corresponding to 95% of the final value, $y(3\tau) \approx 0.95kU_0$.
- For $t = 4\tau$, the output of the system is around 98% of the final value, $y(4\tau) \approx 0.98kU_0$, which is considered the *settling time* for first-order systems (the response remains within 2% of the final value after this time). According to Eq. (3.13), the steady state is achieved mathematically in infinite time, so that in practice it is considered that the system output is in steady state when it reaches 98% of the final value, equal to four time constants.
  Another important feature is that the slope of the time response of the system in $t = t_0$ is kU_0/τ (where t_0 is the instant the step input is applied to the system).

A particular case study of a first-order system called *integrator* will be studied in Sect. 3.6, having its pole at the origin of the s-plane ($s = 0$) and described by

$$\frac{dy(t)}{dt} = ku(t) \to G(s) = \frac{k}{s}. \tag{3.14}$$

This system would be equivalent to a first-order one with $\tau \to \infty$. It is easy to see that when a step input is introduced to an integrator, the output is a ramp and therefore unbounded.

The time response to other inputs (impulse, ramp, etc.) can be analytically obtained by following a similar procedure. The unit impulse response can be calculated taking into account that in this case $U(s) = 1$, so that

$$Y(s) = G(s)U(s) = \frac{k}{\tau s + 1} \to y(t) = \frac{k}{\tau}e^{-t/\tau}.$$

---

[11] Value that the output of the system reaches after all elements of the transient response have declined [4].

[12] In the case of linearized systems, it is the ratio between the change experienced by the output and the amplitude of the step input, considering their steady-state values.

Notice that if $t = 0$, then $y(0) = k/\tau$ and when $t \to \infty$, $y(t \to \infty) = 0$. Thus, the initial value of the response provides the relationship k/τ while for $t = \tau$, $y(\tau) = 0.368k/\tau$, that is, τ is the time the system takes to reach 36.8% of its initial value when evolving toward zero (the time it takes to reach 63.2% of the total change). The slope at the beginning is given by $dy(t)/dt|_{t=0} = -k/\tau^2$.

Notice that from Eq. (3.2), the solution to the normalized ODE (3.11) when the input is an impulse is given by

$$y(t) = \int_{-\infty}^{t} h(t - \xi)u(\xi)d\xi = \int_{-\infty}^{t} \frac{k}{\tau}e^{-\frac{1}{\tau}(t-\xi)}u(\xi)d\xi, \tag{3.15}$$

so that the impulse response is given by $h(t) = (k/\tau)e^{-t/\tau}$, as obtained at the begining of this chapter.

The unit slope ramp response can be also easily obtained as follows:

$$Y(s) = G(s)U(s) = \frac{k}{\tau s + 1}\frac{1}{s^2} \to y(t) = k\left(t - \tau + \tau e^{-t/\tau}\right).$$

For $t = 0$, $y(0) = 0$, while for $t \to \infty$, $e^{-t/\tau} \to 0$, so the system output in steady state in general converges to a ramp with a slope k (kU_0 if the input is $u(t) = U_0 t$). Only in the case $k = 1$, the output converges to a ramp parallel to the input $u(t) = t$ with a difference equal to τ. For $t = \tau$, $y(\tau) = 0.368k\tau$. Notice that $dy(t)/dt = k(1 - e^{-t/\tau})|_{t=0} = 0$ (the slope at $t = 0$ is zero).

3.2.1.3 References Related to this Concept

- [5] Bolzern, P., Scattolini, R., & Schiavoni, N. (2009). *Fundamentos de control automático (Fundamentals of automatic control)*. McGraw-Hill. ISBN: 978-84-481-6640-3. Chapter 4, Sect. 4, paragraph 3, pp. 103–104.
- [6] Shahian, B., & Hassul, M. (1993). *Control system design using MATLAB®*. Prentice Hall. ISBN: 0-13-174061-X. Chapter 1, Sect. 5, paragraph 1, pp. 10–11.
- [7] Franklin, G. F., Powell, J. D., & Emani-Naeni, A. (2015). *Feedback control of dynamic systems* (7th ed.). Pearson. ISBN: 978-0-13-349659-8. Chapter 3, Sect. 3, pp. 123–126.
- [9] Golnaraghi, F., & Kuo, B. C. (2017). *Automatic control systems* (10th ed.). McGraw Hill Education. ISBN: 978-1-25-964384-2.
- [10] Ogata, K. (2010). *Modern control engineering* (5th ed.). Prentice Hall. ISBN: 978-0-13-615673-4. Chapter 5, Sect. 2, pp. 161–164.

Application Interactive tool: t_first_order

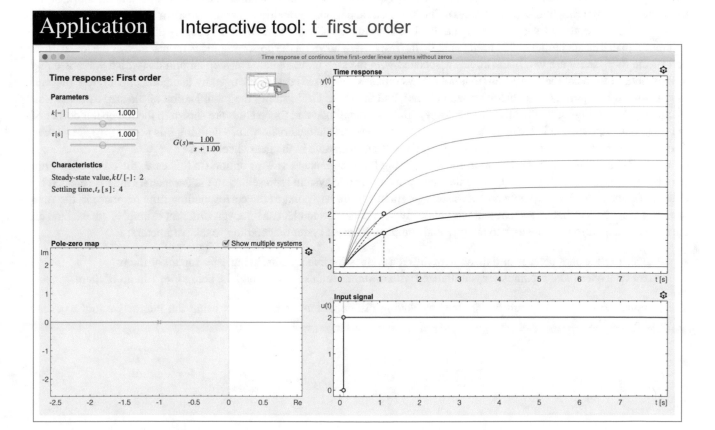

Time Response: First Order

The main objective of this tool is to analyze the step response of a continuous-time LTI first-order system based on the pole location and the static gain values. Thus, the relationships among the s-plane, the transfer function, and the time response are analyzed. The application is fully interactive, in such a way that the modification of a parameter or a representative value in the time response is reflected automatically in the rest of the tool representations.

The main screen is divided into four different areas. The upper-left part of the tool shows the numeric **Parameters** (static gain k and time constant τ) and the transfer function that represents the system under study. Furthermore, some indices summarizing the time response of the selected system (**Characteristics**) are shown in this area. In this case, the output steady-state value and the 2% settling time (4τ) are represented.

The static gain value k can be modified using the textbox that shows its value. After clicking on the textbox, its color changes indicating that a new value can be introduced from the keyboard, which will be evaluated after pressing the enter key or clicking with the mouse pointer on any part of the tool. The static gain value can also be updated by using the slider located below the textbox. Negative static gain values can be introduced through the textbox, and the sign modification will be automatically considered in the slider limits.

The time constant τ can be modified in a similar manner using the corresponding textbox and slider. When the k and τ values are changed, the time response characteristics and the symbolic transfer function representation are automatically updated, as well as all graphical representations.

Pole–zero map: The s-plane (or complex plane) is represented on the lower-left part of the tool, where the pole location of the analyzed system (or the pole location for several systems when the Show multiple systems option is activated, such as described below) is depicted by using ×as representation symbol. The pole location can be modified by clicking and dragging it with the mouse (moving the pole to the left or to the right). When the mouse is over the pole, the pole value is shown in the lower-left corner of the tool. The gearwheel in the graph allows the modification of the scale (Chap. 1, Sect. 1.7.1).

The **Time response** of the different first-order systems are shown in the upper-right part of the tool. When the mouse is located over any point of the plot, the output values defining this point (t, y) are shown over the plot. The graphic includes two circles associated with the time response. The upper circle allows the modification of the final value and the slope of the time response (affecting both k and τ values). The lower circle permits to modify the τ value. When the mouse is located on these circles, the values for the static gain and the time constant are shown in the lower-left corner of the tool.

Input signal: The step input signal used to obtain the responses shown on the upper graphic is represented at the lower-right part of the tool. There are two black circles associated with the step input signal. The step amplitude (which value is equal to 2 by default to make the user aware that not always unit steps are used) can be changed by using the upper circle (as well as clicking and dragging on the black horizontal line), and the lower circle allows the modification of the step time (as well as clicking and dragging on the black vertical line). The step amplitude and time values are shown in the lower-left corner of the tool. When the mouse is located over any point of the step, the characteristic values defining this point (t, u) are shown. The scales of the figures can be modified by using the settings available in the gearwheel icon.

When five system responses are being shown by using the Show multiple systems option (in this case, the color code from the library is used: red, blue, green, magenta, and cyan), the active system represented in the **Parameters** area is selected by clicking on the representative pole on **Pole–zero map** graphic or on any point of the corresponding time response in the **Time response** graphic. The active system will automatically be represented in black. Moreover, different examples are enabled on the Options menu allowing the user to analyze and compare different systems according to the parameters:

- Gain effect: The k parameter is initialized with different values for five models, being $\tau = 1$ in all of them.
- Time constant effect: The τ parameter is initialized with different values for five models, being $k = 1$ in all of them.

The response to other inputs can be obtained by adding differentiators or integrators using the interactive tool t_generic in card 3.6 (see Sect. 2.1.1).

3.2.1.4 Homework

1. Move the system pole along the real axis without leaving the left half s-plane (LHP, shaded in yellow). What change is observed in the time response? What is the effect of moving it to the right? And to the left? Which parameters of the transfer function are affected?

2. Move the system pole inside the right half s-plane (RHP). What kind of response is obtained? Is there any effect in this case if a negative static gain is selected?

3. Place the activation of the input step signal at $t = 5$ and its amplitude to a value equal to 2. Select a value of $k = 1$ and $\tau = 3$ s (a change in the scale of the plots using the associated gearwheel icon will be required, for instance, selecting variable scale settings). How long does it take until the system response reaches 63% of its final value after the introduction of the step? How long does it take for the system response to reach 98% of the steady-state value? Check your answer with the values shown in the **Parameters** and **Characteristics** areas of the tool.

4. Place the activation of the input step signal at $t = 5$ and its amplitude to a value equal to 2. Select a value of $k = -1$ and $\tau = 3$ s. How long does it take until the system response reaches 63% of its final value after the introduction of the step? How long does it take for the system response to reach 98% of the steady-state value? What are the main differences with respect to the previous exercise?

5. How does the location of the pole in the s-plane change when the static gain is modified using the slider? Justify the answer.

6. Let's consider a system with a pole at $s = -3$ and whose steady-state value for a unit step input is 0.5. Include those values in the tool and verify the results. What is the value of the system time constant? And that of its static gain? How long does it take until the output of the system reaches 63 and 98% of its final value after introducing a unit step input? What is the initial slope of the response? What is the transfer function of the system? What is the steady-state value of the output? Check the results using the tool.

7. For the default settings that appear when starting the tool, analyze what happens in the system response when the value of τ is changed to 0.01 and 100 s. What is the value of the initial slope of the response?

8. Selecting the option Show multiple systems, access the Options menu and choose Gain effect. State the value of the static gain and time constant of the five systems represented.

9. Selecting the option Show multiple systems, access the Options menu and choose Time constant effect. State the value of the static gain and time constant of the five systems represented.

10. The time response of two linear dynamical systems when an input step of amplitude equal to 2 is introduced at $t = 0$ is given by

$$y(t) = 6(1 - e^{-\frac{t}{5}}), \qquad \text{for} \quad t \geq 0,$$
$$y(t) = -4(1 - e^{-\frac{t}{2}}), \qquad \text{for} \quad t \geq 0.$$

State the value of the time constant and static gain of each system and its settling time.

11. Consider the differential equation $\dot{y}(t) + 2y(t) = u(t)$.
    If $u(t)$ is constant, then $\dot{y}(t) \approx 0$ when time goes to infinity. What value will $y(t)$ approach as $t \to \infty$ if $u(t) = 5$? Verify using the tool the time constant, static gain, and step time response, as well as the pole location.
    Determine the transfer function relating $U(s)$ and $Y(s)$ for the differential equation above. The solution of this exercise can be found in [11].

3.3 Time Response of Continuous-Time Second-Order Linear Systems without Zeros

3.3.1 Interactive Tool: t_second_order

3.3.1.1 Concepts Analyzed in the Card and Learning Outcomes
- Modeling of LTI dynamical systems using a second-order linear differential equation.
- Obtaining the transfer function of a second-order system from a linear differential equation.
- Analysis of the time response of an LTI second-order dynamical system for a step input.
- Concept of static gain and its effect on the system time response for a step input.
- Concept of relative damping factor and its effect on the step response.
- Concept of undamped natural frequency and its effect on step response.

- Types of dynamic behavior in second-order systems: overdamped, critically damped, underdamped, critically stable, and unstable.
- Stability analysis in second-order linear systems.

3.3.1.2 Summary of Fundamental Theory

Second-order systems, as its name suggests, can be described by a standard second-order differential equation such as

$$\frac{d^2y(t)}{dt^2} + 2\zeta\omega_n\frac{dy(t)}{dt} + \omega_n^2 y(t) = k\omega_n^2 u(t),\tag{3.16}$$

where $y(t)$ and $u(t)$ are the system output and input, respectively.

There are models with "pure" second-order dynamics or formed by the combination of two first-order systems in series (their model is the product of two transfer functions of first order[13]). In [8], several examples are given: the position of a car on the road, satellites, electric systems where energy is stored in two elements, level in two connected tanks, pressure in two connected vessels, pneumatic valves, or simple bicycle models.

The standard transfer function of a second-order system is given by

$$G(s) = \frac{k\omega_n^2}{s^2 + 2\zeta\omega_n s + \omega_n^2},\tag{3.17}$$

wherein the denominator polynomial is called *characteristic polynomial* $a(s) = s^2 + 2\zeta\omega_n s + \omega_n^2$, in which roots (solution of the characteristic equation $a(s) = 0$) are the poles of the transfer function, which in this case can be real or complex conjugates. The parameters that define the transfer function are as follows:

- k is the *static gain* (canonical gain, DC gain, or zero frequency gain). It can be obtained by doing $s = 0$ in the transfer function, $G(0) = k$. In stable systems, it represents the quotient of the amplitude of the steady-state response of the system and the amplitude of the input step.
- ζ is the relative damping factor (ratio or coefficient) of the system (dimensionless), which determines the shape of the transient response. Depending on its value, it can be deduced if the system is unstable ($\zeta < 0$), critically stable or not damped ($\zeta = 0$), underdamped ($0 < \zeta < 1$), critically damped ($\zeta = 1$), or overdamped ($\zeta > 1$).
- ω_n undamped natural frequency [rad/s], which corresponds to the frequency of oscillation of the system if there were no damping ($\zeta = 0$, cosine response).

Obviously, to obtain a bounded response when the input signal has a step shape, the poles of the system must be in the LHP, as commented in the section devoted to first-order systems. If any of the roots is in the RHP, the system will be unstable. If the system characteristic equation ($a(s) = s^2 + 2\zeta\omega_n s + \omega_n^2 = 0$) has its roots on the imaginary axis ($j\omega$-axis), the output in steady state when the input is a step signal will be of cosine type (steady oscillations or limit cycles). When the input is a sine wave whose frequency is equal to the magnitude of the roots of the $j\omega$-axis, the output will be unbounded.[14] Such a system is called *marginally stable*, because only some bounded inputs (sinusoids with the same frequency of the poles) will produce unbounded outputs. The time response when the input has a step shape of amplitude U_0 ($U(s) = U_0/s$) can be obtained from $Y(s) = G(s)U(s)$ by applying the inverse Laplace transform, $y(t) = \mathcal{L}^{-1}\{Y(s)\}$, or by solving the differential equation with

$$u(t) = \begin{cases} 0, & t < 0, \\ U_0, & t \geq 0, \end{cases}$$

making it necessary to distinguish between different cases depending on the value of ζ:

- **Underdamped system:** In the case $0 < \zeta < 1$ the two poles of the system (roots of the characteristic equation) are complex conjugate, ($p_1 = -\zeta\omega_n + j\omega_n\sqrt{1-\zeta^2} = -\sigma + j\omega_d$ and $p_1^* = -\zeta\omega_n - j\omega_n\sqrt{1-\zeta^2} = -\sigma - j\omega_d$), where the product $\sigma = \zeta\omega_n$ is called *damping factor* or *decay rate*, which is a constant that determines the damping properties

---

[13] As treated in Chap. 7, the equivalent transfer function of two systems in series $G_1(s)$ and $G_2(s)$ is the product $G_1(s)G_2(s)$, as a consequence of applying the properties of the Laplace transform.

[14] In such case, $Y(s) = k\omega_n^3/(s^2 + \omega_n^2)^2$, providing $y(t) = (k/2)(\sin(\omega_n t) - \omega_n t \cos(\omega_n t))$ from the application of the inverse Laplace transform (line 15 in Table 3.2).

of a system, and $\omega_d = \omega_n\sqrt{1-\zeta^2}$ is the *natural damped frequency*. The reader can easily check that $s^2 + 2\zeta\omega_n s + \omega_n^2 = (s-p_1)(s-p_1^*) = (s+\sigma)^2 + \omega_d^2$.

The time response for a step input can be obtained using the inverse Laplace transform:

$$Y(s) = G(s)U(s) = \frac{k\omega_n^2}{s^2 + 2\zeta\omega_n s + \omega_n^2}\frac{U_0}{s} = \frac{c_1 s + c_2}{s^2 + 2\zeta\omega_n s + \omega_n^2} + \frac{c_3}{s}, \tag{3.18}$$

where c_i, $i = 1\ldots 3$ are the residues of $Y(s)$ at $s = p_i$. By multiplying (3.18) by $(s-p_1)$ and making $s = p_1$,

$$Y(s) = \frac{kU_0\omega_n^2(s-p_1)}{(s-p_1)(s-p_1^*)s}\bigg|_{s=p_1} = \frac{(c_1 s + c_2)(s-p_1)}{(s-p_1)(s-p_1^*)}\bigg|_{s=p_1} + \frac{c_3(s-p_1)}{s}\bigg|_{s=p_1},$$

so that $p_1(c_1 p_1 + c_2) = kU_0\omega_n^2$, with $p_1 = -\sigma + j\omega_d$. Operating and equaling both real and imaginary parts, it is obtained that $c_1 = -kU_0$ and $c_2 = 2\sigma c_1$. In the same way, by multiplying (3.18) by s and making $s = 0$, it results that $c_3 = kU_0$. Therefore,

$$Y(s) = \frac{c_1 s + c_2}{s^2 + 2\zeta\omega_n s + \omega_n^2} + \frac{c_3}{s} = kU_0\left(\frac{1}{s} - \frac{(s+2\sigma)}{(s+\sigma)^2 + \omega_d^2}\right) = kU_0\left(\frac{1}{s} - \left[\frac{(s+\sigma)}{(s+\sigma)^2 + \omega_d^2} + \frac{\sigma}{(s+\sigma)^2 + \omega_d^2}\right]\right).$$

By using lines 2, 8, and 9 of Table 3.2 (main Laplace transforms),

$$\mathscr{L}^{-1}\left(\frac{s+\sigma}{(s+\sigma)^2 + \omega_d^2}\right) = e^{-\sigma t}\cos(\omega_d t),$$

$$\mathscr{L}^{-1}\left(\frac{\omega_d\frac{\sigma}{\omega_d}}{(s+\sigma)^2 + \omega_d^2}\right) = \frac{\sigma}{\omega_d}e^{-\sigma t}\sin(\omega_d t), \quad \text{with } \frac{\sigma}{\omega_d} = \frac{\zeta}{\sqrt{1-\zeta^2}}.$$

The step response is obtained as follows:

$$y(t) = kU_0\left[1 - e^{-\zeta\omega_n t}\left(\cos(\omega_d t) + \frac{\zeta}{\sqrt{1-\zeta^2}}\sin(\omega_d t)\right)\right], \quad t \geq 0. \tag{3.19}$$

It can be observed how the complex component of the poles produces a time response with the presence of sines and cosines that gives rise to oscillations which are damped by the exponential envelope with the representative time constant $\tau = 1/\sigma = 1/(\zeta\omega_n)$. Notice that ω_d determines the oscillation frequency of the step response of an underdamped second-order system.

- **Critically stable system:** As can be seen in Eq. (3.18), when $\zeta = 0$ the two complex conjugate poles are located on the imaginary axis (with zero real part) and the response shows a maintained oscillation given by

$$Y(s) = \frac{k\omega_n^2}{s^2 + \omega_n^2}\frac{U_0}{s} \rightarrow y(t) = kU_0(1 - \cos(\omega_n t)), \quad t \geq 0. \tag{3.20}$$

- **Overdamped system:** When the relative damping factor $\zeta > 1$, the poles of the second-order transfer function are real ($p_1 = -1/\tau_1 = -\zeta\omega_n - \omega_n\sqrt{\zeta^2-1}$ and $p_2 = -1/\tau_2 = -\zeta\omega_n + \omega_n\sqrt{\zeta^2-1}$). The transfer function in this case is given by

$$G(s) = \frac{k}{(\tau_1 s + 1)(\tau_2 s + 1)}, \tag{3.21}$$

where τ_2 is the time constant associated with the pole closest to the imaginary axis, which causes the slower exponential response ($\tau_2 > \tau_1$), and the time response can be obtained as a superposition of that given by two first-order systems in series:

$$Y(s) = G(s)U(s) = \frac{k}{(\tau_1 s + 1)(\tau_2 s + 1)} \frac{U_0}{s} = \frac{c_1}{\tau_1 s + 1} + \frac{c_2}{\tau_2 s + 1} + \frac{c_3}{s},$$

$$c_1 = Y(s)(\tau_1 s + 1)|_{s=-1/\tau_1} = kU_0 \frac{\tau_1^2}{\tau_2 - \tau_1},$$

$$c_2 = Y(s)(\tau_2 s + 1)|_{s=-1/\tau_2} = -kU_0 \frac{\tau_2^2}{\tau_2 - \tau_1}, \quad c_3 = Y(s)s|_{s=0} = kU_0,$$

so that, by using lines 2 and 5 of Table 3.2 of representative Laplace transforms, it results in

$$y(t) = kU_0 \left(1 - \frac{\tau_2}{\tau_2 - \tau_1} e^{-\frac{t}{\tau_2}} + \frac{\tau_1}{\tau_2 - \tau_1} e^{-\frac{t}{\tau_1}} \right), \quad t \geq 0. \tag{3.22}$$

Notice that each term $c_i e^{p_i t}$ provided by the inverse Laplace transform is called *natural mode of the system* and it is said to be asymptotically stable if $p_i < 0$ (the exponential decays to zero as time increases).

- When $\zeta = 1$, both real roots are equal ($\tau_1 = \tau_2$) and the system is called *critically damped*. Its transfer function is given by

$$G(s) = \frac{k}{(\tau s + 1)^2}. \tag{3.23}$$

For a step input signal, the residues of the partial-fraction decomposition for inverse Laplace transform have to be calculated as follows:

$$Y(s) = G(s)U(s) = \frac{k}{(\tau s + 1)^2} \frac{U_0}{s} = \frac{c_1}{(\tau s + 1)} + \frac{c_2}{(\tau s + 1)^2} + \frac{c_3}{s},$$

$$c_2 = Y(s)(\tau s + 1)^2|_{s=-1/\tau} = -kU_0\tau, \quad c_3 = Y(s)s|_{s=0} = kU_0,$$

$$\frac{d}{ds} \left[Y(s)(\tau s + 1)^2 \right] \Big|_{s=-1/\tau} = -\frac{kU_0}{s^2} \Big|_{s=-1/\tau} = c_1 \tau \rightarrow c_1 = -kU_0\tau,$$

and using lines 2, 5, and 10 of Table 3.2, the step response has the following analytical expression:

$$y(t) = kU_0 \left(1 - e^{-\frac{t}{\tau}} - \frac{t}{\tau} e^{-\frac{t}{\tau}} \right), \quad t \geq 0. \tag{3.24}$$

Here it is easy to see (by computing $dy(t)/dt|_{t=0}$) that the initial slope of the response is zero.

- **Unstable system:** If $\zeta < 0$, the system will be unstable, with two complex conjugate poles with real part within the RHP if $-1 < \zeta < 0$ (unstable oscillatory response) or either two real poles in the RHP if $\zeta \leq -1$ (unstable response of exponential type).

For both the overdamped and underdamped cases, an interesting analysis is to study the location of the poles of a second-order system as a function of the characteristic parameters of the transfer function. From these parameters, several relationships with certain specific characteristics of the time response of the system can be found (which, as discussed in Chap. 8, can also be used as closed-loop performance specifications in control system design). For stable systems, the best-known features of the time response are as follows (see Fig. 3.2):

- **Peak time** (t_p [s]): It is the time that the system response takes to reach its maximum (peak) value measured from the moment a step input is applied. By applying the time derivative to Eq. (3.19) and equating it to zero, it is possible to obtain the peak time as follows:

$$t_p = \frac{\pi}{\omega_d} = \frac{\pi}{\omega_n \sqrt{1 - \zeta^2}}. \tag{3.25}$$

- **Overshoot** (*OS*), maximum overshoot or peak overshoot: It represents the difference between the maximum peak value of the response and its steady-state value, relative to that steady-state value (when it is represented in %, it is called maximum percentage overshoot). For the underdamped case, applying the time derivative to Eq. (3.19) and equating it to zero, the maximum value of $y(t_p)$ which defines the peak overshoot can be obtained using the value of the peak time. The values of the maximum overshoot and maximum percentage overshoot are given by

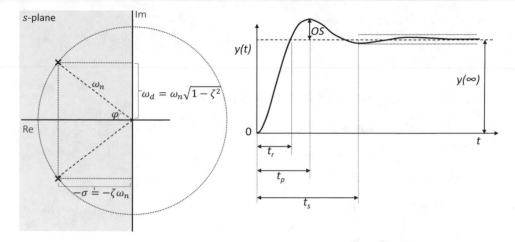

Fig. 3.2 Parameters characterizing the time response and the complex plane representation of an underdamped second-order system

$$OS = \exp\left(\frac{-\zeta\pi}{\sqrt{1-\zeta^2}}\right); \quad OS\,[\%] = 100\,OS. \tag{3.26}$$

This definition makes sense only in the underdamped case.

- **Rise time** (t_r [s])**:** In the underdamped case, it is the time elapsed since the output of the system begins to evolve until it reaches the steady-state value for the first time. It can be obtained by making $y(t_r) = 1$ in Eq. (3.19):

$$t_r = \frac{\pi - \varphi}{\omega_d}, \tag{3.27}$$

with $\zeta = \cos\varphi$, φ being the angle formed by the complex conjugate poles with the x-axis. In overdamped systems, it is defined as the time the system response takes to evolve from 10 to 90% of its steady-state value.

- **Settling time** (t_s [s])**:** It is the time elapsed from the moment the output of the system begins to evolve until it remains at around 2% of the steady-state value. An upper bound calculated from the envelope curves of the transient response (with a time constant $1/(\zeta\omega_n)$) is given by

$$t_s \approx \frac{4}{\zeta\omega_n}. \tag{3.28}$$

For the 5% case it is given by $t_s \approx \frac{3}{\zeta\omega_n}$. These are valid for $\zeta < 0.9$. Notice that the rise time can be obtained by making $y(t_s) = 0.98$ or $y(t_s) = 0.95$ in (3.19). The relationship between ζ and t_s is discontinuous, thus being the reason for approximating the relationship.

In the overdamped case ($\zeta \geq 1$), its value is obtained as $t_s \approx 4(\tau_1 + \tau_2)$.

It is also possible to analyze the effect of varying the location of the poles of an underdamped second-order system using an expression in terms of the real and imaginary parts of the roots [6] (see Fig. 3.2), where $\sigma = \zeta\omega_n$ and $\omega_d = \omega_n\sqrt{1-\zeta^2}$. The time constant of the exponential envelope of the time response of an underdamped second-order system is $\tau = 1/\sigma$:

$$G(s) = \frac{\omega_n^2}{s^2 + 2\zeta\omega_n s + \omega_n^2} = \frac{\omega_d^2 + \sigma^2}{s^2 + 2\sigma s + (\omega_d^2 + \sigma^2)}.$$

Case 1: Effects of increasing σ (with constant ω_d)

- The imaginary part of the poles remains constant and the real part (in absolute value) increases.
- The settling time decreases.
- The rise time decreases because the distance from the poles to the origin increases.
- The peak overshoot is reduced because ζ increases.

- The peak time remains constant because ω_d has been fixed.
- The bandwidth[15] increases because it is proportional to ω_n.

Case 2: Effects of increasing ω_d (with constant σ)

- The real part of the poles remains constant while the imaginary part increases.
- The settling time remains constant.
- The peak overshoot and bandwidth increase.
- The peak time and rise time decrease.

Case 3: Effects of increasing ω_n (with constant ζ)

- The poles are moved radially away from the origin.
- The peak overshoot remains constant.
- The rise, peak, and settling times decrease.
- The bandwidth increases.

Case 4: Effects of increasing ζ (with constant ω_n)

- The rise time increases.
- The peak overshoot and settling time decrease.
- The peak time increases.

The same procedure as in the case of first-order systems can be followed to obtain the impulse and ramp responses. The stable cases of the impulse response are summarized here as follows:

$$0 < \zeta < 1, \quad y(t) = \frac{\omega_n}{\sqrt{1 - \zeta^2}} e^{-\zeta \omega_n t} \sin(\omega_d t),$$
$$\zeta = 0, \quad y(t) = \omega_n \sin(\omega_n t),$$
$$\zeta = 1, \quad y(t) = \omega_n^2 t e^{-\omega_n t},$$
$$\zeta > 1, \quad y(t) = \frac{\omega_n}{2\sqrt{\zeta^2 - 1}} (e^{-t/\tau_2} - e^{-t/\tau_1}).$$

3.3.1.3 References Related to this Concept

- [4] Dorf, R. C., & Bishop, R. H. (2011). *Modern control systems* (12th ed.). Prentice Hall. ISBN: 978-0-13-602458-3. Chapter 5, Sect. 3, pp. 308–314.
- [5] Bolzern, P., Scattolini, R., & Schiavoni, N. (2009). *Fundamentos de control automático (Fundamentals of automatic control)*. McGraw-Hill. ISBN: 978-84-481-6640-3. Chapter 4, Sect. 4, paragraph 4, pp. 105–111.
- [6] Shahian, B., & Hassul, M. (1993). *Control system design using MATLAB®*. Prentice Hall. ISBN: 0-13-174061-X. Chapter 1, Sect. 5, paragraph 2, pp. 11–16.
- [7] Franklin, G. F., Powell, J. D., & Emani-Naeni, A. (2015). *Feedback control of dynamic systems* (7th ed.). Pearson. ISBN: 978-0-13-349659-8. Chapter 3, Sect. 4, pp. 131–136, Chapter 6, Sect. 1, pp. 314–317.
- [9] Golnaraghi, F., & Kuo, B. C. (2017). *Automatic control systems* (10th ed.). McGraw Hill Education. ISBN: 978-1-25-964384-2. Chapter 3, sections 4.2 and 4.3; Chapter 7, Sect. 6.
- [10] Ogata, K. (2010). *Modern control engineering* (5th ed.). Prentice Hall. ISBN: 978-0-13-615673-4. Chapter 5, Sect. 3, pp. 164–179.

---

[15] A concept that is discussed in Chap. 4 related to the speed of response of the (usually closed-loop) system (the higher the bandwidth, the faster the response).

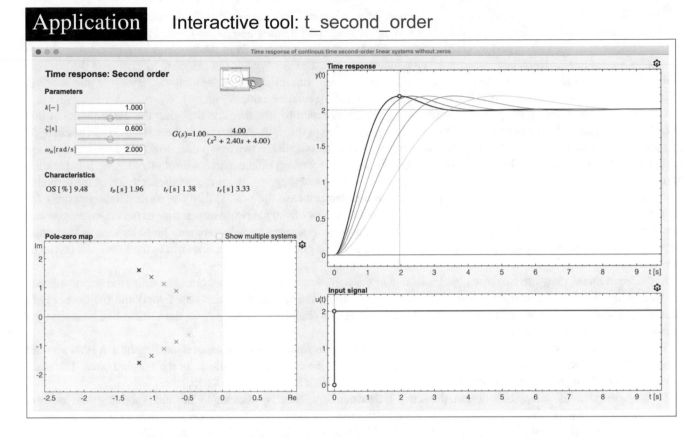

Time Response: Second Order

The main objective of this tool consists of analyzing the step response of continuous-time second-order LTI systems based on their characteristic parameters. Comparisons between different systems can be made by setting some of their characteristic parameters ζ, ω_n, σ, ω_d.

The upper-right part of the tool is dedicated to the **Parameters** definition k, ζ, and ω_n, determining the system transfer function (shown in a normalized format). The values of the parameters can be modified using the associated textbox (introducing the new value from the keyboard) or the corresponding slider.

The time response **Characteristics** are shown below the **Parameters** area. The time response of an underdamped second-order system is characterized by the following parameters: Percentage overshoot OS [%], peak time t_p [s], rise time t_r [s], and settling time t_s [s].

For values of $\zeta \geq 1$, the **Parameters** area is changed to describe an overdamped response defined by Eq. (3.22) where τ_1 and τ_2 represent the characteristic time constants of the two real poles of the system. The static gain k keeps the same meaning. The symbolic representation of the transfer function is also updated, as well as the **Characteristics** area, where now only the settling time value is shown ($4(\tau_1 + \tau_2)$).

The **Pole–zero map** is located at the lower-left part of the tool, representing the s-plane or complex plane. This graphic allows the analysis of the dynamic behavior of the system according to changes in the pole location (which are represented by the symbol $\times$). When the mouse is placed over the poles, the pole location is shown on the lower-left corner of the tool. Using this graphic, it is possible to move from underdamped to overdamped responses (and vice versa) by clicking and dragging the poles to the real axis or moving them away from it. The tool shows only a unique second-order system by default, but it is possible to show several systems (five systems) simultaneously by selecting the option Show multiple systems located on the **Pole–zero map** graphic. When several systems are shown, the active system is selected by clicking on its poles or its time response. Then, the information of the active model will be represented in black and in thick style. The rest will be drawn using the colors defined in the library. The gearwheel icon allows the modification of the scale.

The right part of the tool represents the step response of the system. The system output is shown at the upper part in the **Time response** plot, and the step input signal is defined on the lower part by the (**Input signal**) plot. The relationship between the time response of the system and its characteristic parameters can be analyzed from the **Time response** plot, being able to change its values interactively. For instance, it is possible to modify the overshoot of the response by changing the position of the black circle located at the maximum value of the time response, thus affecting the damping factor and the pole location (there exists a biunivocal relation between the peak overshoot and the relative damping factor).

On the other hand, the undamped natural frequency can be modified by clicking and dragging the dashed vertical line (which passes through the maximum value of the response) to the left or the right. Furthermore, the static gain can be changed by moving the dashed horizontal line in a vertical direction. When the system has two real poles, two dashed vertical lines are available in the **Time response** plot to interactively modify the corresponding time constants. From this figure, it is possible to move from the underdamped to the overdamped cases by clicking and dragging down the maximum value of the response (the critically damped case is reached for $\zeta = 1$, and the overdamped case for $\zeta > 1$, then the corresponding options for the time constants will be available). However, it is not possible to move from the overdamped case to the underdamped one (multi-valued problem), making it necessary to do it from the **Parameters** area or **Pole–zero map** graphic, such as discussed above. When the mouse is located over any interactive element in the graphics, its value is shown in the lower-left corner of the tool.

In the **Input signal**, there are two black circles related to the step parameters. The upper circle modifies the step amplitude (as well as by clicking and dragging on the horizontal line defining the amplitude of the step signal) and the lower circle allows the modification of the step time (that can also be modified by clicking and dragging on the vertical line located at the time value where the step value changes).

When the mouse is placed over any point of the plots shown in the **Time response** or **Input signal** graphics, a label appears showing the time instant (t) and the output (y) or input (u) values (in the corresponding plot), for the selected point. The scale can be modified using the settings available in the gearwheel icon, such as described in Chap. 1.

By using the Show multiple systems option, the active system can be selected by clicking on the corresponding pole location (on the **Pole–zero map**) or on the desired time response (at the **Time response**). The **Parameters** area shows the information of the active system. When multiple systems are shown, the following examples appear in the Options menu:

- Damping factor: Systems with the same ω_n are compared. The **Pole–zero map** plot includes a circle representing the geometric locus for the points with constant ω_n value. The poles for the five systems are located on this circle. In this way, it is possible to analyze, taking a quick look, how the time response can be affected based on the relative damping factor ζ. The pole location can be modified along the circle to visualize the effect of this modification on the rest of the system parameters and on the time response (the active system is shown in black and in thick style). The ω_n value describing the geometric locus can be modified by clicking and dragging on the circle (dashed line) toward the inside of the circle (ω_n decrease) or the outside of the circle (ω_n increase). Obviously, the new value of ω_n is updated in the **Parameters** area and the symbolic transfer function representation of the active system, updating the **Time response** plot as well. This example allows the analysis of the case study IV described in the theory summary of this section.
- Undamped natural frequency effect: Second-order systems with the same ζ value and with different values of ω_n are compared (corresponding with the case study III of the theory section). In this case, two dashed lines starting in the origin appear in the **Pole–zero map** plot, which define the geometric locus for the points with a constant value for ζ (notice that the φ angle formed by these two lines with the abscissa axis is related to ζ by $\cos \varphi = \zeta$). Once one of the five systems is selected, its poles can be moved along the lines to observe that the natural frequency is modified and its effect on the time responses. The φ angle can be modified interactively by clicking and dragging the dashed lines, thus changing the ζ value and updating the five time responses and the values of the parameters associated with the active system shown in the **Parameters** area.
- Constant imaginary part: Second-order systems with the same imaginary part of their roots and different real parts are compared. When this option is selected, two horizontal dashed lines appear on the **Pole–zero map** plot, representing the geometric locus of the poles with the same imaginary part. This value can be modified by vertically clicking and dragging these two dashed lines. Furthermore, the poles can be moved along these lines automatically updating the values of the different parameters and the time responses in the rest areas of the tool. This example is associated with the case study I discussed in the theoretical summary of this section.
- Constant real part: Second-order systems with the same real part of the poles and different imaginary parts are compared (case study II of the theoretical summary). The geometric locus of the roots with the same real part is drawn in the **Pole–zero map** plot by using a vertical dashed line, which can be modified by moving the line to the left or to the right. The poles

can be vertically dragged to analyze the effect the displacement has on the time response (**Time response** plot) and on the representative system parameters, always showing the active system in black and in thick style.

3.3.1.4 Homework

1. With the help of the tool, try to analyze the following observations [12]:
   a. In critically damped and overdamped systems there is no overshoot.
   b. t_r has a direct relation with ζ and an inverse relation with ω_n.
   c. t_s has an inverse relation with ζ and ω_n.
   d. t_s is discontinuous in ζ.
2. Using the tool t_second_order, examine and describe the effect of typical parameters of a second-order system on the step response:
   a. Select $\zeta = 0.1$, $k = 1$ and comparatively analyze which values are obtained for the peak overshoot, rise time, peak time, and settling time when $\omega_n = 1$, $\omega_n = 2$, and $\omega_n = 4$ rad/s. Locate the position of the poles of the system in each case, relating them with the values of ζ and ω_n.
   b. Choose the settings $\omega_n = 1$ rad/s, $k = 1$ and comparatively analyze what values are obtained for the peak overshoot, rise time, peak time, and settling time when $\zeta = 0.1$, $\zeta = 0.5$, and $\zeta = 0.8$. Also indicate the location of the system poles in each case, relating them with the values of ζ and ω_n.
3. For a second-order system with static gain equal to one ($k = 1$), $\zeta = 0.4$ and $\omega_n = 2$ rad/s, estimate using the time response shown in the **Time response** plot the associated values of the peak overshoot, peak time, rise time, and settling time and compare these values with those shown in the **Characteristics** area of the tool.
4. The step response of a system shows static gain equal to one ($k = 1$), 12% peak overshoot, and a peak time of 2.5 s. Analytically obtain the characteristic parameters of this system and corroborate the expected time behavior using the tool.
5. For a value of $\omega_n = 2$ rad/s and $k = 1$, consider the time response and the location of the poles of a second-order system for the values of $\zeta = -2$, $\zeta = -0.3$, $\zeta = 0$, $\zeta = 0.3$, $\zeta = 0.7$, $\zeta = 1$, and $\zeta = 2$. Indicate the type of response obtained and describe its relationship to the location of the poles of the system.
6. As noted in the previous exercise, the value of parameter ζ greatly influences the time response of the system. There is a value above which the system leaves to be oscillatory and below which an oscillatory response is obtained. Using the tool, try to determine that value.
7. Similarly, for $\zeta = 0$ check that the oscillation frequency coincides with the natural undamped frequency ω_n using the tool.
8. The options in the interactive tool allow analyzing that, for constant values of ω_n, by varying ζ in the range $0 < \zeta < 1$, it changes the angle formed by the location of the complex poles with the x-axis. Likewise, for constant values of ζ, by varying ω_n, it can be seen how the distance from the complex poles to the origin is modified (hypotenuse of the triangle formed by each of the complex poles and the origin). Therefore, given a particular value of the two complex poles $p_1 = -\sigma + j\omega_d$ and $p_1^* = -\sigma - j\omega_d$, try to relate the real and imaginary parts of the poles with the parameters ζ and ω_n of a second-order system. Use the tool to understand and analyze the results.
9. In the overdamped case ($\zeta > 1$), leaving one of the poles in a fixed location, analyze the effect on the transient response by moving the other pole along the real axis.
10. Select the option Constant imaginary part on the menu Options. With the default configuration that appears when starting the tool, obtain the values of the peak overshoot, peak time, rise time, and settling time of the five systems. Keeping the same imaginary part, which damping factor ($\sigma = \zeta\omega_n$) must the roots have so that the percentage peak overshoot is of 2%? What is the value of the settling time in that case? Does the peak time change in this study?
11. Select the option Constant real part on the menu Options. With the default configuration that appears when starting the tool, obtain the values of the peak overshoot, peak time, rise time, and settling time of the five systems. Keeping the same real part, what natural damped frequency (ω_d) must the roots have so that the percentage peak overshoot is of 2%? What is the value of the settling time in that case? Does the settling time change when the imaginary part of the roots is modified? Justify the answer.
12. Select the option Damping factor effect on the menu Options. With the default configuration that appears when starting the tool, obtain the values of the peak overshoot, peak time, rise time, and settling time of the five systems. Keeping the same natural undamped frequency, what location must the complex poles have so that the percentage peak overshoot is of 1%? What is the value of the settling time in that case?
13. Select the option Natural frequency effect on the menu Options. With the default configuration that appears when starting the tool, obtain the values of the peak overshoot, peak time, rise time, and settling time of the five systems. Keeping the same

relative damping factor, what location must the complex poles have so that the settling time is of 2 s? What is the value of the peak overshoot? Does it change in this exercise? Justify the answer.

14. Introduce the following transfer function:

$$G(s) = \frac{2}{(s+1)^2},$$

and enter a unit step at the input. Analytically check the following (and corroborate it using the tool):

a. The output starts to evolve with zero slope.

b. When $t = \tau$ the output reaches the value $y(\tau) = 0.5284$.

15. Below, differential equations that describe dynamical systems are given together with system inputs and initial conditions. Use the de Laplace transform to determine the system outputs and compare the obtained results with those given by the interactive tool.

$$\frac{d^2 y(t)}{dt^2} + 3\frac{dy(t)}{dt} + 2y(t) = u(t),$$

$$u(t) = \begin{cases} 0, & t < 0 \\ 1, & t \geq 0 \end{cases}, \qquad \frac{dy(t)}{dt}(0) = y(0) = 0.$$

The solution of this exercise can be found in [11].

3.4   Effect of a Zero on the Time Response of Continuous-Time First-Order Linear Systems

3.4.1   Interactive Tool: t_first_order_zero

3.4.1.1 Concepts Analyzed in the Card and Learning Outcomes
- Effect of a zero in the step response of an LTI first-order system.
- Effect of an RHP zero. Inverse response and nonminimum phase (NMP) systems.

3.4.1.2 Summary of Fundamental Theory
As discussed in Sect. 3.2, the transfer function and the time response of a first-order system to a step input are given by[16]

$$G_1(s) = \frac{Y_1(s)}{U(s)} = \frac{k}{\tau s + 1} \rightarrow Y_1(s) = \frac{k}{\tau s + 1}\frac{U_0}{s},$$

$$y_1(t) = kU_0 \left(1 - e^{-\frac{1}{\tau}t}\right), \tag{3.29}$$

where $y_1(t)$ and $u(t)$ are the system output and input, respectively, k is the static gain, and τ the time constant (see Sect. 3.2). If the output not only depends on the input but also on the input derivative, the describing differential equation is given by

$$\tau \frac{dy(t)}{dt} + y(t) = k \left(\beta \frac{du(t)}{dt} + u(t)\right), \tag{3.30}$$

where $y(t)$ and $u(t)$ are the system output and input, respectively. The transfer function of this case is

$$G(s) = \frac{Y(s)}{U(s)} = k\frac{\beta s + 1}{\tau s + 1}. \tag{3.31}$$

This transfer function has a polynomial of degree 1 both in the numerator and the denominator. The root of the characteristic equation ($s = -1/\tau$) is the *pole* of the system and that of the numerator ($s = -1/\beta$) the *zero*, which comes from the presence of the derivative of the input in the differential equation. Assimilating the variable s to the derivative operator[17]

---

[16] Subscript 1 is used both in the output and the transfer function because it facilitates the explanation.

[17] Property of derivation in t of the Laplace transform (Table 3.1).

($\mathcal{L}(du(t)/dt) = sU(s) - u(0)$, with $u(0) = 0$), the time response of the system when a step input of amplitude U_0 is introduced at $t = 0$ ($U_0 = 1$ for a unit step) can be written as the sum of the system response with $\beta = 0$ (that of the system without zero described by Eq. (3.29)), plus the derivative of that response weighted by β:

$$y(t) = \mathcal{L}^{-1}\left\{ G(s)\frac{U_0}{s}\right\} = y_1(t) + \beta\frac{dy_1(t)}{dt} = kU_0\left(1 - e^{-\frac{1}{\tau}t} + \frac{\beta}{\tau}e^{-\frac{1}{\tau}t}\right),$$

$$y(t) = kU_0\left(1 + \left(\frac{\beta}{\tau} - 1\right)e^{-\frac{1}{\tau}t}\right), \tag{3.32}$$

for $t \geq 0$. The same expression can be obtained by decomposing $Y(s)$ into partial fractions and computing the residues:

$$y(t) = \mathcal{L}^{-1}\left\{k\frac{\beta s + 1}{\tau s + 1}\frac{U_0}{s}\right\} = kU_0\left(\mathcal{L}^{-1}\left\{\frac{1}{s}\right\} + \mathcal{L}^{-1}\left\{\frac{\beta - \tau}{\tau s + 1}\right\}\right) = kU_0\left(1 + \left(\frac{\beta}{\tau} - 1\right)e^{-\frac{t}{\tau}}\right), t \geq 0.$$

As can be seen, the zero does not affect the system stability, as β does not appear in the exponential of (3.32) and it only modifies the shape of the transient response. In particular, if $\tau > 0$, the following properties hold:

- **Steady-state response:** When $t \to \infty$ the output of the system tends to kU_0, regardless the value of β and τ. The *settling time* in this case can be approximated (by substituting $y(t_s) = 0.98kU_0$ in (3.32)) as

$$t_s \approx \tau \ln\left(\frac{50|\beta - \tau|}{\tau}\right). \tag{3.33}$$

This time is greater than 4τ (settling time of a first-order system without zero) when $\beta > 2\tau$. In the case that $\beta \in [0, 2\tau]$, there is a significant cancellation of the effect of the pole and zero.
- **Initial value:** At $t = 0$ the output of the system given by (3.32) takes the value $kU_0(\beta/\tau)$, which it is not therefore equal to zero. The reason is that the system is not strictly proper or causal, since the degree of the polynomial in the numerator ($m = 1$) is equal to that of the denominator ($n = 1$), and there is a direct transmission of the input signal to the output (static component). The following cases can be distinguished:
  - $\beta > 0$, the response has a discontinuity at $t = 0$ and the value taken by the response has the same sign as the value in steady state.
  - $\beta < 0$, the response has a discontinuity at $t = 0$ and the value taken by the response has the opposite sign to that of the steady-state value. This type of response is called *inverse response* or *nonminimum phase* (NMP) response (because the interpretation in the frequency domain, Chap. 4), characteristic of many systems, such as water level control in a boiler [13].
  - $\beta = 0$, then the system is a first-order one, already treated in card 3.2.

If $\tau = 0$ the model should not be realizable, as it would consist of a first-order polynomial in the numerator. If $\tau < 0$, the system is unstable.

3.4.1.3 Reference Related to this Concept

- [6] Shahian, B., & Hassul, M. (1993). *Control system design using MATLAB®*. Prentice Hall. ISBN: 0-13-174061-X. Chapter 1, Sect. 5, paragraph 3, pp. 16–19.

Application	Interactive tool: t_first_order_zero

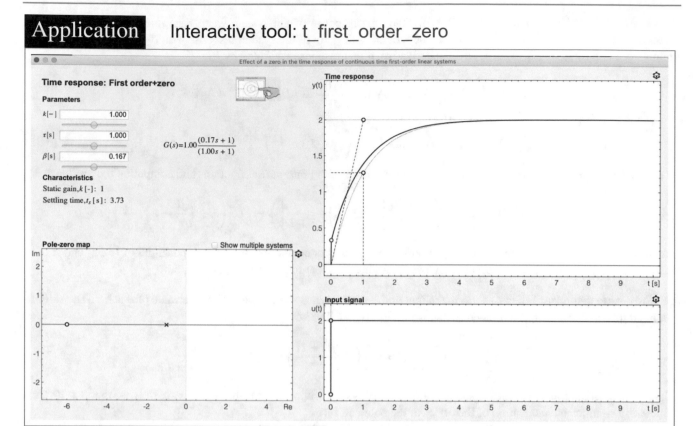

Time Response: First Order + Zero

> The main objective of this tool consists of analyzing the step response of a continuous-time first-order LTI system with a zero based on its characteristic parameters. Special attention will be paid to the initial value of the step response and to understand the inverse response when the zero of the transfer function is in the RHP.

The numeric **Parameters** and the transfer function representation are shown on the upper-left part of the tool. Furthermore, different indices are shown describing the time response of the system (**Characteristic**), being in this case the static gain and the 2% settling time (given by Eq. (3.33)). Such as described for the first-order systems without zero, the values of the characteristic parameters can be modified by using the textboxes or by using the sliders. In this tool, in addition to the static gain k and the time constant τ, the zero time constant β is also included as an interactive parameter.

Pole–zero map (complex s-plane) is shown on the lower-left part of the tool, where the pole ($\times$) and zero ($\circ$) are drawn (five zeros if the Show multiple systems option is selected). The pole and zero locations (shown on the lower-left corner of the tool) can be modified by clicking and dragging on them along the abscissa axis with the mouse. The scale can be modified using the settings available in the gearwheel icon.

The **Time response** for the different represented systems is shown on the upper-right part of the tool. The response with zero is depicted by a solid black line, while the response without zero is represented by a dotted black line. The interactive elements are as follows:

- The dashed horizontal line allows the modification of the static gain of the transfer function.
- The circle located on this line allows the modification of the initial slope of the response without zero (and thus affecting k and τ).
- The lower circle on this same line allows the modification of τ (as for the case of the first-order system without zero).
- Finally, the modification of β can be done by using a blue circle ($\circ$) that can be moved in a vertical way to represent direct (minimum phase) and inverse (nonminimum phase) responses. The vertical displacement of this circle affects the zero location in the **Pole–zero map** plot, and also the value of β in the **Parameters** area.

When the Show multiple systems option is selected, the only change appearing with respect to the previous description is that the blue circle (∘) allows the modification of the zero location of the active system (that can be selected choosing the zero in **Pole–zero map** plot or selecting the desired response in the **Time response** plot). Notice that all the systems share the same pole.

The **Input signal** with step shape introduced to the system is shown on the lower-right part of the tool (the default value of the amplitude is 2). It has two black (∘) circles associated with the step time and step amplitude.

All these graphics are completely interactive. The pole and zero locations can be modified in the s-plane; the initial slope, the initial value, and the final value of the time response can be modified from the **Time response** plot; the step amplitude and the step time can be modified from the input signal graphic; and the systems parameters can be also set quantitatively. Furthermore, the different graphic scales can be changed using the settings available in the gearwheel icon.

3.4.1.4 Homework
1. Selecting a value of $k = 1$ and $\tau = 1$ s, analyze the response obtained with values of $\beta = 0.25, 0.5, 1, 2$, and -2 s, respectively. In each case indicate the location of the zero in the s-plane (**Pole–zero map**) and the settling times obtained. What happens if $\beta \approx \tau$? Discuss the answer by analyzing the time evolution of the output.
2. Repeat the previous exercise by selecting the option Show multiple systems and using the location of the zeros that appear by default.
3. What happens if β takes a very large value compared to τ? (e.g. $\beta = 100$ and $\tau = 1$ s). Discuss the obtained response and what would be expected if $\beta \to \infty$.
4. What happens if τ takes a very large value compared to β? (e.g. $\beta = 1$ and $\tau = 100$ s). Discuss the obtained response and what would be expected if $\tau \to \infty$.
5. Which value of β provides the response of the first-order system with a zero similar to that of a first-order system without zero? Check the answer using $k = 1$ and the following configurations of the pair (τ, β) [s]: $(1, 0.25)$, $(2, 1)$, $(2, -1)$.
6. For $k = 1$ and values of $\tau = 0.1, 1, 10$ s, respectively, indicate the ranges of β in which the following occurs:
   a. The settling time of the system with a zero is less than that of the system without a zero.
   b. The settling time of the system with a zero is greater than that of the system without a zero.
7. Simulate, using the tool, the unit step response for two first-order systems with a zero whose output is given by

$$y(t) = 2(1 - 0.9e^{-t}), \quad t \geq 0,$$
$$y(t) = 2(1 + 9e^{-10t}), \quad t \geq 0.$$

Determine the value of the parameters of the transfer function in both cases.
8. Using the option Show multiple systems, find suitable combinations for the parameters k, τ, and β to provide a single starting point of the time response in $t = 0$ when the input is a unit step.

3.5 Effect of a Zero on the Time Response of Continuous-Time Second-Order Linear Systems

3.5.1 Interactive Tool: t_second_order_zero

3.5.1.1 Concepts Analyzed in the Card and Learning Outcomes
- Effect of a zero in the time response to a step input of a linear time-invariant second-order system.
- Effect of an RHP zero and inverse response.

3.5.1.2 Summary of Fundamental Theory
The main objective of this card is to analyze and understand the influence exerted by the presence of a zero on the step response of a second-order system. The generic transfer function of a second-order system is given by[18]

---

[18] Subscript 2 is used both in the output and the transfer function because it facilitates the explanation.

$$G_2(s) = \frac{Y_2(s)}{U(s)} = \frac{k\omega_n^2}{s^2 + 2\zeta\omega_n s + \omega_n^2}, \tag{3.34}$$

where k is the *static gain*, ζ is the *relative damping factor* of the system, and ω_n is the *natural undamped frequency* (see Sect. 3.3).

When considering in the above transfer function (3.34) the presence of a zero with related time constant β (and therefore located in the complex plane at $s = -1/\beta$), the following transfer function is obtained:

$$G(s) = \frac{Y(s)}{U(s)} = \frac{k\omega_n^2(\beta s + 1)}{s^2 + 2\zeta\omega_n s + \omega_n^2}. \tag{3.35}$$

Analyzing this new transfer function, $G(s)$, it can be observed that it can be separated into two terms based on the transfer function of the system without the zero, $G_2(s)$:

$$G(s) = G_2(s)(1 + \beta s) = G_2(s) + \beta s G_2(s) = \frac{k\omega_n^2}{s^2 + 2\zeta\omega_n s + \omega_n^2} + \beta s \frac{k\omega_n^2}{s^2 + 2\zeta\omega_n s + \omega_n^2}. \tag{3.36}$$

Considering that the time response when the input signal is a step of amplitude U_0 for the system represented by $G_2(s)$, the step response of the system described by $G(s)$ can be obtained from $Y_2(s) = G_2(s)U(s)$ as follows:

$$Y(s) = G(s)\frac{U_0}{s} = \left(G_2(s) + \beta s G_2(s)\right)\frac{U_0}{s} = G_2(s)\frac{U_0}{s} + \beta s G_2(s)\frac{U_0}{s} = Y_2(s) + \beta s Y_2(s), \tag{3.37}$$

$Y_2(s)$ being the Laplace transform of $y_2(t)$, which is the step response of the system without the zero, represented by Eq. (3.34). Applying the inverse Laplace transform and taking into account that $y_2(0) = 0$,

$$y(t) = y_2(t) + \beta\frac{dy_2(t)}{dt}. \tag{3.38}$$

That is, based on the result obtained in (3.38), it can be interpreted that the influence of a zero in the time response of a second-order system is given by the original time response plus the time derivative of that response weighted by a constant β when the input signal is a step.

If the system is overdamped, the representation of the transfer function of the second-order system with a zero is

$$G(s) = \frac{k(\beta s + 1)}{(\tau_1 s + 1)(\tau_2 s + 1)}, \tag{3.39}$$

where τ_1 and τ_2 are the time constants associated with the real poles (see Sect. 3.3). In this case, the response is affected by the relative position of the zero with respect to the two real poles. When a zero is located near a pole, the effect of the two elements in the response is largely canceled. If $\tau_1 \neq \beta$, $\tau_2 \neq \beta$, and $\tau_2 > \tau_1 > 0$, the response to a step input can be obtained by applying the inverse Laplace transform to $Y(s) = G(s)\frac{U_0}{s}$:

$$y(t) = kU_0\left(1 - \frac{\tau_2 - \beta}{\tau_2 - \tau_1}e^{-\frac{t}{\tau_2}} + \frac{\tau_1 - \beta}{\tau_2 - \tau_1}e^{-\frac{t}{\tau_1}}\right), \quad t \geq 0. \tag{3.40}$$

Both in the underdamped case and in the overdamped one, the derivative of the output is not zero at $t = 0$, a fact that does occur in second-order systems without zeros. The responses influenced by the presence of a zero are classified into two types according to the sign of the zero. When $\beta > 0$, the zero is in the LHP and the response is known as *minimum phase* (MP), where the time response is affected in the form of an increase in the response speed and in the overshoot. On the other hand, when $\beta < 0$, the zero is located on the RHP, and the response is known as *nonminimum phase* (NMP), where now the term that accompanies the derivative $dy_2(t)/dt$ in (3.38) is subtracted from the time response $y_2(t)$, causing an inverse response during a given period of time. In [5], the following cases are distinguished:

1. $\beta < 0$: The system has an *inverse response*, which is more pronounced the closer the zero $-1/\beta$ is to the origin of the complex plane.

2. $\beta > \tau_2 > \tau_1$: The response presents an *overshoot* even more pronounced the closer the negative zero is to the origin with respect to the location of the poles.

3. $\beta \approx \tau_2 \gg \tau_1$: In this case, the output evolution can be approximated by $y(t) \approx k(1 - \exp(-t/\tau_1))$, $t \geq 0$, and the system can be treated like a first-order one, although due to the disregarded pole–zero pair, a transient behavior of small magnitude is generated, producing a slow drift of the output to the steady-state value.

4. $\tau_2 > \beta > \tau_1$: The presence of the zero tends to speed the response when compared to the case without zero. If the zero is near the pole furthest from the origin, the response increasingly approaches that of a first-order system with a time constant τ_2. In this case, there is no slow drift of the output toward the steady-state value because the disregarded dynamics vanish quickly.

5. $\tau_2 > \tau_1 > \beta$: When moving the zero away from the origin of the complex plane (and from the poles), the response tends to be that of a second-order system with the same poles but without zero.

3.5.1.3 References Related to this Concept

- [1] Åström, K. J., & Murray, R. M. (2014). *Feedback systems: An introduction for scientists and engineers* (2nd ed.). Princeton University Press. ISBN: 9780691193984. Chapter 6, exercise 6.14, pp. 6–34.
- [4] Dorf, R. C., & Bishop, R. H. (2011). *Modern control systems* (12th ed.). Prentice Hall. ISBN: 978-0-13-602458-3. Chapter 5, Sect. 4, pp. 314–317.
- [5] Bolzern, P., Scattolini, R., & Schiavoni, N. (2009). *Fundamentos de control automático (Fundamentals of automatic control)*. McGraw-Hill. ISBN: 978-84-481-6640-3. Chapter 4, Sect. 4, paragraph 4, pp. 105–111.
- [7] Franklin, G. F., Powell, J. D., & Emani-Naeni, A. (2015). *Feedback control of dynamic systems* (7th ed.). Pearson. Chapter 3, Sect. 5, pp. 137–146.
- [12] Bavafa-Toosi, Y. (2017). *Introduction to linear control systems*. Academic Press-Elsevier. ISBN: 978-0-12-812748-3. Chapter 4, Sect. 4.9, pp. 290–291.

Application   Interactive tool: t_second_order_zero

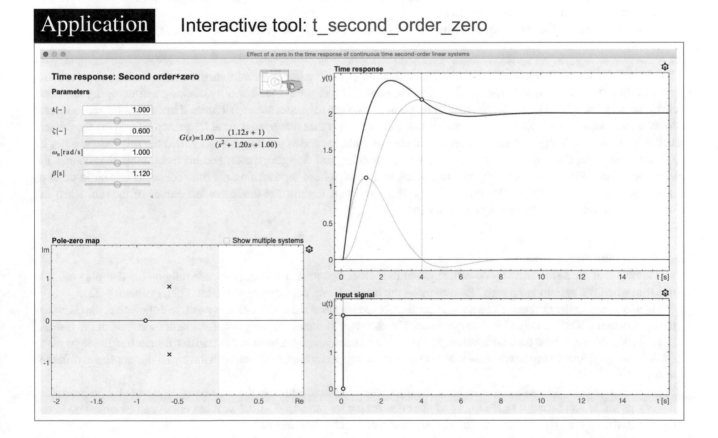

Time Response: Second Order + Zero

> The main objective of this tool consists of studying and analyzing the effect of a zero on the time response of a continuous second-order system in an interactive way. It is very useful to analyze the inverse response concept.

The left part of the tool is dedicated to defining the **Parameters** that define the transfer function of the system, for the underdamped (k, ζ, ω_n, and β) and for the overdamped (k, τ_1, τ_2, and β) cases. The values of these parameters can be modified numerically by using the textboxes or the sliders appearing in this area. These values are also affected by the modification of the poles and/or zero locations from the **Pole–zero map** plot. Moreover, the **Parameters** section includes a symbolic representation of the system transfer function (according to the nomenclature used in Eq. (3.35)), whose coefficients are updated after any change produced in the tool.

The tool loads only one second-order system with zero by default. However, it is possible to load five transfer functions with different values of the zero by using the Show multiple system option. The zeros are represented by circles of different colors (the active system being represented in black, o), while the two poles are represented using the symbol ×. The tool includes two conjugated poles by default, but these poles can be changed to be real poles moving their location to the abscissa axis.

The system parameters can also be interactively modified using the **Time response** plot. When the mouse is located over the plot, a label appears showing the time instant and the output value for this point of the response. These graphics include three plots:

- One represented by a black dashed line corresponding to the response without zero (transfer function (3.34) or that described by an underdamped second-order system without zero (3.21)).
- A plot with a solid line and in black represents the system response with zero (transfer function (3.35) for underdamped systems and (3.39) for overdamped systems).
- The third plot is dashed blue line corresponding to the derivative of $y_2(t)$ weighted by the β value, ($\beta \, dy_2(t)/dt$), described by (3.38).

The blue circle (o) on this last plot allows the modification of the β parameter and thus the corresponding zero location. Moreover, it is possible to modify the relative damping factor, and thus the overshoot, by clicking and dragging the black circle of the original system located at the maximum point of the time response, such as described in the tool for second-order systems without zero (Sect. 3.3). Furthermore, it is possible to modify the undamped natural frequency by moving the vertical dashed line associated with the system response without zero and located at the time instant where the response reaches its maximum value. The static gain can also be modified by clicking and dragging the dashed horizontal line that defines the steady-state value. When the system has two real poles, two vertical lines representing the time constants appear in order to modify their values interactively. The value of these time constants is shown at the lower-left corner of the tool when the mouse is located over one of these interactive lines.

3.5.1.4 Homework

1. It is known that for a standard second-order system the percentage peak overshoot depends only on the damping ratio. Is this true when the system has a zero? Reason using the tool and verify it theoretically (Ref. [12], Problem 4.22).
2. Study and describe the difference observed in the time response when the system zero is positive or negative, that is, when the zero is in the RHP or in the LHP, respectively. Use the Show multiple systems option for a better analysis of the results.
3. In [12] (p. 290) it is stated that the addition of a zero to an underdamped second-order transfer function at the right place ($\beta \approx 1/(4\zeta\omega_n)$) speeds up the response without much affecting performance. Check with the tool if this applies to different settings.
4. Enter an example in the tool using a value of the time constant associated with the zero equal to $\beta = 0.1$ s. Interactively modify its value increasing it to $\beta = 4$ s and after decreasing its value to $\beta = 0.1$ s. What effect can be observed in the time response of the system? How do these changes affect the blue dashed line?
5. Enter an example in the tool with $\beta = -0.2$. Interactively decrease its value to $\beta = -1$ and then increase it to $\beta = -0.2$. What effect can be observed in the time response of the system? How do these changes affect the blue dashed line?

6. For the configuration of poles and zeros that appears by default when starting the tool, analyze (in relation to the real part of the poles of the system given by $-\sigma = -\zeta\omega_n$) from which value of β the effect of the zero is not appreciable. Perform this analysis for different configurations of poles and zeros.

7. Using the configuration of poles and zeros that appears by default when starting the tool, indicate what value of β makes the peak overshoot twice that of the original system without the zero? Is the peak time affected? Justify the answer. Perform this analysis for different configurations of poles and zeros.

8. Using the pole–zero configuration loaded by default and only changing the value of $\beta = 1$, justify what condition defines the point in which there is an intersection between the original system response without the zero (dashed line) and that of the system including the zero (thick solid line) for the first time. Justify the answer.

9. In the case of inverse response (NMP systems), justify in the underdamped case if it is possible to obtain a peak overshoot greater than that of the original system without a zero. If so, does the maximum overshoot take place at a peak time greater or less than that of the original system?

10. For a configuration with two real poles (corresponding to $\tau_1 = 0.5$ s and $\tau_2 = 1$ s), find the value of β from which the response is similar to that of the original system without a zero. Modify the location of the poles and perform the analysis again. Justify the obtained results.

11. For a configuration with two real poles (corresponding to $\tau_1 = 0.5$ s and $\tau_2 = 1$ s), comment on the response obtained when $\beta = 0.5$ s and when $\beta = 1$ s. Justify the answer.

12. Check the analysis performed in the summary of theory with Eq. (3.40) with the interactive tool. Take adequate values of β, τ_1, and τ_2 that allow the analysis of the step response in terms of the relative position of the poles and the zero:
    a. $\beta < 0$: $\beta = -1$.
    b. $\beta > \tau_2 > \tau_1$: $\beta = 3$, $\tau_2 = 2$, $\tau_1 = 1$.
    c. $\beta \approx \tau_2 \gg \tau_1$: $\beta = 2$, $\tau_2 = 1$, $\tau_1 = 0.1$.
    d. $\tau_2 > \beta > \tau_1$: $\tau_2 = 10$, $\beta = 5$, $\tau_1 = 1$.
    e. $\tau_2 > \tau_1 > \beta$: $\tau_2 = 10$, $\tau_1 = 5$, $\beta = 1$.
    Verify that the comments made in the summary of the theory are true.

13. Using the default configuration when starting the tool, calculate the value of β that produces the following:
    a. A percentage overshoot of 50%.
    b. A peak time of 2 s.
    c. An inverse response whose maximum absolute value coincides with the maximum of the direct response.

14. Consider the following second-order system (Ref. [1], pp. 6–34):

$$\frac{d^2 y(t)}{dt^2} + 0.5 \frac{dy(t)}{dt} + y(t) = \beta \frac{du(t)}{dt} + u(t),$$

with zero initial conditions.
    a. Demonstrate that the initial slope of the step response is β. Explain what $\beta < 0$ means.
    b. Show that there are points in the step response that are invariant to the value of β. Qualitatively discuss the effect of parameter β in the solution (the use of the Show multiple systems option can help for this).
    c. Simulate the system with the tool and explore the effect of β on the rise time and peak overshoot.

3.6 Time Response of Generic Continuous-Time Linear Systems

3.6.1 Interactive Tool: t_generic

3.6.1.1 Concepts Analyzed in the Card and Learning Outcomes
- High-order LTI systems.
- Effect of the inclusion of new zeros, poles, integrators, and differentiators on the step response.
- Effect of time delay on the open-loop response.
- Time response to other inputs (impulse, ramp, and parabola) as an application of the properties of linear systems.

3.6.1.2 Summary of Fundamental Theory

In the previous sections, the concepts associated with the step response of LTI first-order and second-order systems have been studied, as well as the effect of adding a zero to the transfer function. This section also analyzes the presence of the following:

- **Integrators:** Poles at the origin $s = 0$. As discussed in Sect. 3.2, a particular case of first-order system is an integrator, which is represented by a transfer function $G(s) = 1/s$.
- **Differentiators:** A differentiator is a system that has a transfer function $G(s) = s$, i.e. a zero at the origin $s = 0$.
- A pure **time delay**, t_d: Events occurring at time t at one point in the system occur at another point in the system at a later time $t + t_d$ [4]. Time delay is a shift in the effect of an input on an output dynamic response. The case of input–output time delay is analyzed in this card, so that the effect of an input step at time t is seen at the output from time $t + t_d$.

Also, when models of physical systems are obtained or when closed-loop control systems are implemented, high-order systems may be obtained (the order of the model is the degree of the denominator polynomial of the transfer function, called *characteristic polynomial*). In many cases, the response of high-order models can be assimilated to that of simpler ones (increasing modeling errors) to facilitate, for example, the design of controllers. In these cases, it can be seen that the response of high-order systems can be built from the known response of first-order and second-order systems.

Most textbooks treat the case of analyzing the effect of adding a pole or a zero to a second-order system. The case of the zero has been analyzed in the previous section. When adding a pole keeping the original static gain, if the pole is at infinity it does not affect the original system. As the pole moves toward the origin the rise time increases and the maximum overshoot decreases. Anyway, all these cases can be easily analyzed with the interactive tool presented in this card.

A generic system can be represented by the following differential equation of order n:

$$a_n \frac{d^n y(t)}{dt^n} + a_{n-1} \frac{d^{n-1} y(t)}{dt^{n-1}} + \cdots + a_1 \frac{dy(t)}{dt} + a_0 y(t) = \tag{3.41}$$

$$= b_m \frac{d^m u_d(t)}{dt^m} + b_{m-1} \frac{d^{m-1} u_d(t)}{dt^{m-1}} + \cdots + b_1 \frac{du_d(t)}{dt} + b_0 u_d(t),$$

where $y(t)$ is the system output, $u(t)$ the input, and $u_d(t) = u(t - t_d)$. Applying the Laplace transform with zero initial conditions, the following transfer function is obtained:

$$G(s) = \frac{Y(s)}{U(s)} = \frac{b_m s^m + b_{m-1} s^{m-1} + \cdots + b_1 s + b_0}{a_n s^n + a_{n-1} s^{n-1} + \cdots + a_1 s + a_0} e^{-t_d s}; \quad (m \leq n), \tag{3.42}$$

where the exponential term represents the time delay (from the application of the *time shifting* property of the Laplace transform, Table 3.1). The condition $m \leq n$ is necessary to ensure causality and that the model has the physical sense (otherwise the model output at the current instant would depend on the future). This transfer function can be factorized in ZPK format:

$$G(s) = \kappa \frac{(s - z_1)(s - z_2) \cdots (s - z_m)}{(s - p_1)(s - p_2) \cdots (s - p_n)} e^{-t_d s}, \tag{3.43}$$

where z_ℓ ($\ell : 1 \ldots m$) are the zeros and p_i ($i : 1 \ldots n$) the poles.

If the final value theorem is applied to Eq. (3.43), the static gain of $G(s)$ is obtained as follows: $G(0) = k = \kappa \left(\prod_{\ell=1}^{m} (-z_\ell) / \prod_{i=1}^{n} (-p_i) \right)$.

In what follows, the most intuitive and simple cases [10] are analyzed. For the sake of simplicity, it is considered that the system has no time delay or zeros. The unit step response when only distinct real poles are considered in (3.43) is given by

$$Y(s) = \frac{k}{s} + \sum_{i=1}^{n} \frac{c_i}{s - p_i}. \tag{3.44}$$

In (3.44), c_i is the residue of the pole at $s = p_i$ (if the system has multiple poles, then $Y(s)$ will involve multiple pole terms). In the stable and strictly causal case (all poles lying in the LHP, $p_i < 0$, and $n \geq m$), the residues determine the relative importance of the components in the expanded form of the output [10]. If the real part of a pole is located far from the imaginary axis, its residue can be small (depending on the relative positions of all poles), indicating that its influence over the output is small and lasts a short time (having in this case a "non-dominant" pole). Thus, the *dominant poles* are those in which the real part is located closer to the imaginary axis, making it possible to reduce the order of the obtained models by not considering the effect of the poles farther from the imaginary axis. This analysis was clear when dealing with second-order overdamped systems in Sect. 3.3.

The poles will then be real or pairs of complex conjugates (that produce second-order terms in s), being possible to obtain a factorized form of the characteristic equation in first-order and second-order terms [10], so that the system output to a unit step input is given by

$$Y(s) = \frac{k}{s} + \sum_{i=1}^{q} \frac{c_i}{s - p_i} + \sum_{\ell=1}^{r} \frac{d_\ell(s + \zeta_\ell \omega_{n\ell}) + f_\ell \omega_{n\ell} \sqrt{1 - \zeta_\ell^2}}{s^2 + 2\zeta_\ell \omega_{n\ell} s + \omega_{n\ell}^2}; \quad (q + 2r = n), \quad (3.45)$$

where it is supposed that all the poles are different and d_ℓ and f_ℓ are coefficients obtained from the decomposition into simple fractions. From this equation, it can be inferred that the response of a high-order system without delay and without zeros to a unit step input is the superposition of responses of first- and second-order systems:

$$y(t) = k + \sum_{i=1}^{q} c_i e^{p_i t} + \sum_{\ell=1}^{r} d_\ell e^{-\zeta_\ell \omega_{n\ell} t} \cos\left(\omega_{n\ell} \sqrt{1 - \zeta_\ell^2} t\right) + \sum_{\ell=1}^{r} f_\ell e^{-\zeta_\ell \omega_{n\ell} t} \sin\left(\omega_{n\ell} \sqrt{1 - \zeta_\ell^2} t\right); \quad t \geq 0. \quad (3.46)$$

If the system is stable, the response signal is the sum of exponential and damped sinusoidal signals, and all the poles will have negative real parts, so that in steady state, the unit step response is $y(t \to \infty) = k$.

In systems with zeros and/or time delay, similar expressions and conclusions can be drawn, but it is preferred to tackle the previous cases because they are more intuitive and easy to understand. The effect of zeros is the same as explained in Sects. 3.3 and 3.5, increasing the speed of the response and eventually the amplitude of the oscillations when they are located in the LHP and producing inverse response if they are located in the RHP. What has been said about dominance can also be applied to the zeros, but possible pole–zero cancellations effects may occur. In open-loop systems, the time delay only causes the output to start evolving t_d time units after the input changes. Time delay has negative effects on the stability and behavior of closed-loop systems, as will be analyzed in Chaps. 7 and 8.

The main idea of this card is to analyze high-order systems, including multiple poles, zeros, and delay time, by using an interactive tool, with the aim of facilitating the learning of the concepts involved without the need to resort to long mathematical formulations.

This card also allows the user to analyze the response to other types of inputs, thanks to the properties of linear systems. For instance, to analyze the unit impulse response (the impulse is the derivative of the unit step), a pure differentiator can be added to the system using the pole–zero editor. Although in reality this differentiator does not belong to the plant represented by $G(s)$, the overall effect is like if the derivative of the step input is introduced. The same happens with the ramp (adding an integrator), the parabola (adding a double integrator), and so on.

Another interesting aspect is that this tool helps to analyze the effect of approximating a time delay t_d by rational functions. For instance, the two most approximations used in this text in Sect. 8.6 are based on the following:

- The truncated Taylor/McLaurin series expansion of the exponential function $e^{-t_d s} \approx (1 - t_d s)$, which helps to replace the exponential (irrational term) with an RHP zero (rational term), so that the inverse response produces the effect of a time delay.
- A Padé approximation [14] to the time delay: $e^{-t_d s} \approx \frac{(1 - s t_d/2)}{(1 + s t_d/2)}$, so that the exponential is approximated by a quotient of first-order polynomials, with an RHP zero and an LHP pole defined by half of the value of t_d.

More examples can be found in [12] (Sect. 2.3.13).

The concept of *dominance ratio* or relative dominance will be dealt with in more detail in the next section.

3.6.1.3 References Related to this Concept

- [4] Dorf, R. C., & Bishop, R. H. (2011). *Modern control systems* (12th ed.). Prentice Hall. ISBN: 978-0-13-602458-3. Chapter 5, Sect. 4, pp. 314–317.

- [6] Shahian, B., & Hassul, M. (1993). *Control system design using MATLAB®*. Prentice Hall. ISBN: 0-13-174061-X. Chapter 1, Sect. 5, paragraph 3, pp. 16–19.
- [7] Franklin, G. F., Powell, J. D., & Emani-Naeni, A. (2015). *Feedback control of dynamic systems* (7th ed.). Pearson. ISBN: 978-0-13-349659-8. Chapter 3, Sect. 1, pp. 107–109, example 3.30, pp. 143–146.
- [9] Golnaraghi, F., & Kuo, B. C. (2017). *Automatic control systems* (10th ed.). McGraw Hill Education. ISBN: 978-1-25-964384-2. Chapter 7, Sect. 7.1.
- [10] Ogata, K. (2010). *Modern control engineering* (5th ed.). Prentice Hall. ISBN: 978-0-13-615673-4. Chapter 5, Sect. 4, pp. 179–181.
- [12] Bavafa-Toosi, Y. (2017). *Introduction to linear control systems*. Academic Press-Elsevier. ISBN: 978-0-12-812748-3. Chapter 2, Sect. 2.3.13, pp. 115–121; Chapter 4, Sect. 4.11, pp. 293–295.
- [15] Barrientos, A., Sanz, R., Matía, F., & Gambao, E. (1996). *Control de sistemas continuos. Problemas resueltos (Control of continuous systems. Problems solved)*. McGraw-Hill. ISBN: 84-481-0605-9. Chapter 4, Sect. 4, pp. 142–144; Sect. 11, pp. 172–176.

Application Interactive tool: t_generic

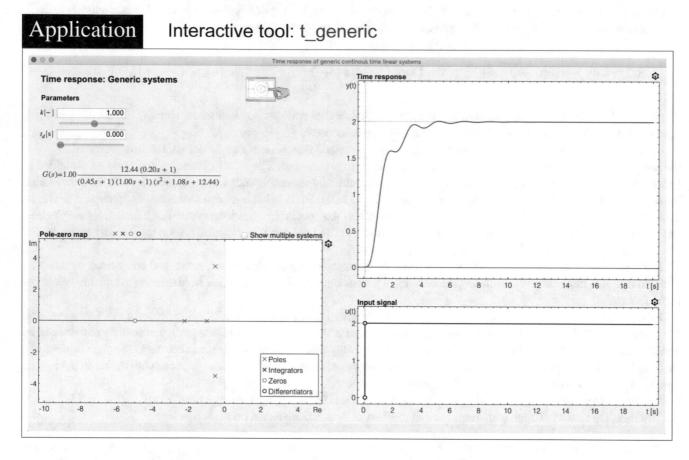

Time Response: Generic System

The main purpose of this tool is to analyze the time response of generic LTI systems. Generic means that the system has an arbitrary number of poles (including integrators), an arbitrary number of zeros (including differentiators), and time delay. The tool ensures that the transfer functions are causal (meaning that the denominator polynomial degree must be greater than that of the numerator). A limitation on the number of poles and zeros has been included in the tool to preserve computational efficiency. This tool is also useful to analyze the time response to other inputs different to steps (impulse, ramp, parabola, etc.), using the properties of linear systems and the possibility of including integrators and differentiators in the transfer functions.

As in previous tools, this tool is completely interactive and it is divided into four main areas. In this case, new features have been included regarding interactive elements.

Parameters: The upper-left area contains a symbolic representation of the transfer function (red color). Over it, there are two sliders that allow the modification of the values of the static gain k and the time delay t_d. The values of both parameters can also be set through textboxes. If the Show multiple systems option is activated on the **Pole–zero map** graph, this setup is replicated for the second transfer function (blue color).

Pole–zero map: The main novelty introduced in this tool is the possibility of using a *pole–zero editor* to configure the system transfer function to be studied. In the graph there is a legend stating that the poles are drawn using the cross symbol ×, the integrators using a bold cross **x**, the zeros are drawn using a circle symbol ∘, and the differentiators using a bold circle ∘. In the default setup, the selected system is shown in red. All this setup is replicated in blue color when activating Show multiple systems option, in such a way that two legends are shown in the **Pole–zero map** plot. The model configuration can be defined by interacting with the elements placed in the repository (×, **x**, ∘, ∘) over the **Pole–zero map**. These elements can be dragged into the s-plane, so that if they are placed over the real axis they correspond to a real pole or zero; otherwise, they are imaginary poles or zeros. Clicking and dragging an integrator or a differentiator on any area of the graph, it is automatically placed at the origin of the s-plane ($s = 0$). The elements can be removed from the system by double-clicking on them or by dragging the element back toward the repository. A maximum number of four poles is allowed but there is no limitation on the number of integrators, and the maximum number of zeros and differentiators is that ensuring system causality. This condition is checked even when elements are removed (poles cannot be removed if that action causes causality problems). Above this figure, there is a synchronization button (Synchro) which, when the Show multiple systems option is active, causes the blue model to be exactly equal to the red one. This allows a direct analysis of the effect of changing the parameters of one of the two systems in relation to the initial configuration. When placing the mouse over a pole or a zero, its position is indicated on the lower-left corner of the application.

Time response: The represented systems step response is shown on the upper-right area of the application. The scale can be modified using the settings available in the gearwheel icon. The systems' static gain can be modified through the dashed horizontal line with the corresponding color. In the same way, the time delay can be changed by moving the dashed vertical line with the color related to the selected system. When the mouse is placed over one of the interactive lines, the corresponding value (gain or time delay) is displayed on the lower-left corner. If the mouse is placed over any point of the time response, a label is shown containing the coordinates time (t) and output (y).

Input signal: In the lower-right area, the input step signal (with amplitude 2 by default) which generates displayed output at the upper-right area is shown. Placing the mouse over the top black circle (or over the horizontal line), the amplitude of the step can be adjusted, while the lower circle (or dragging the vertical line) enables changing the activation time of the signal. The step instant and its amplitude on the lower-left corner of the tool are also displayed.

In the Options menu, the possibility to introduce an arbitrary transfer function using (NUM, DEN) or ZPK formats is included. Three illustrative examples have been included as follows:

- Example 1: Displays two systems, corresponding to transfer functions $G_1(s)$ and $G_2(s)$, to study the concept of the dominant pole in the most simple configuration:

$$G_1(s) = \frac{1}{(1 + 0.5s)(1 + s)}, \quad G_2(s) = \frac{1}{(1 + s)}.$$

- Example 2: System with a pole at $s = -2$, another at $s = -5$, and a zero at $s = -10$. This example is useful to analyze the effect of zeros in an overdamped system, to introduce the pole–zero cancellation concept, and to analyze the effect of a zero in the dominance of the poles.
- Example 3: Defines a system with three real poles at $s = -1$ and two complex zeros with real part at $s = -4$. This example helps to analyze the effect of complex zeros in the system time response. It is also useful to analyze the effect of placing the complex zeros in the RHP near the abscissa axis, as a possible polynomial approximation to a time delay.

Furthermore, the option (Analyze dominance) has been included. This option will be used in the following card.

3.6.1.4 Homework

1. Include only an integrator in the **Pole–zero map**. Comment on the type of response obtained by a step input and indicate a characteristic parameter of the response. As a pure integrator is a special case of the first-order model (Sect. 3.2), try to establish the relationships between the parameters describing a pure integrator and those of a first-order transfer function and reason about the differences observed in the step response.
2. Maintaining the previous integrator, include a real pole. How is the time response affected? What is the effect of moving the pole away from the origin of the s-plane? Perform the exercise both in the case of placing the pole on the LHP and

the RHP. Notice that the obtained response is like the ramp response of a first-order system, as the integrator acts as the integral of the input step.

3. Is it possible to include only a differentiator in the **Pole–zero map**? Justify the answer. Enter a pole and a differentiator and comment on the effect of bringing the pole closer to the differentiator. Perform the exercise by placing the real pole both on the LHP and on the RHP. Notice that the obtained response is like the impulse response of a first-order system, as the differentiator acts as the derivative of the step input.

4. As a generalization of the two previous exercises, using the tool, analyze the impulse, ramp, and parabola response of a first-order system, previously configuring the step amplitude to 1 (unit step).

5. Introduce a second-order system with unit gain and two poles in $s = -2$. If a time delay is introduced, indicate what areas of the tool are affected and justify the answer. Select $t_d = 2$ and the Show multiple systems option, clicking on the Synchro button. Remove the delay in the blue system by the corresponding slider or textbox. Enter a zero in the RHP and find the value of its corresponding time constant that better approximates the time delay. Is that a good approximation? Does it fit what is expected from approximating the exponential by a first-order Taylor series expansion? Now set $t_d = 0.5$ in the original (red) system and repeat the approximation. Draw conclusions on the validity of the approximation in relation to the relative value of t_d when compared with the fundamental time constants of the model.

6. Repeat the previous exercise using a Padé approximation of the delay (adding an RHP zero and an LHP pole with time constants equal to half of the delay).

7. Select Example 1. How far should the origin of the s-plane from the pole be at $s = -2$ so that the response obtained from $G_1(s)$ is almost similar to that obtained from $G_2(s)$?

8. Select Example 2. Gradually bring the zero closer to the origin of the s-plane and comment on what happens when the zero is over the poles. Use the option Show multiple systems to enter a first-order model with its pole placed at $s = -2$ and after at $s = -5$ (and the same static gain of the original model). These two configurations help to understand and comment on the answer.

9. Select Example 2 from the Options menu. Analyze the influence of the zero in the time response when compared to the response of the original system without the zero. To perform the comparison, use the option Show multiple systems when entering the poles and click on the button Synchro, removing the zero of one of the systems from the **Pole–zero map**. Move the zero from its initial position to the same position in the RHP. It is possible to approximate the effect of an RHP zero to a time delay?

10. Select Example 3 from the Options menu and place two of the poles to be complex conjugates. Analyze the effect of modifying the location of the complex conjugate zeros in relation to the other dynamic elements. What happens if two complex conjugate zeros are placed on the LHP? Can their effect be assimilated to a time delay? Place the zeros on a location so that their effect on the time response is similar to that of a time delay of 2 s.

11. Using the option that allows the entry of a transfer function in (NUM, DEN) format, represent the unit step response of the following systems (Ref. [15]):

$$G_1(s) = \frac{-s}{s+2}, \ G_2(s) = \frac{s-2}{s+4}, \ G_3(s) = \frac{1.25}{s^2+s+2.5}, \ G_4(s) = \frac{1}{s^2+2s-1}.$$

12. Given the system

$$G(s) = \frac{5}{(s+c)(s^2+s+4.25)},$$

compare its response when the input is a unit step for $c = 5$, $c = 1$, and $c = 0.1$ (Ref. [15]).

13. For the system

$$G(s) = \frac{5(s+b)}{(s+2)(s^2+s+4.25)},$$

compare its unit step response for $b = 5$, $b = 1$, $b = 0.1$, and $b = -1$.

14. The following system (modified from [16]) relates an aircraft altitude $y(t)$ with the elevator (the control signal $u(t)$ is considered the integral of the actions performed over the elevator helm). The transfer function is given by

$$G(s) = \frac{5(s-6)}{(s^2+4s+13)}.$$

Plot and analyze the response to a unit step input.

15. Consider the following plants (Ref. [12], Example 4.21, p. 294):

$$G_1(s) = \frac{(-s+1)(s+0.9)}{(s+1)^2}, \quad G_2(s) = \frac{(-s+1)(s+0.9)}{(s+1)^3},$$

$$G_3(s) = \frac{(-s+1)(s+0.9)}{(s+1)^3}, \quad G_4(s) = \frac{(-s+1)^2(s+0.9)}{(s+1)^4},$$

$$G_5(s) = \frac{(-s+1)^2(s-0.9)}{(s+1)^3}, \quad G_6(s) = \frac{(-s+1)^2(s-0.9)}{(s+1)^4}.$$

Using the tool, observe the step response of $G_i(s)$ and $-G_i(s)$ for all $i = 1 \dots 6$. Check that the so-called inverse response phenomenon sometimes shows itself as the initial inverse-sign response.

16. MP systems can show a kind of inverse response (Ref. [12], Example 4.22, p. 294). Enter in the tool the transfer function:

$$G(s) = \frac{s^2 + 0.4s + 2}{s^2 + 1.2s + 1},$$

notice that this is not a strictly causal system, so that there is direct transmission from the input to the output so that the output starts to evolve from $y(0) = 1$ initially moving in the inverse direction of the steady state (but with the same sign). However, this is different from the meaning of the NMP transfer function. For NMP systems the output has the opposite sign of its final value for a duration of time, but for MP systems the output can never have the opposite sign of its final value, as shown in this example.

17. Analyze the step response of the systems represented by the following transfer functions (Ref. [12], Example 4.23, p. 295):

$$G_1(s) = \frac{(-s+1)^2}{(s+1)^3}, \quad G_2(s) = \frac{4(-s+1)(-s+2)(s+3)}{(s+1)(s+2)(s+3)(s+4)},$$

and find out the reason of the sign of the final value in this case. This is related to the comment in [12] (Sect. 4.11) that "it is folk knowledge that if the number of NMP zeros is odd then the step response first moves in the opposite direction of the input and takes values of opposite sign, and if it is even then it moves in the direction of the input but then changes its direction. The above folk knowledge is correct only in certain situations. The exact phenomenon is somewhat more involved".

3.7 Dominance in the Time Domain

3.7.1 Interactive Tool: t_dominance

3.7.1.1 Concepts Analyzed in the Card and Learning Outcomes
- Dominant poles.
- Dominance ratio or relative dominance in the time domain.

3.7.1.2 Summary of Fundamental Theory
In the previous card, devoted to studying the time response of generic linear systems (Sect. 3.6), it has been stated that the farther apart the real part of the poles of the imaginary axis is, the faster their associated exponential terms go to zero. Therefore, their influence on the transient response is almost irrelevant if there are poles whose real part is closer to the imaginary axis (the latter poles are called *dominant poles*). Notice that this is true always that the steady-state value is not affected. In other words, each pole contributes to the output both with the corresponding exponential term (that rapidly goes to zero) and also through its coefficient (residue), which could affect steady-state response. Let us consider the output to an input step response considering a transfer function with only real poles with simple multiplicity:

$$Y(s) = \frac{b(s)}{a(s)}\frac{1}{s} = \frac{\kappa \prod_{\ell=1}^{m}(s - z_\ell)}{\prod_{i=1}^{n}(s - p_i)}\frac{1}{s} = \frac{c_0}{s} + \frac{c_1}{s - p_1} + \frac{c_2}{s - p_2} + \cdots + \frac{c_n}{s - p_n},$$

$$y(t) = c_0 + \sum_{i=1}^{n} c_i e^{p_i t}, \tag{3.47}$$

where c_i, $i = 1 \ldots n$, are the residues (constants) associated with each pole, which can be obtained as follows:

$$c_i = \left(\frac{b(s)}{a(s)}(s - p_i)\right)_{s=p_i} = \frac{\kappa}{|p_i|}\frac{\prod_{\ell=1}^{m}(p_i - z_\ell)}{\prod_{\iota=1, \iota \neq i}^{n}(p_i - p_\iota)}. \tag{3.48}$$

In many textbooks, it is supposed that less dominant poles should also contribute with small residues, but this is not always true, as treated in [12]. In fact, it is very important to analyze the value achieved by c_i in (3.48). Several conclusions can be drawn as follows [17]:

1. Poles are in the denominator of c_i and therefore have more impact than the zeros.
2. The term $|1/p_i|$ indicates that poles that are closer to the origin are more dominant than poles that are further away.
3. A first exception to 2 occurs if two poles p_i and p_l are very close to each other, since then the difference $|p_i - p_l|$ can become very small and therefore these poles can be dominant even if other poles are closer to the origin.
4. A second exception to 2 also occurs if a pole p_i and a zero z_ℓ are very close to each other, as the difference $|p_i - z_\ell|$ can become very small and then the p_i pole has almost no impact on the response to a step input.

From these considerations, it can be concluded that, if there are no poles close to each other (condition 3), poles close to the origin are dominant unless a zero is close to a pole (condition 4). In the latter case, this pole would not be dominant. Furthermore, poles that are strictly in the RHP (unstable poles) are always dominant, since, in the development of simple fractions, they contribute to the response of the system with a positive exponential.

The poles determine the individual response modes and the zeros determine the relative influence of the functions of the individual modes (shape of the response). Moving a zero closer to a specific pole will reduce the relative contribution of the shape function corresponding to that pole. The *relative dominance* or *dominance ratio* of the poles can be determined by the quotient of their real parts. If the ratios of the real parts are greater than a certain value (which is usually between 2 and 10 depending on the author consulted) and there are no zeros nearby, the poles with the real part closest to the origin will dominate in the transient response behavior (slowest components of the response) [10]. Given a transfer function $G(s)$ of a stable system, once considered eventual pairs of poles and zeros close together in the complex plane (that to some extent *cancel* their effects), the *dominant poles* are those whose real part is located closer to the imaginary axis than that of the other poles of the system always that condition 3 does not fulfill.

The step response of a system with dominant poles can be approximated by that of a system with a transfer function that contains only those dominant poles and with the same static gain that the original system. Therefore, if a system has only one or two dominant poles, its step response can be determined from the known results for first-order or second-order systems [5]. Naturally, to perform the approximation, the presence of zeros which have a comparable distance or less than that of the dominant poles from the imaginary axis should be taken into account. For instance, for a third-order system with a real pole and two complex conjugate ones, the relative dominance of the complex roots is determined by the ratio between the real root and the real part of the complex roots, so that such a system could be approximated by a second-order one with the same pair of complex conjugate roots if the ratio is greater than 5 [4].

A *reduced equivalent system* is one that has fewer poles and/or zeros than the original system and shows a similar time response. The simple rules usually followed to obtain a reduced equivalent system are as follows [15]:

1. Never neglect the effect of an unstable pole.
2. Neglect the effect of the poles and/or zeros that are relatively far away from the imaginary axis.

3. Simplify the pole–zero pairs in the LHP that are relatively close together.
4. The original and reduced system must have the same equivalent static gain.

A more formal approach for obtaining reduced equivalent systems can be found in [9]. There are specialized model reduction methods summarized for instance in [12].

3.7.1.3 References Related to this Concept

- [4] Dorf, R. C., & Bishop, R. H. (2011). *Modern control systems* (12th ed.). Prentice Hall. ISBN: 978-0-13-602458-3. Chapter 5, Sect. 4, pp. 314–317.
- [5] Bolzern, P., Scattolini, R., & Schiavoni, N. (2009). *Fundamentos de control automático (Fundamentals of automatic control)*. McGraw-Hill. ISBN: 978-84-481-6640-3. Chapter 4, Sect. 4, paragraph 6, pp. 111–112.
- [6] Shahian, B., & Hassul, M. (1993). *Control system design using MATLAB®*. Prentice Hall. ISBN: 0-13-174061-X. Chapter 1, Sect. 5, paragraph 3, pp. 16–19.
- [9] Golnaraghi, F., & Kuo, B. C. (2017). *Automatic control systems* (10th ed.). McGraw Hill Education. ISBN: 978-1-25-964384-2. Chapter 7, Sect. 7.1.
- [10] Ogata, K. (2010). *Modern control engineering* (5th ed.). Prentice Hall. ISBN: 978-0-13-615673-4. Chapter 5, Sect. 4, p. 182.
- [12] Bavafa-Toosi, Y. (2017). *Introduction to linear control systems*. Academic Press-Elsevier. ISBN: 978-0-12-812748-3. Chapter 4, Sect. 4.7, pp. 284–289.
- [15] Barrientos, A., Sanz, R., Matía, F., & Gambao, E. (1996). *Control de sistemas continuos. Problemas resueltos (Control of continuous systems. Problems solved)*. McGraw-Hill. ISBN: 84-481-0605-9. Chapter 4, pp. 130–131 and 165–166.

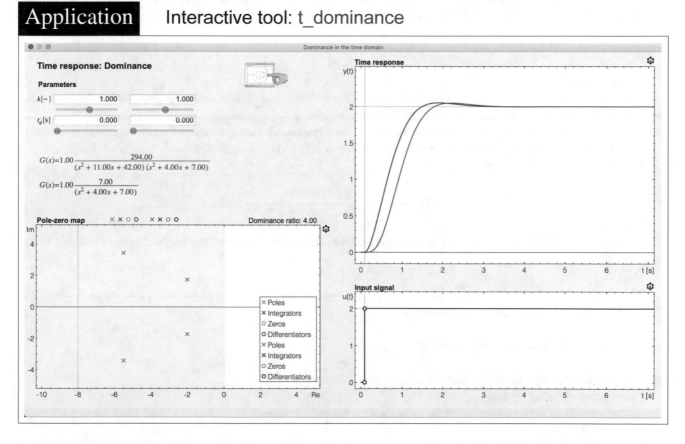

| Application | Interactive tool: t_dominance |

Time Response: Dominance

This tool deals with the dominance and dominance ratio concepts in high-order systems, including representative examples. This is very useful to obtain reduced equivalent models of high-order transfer functions.

This tool is based on the one developed in Sect. 3.6, so its basic description is not repeated. The main novelty is that the option Analyze dominance in the Options menu is active by default. The components (poles/zeros) of the transfer function shown in blue are shared by both systems while the components shown in red are only included in the corresponding transfer function.

When the Analyze dominance option is active, three new options appear in the Options menu:

- Dominance ratio: It shows a textbox that allows the introduction of the dominance ratio in the range between 2 and 10, shown as a dotted black vertical line which indicates the dominance relation boundary. This line is interactive and allows the modification of the dominance relationship. As an example, if a dominance ratio of 4 is fixed, the poles closer to the imaginary axis can be considered dominant if the rest of the poles of the system are placed to the left of the line. The vertical line interactively moves according to the real part of the dominant poles.
- Dominance example 1: This configuration corresponds to the default option. It corresponds to a fourth-order system with two complex poles and a dominance ratio of 4. In the parameters area, a symbolic representation of the transfer function is shown in red. Also, the system containing only the dominant poles and the same static gain and delay is shown in blue. In the slider over the symbolic representation, the gain k and delay t_d can also be modified. On the lower-right area, the step input signal can be seen, while on the upper-right area, the step response for both the complete and the reduced equivalent models can be seen. On the **Pole–zero map**, both the dominant poles and non-dominant ones can be dragged. Also, the vertical line defining the dominance ratio is automatically moved according to the real part of the dominant poles. It is important to notice that if non-dominant poles are moved to the left of the vertical line, both time responses become very similar.
- Dominance example 2: In this case, the original system corresponds to an overdamped second-order system while the reduced-order one is a first-order system with a pole equal to the original system dominant pole. By clicking and dragging both poles, the relative position between the different poles can be changed. This modifies the dominance ratio and the related time response.

3.7.1.4 Homework

1. Select Dominance example 1. Indicate the dominance ratio of the system. Place the less dominant poles so that the dominance ratio is 2. Is there a large difference from the previous case?
2. Select Dominance example 2. Indicate the dominance ratio of the system. Comment on the differences in the time response when the less dominant pole is placed so that the dominance ratio is 2 and 10, respectively.
3. Obtain a reduced equivalent system of [15]

$$G(s) = \frac{0.05(s+2)}{(s+3)(s+0.43)(s+1.97)}.$$

Comment on the differences that can be appreciated between the time response of the original system and that of the reduced equivalent one.

3.8   Model Fitting in the Time Domain

3.8.1   Interactive Tool: t_model_fitting

3.8.1.1 Concepts Analyzed in the Card and Learning Outcomes
- Model parameters fitting by trial and error.
- Obtaining models of first- and second-order systems with delay from experimental data obtained from the step response of a dynamical system.
- Performance indices of the time response of a dynamical system.

3.8.1.2 Summary of Fundamental Theory
As control systems are equipped with *sensors* and *actuators*, it is possible to obtain dynamic models of a process from experimental data.

The simplest way to determine the dynamics of a process is to observe the response to a unit step signal at its input when the process is in a steady state defining the operating point. Under these conditions, a step input is introduced, observing the dynamic behavior until the system reaches a new steady-state situation. This test, known as *reaction curve method*, provides (through the shape of the response) relevant information on the dynamics of the process, for instance, typical response time or kind of response (underdamped or overdamped) [18]. This test is used for obtaining linear dynamical models of the behavior of the system about the operating point at which the input step is introduced, where the input and output variables of said models represent deviation variables from the operating point (incremental variables starting from zero, so that the conditions for obtaining the representative transfer function of the process apply).

As can be seen in Fig. 3.3a, when the reaction curve method is applied on a process and the shape of the response is of sigmoidal or exponential type, it is easy to estimate, by a visual analysis, the characteristic parameters of the model of a first-order system with time delay (FOTD), given by

$$G(s) = \frac{k}{\tau s + 1} e^{-t_d s}. \tag{3.49}$$

- The static gain k is obtained as the quotient between the change that the output of the process experiments in steady state after the introduction of a step input and the amplitude of the input step (change in the input). If unit step input is used, k is directly the change experimented by the process output in steady state.
- The time delay t_d is computed as the time elapsed from the instant the input step is introduced until the output of the process starts to evolve.
- The time constant τ is estimated as the time elapsed from the instant the output of the process starts to evolve (after the time delay) until it reaches 63% of the final steady-state value. Remember that this percentage comes from making $t = \tau$ in the solution of the first-order differential equation that defines an LTI first-order system, when a step input is used (see Sect. 3.2).
- Notice that the unit step response can be also characterized by the parameters a and t_d in Fig. 3.3a, which are the intercepts of the steepest tangent S of the step response with the coordinate axes. The parameter t_d is an approximation of the time delay of the system and a/t_d is the steepest slope S of the step response [1]. The relationship with parameters k and τ is also included in the figure ($a \approx k t_d / \tau$).

Similarly, when the step response of the system shows a behavior like the one shown in Fig. 3.3b, such response can be modeled as that of an underdamped second-order system with a time delay (SOTD):

$$G(s) = \frac{k \omega_n^2}{s^2 + 2\zeta \omega_n s + \omega_n^2} e^{-t_d s}. \tag{3.50}$$

The static gain k and the time delay t_d are estimated in the same way explained in the previous paragraph for a first-order model.

- The relative damping factor ζ is obtained by measuring the peak overshoot (OS) in the step response and applying the formula that links the peak overshoot with ζ (Eq. (3.26) in Sect. 3.3):

$$OS = \frac{y(t_p) - y(t \to \infty)}{y(t \to \infty)} = \exp\left(\frac{-\zeta \pi}{\sqrt{1 - \zeta^2}}\right) \to \zeta = \sqrt{\frac{[\ln(OS)]^2}{[\ln(OS)]^2 + \pi^2}}. \tag{3.51}$$

- Once ζ has been computed, the value of the natural undamped frequency ω_n can be obtained from one of the expressions of the three characteristic times of the step response of second-order systems shown in Fig. 3.3b: peak time (Eq. (3.25)), rise time (Eq. (3.27)), or 2% settling time (Eq. (3.28)). For instance, if the peak time is measured on the response plot,

$$t_p = \frac{\pi}{\omega_d} = \frac{\pi}{\omega_n \sqrt{1 - \zeta^2}} \to \omega_n = \frac{\pi}{t_p \sqrt{1 - \zeta^2}}. \tag{3.52}$$

The goal of model fitting in the time domain (where the model is usually described by a linear differential equation or the related transfer function) is to obtain, by modifying its descriptive parameters, a model output $y_m(t)$ as close as possible to that of the real process $y(t)$. There exist different methods for automatic model fitting, the most common seeking to obtain a

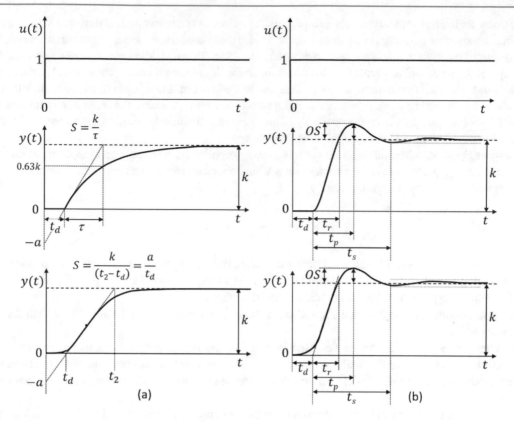

Fig. 3.3 Characterization of the unit step response as: **a** FOTD, **b** SOTD

model that minimizes the integral square error (ISE) between the model output and the actual system output in a time interval $[t_0,\ t_f]$[19]:

$$\theta^* = \arg \min_\theta \int_{t_0}^{t_f} e_m^2(t)dt = \arg \min_\theta \int_{t_0}^{t_f} (y(t) - y_m(t, \theta))^2 dt, \tag{3.53}$$

where θ is the vector of model parameters considered in the identification process and $e_m(t)$ is the difference between the system output and the model output at time t. θ^* is the vector of *optimal* parameters of the model, that is, the set of parameters that minimizes the integral square error as a product of the identification stage, t_0 is the initial time, and t_f the final interval (in the usual definition of the index $t_f \to \infty$, but in practice it is selected to be the settling time).

When adjusting the output of a model to that of a real process, what is available is a discrete-time set of input and output data of the process (obtained by sampling the signals), so that in the previous expression instead of using the integral, a sum of errors measured at certain instants of time is used, leading to the so-called *least squares identification problem* [19].

The transfer function is a description of the system in continuous time, so that the use of interactive tools permits to roughly fit a process model by modifying the characteristic parameters of the transfer function, trying that the response of the model closely matches that of the process, that usually proceeds from a sampled system, replacing in this case the minimization of a cost function by a visual inspection of the results. The interactive tool has been limited to the fitting of responses that can be modeled through first-order or second-order models, with or without delay and with one zero to facilitate the understanding of the basic ideas.

---

[19] Other indices can be used as the integral of absolute error (IAE) defined by $\int_{t_0}^{t_f} |e_m(t)|dt$ or the integral of time multiplied by absolute error (ITAE), given by $\int_{t_0}^{t_f} t|e_m(t)|dt$. In practice, these indices are not only used for model identification purposes, but for measuring the performance of closed-loop systems. ISE discriminates between excessively overdamped and excessively underdamped systems. The minimum value of the integral occurs for a compromise value of damping. Therefore, this criterion gives a higher weighting when there is a large error and a soft weighting when the error is small. IAE is one of the easiest implementable indices. It provides optimal results when dealing with reasonable damping and satisfactory transient response. However, this performance index is not easy to evaluate by analytical means. ITSE has the characteristic that with a unit step input the response has a large initial error because it has a small weight, but as time progresses, the error is penalized more heavily.

3.8.1.3 Reference Related to this Concept

- [16] Franklin, G. F., Powell, J. D., & Emani-Naeni, A. (2010). *Feedback control of dynamic systems* (6th ed.). Pearson. ISBN: 978-0-13-500150-9. Chapter 3, Sect. 7, paragraph 1, pp. 158–164.

Application	Interactive tool: t_model_fitting

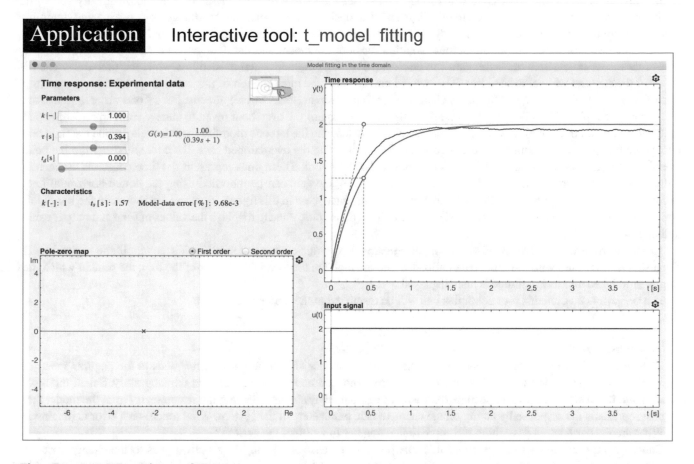

Time Response: Experimental Data

> This tool is devoted to fitting experimental data to linear FODT and SODT model structures. It serves as a simple introduction to model identification to verify that low-order models can be easily calibrated using step response data, but the complexity increases as the order of the model does.

This tool is divided into four main areas. Data and responses representing the model are drawn in red, while real process data are shown in black.

Parameters: Similar to previous tools, in this area the FODT or SODT transfer function is displayed, together with representative parameters: static gain k, time constant τ, and time delay t_d for FODT models; and static gain k, undamped natural frequency ω_n, relative damping factor ζ, and time delay t_d for SODT models. The system order can be selected through the available radio buttons over the **Pole–zero map**. The parameters can be freely modified through the sliders or textboxes (when the static gain sign or the gain range wants to be changed the second option must be used). Notice that if $\zeta \geq 1$ is introduced, the representation automatically changes to the one showing the time constant of each pole (τ_1 and τ_2). This change can also be activated in the **Time Response** figure which will be introduced later. Additionally, some time response **Characteristics** from the selected systems are shown as follows.

- In the First-order model case, the characteristics are the static gain k, the 2% settling time ($t_e \approx 4\tau$), and the fitting quadratic error (Model-data error [%]), defined by $\frac{1}{n_d}\sum_{i=1}^{n_d} e_m(t_i)^2$, with $e_m(t_i) = 100\left((y(t_i) - y_m(t_i, \theta))/y(t_i)\right)$, with $y_m(t_i, \theta)$ being the selected model output, $y(t_i)$ the real process output at time t_i, and n_d the total number of available time samples from the real system (in the exercise section some preloaded examples will be discussed).

- In the Second-order systems case, the characteristics are the percentage peak overshoot (OS [%]), peak time (t_p [s]), rise time (t_r [s]), settling time (t_s [s]), and the fitting quadratic error (Model-data error [%]), defined above.

In the **Pole–zero map**, a first- or second-order model can be selected while its associated poles are automatically shown in the figure (time delay is not represented in this diagram). The position of the poles can be changed by clicking and dragging them. The tool allows a change of the scale by changing the settings available in the gearwheel icon.

The **Time response** figure shows the output evolution for both the model and the data gathered from a real process. If a first-order model is selected, the static gain can be changed through the horizontal dashed line while the time delay can be changed through the vertical dashed line. The two red circles allow the modification of the initial slope of the response (top circle), affecting both the gain and the time constant or the time constant (lower circle) directly. For second-order systems, the two circles are replaced by one circle which allows the modification of the overshoot (even to change from an underdamped model to an overdamped one). Two vertical lines are also included, and the left one modifies the time delay while the right one allows the fixing of the undamped natural frequency. When choosing the overdamped case, the circle representing the peak overshoot becomes two dashed vertical lines containing the × symbol. These lines represent the time constants associated with each one of the real poles. Similar to the first-order model, the gain can be modified using the dotted horizontal line. When the cursor is placed over any of the interactive elements that appear in this figure, its values are shown on the lower-left corner of the tool. When the mouse is placed over the time response plot, one tag displays the values of time and output value at that point.

The lower-right figure corresponds to the **Input signal**, which in this case is not modifiable (the amplitude is fixed to 2) because it is associated with the input data preloaded into the tool. On the lower-left corner of the tool, the time in which the step is introduced and its amplitude is also displayed.

The Options menu contains four examples that will be discussed in the homework section.

3.8.1.4 Homework

1. Choose from the Options menu Experiment 1. The process response shown in black corresponds to the angular speed of a DC motor when a step input of amplitude 2 is introduced at time $t = 0$, measured in the laboratory. Select the most suitable model structure (first or second order with or without delay). After selecting the structure, try to fit the model that best represents the response by acting on the characteristic parameters of the system. What model-data error is obtained? What are the parameters describing the approximate transfer function of the system?

2. Choose from the Options menu Experiment 2. The process response shown in black corresponds to the change that the temperature of a room experiments when a window is opened 2% at time $t = 0$ (note that it is an incremental variable around a particular operating point and so it evolves from zero). Repeat the steps in the previous exercise for this example.

3. Choose from the Options menu Experiment 3. The process response shown in black corresponds to the change that the temperature of a room experiments when the heating system is activated at time $t = 0$. Repeat the steps in the previous exercise for this example.

4. Choose from the Options menu Experiment 4. The process response shown in black corresponds to the angular position of a DC motor controlled in a closed loop with a proportional controller (these kinds of controllers will be analyzed in Chap. 7 and subsequent chapters), where the input signal in this case is the reference to the closed loop. Repeat the steps of the previous exercise for this example.

References

1. Åström, K. J., & Murray, R. M. (2014). *Feedback systems: An introduction for scientists and engineers* (2nd ed.). Princeton University Press.
2. Ollero, A. (1991). *Control por computador: Descripción interna y diseño óptimo (Computer controlled systems: internal description and optimal design)*. Marcombo.
3. Chen, C.-T. (1999). *Linear system theory and design* (3rd ed.). Oxford University Press.
4. Dorf, R. C., & Bishop, R. H. (2011). *Modern control systems* (12th ed.). Prentice Hall.
5. Bolzern, P., Scattolini, R., & Schiavoni, N. (2009). *Fundamentos de control automático (Fundamentals of automatic control)*. McGraw-Hill.
6. Shahian, B., & Hassul, M. (1993). *Control system design using MATLAB®*. Prentice Hall.
7. Franklin, G. F., Powell, J. D., & Emani-Naeni, A. (2015). *Feedback control of dynamic systems* (7th ed.). Pearson.
8. Åström, K. J. (2004). *Introduction to control*. Department of Automatic Control, Lund Institute of Technology, Lund University.
9. Golnaraghi, F., & Kuo, B. C. (2017). *Automatic control systems* (10th ed.). McGraw Hill Education.

10. Ogata, K. (2010). *Modern control engineering* (5th ed.). Prentice Hall.
11. KTH (2016). Royal Institute of Technology and Linköpings Universitet, Sweden. Reglerteknik ak med utvalda tentamenstal (automatic control exercises: Computer exercises, laboratory exercises). Retrieved July 01, 2011, from https://cutt.ly/jYkcFZV.
12. Bavafa-Toosi, Y. (2017). *Introduction to linear control systems*. Academic Press-Elsevier.
13. Albertos, P., & Mareels, I. (2010). *Feedback and control for everyone*. Springer.
14. Baker, G. A., & Graves-Morris, P. (1996). *Padé approximants*. Cambridge University Press.
15. Barrientos, A., Sanz, R., Matía, F., & Gambao, E. (1996). *Control de sistemas continuos. Problemas resueltos (Control of continuous systems. Problems solved)*. McGraw-Hill.
16. Franklin, G. F., Powell, J. D., & Emani-Naeni, A. (2010). *Feedback control of dynamic systems* (6th ed.). Pearson.
17. Díaz, J. M., Costa-Castelló, R., & Dormido, S. (2021). An interactive approach to control systems analysis and design by the root locus technique. *Revista Iberoamericana de Automática e Informática Industrial, 18*(2), 172–188.
18. Åström, K. J., & Hägglund, T. (2006). *Advanced PID control*. ISA - The Instrumentation, Systems and Automation Society.
19. Ljung, L. (1998). *System identification: Theory for the user* (2nd ed.). Pearson.

Frequency Response

4.1 Introduction

In this chapter, the concepts associated with frequency response and its different graphic representations will be dealt with using interactive tools and related learning cards. The basic requirements for addressing the study of frequency response are basic knowledge in the complex variable theory, the Laplace transform, and transfer functions, which have been discussed in the Introduction to Chap. 3.

Frequency response is a complementary method to time response for analyzing LTI systems through their steady-state time response to sinusoidal input signals[1] (commonly referred to as the steady-state regime in this scope) of different frequencies [1]. It is of great interest, as any periodic signal can be decomposed as a linear combination of sinusoidal signals (e.g. using Fourier series[2]).

Since $s = \sigma + j\omega$ is a complex variable, transfer functions such as $G(s)$ (understood as a generic transfer function) are also complex functions. As summarized in the terminology and abbreviations chapter, complex numbers can be expressed in polar form, so the magnitude and phase diagrams as a function of s are three-dimensional. Usually, only imaginary values of s are taken into account. This provides the *permanent (steady-state) sinusoidal response*, which characterizes the system. There are several ways of representing the frequency response, the most important being the *Bode diagram*. Its importance lies in the fact that by taking logarithms of magnitude and frequency, the scaling achieved allows a wide spectrum of frequencies to be analyzed. In addition, it has the advantage that the product of gains is transformed into a sum by taking logarithms. Bode diagrams require two different graphs. If the frequency is not explicitly considered and the imaginary part of $G(j\omega)$ is drawn on the ordinate axis and the real part of $G(j\omega)$ on the abscissa axis, the *Nyquist diagram* is obtained, while the *Nichols diagram* represents the logarithmic magnitude (ordinate axis) versus the phase in degrees (abscissa axis). The fundamental advantage of the Bode, Nyquist, and Nichols diagrams is that they allow the stability of the closed-loop system to be analyzed by examining the frequency response of the open loop (loop transfer function), as discussed in Chap. 6. Frequency domain also plays a relevant role in the design of closed-loop systems, where high-order systems are easily handled than in time-domain analysis.

After introducing the concept of frequency response and representative plots, the brief review of concepts presented in this chapter follows with first and second-order systems and later extends to higher-order systems, as the frequency response plots of generic systems can be obtained as a combination of those of first and second-order systems.

---

[1] As pointed out in Chap. 2, a signal, in a broad sense, is any physical quantity that evolves over time.

[2] Infinite series converging punctually to a piecewise periodic and continuous function. Fourier series is the basic mathematical tool of Fourier analysis used to analyze periodic functions through the decomposition of said function into an infinite sum of much simpler sinusoidal functions (as a combination of sines and cosines with integer frequencies).

Supplementary Information The online version contains supplementary material available at https://doi.org/10.1007/978-3-031-09920-5_4.

4.2   Frequency Response Concept

4.2.1   Interactive Tool: f_concept

4.2.1.1 Concepts Analyzed in the Card and Learning Outcomes
- Frequency response concept.
- Logarithmic scales, decibels (dB), and decades.
- Graphical representations of frequency response: Bode, Nyquist, and Nichols diagrams.
- Basic frequency response factors.
- Relationship of the frequency response to the pole–zero representation and the parameters of the transfer function.
- Low-pass filters, cutoff frequency, and bandwidth.
- Minimum phase (MP) and nonminimum phase (NMP) systems.
- Influence of time delay on frequency response.

4.2.1.2 Summary of Fundamental Theory

A key concept in LTI systems is the *frequency response*. The term frequency response represents the steady-state (steady-state regime) response of a system to a sinusoidal input. The steady-state response of an LTI system to a sinusoidal input is another sinusoidal signal of the same frequency as that of the input, but with different magnitude and phase, such difference being a function of the input frequency and the system characteristics. An analysis of the frequency response of a system can be made from the transfer function $G(s)$ when $s = j\omega$. This allows the relationship between $G(j\omega)$ and ω to be analyzed in different graphical representations, such as the Bode, Nyquist, and Nichols plots which will be discussed later. The frequency response is usually obtained by introducing sinusoidal input signals of different frequencies and analyzing the resulting response in terms of amplitude and phase shift. This is one of the advantages of the method, since data obtained directly from measurements on the physical system can be used for analysis without having to deduce its mathematical model. The frequency response is important to understand the behavior of the open-loop system, in addition to being a fundamental tool for studying the stability and performance of closed-loop systems based on open-loop information, as will be seen in Chaps. 6 and 7.

Consider an LTI system described by its transfer function $G(s) = Y(s)/U(s)$, where $Y(s)$ and $U(s)$ are, respectively, the Laplace transforms of the output and input with zero initial conditions (anyway, the steady-state response of that system to a sinusoidal input does not depend on the initial conditions and therefore it can be assumed that the initial conditions are zero). According to the definition of frequency response, for a sinusoidal input $u(t) = U_0 \sin(\omega t)$, the evolution of the output can be obtained through the inverse Laplace transform,[3] where

$$U(s) = U_0 \frac{\omega}{s^2 + \omega^2} \quad \text{and} \quad Y(s) = G(s)U_0 \frac{\omega}{s^2 + \omega^2}.$$

For the sake of simplicity and without loss of generality, it is considered that the poles of $G(s)$ are different, so that the following decomposition into partial fractions can be made:

$$Y(s) = G(s)U(s) = G(s)\frac{U_0\omega}{s^2 + \omega^2} = \frac{c_1}{s - p_1} + \cdots + \frac{c_n}{s - p_n} + \frac{d_o s + d_1}{s^2 + \omega^2} =$$

$$= \underbrace{\frac{c_1}{s - p_1} + \cdots + \frac{c_n}{s - p_n}}_{\text{transient regime}} + \underbrace{\frac{d}{s + j\omega} + \frac{d^*}{s - j\omega}}_{\text{steady-state regime}}. \tag{4.1}$$

The transient regime is a consequence of the poles of $G(s)$ and the steady-state regime comes from the sinusoidal excitation function. Thus, the residues that play a role in the steady-state regime are

$$d = G(s)\frac{U_0\omega}{s^2 + \omega^2}(s + j\omega)\Big|_{s=-j\omega} = \frac{-U_0 G(-j\omega)}{2j},$$

$$d^* = G(s)\frac{U_0\omega}{s^2 + \omega^2}(s - j\omega)\Big|_{s=j\omega} = \frac{U_0 G(j\omega)}{2j}. \tag{4.2}$$

---

[3] See Table 3.2 at the beginning of Chap. 3.

Indeed, the inverse Laplace transform of (4.1) provides:

$$y(t) = c_1 e^{p_1 t} + \cdots + c_n e^{p_n t} + d e^{-j\omega t} + d^* e^{j\omega t}. \tag{4.3}$$

If the system is stable, (all $\text{Re}(p_i) < 0$, $i = 1 \ldots n$), the exponential terms drop to zero when $t \to \infty$. In that case

$$y(t \to \infty) = d e^{-j\omega t} + d^* e^{j\omega t}. \tag{4.4}$$

If the system is unstable, the terms corresponding to the transient regime will grow with time without limit, the use of the term steady-state regime having no meaning in those cases.

Recalling the value of d and d^* in (4.2), if $G(j\omega) = \rho e^{j\phi}$, where $\rho = |G(j\omega)|$ is the magnitude and $\phi = \lfloor G(j\omega) = \arctan\left(\frac{\text{Im}(G(j\omega))}{\text{Re}(G(j\omega))}\right)$ the phase of $G(j\omega)$:

$$d = \frac{-U_0 G(-j\omega)}{2j} = \frac{-U_0 \rho e^{-j\phi}}{2j} \; ; \; d^* = \frac{U_0 G(j\omega)}{2j} = \frac{U_0 \rho e^{j\phi}}{2j},$$

then Eq. (4.4) can be written as[4]

$$y(t \to \infty) = U_0 \rho \frac{-e^{-j(\omega t + \phi)} + e^{j(\omega t + \phi)}}{2j} = U_0 \rho \sin(\omega t + \phi). \tag{4.5}$$

As can be seen, the output is a sine wave signal of the same frequency as the input, but in general with different amplitude (given by the input sine amplitude multiplied by ρ) and phase shift given by ϕ. Therefore, $|G(j\omega)|$ represents the ratio of the output sinusoidal amplitude to the input sinusoidal amplitude and $\lfloor G(j\omega)$ represents the phase shift of the output signal relative to the input signal. For this reason, the function $G(j\omega)$ is called the *frequency-domain transfer function* or *sinusoidal transfer function* and for LTI systems is obtained by substituting $s = j\omega$ in the system transfer function. The frequency response function and the transfer function are directly related to each other. If one is known, the other can also be known.

The previous features of the sinusoidal transfer function facilitate the use of three graphical representations:

1. **Bode or logarithmic plot (diagram):** Formed by two graphs:

   - One represents on the ordinate axis the absolute value (magnitude) of the sinusoidal transfer function in linear scale in decibels [dB], which are the units of logarithmic gain in base 10, given by $20 \log(|G(j\omega)|)$ (this logarithm applies to the ratio of two numbers representing the output and the input of the plant which do not necessarily have the same units) versus the frequency on the abscissa axis in logarithmic scale [rad/s]. This is called the *magnitude plot*.
   - The other represents the phase angle in degrees ($\lfloor G(j\omega)$) in linear scale versus the frequency in logarithmic scale on the abscissa axis [rad/s]. This is called the *phase plot*.

Therefore, both curves are drawn on semilogarithmic diagrams, with logarithmic scale for frequency (frequency ω represents a power of 10) and linear scales for magnitude in dB and phase angle in degrees. In the frequency axis (logarithmic scale), the distance between a frequency and its ten times more or less (e.g. 1 and 10 or 0.1), is divided in length proportional to $\log(1) = 0$; $\log(2) = 0.3010$; $\log(3) = 0.4711$; $\log(4) = 0.6020$; $\log(5) = 0.699$; ... ; $\log(9) = 0.9542$; $\log(10) = 1$ rad/s. Logarithmic scales are used when data of different orders of magnitude are to be represented on the same axis, separating them into decades. A decade is a factor of ten in frequency, that is the distance between any two frequencies whose larger one is 10 times the smaller one. The following property of decimal logarithms is thus used: $\log(k\,10^n) = \log(k) + n$, so that the order of magnitude (n) establishes a shift, separating one decade ($n = i$) from the next ($n = i + 1$) and the points corresponding to the same order of magnitude (decade) have the same space to be represented as those belonging to a higher decade.

The main advantage of this diagram is that the multiplication of magnitudes is converted into the sum (thanks to the properties of the logarithms) and it is also easy to find asymptotic approximations (straight lines) to the frequency response plots. The disadvantage is that it is not possible to draw the curves at zero frequency, due to the use of a logarithmic scale on the abscissa axis.

---

[4] Recall that $e^{j\phi} = \cos\phi + j\sin\phi$.

Table 4.1 Low- and high-frequency values of the Nyquist diagram. $\aleph$ is the system type, $(n - m)$ the relative degree

Frequency		Magnitude	Phase
0^+	$\aleph > 0$	∞	$-90\aleph$
	$\aleph = 0$	$|G(j0)|$	$-90\aleph$
	$\aleph < 0$	0	$-90\aleph$
∞	$n > m$	0	$-90(n - m)$
	$n = m$	$|G(j\infty)|$	$-90(n - m)$
	$n < m$	∞	$-90(n - m)$

2. **Nyquist or polar plot (diagram):** The projections of $G(j\omega)$ on the real and imaginary axes are its real and imaginary components. The Nyquist diagram represents the real part of $G(j\omega)$ ($\mathrm{Re}(G(j\omega))$) on the abscissa axis versus its imaginary part ($\mathrm{Im}(G(j\omega))$) on the ordinate axis, parameterized according to the frequency ω. Therefore, it constitutes the geometrical locus of the vectors ($|G(j\omega)|$, $\lfloor G(j\omega)$) in polar coordinates when ω varies from zero to infinity. In polar plots, phase angles are positive when measured counterclockwise from the positive real axis. The main advantage of this polar plot is that it shows in a single graph the frequency response over the entire frequency range, but its disadvantage is that the dependence on frequency is not explicit, placing a problem in engineering applications and, moreover, it does not allow the individual contribution of the transfer function terms to be easily visualized. Each point on the polar diagram of $G(j\omega)$ represents the terminal point of a vector at a given value ω. In order to build Nyquist diagrams, as in other cases, it is necessary to substitute $s = j\omega$ in the transfer function, computing as a function of the frequency the real and imaginary parts. It also helps to compute the magnitude and phase as in the Bode diagram case, mostly to be aware of the values of the phase for low and high frequencies. Then, selecting several values of frequencies (sometimes those related to the poles and zeros location, as well as those obtained from the crossings with the abscissa and ordinate axes by making the imaginary and real parts equal zero) a sketch of the diagram can be obtained. Of special interest are the low- and high-frequency values, obtained, respectively, by making $\omega = 0$ and $\omega \to \infty$. Table 4.1 summarizes the values depending on the number of poles n, number of zeros m, and system type $\aleph$ (number of poles at the origin of the s-plane) [2]. Although in some cases for systems with $\aleph > 0$, the Nyquist plot is not close to the axes [2], they are usually drawn close to them at $\omega = 0^+$ as an interpretation of the phase condition in Table 4.1, because the absolute value of the real or imaginary parts in each case are much greater than the other. For instance, for $G(s) = (s + 1)/s^2 \to G(j\omega) = \frac{-1}{\omega^2} - j\frac{1}{\omega}$. As pointed out by [2] (Example 6.6), it corresponds to a parabola in the third quadrant, but from Table 4.1, as $\aleph = 2$, for $\omega = 0^+$ the phase should be close to the $-180°$ axis, what is not correct, but for very low values of ω, the (negative) real part is much greater in absolute value than the imaginary part, that being the reason for drawing the plots close to the axis, but anyway it is important for students to know this fact.

3. **Nichols plot (diagram) or logarithmic magnitude diagram versus phase:** It represents the magnitude $20 \log (|G(j\omega)|)$ on the ordinate axis versus the phase $\lfloor G(j\omega)$ on the abscissa axis as a function of the frequency ω. In this type of plot, a change in the static gain of the system shifts the plot up (gain increase) or down (gain decrease), but the shape of the curve remains the same. The Nichols plot usually incorporates an abacus which includes geometric loci of constant magnitude and phase of the closed-loop system, so that from the points of the open-loop frequency response the closed-loop frequency response can be built. This feature is not analyzed in this text, as only the Bode and Nyquist diagrams are used, but it is very useful in control techniques such as *Quantitative Feedback Theory* (QFT) [3].

When working with Bode diagrams, the so-called *basic factors* are usually distinguished: The gain k, the integral factors $1/(j\omega)$, the derivative factors $j\omega$, the first-order factors $(j\omega\tau + 1)^{\pm1}$ and the second-order factors $((j\omega/\omega_n)^2 + 2\zeta(j\omega/\omega_n) + 1)^{\pm1}$. These basic factors, as well as the frequency response associated with a time delay, are discussed in the following cards. Notice that they apply to logarithmic diagrams, not to polar ones, because the Nyquist plot of a transfer function does not have any tangible relation with the Nyquist plots of its constituent components.

There are several important concepts associated with representations in the frequency domain, which will be analyzed in this chapter (see Fig. 4.1):

- **Corner frequency** (ω_{cf}, rad/s): In this text, it is considered as the frequency (or frequencies) at which the asymptotic approximation of the frequency response of a pole (or zero) changes slope (intersection of the asymptotes, also known as break frequency).

- **Low-pass filters and *cutoff frequency*** (ω_c, rad/s): The ideal low-pass filters are systems where the Bode magnitude plot associated with them is constant up to a certain frequency (the cutoff frequency ω_c) and is worth $-\infty$ for $\omega > \omega_c$, being

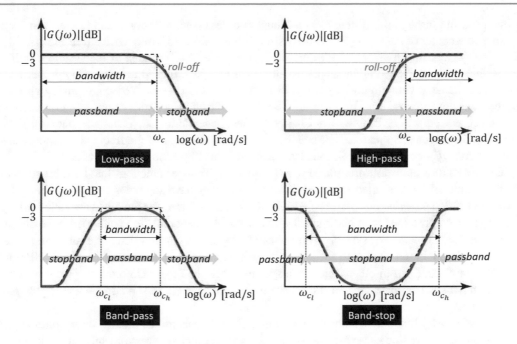

Fig. 4.1 Schematic magnitude Bode diagrams to visualize the frequency characteristic of different filters

its phase plot zero at least up to ω_c. As its realization is impossible, the real low-pass filters are defined as those systems characterized by a frequency response $G(j\omega)$ with an almost constant low-frequency magnitude and decreasing magnitude for $\omega > \omega_c$. Usually, low-pass filters are those that meet the following conditions:

$$\frac{1}{\sqrt{2}} \le \frac{|G(j\omega)|}{|G(j0)|} \le \sqrt{2}, \quad \forall \omega \le \omega_c : \quad \frac{|G(j\omega)|}{|G(j0)|} < \frac{1}{\sqrt{2}}, \quad \forall \omega > \omega_c.$$

Note that in the definition of low-pass filter, the phase does not play any role, although its contribution is usually significant. Regarding the *cutoff frequency*, a more common definition for systems that have a finite and non-zero zero frequency gain, is the frequency at which the magnitude of the system's frequency response is 3 dB (70.7%) below its zero frequency value. In other words, if the low-frequency magnitude is 0 dB, the cutoff frequency is defined as the frequency at which the ratio (output/input) has a magnitude of 0.707 (-3dB).[5]

- **Bandwidth** (BW, rad/s): The frequency range below the cutoff frequency, $0 \le \omega \le \omega_c$, in which the output magnitude does not fall below $1/\sqrt{2}$ that of the input (is reduced by more than 3 dB below its reference value) is called the system *bandwidth* (BW) (also called the *pass band*). The bandwidth of a low-pass system is defined as the frequency where the magnitude drops by a factor of $1/\sqrt{2} = 0.707 = -3$ dB of its DC value (gain at zero or low frequency). For systems that have a finite and non-zero zero-frequency gain, the bandwidth coincides with the cutoff frequency (in strictly causal systems with no poles or zeros at the origin of the s-plane). In systems that attenuate signals at low frequencies but let high-frequency signals pass through (high-pass filters), the reference gain is chosen as the high-frequency gain. In systems that attenuate both low and high frequencies (band-pass filters[6]), the bandwidth is the difference between the frequencies where their

---

[5] This is simplest definition for low-pass filters, but there are other definitions depending on the filtering characteristics (band-pass, high-pass) of the system.

[6] Band-stop filters passes most frequencies unaltered, but attenuates those in a specific range to very low levels. It is the opposite of a band-pass filter.

attenuation when passing through the system remains equal to or less than 3 dB compared to the main frequency, which is usually taken as the center of the band [1]. Band-pass filters have two cutoff frequencies (see Fig. 4.1). The bandwidth indicates how well the system filters or modifies the input sine wave (it is an index of the system's ability to attenuate high-frequency signals, as physical systems should usually act as low-pass filters due to causality properties, see Chap. 3). Bandwidth is also associated with closed-loop behavior (is a typical closed-loop frequency) giving an indication of the properties[7] of the transient response of a control system, as well as the characteristics of noise filtering and robustness[7] of the system (all these aspects will be dealt with in Chap. 6). Often, bandwidth alone is not adequate to indicate a system's ability to filter out noise signals. Sometimes it is necessary to look at the slope of the frequency response around the BW, which is called the *roll-off* or *cutoff ratio* [4], and also plays a role in closed-loop performance.

- **Minimum phase (MP) and nonminimum phase (NMP) systems:** Transfer functions that have no poles or zeros at the open RHP, neither time delays, are called *minimum phase* (MP). If they have any poles or zeros in the open RHP (and/or time delay), they are called *nonminimum phase* (NMP). Systems with MP transfer functions are called MP systems, while NMP systems are those represented with NMP transfer functions. The denomination comes from the difference between the two in the frequency domain: For systems with the same magnitude characteristic, the range in phase angle of the MP transfer function is minimum among all such systems, while the range in phase angle of any NMP transfer function is greater than this minimum [5]. In other words, if a system is MP then there is no other system which has the same magnitude $\forall \omega$ but has a smaller phase lag. An NMP system does not have the least phase lag that is possible for a system with that magnitude [2].

  For MP systems, the transfer function is uniquely determined from the magnitude plot (if the magnitude plot is specified over the whole frequency range, the phase plot is uniquely determined, and vice versa), while this is not the case for NMP systems. The problem sometimes is that one does not known a priori if the system is MP or not.

- **Influence of time delays on frequency response:** Time delays have a great influence on the frequency response of a system, since they constitute an NMP element. A pure delay in the form of a transfer function is given by $G(s) = e^{-t_d s}$. In the frequency domain $G(j\omega) = e^{-j\omega t_d}$, $\rho = 1$, $\phi = -\omega t_d$. Therefore, the time delays do not affect the magnitude plot, but they introduce a linear phase shift with frequency ($-\omega t_d$). In fact, for $\omega = 1/t_d$, a pure delay term introduces a lag of $-57.3°$.

  The polar diagram associated with the time delay is formed by a circle of unit radius traveled an infinite number of times clockwise from the positive semi-real axis. The modulus is constant and equal to 1 for all ω, while the phase is decreasing with ω. This means that when a causal dynamical system includes a time delay term, as the frequency increases the Nyquist diagram of the system runs an increasing number of turns in a clockwise direction around the origin of the complex plane approaching it in the case of strictly causal systems (logarithmic spiral).

4.2.1.3 References Related to this Concept

- [2] Bavafa-Toosi, Y. (2017). *Introduction to linear control systems*. Academic Press-Elsevier. ISBN: 978-0-12-812748-3. Chapter 7, Sect. 7.4, p. 547.
- [4] Golnaraghi, F., & Kuo, B. C. (2017). *Automatic control systems* (10th ed.). McGraw Hill Education. ISBN: 978-1-25-964384-2. Chapter 10.
- [5] Ogata, K. (2010). *Modern control engineering* (5th ed.). Prentice Hall. ISBN: 978-0-13-615673-4. Chapter 7, pp. 398–566, 2010.
- [6] Bolzern, P., Scattolini, R., & Schiavoni, N. (2009). *Fundamentos de control automático (Fundamentals of automatic control, in Spanish)*. McGraw-Hill. ISBN: 978-84-481-6640-3. Chapter 6, pp. 135–169.
- [7] D'Azzo, J. J., Houpis, C. H., & Sheldon, S. N. (2003). *Linear control system analysis and design with MATLAB®* (5th ed.). Marcel Dekker Inc. ISBN: 0-8247-4038-6. Chapter 8, pp. 301–330.
- [8] Dorf, R. C., & Bishop, R. H. (2011). *Modern control systems* (12th ed.). Prentice Hall. ISBN: 978-0-13-602458-3. Chapter 8, pp. 554–603.
- [9] Franklin, G. F., Powell, J. D., & Emani-Naeni, A. (2010). *Feedback control of dynamic systems* (6th ed.). Pearson. ISBN: 978-0-13-500150-9. Chapter 6, Sect. 1, pp. 315–334.
- [10] Rohrs, C. E., Melsa, J. L., & Schultz, D. G. (1994). *Sistemas de control lineal (Linear control systems)*. McGraw-Hill. ISBN: 970-10-0411-6. Chapter 5, pp. 269–320.

---

[7] Attribute of a control system that shows low sensitivity to effects that were not considered in the analysis and design phase of the system, e.g. disturbances, unmodeled dynamics, noise, etc.

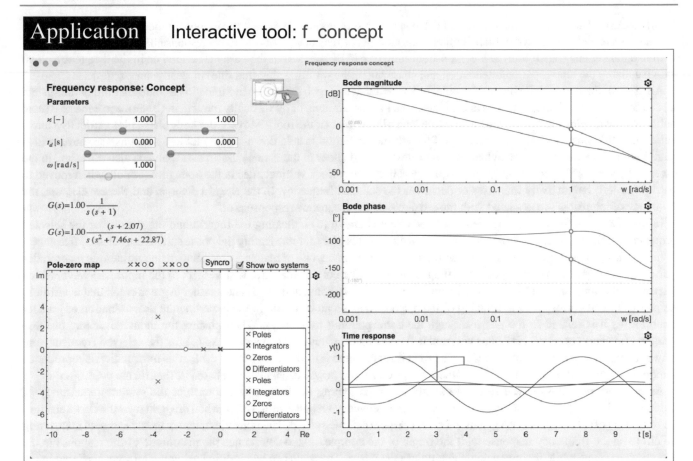

Frequency Response: Concept

The main objective of this tool is to analyze the frequency response concept and related graphical representations (Bode, Nyquist, and Nichols diagrams). The time-domain response to sinusoidal inputs of different frequencies in steady-state regime is shown together with those graphical representations, to facilitate the interpretation and understanding of the main concept.

The layout of the tool includes elements in the left area that have been used in the tools devoted to analysis in the time domain (Chap. 3), replacing the time response graphs by frequency plots in the right area, although in this card a time response graph is preserved to visualize the concept of frequency response as a response to sinusoidal inputs in steady-state regime (the transient response is not shown in the plot).

The **Parameters** area shows by default the transfer function of a second-order system without delay and with an integrator. Transfer functions are represented in this tool in ZPK format (see Eq. (3.6)) and the values of the gain term κ and the time delay t_d can be modified through corresponding textboxes and sliders, as can be done with the frequency of interest ω to be analyzed in the plots of the frequency response. Below, the selected transfer functions (two if the Show two systems option is selected) are displayed. The value of the other dynamic elements (poles and zeros) of the transfer function can be modified in the **Pole–zero map**, where the repository and editor of poles and zeros allows to change the structure of the selected system, adding or removing poles, zeros, differentiators, and integrators, by dragging inside the graph or double-clicking on them or dragging them outside (into the repository) the symbols × × ∘ ∘ located above the graph. The (Synchro) button generates a copy of the first transfer function to facilitate comparative analysis starting from the same configuration of poles and zeros. The scale can be modified using the settings available in the gearwheel icon.

On the right side, different representations of the frequency response are included, selectable through the drop-down Options menu located in the upper frame of the application. By default, the Bode diagram appears, including the 0 dB and −180° lines. The Nyquist diagram and Nichols diagram can also be selected, or the Complete view option, which represents all the diagrams simultaneously. The plots include the gearwheel icon which settings enable scaling the graphs. In all cases, in the lower-right part the **Time response** is drawn, which facilitates the analysis of the concept of frequency response based on the introduction of sinusoidal signals (black line) at a certain frequency (it can be selected in the **Parameters** area or in the frequency plots) and visualizing the amplification/attenuation and phase lag of the resulting sine (red line) in steady-state regime.

All areas of the tool include interactive objects that are related to each other. The modification of the gain term of the system (κ) can be done through the textboxes and sliders in the **Parameters** area, or by vertically moving the **Bode magnitude** plot (note that moving it displays the gain value also in the lower-left corner of the tool). Moving the **Bode phase** plot vertically causes a change in the time delay (which increases if the left mouse button is held down on the plot and a downward movement is made), which can alternatively be changed via the textbox and slider in the **Parameters** area. Its value is also refreshed in the lower-left corner of the tool. Finally, the frequency is linked to a black vertical lines in the **Bode** plots which, when moved to the right or left, interactively increases or decreases the selected frequency. In the **Nyquist diagram** and **Nichols diagram**, the frequency of interest ω is associated with the red circle on the frequency response plot.

In the **Bode magnitude** plot, a magenta colored circle is drawn (o) indicating the logarithmic magnitude for the selected frequency. Similarly, in the **Bode phase** plot, a green circle (o) is placed, representing the phase shift at the selected frequency.

In the **Time response** plot, two non-interactive segments are drawn, one of magenta color, which determines the magnitudes ratio at the selected frequency (quotient of the magnitude of the output sine wave with respect to the input sine wave, which determines the magnitude of the response in the Bode diagram one converted to dB) and another in green color, that determines the phase shift (it corresponds to time lag of the responses, as it can be useful in the experimental determination of transfer functions, as it is related to the phase through the corresponding frequency). When placing the mouse over any of these elements (magenta or green circles or segments), the values of magnitude (m) and phase (p) for the selected frequency are shown in all plots (in the case of time response what is represented is the time lag). In all cases, this information is reproduced in more detail in the bottom left corner of the tool, specifying: w [rad/s], Magnitude [dB], Phase [°], Time lag [s].

As pointed out, the **Nyquist diagram**, in addition to representing the frequency response of the system under analysis, includes a red circle (o) as an interactive element on the frequency response plot, which when dragged over the curve changes the frequency of interest and consequently the **Time response** of the system. The non-interactive magenta segment connecting the circle on the frequency response with the origin of the complex plane determines the magnitude of the response at the selected frequency, while the green angle determines the phase. The scale can be modified using the settings available in the gearwheel icon.

In the same way, in the **Nichols diagram**, the user can interactively modify the frequency corresponding to the magnitude and phase of the frequency response under study by accessing the red circle (o) over the frequency response. The green segment indicates the phase shift in this plot, while the magenta represents the magnitude in dB.

In both **Nyquist** and **Nichols** diagrams, by placing the mouse over the circles located on the frequency response (o), information about the magnitude (m) and phase (p) linked to that point appears in the graphs, while a more detailed description is included in the lower-left corner of the tool, as previously indicated.

Three examples with different pole, zero, and time delay structures can be selected from the Options menu. In addition, an arbitrary transfer function can be entered using the MATLAB® formats explained in Table 1.2 of Chap. 1:

- Example 1 ($P_{11}(s)$): Uses two systems of third and fifth-order, respectively, whose frequency responses are very similar in phase in the low-frequency range (notice that the default scale has to be changed using the gearwheel icon). This is a very illustrative example of the influence of two additional complex conjugated poles and zeros in the blue system and their effect on the frequency response (effect of the less dominant poles and influence of complex zeros).
- Example 2 uses as base system the transfer function $P_4(s)$, formed by a pole and a zero (in this case in the RHP, NMP) and compares it with a first-order system, to analyze the different behaviors at low, medium, and high frequency.
- Example 3 ($P_2(s)$) uses a first-order system. This example serves as introduction to the next card, and its objective is to analyze the different graphical representations of the frequency response of a first-order system.

4.2.1.4 Homework

1. Given $G(s) = 6/(s + 2)$ find the steady-state time response for $u(t) = 4\sin(4t)$ (Ref. [10], Exercise 5.2-1, p. 275).
2. Indicate the slope of the **Bode magnitude** plot when $\omega \to \infty$ of the following transfer functions. Relate the results obtained to the difference between the relative degree between that of the numerator polynomial and that of the denominator (proper or causal systems and strictly causal systems):

$$G_1(s) = \frac{1}{(s+1)}, \; G_2(s) = \frac{(s+1)}{(s+2)}, \; G_3(s) = \frac{(s+2)}{(s+1)}, \; G_4(s) = \frac{1}{s},$$

$$G_5(s) = \frac{(s^2+s+1)}{(s+1)(s+2)(s+3)}, \; G_6(s) = \frac{(s^2+s+1)(s+0.5)}{(s+1)(s+2)(s+3)}.$$

3. Select Example 3 ($P_2(s)$) in the Options menu of the tool. Set the transfer function to ($G(s) = 1/(s+1)$):
   a. Analytically obtain the sinusoidal response in steady-state regime, indicating the value of $|G(j\omega)|$ and $\lfloor G(j\omega)$. For different values of the frequency ω (from low to high frequencies, including the frequencies $\omega = 0.01, 1, 2$, and 10 rad/s), measure in the **Time response** the ratio of amplitudes between the output and input sine wave signal, as well as the associated time lag. Based on these data (transforming the amplitudes ratio to magnitude in dB and the time lag to phase in degrees), construct a Bode plot and compare it with the one drawn in the tool. What is the value of the magnitude and phase at low frequencies? And at high frequencies? Calculate the value of the cutoff frequency and the bandwidth of the system and write the equation of the output signal (including the value of the output sine wave amplitude and its phase lag) from the input for that cutoff frequency.
   b. Determine the value and location of the cutoff frequency in the three graphic representations of the frequency response, justifying the results. What shape is obtained in the **Nyquist diagram**? (note the scale on the graph). Determine on the **Nyquist diagram** and on the **Nichols diagram** the points corresponding to $\omega = 0$, $\omega = 1$ and $\omega = \infty$ rad/s.
   c. Use the same value of the gain but with a negative sign. Discuss the changes in the Bode plots and the sine response graph and justify those results. Repeat the procedure using the **Nyquist diagram** and **Nichols diagram** as representative plots.
4. Select Example 1 ($P_{11}(s)$) in the Options menu. In the **Pole–zero map** plot, change the scales to Variable using the gearwheel icon and drag the two conjugated complex poles furthest from the imaginary axis of the second system (blue) closer to the two conjugated complex zeros closer to the imaginary axis. Discuss what happens to the frequency responses of the two systems being compared and comment on the results.
5. Select Example 2 ($P_4(s)$) in the Options menu. In the **Pole–zero map** plot, move the zero of the first system and indicate its location so that in $\omega = 1$ rad/s the two systems have the same magnitude. Repeat the procedure to analyze the position of the zero that makes the two systems having the same phase for that frequency $\omega = 1$ rad/s. Change the scales using the gearwheel icon when required.
6. Select a system with one pole and one zero. Analytically calculate the steady-state output to a sinusoidal input, obtaining the expression of the magnitude and the phase shift. Indicate what relationship must exist between the position (or equivalently the time constants) associated with the zero and the pole of the system for the phase to increase or decrease with frequency. Check the analytical calculations using the interactive tool and indicate the phase and magnitude values in each case for low frequencies, for the cutoff frequency, and for very high frequencies as a function of the relative position between the pole and the zero (perform tests for different relative positions). Notice that the transfer function is here represented in ZPK format, so that κ is not the static gain of the system.
7. Select an integrator system with a transfer function $G(s) = k/s$ (for example using the editor for poles and zeros, notice that in this case $k = \kappa$). Justify the magnitude and phase plot shapes obtained in the **Bode diagram** by providing the formulas that describe them. What changes if the integrator's gain is negative (variable axis scale should be selected through the gearwheel in one of the plots)? In both cases, describe the behavior observed in the sinusoidal time response in steady-state regime. For the case with positive static gain, what happens if time delay is included? Does the phase plot cut the horizontal line of $-180°$? Repeat the procedure using the **Nyquist diagram** (polar diagram) and the **Nichols diagram** as the representation graph. What is the polar diagram of $G(j\omega) = 1/j\omega$? And its Nichols diagram? Justify the answers.
8. Select, using the pole–zero editor and repository, a system with a pole far from the imaginary axis (for example, at $s = -30$, changing the scale as necessary in all graphs) and a differentiator (zero at $s = 0$), and adjust κ so that the static gain is 0.1. Justify the shapes of the **Bode magnitude** and **Bode phase** plots obtained in the frequency range $0 \leq \omega \leq 100$ rad/s. What changes if the gain of the system is negative? In both cases, describe the behavior observed in the sinusoidal time response in steady-state regime. For the case with positive gain, what happens if time delay is included? For which value of the time delay the phase plot cuts the horizontal line of $-180°$? Repeat the procedure using the **Nyquist diagram** (polar plot) and the **Nichols diagram** as the representation graph. What is the polar diagram of $G(j\omega) = j\omega$, and its Nichols diagram? Justify the answers.

9. Select two systems with a pole at approximately $s = -1$ and a zero at approximately $s = -2$ (use the Synchro option to place the pole and the zero in the same location). Move one of the zeros at $s = 2$ location on the right half plane and keep the same static gain modifying κ adequately. Selecting appropriate scales through the gearwheel icon, discuss the differences observed between the system that has its pole and zero on the LHP (MP system) and the one that has zero on the RHP (NMP system), using all possible representations of the frequency response and also reasoning on the **Time response** plot.

10. Repeat the previous exercise using a configuration with the poles at $s = -2$ and the zeros at $s = -1$ and $s = 1$, respectively. Comment on the results as in the previous exercise.

11. Build a second-order system with a transfer function

$$G(s) = \frac{3}{s^2 + s + 3},$$

using the Introduce plant 1 (red) (NUM,DEN) option in the Options menu (3, [1, 1, 3]). Analyze the shape of the Bode, Nyquist, and Nichols plots (**Complete view** and adjusting the scales). On these diagrams, determine the value at which the maximum amplitude of the response is obtained and the corresponding frequency (it is also very interesting to analyze what happens in the time response plot as frequency changes). Also, indicate the location of the zero and infinite frequency points in the **Nyquist diagram**.

12. Consider the mass-spring-damper shown in the figure. Obtain the transfer function of the system considering linear dynamics: $m\ddot{x} + b_m\dot{x} + kx = f(t)$. If $f(t) = F \sin(\omega t)$, find the complete response of the system and that of steady-state regime (Ref. [2], Example 2.11, Problem 4.26).

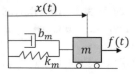

13. Select $\omega = 2$ rad/s in the **Parameters** area. Using the Introduce plant 1 (red) (NUM,DEN) option in the Options menu, enter the transfer function

$$G(s) = \frac{1}{s^2 + 0.01s + 1}.$$

Indicate the position of the poles of this system and their associated magnitude. Using the slider, slowly start to reduce the frequency and analyze the behavior observed in the **Time response** plot. Is there a value of the frequency that makes the output unbounded? Relate your answer to the concept of *marginal stability* discussed in Sect. 3.3.

4.3   Frequency Response of Continuous-Time First-Order Linear Systems Without Zeros

4.3.1   Interactive Tool: f_first_order

4.3.1.1 Concepts Analyzed in the Card and Learning Outcomes
- Frequency response analysis of an LTI first-order system through the Bode diagram of its representative transfer function.
- Asymptotic approximations to the magnitude and phase plots of the Bode diagram of a first-order transfer function.
- Relationship between static gain and magnitude plot at low frequency.
- Relationship between the time constant and the cutoff and corner frequencies.
- Low-pass filtering characteristics of a first-order transfer function.

4.3.1.2 Summary of Fundamental Theory
This card analyzes the frequency response of LTI continuous systems which representative transfer function is of first-order without zeros. The Bode diagram is used as a graphical representation of the frequency response, which, as seen in the previous card, consists of two graphs, one in which the magnitude of the sinusoidal transfer function is represented on the ordinate axis in decibels ($20 \log(|G(j\omega)|)$) versus the logarithmic scale frequency on the x-axis (in rad/s), and another representing the phase angle in degrees ($\angle G(j\omega)$) versus logarithmic scale frequency on the x-axis (in rad/s). The frequency response of a first-order LTI system can be obtained from its transfer function by substituting $s = j\omega$ (Eq. (4.7)).

When working with the Bode diagram, the gain only contributes to the magnitude plot as $20 \log |k|$ dB (horizontal line), while the phase lag contribution is zero (the modification of the static gain does not affect the phase lag), unless the static gain is negative, in which case it contributes a constant lag of $180°$ (which is the one chosen in this book, although depending on the selected angle criterion, $-180°$ could also be considered). Notice that a static gain greater than unity has a positive value in dB, while a static gain smaller than unity has a negative value. Some typical examples can be considered If an integral

k	0.005	0.01	0.1	0.5	1	2	10	100	200
dB	-46	-40	-20	-6	0	6	20	40	46

factor is analyzed (particular case of a first-order transfer function)

$$G(s) = \frac{1}{s} \rightarrow G(j\omega) = \frac{1}{j\omega} = \frac{-j}{\omega} \rightarrow \begin{cases} |G(j\omega)| = \frac{1}{|\omega|}, \\[2mm] \lfloor G(j\omega) = \arctan\left(\frac{-1/\omega}{0}\right) = -90°. \end{cases} \tag{4.6}$$

It can easily be seen that the logarithmic magnitude of $1/j\omega$ in dB is $-20 \log (|\omega|)$ and its phase angle is constant and equal to $-90°$. In a Bode diagram, the magnitude $-20 \log (|\omega|)$ is a line of slope -20 dB/decade passing through the point (0 dB, $\omega = 1$ rad/s),[8] as the frequency axis is in decimal logarithmic scale.

If the frequency response of a first-order transfer function is analyzed

$$G(s) = \frac{k}{\tau s + 1} \rightarrow G(j\omega) = \frac{k}{j\omega\tau + 1} = \frac{k}{j\omega\tau + 1}\frac{(-j\omega\tau + 1)}{(-j\omega\tau + 1)} = \frac{k(1 - j\omega\tau)}{\tau^2\omega^2 + 1},$$

$$|G(j\omega)| = \frac{|k|}{|\sqrt{\tau^2\omega^2 + 1}|}, \quad \phi = \lfloor G(j\omega) = \arctan\frac{(-\omega\tau)}{1} = -\arctan(\omega\tau), \tag{4.7}$$

it can be seen how the term $(j\omega\tau + 1)^{-1}$ contributes to both the magnitude and phase plots. In the case of the magnitude plot

$$20 \log\left(\left|\frac{1}{j\omega\tau + 1}\right|\right) = -20 \log\left(|\sqrt{\omega^2\tau^2 + 1}|\right) \text{ [dB].} \tag{4.8}$$

It is easy to verify that the *low-frequency asymptote* ($\omega\tau \ll 1$) is 0 dB, while the *high-frequency asymptote* ($\omega\tau \gg 1$) is $-20 \log (|\omega\tau|)$ dB, which in the Bode plot is a straight line of slope -20 dB/decade that intersects the 0 dB line at $\omega_{cf} = 1/\tau$, where ω_{cf} is the corner frequency. Notice that the maximum error provided by the Bode magnitude asymptotes occurs at ω_{cf}, where both asymptotes intersect. For $\omega = 1/\tau$, the magnitude of the frequency response (4.7) is $-20 \log(|\sqrt{2}|) = -3$ dB, while the asymptotes intersect at 0 dB. Notice that by definition, $\omega = 1/\tau$ is also the cutoff frequency ω_c of a first-order system ($\omega_c = \omega_{cf}$ in this case). By looking at the Bode magnitude plot, it can be seen that it represents a first-order low-pass filter, as it passes signals with a frequency lower than the cutoff frequency and attenuates signals with frequencies higher than the cutoff frequency.

The phase ϕ of the factor $(j\omega\tau + 1)^{-1}$ is $\phi = -\arctan(\omega\tau)$. For $\tau > 0$, at zero frequency, the phase is $0°$, at high frequencies ($\omega \rightarrow \infty$), the phase angle is $-90°$, while at the cutoff (and corner) frequency, it is $-45°$. In the case of the phase plot, the low- and high-frequency asymptotes are traditionally defined as horizontal lines to the corresponding phases that are joined by a vertical line that intersects the phase plot at the point where it has changed $\pm45°$ (depending on the sign of the time constant) from its initial value (*step* shape). There are other asymptotic approaches to the phase plot, such as *linear*, where the low- and high-frequency asymptotes are horizontal, but incorporate a linear approximation from the previous decade to the next one by cutting the phase plot at the point where it has changed $45°$ (corresponding to the corner frequency ω_{cf}). The error provided by this linear approximation is less than $6°$ for all frequencies. Another asymptotic approach to phase may be the so-called *optimum* approach, where the slope of the line is adjusted to that of the phase plot to minimize the error between them and to coincide at the point of interest (cutoff frequency), thus not starting from the previous decade or ending in the next decade.

As indicated in the summary of fundamental theory in Sect. 4.2, the bandwidth (BW) is defined for low-pass filters as the frequency at which the frequency response has deviated -3 dB from its low-frequency value. The bandwidth of a first-order

---

[8] The modulus operator is kept for $|\omega|$, but it could be removed as only positive frequencies have physical meaning. Negative values of the frequency will be used to build Nyquist diagrams in the framework of the Nyquist stability criterion in Chap. 6. Also, notice that if the transfer function contains the factor $(j\omega)^{-n}$, the log magnitude becomes $-20n \log (|\omega|)$.

transfer function thus coincides with the absolute value of the pole $BW = |p| = |1/\tau| = \omega_c = \omega_{cf}$. It is a measure of the system's ability to faithfully reproduce an input signal. Generally, the system response for frequency values above the bandwidth will be attenuated. This can be easily determined by the definition of frequency response. The output to a sinusoidal input $u(t) = U_0 \sin(\omega t)$ is given by Eq. (4.5):

$$y(t) = \frac{kU_0}{\sqrt{1 + \tau^2\omega^2}} \sin(\omega t - \arctan(\omega\tau)) = \frac{kU_0 p}{\sqrt{p^2 + \omega^2}} \sin(\omega t - \arctan(\omega/p)).$$

It can be seen that at $\omega = p = 1/\tau$, the output magnitude reduces to $1/\sqrt{2}$ times that of the input. The bandwidth is a concept usually associated with the closed-loop response of a control system (Chaps. 6 and 7). As the bandwidth increases, the speed of response also increases (the higher the bandwidth of the system, the faster the response will be). This is very easy to understand in a first-order transfer function, as $BW = |1/\tau|$, so if BW increases, the time constant decreases. Bandwidth is also a direct measure of the system's sensitivity to noise (very large bandwidth indicates that the system is sensitive to high-frequency noise). These considerations (trade-off between speed of response and noise rejection) generally hold for higher-order systems and are dealt with in closed-loop analysis.

4.3.1.3 References Related to this Concept

- [2] Bavafa-Toosi, Y. (2017). *Introduction to linear control systems*. Academic Press-Elsevier. ISBN: 978-0-12-812748-3. Chapter 4, Sect. 4.6.1, pp. 274–275.
- [5] Ogata, K. (2010). *Modern control engineering* (5th ed.). Prentice Hall. ISBN: 978-0-13-615673-4. Chapter 7, sections 2–4, pp. 403–445.
- [6] Bolzern, P., Scattolini, R., & Schiavoni, N. (2009). *Fundamentos de control automático (Fundamentals of automatic control, in Spanish)*. McGraw-Hill, ISBN: 978-84-481-6640-3. Chapter 6, Sect. 6, pp. 146–148.
- [8] Dorf, R. C., & Bishop, R. H. (2011). *Modern control systems* (12th ed.). Prentice Hall. ISBN: 978-0-13-602458-3. Chapter 8, Sect. 2, pp. 556–557.
- [11] Shahian, B., & Hassul, M. (1993). *Control system design using MATLAB®*. Prentice Hall. ISBN: 0-13-174061-X. Chapter 1, Sect. 5, paragraph 1, pp. 10–11.

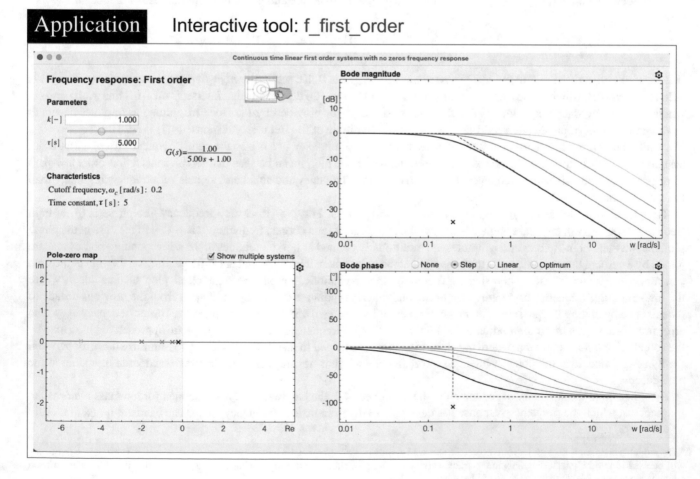

Application   Interactive tool: f_first_order

Frequency Response: First Order

This interactive tool is devoted to analyzing the frequency response of first-order transfer functions. The transfer function is represented in time constant format and the tool helps to analyze the effect of static gain and time constant (or equivalently, the pole location) on the Bode magnitude and phase diagrams. Although there is a tool focusing on NMP systems, this tool also allows the introduction of negative static gains and time constants (pole in the RHP) to verify the effect in the phase plot.

Parameters: The upper-left side shows the transfer function (represented in time constant format) and its representative parameters: static gain k and time constant τ. Indices that describe the frequency response of the selected system are displayed in the (**Characteristics** area, in this case, the Time constant and Cutoff frequency). The value of k and τ can be modified via the textbox or the slider. Negative values can be entered by typing their value in the textbox (and the scroll bar limits will automatically take into account the new sign of the parameter). When the values of k and τ are changed, the representation of the transfer function and the frequency characteristics are automatically refreshed in this area, as well as the data shown in the other graphical representations.

Pole–zero map: It shows the location of the pole of the system under analysis (or the position of the poles of the systems under study if the Show multiple systems) option is selected. As in the previous cards, the pole is represented by the symbol $\times$, which can be clicked and dragged to any location on the real axis of the complex plane (its location is always displayed in the lower-left corner of the tool). The scale can be modified using the settings available in the gearwheel icon.

The **Bode magnitude** and **Bode phase** plots show, respectively, the magnitude of the frequency response in dB and the phase in degrees of the transfer functions represented, as a function of frequency ω in rad/s (in logarithmic scale). Both diagrams can be enlarged or reduced in scale using the settings available in the gearwheel positioned in the upper right corner of the charts. In both graphs, there is a cross symbol ($\mathbf{x}$) which corresponds to the location of the cutoff frequency, which value can be modified by clicking and dragging it to the right or to the left, affecting the **Parameters** and **Pole–zero map** representations. When the mouse pointer if placed over it, the value of the corresponding frequency is displayed in the lower-left corner of the tool.

In addition to the frequency response, the corresponding asymptotes are shown. In the case of the magnitude plot, the asymptotes are drawn as black dashed lines. The low-frequency one is horizontal and the high-frequency one is a line with a -20 dB/decade slope which cuts the low-frequency one at the cutoff frequency ω_c, which coincides with the corner frequency ω_{cf}. It should be noted that the change in low-frequency magnitude (and thus in static gain) is done in these plots by clicking and dragging up/down the thick black line located on the right side of the **Bode magnitude** plot.

For the phase plot, the low- and high-frequency asymptotes are horizontal dashed lines to the corresponding phases. A vertical dashed line is drawn at the cutoff frequency, intersecting the phase plot at the point where it has changed $\pm 45°$ (depending on the sign of time constant) from its low-frequency value (step shape). It is possible to represent both plots without asymptotes or to choose between the different representations of the phase asymptotes by activating the radio buttons above the phase plot: ⊙ None ⊙ Step ⊙ Linear ⊙ Optimum.

By selecting the Show multiple systems checkbox in the **Pole–zero map**, different options are made available to initialize the location of the poles and to let comparing the five transfer functions according to their parameters (each with an associated color according to the color library explained in Chap. 1) in the Options menu on the top frame of the tool:

- Time constant effect: It starts the τ attribute of the transfer functions of five systems with different values, keeping k constant ($k = 1$) in all of them.
- Gain effect: It starts the k attribute with different values, keeping constant the value of $\tau = 1$.

The active system in the **Parameters** area is chosen by clicking on the location of its pole in the **Pole–zero map** plot or over any point of the selected frequency response, either in the **Bode magnitude** plot or in the **Bode phase** one, which are automatically highlighted in bold and black color.

It should be noted that when the pointer is placed over the frequency response plots of the active system, the frequency (w) and magnitude (m) or phase (p) values corresponding to the selected point on the curve are shown in a legend. Similarly, this information is displayed in the lower-left corner of the tool.

4.3.1.4 Homework

1. Using the default configuration when opening the tool, move the low-frequency asymptote on the magnitude plot of the Bode diagram vertically. Which elements of the interactive tool are modified? Justify the answer.

2. Select a first-order transfer function with $k = 1$ and $\tau = 1$ s. What are the values of the magnitude and phase at the cutoff frequency? Estimate the errors that occur at different frequencies between the magnitude plot and the asymptotic lines. What is the maximum value of that error and at what frequency does it occur?

3. Move the symbol **x** present on the magnitude and phase plots of the tool to the right and left. Which elements of the interactive tool are modified? If it is dragged to the right, does it increase or decrease the system bandwidth? How does the bandwidth relate to the time constant? And to the static gain? And to the pole location?

4. Can a negative time constant be achieved by shifting the **x** symbol that appears on the magnitude and phase plots of the tool? Justify the answer.

5. If the pole is dragged away from the origin of the s-plane, does the system's bandwidth increase or decrease? And the cutoff frequency? And the speed of response?

6. Try to place the pole at or near the origin $s = 0$. Comment on and justify the shape of the magnitude and phase plots (this case can be studied in detail using the interactive tool f_generic, card 4.7).

7. Enter a negative value for the static gain. Explain and justify what happens to the magnitude and phase plots in the Bode diagram.

8. Select a time constant $\tau = -1$ s. Justify the low-frequency and high-frequency values of the phase. Using the Show multiple systems option, set the pole of one of them at $s = -1$ and the pole of another at $s = 1$. Compare the magnitude and phase plots and comment on the differences. Which of the two systems could be called nonminimum phase? Move the pole from $s = 1$ to $s = -1$ and analyze and comment on what happens with the magnitude and phase plots.

9. By activating the Show multiple systems option, select from the Options menu Time constant effect. Analyze the effect on the Bode diagram the left to right shift of the pole furthest from the imaginary axis, including positive values for that pole.

10. By activating the Show multiple systems option, select from the Options menu Gain effect. Why do all the frequency responses have the same phase diagram? What is the value of the cutoff frequency of each of the represented systems?

11. Choose a system with $k = 1$ and $\tau = 0.5$ s. Indicate the value of the cutoff frequency and its relationship to the system time constant. Then analyze the accuracy of the different asymptotic phase approximations, indicating for each one the frequency at which the greatest difference between the phase plot and its asymptotic approximation is obtained and the value of that difference.

12. By activating the Show multiple systems check button, select from the Options menu Time constant effect. For the default poles configuration, calculate the bandwidth in each case. Using the interactive tool t_first_order (card 3.2), simulate these examples and relate the system's speed of response to the bandwidth.

13. Given the transfer function $G(s) = 1/(0.2s + 1) = 5/(s + 5)$ and using the interactive tool f_concept (card 4.2), analyze the input and output in the **Time response** plot for different frequencies: 0.5, 2, 5, and 8 rad/s. Note that for frequency 5 rad/s the magnitude of the output is about 0.707 that of the input, which is the definition of bandwidth (Ref. [2], Example 4.6).

4.4   Frequency Response of Continuous-Time Second-Order Linear Systems Without Zeros

4.4.1   Interactive Tool: f_second_order

4.4.1.1 Concepts Analyzed in the Card and Learning Outcomes

- Frequency response analysis of an LTI second-order system through the Bode diagram of its representative transfer function.
- Asymptotic approximations to the magnitude and phase plots of the Bode diagram of a second-order transfer function.
- Relationship between static gain and low-frequency magnitude plot.
- Relationship between relative damping factor, resonant frequency, and resonant peak.
- Relationship between undamped natural frequency and relative damping factor to gain cutoff frequency. Bandwidth in second-order systems.

4.4.1.2 Summary of Fundamental Theory

Although the frequency response of a system qualitatively represents the transient response, the correlation between frequency and transient responses is approximate, except in the case of second-order systems. Therefore, it is very important to understand the frequency response characteristics of these kinds of systems. The standard transfer function of second-order systems is:

$$G(s) = \frac{k\omega_n^2}{s^2 + 2\zeta\omega_n s + \omega_n^2} \rightarrow G(j\omega) = \frac{k}{\left(j\frac{\omega}{\omega_n}\right)^2 + 2\zeta\left(j\frac{\omega}{\omega_n}\right) + 1}, \tag{4.9}$$

where k is the *static gain*, ζ is the *relative damping factor* and ω_n is the *undamped natural frequency* of the system. The poles are placed at $p_1 = -\zeta\,\omega_n + j\omega_n\sqrt{1 - \zeta^2}$ and $p_1^* = -\zeta\,\omega_n - j\omega_n\sqrt{1 - \zeta^2}$. In the stable case, those poles can be real (overdamped system, $\zeta > 1$), real and equal (critically damped system, $\zeta = 1$) or complex conjugates (underdamped system, $0 < \zeta < 1$). For $\zeta = 0$, two complex conjugate poles on the j-axis are obtained (critically stable system) and for $\zeta < 0$ the system is unstable.

The magnitude and phase plot are obtained as

$$|G(j\omega)|_{dB} = 20\log\left(|k|\right) - 20\log\left(\left|\sqrt{\left(1 - \frac{\omega^2}{\omega_n^2}\right)^2 + \left(2\zeta\frac{\omega}{\omega_n}\right)^2}\right|\right), \tag{4.10}$$

$$\phi = \arctan\left(\frac{2\zeta\frac{\omega}{\omega_n}}{1 - \left(\frac{\omega}{\omega_n}\right)^2}\right). \tag{4.11}$$

In what follows $k = 1$ without loss of generality, as the effect of the gain on the Bode diagram is a vertical displacement of the whole magnitude plot without adding any contribution to the phase for positive gains (180° for negative gains).

In the case where the transfer function has its poles in the LHP and taking into account equations (4.10) and (4.11), for low frequencies ($\omega/\omega_n \ll 1$), the logarithmic magnitude is 0 dB[9] and the phase 0° (low-frequency asymptotes), while for high frequencies ($\omega/\omega_n \gg 1$), the magnitude asymptote is a straight line with slope -40 dB/decade ($|G(j\omega)|_{dB} \approx -40\log(\omega/\omega_n)$) and the phase is $\lfloor G(j\omega) \approx \lfloor (j\omega/\omega_n)^2 = -180°$. As explained in Sect. 4.3, there are different asymptotic approaches to the phase plot[10]:

- **Step:** Low-frequency (0°) and high-frequency ($-180°$) asymptotes are joined by a vertical line that cuts to the actual phase plot at the point where it has changed $-90°$.
- **Linear:** The low- and high-frequency asymptotes are joined by a line that goes from the previous decade to the subsequent decade, cutting to the phase plot at the point where it has changed $-90°$.
- **Optimum:** The slope of the linear asymptote adjusts to the real curve minimizing the error between them and coinciding at the midpoint ($-90°$).

On the magnitude graph, the high-frequency asymptote cuts to the low-frequency asymptote in $\omega = \omega_{cf} = \omega_n$. For this reason, asymptotic approximations do not provide very accurate results for low values of ζ in the range $0 \le \zeta \le 0.707$, those for which Eq. (4.10) has a maximum.

In the underdamped case, the Bode plot presents a *resonant peak* M_r (maximum frequency response amplitude) near ω_n (in ω_r, called the *resonant frequency*, which is the frequency at which the maximum value of the frequency response of the complex conjugate pole pair is achieved). The following expressions can be obtained deriving (4.10) with respect to ω and equalling zero:

$$\omega_r = \omega_n\sqrt{1 - 2\zeta^2}, \quad M_r = \frac{1}{2\zeta\sqrt{1 - \zeta^2}}, \quad \forall \zeta \le \frac{1}{\sqrt{2}}, \quad M_r \text{ not in dB.} \tag{4.12}$$

It is easy to see that [6]

$$|G(j\omega_n)| = \frac{1}{2\zeta}, \tag{4.13}$$

---

[9] $20\log(|k|)$ if $k \neq 1$.

[10] The profile of the phase plot and that of the asymptotes could change if the transfer function has negative static gain (adding 180°) and/or RHP poles. This can be analyzed in Sect. 4.8 devoted to NMP systems.

by substituting $\omega = \omega_n$ in (4.10). It is important to note that the peak magnitude only depends on ζ. As the relative damping factor ζ tends to zero, $\omega_r \to \omega_n$ and $M_r \to \infty$. For $\zeta = 0.707$, there is no resonant peak and $M_r = 1$.

The response of an underdamped second-order system to a sinusoidal input $u(t) = U_0 \sin(\omega t)$ is given by

$$Y(s) = \frac{k\omega_n^2}{s^2 + 2\zeta\omega_n s + \omega_n^2} \frac{U_0 \omega}{s^2 + \omega^2},$$

by decomposing into simple fractions and applying the inverse Laplace transform [2]

$$y(t) = \frac{kU_0 \omega \omega_n^2}{(\omega_n^2 - \omega^2)^2 + 4\zeta^2 \omega_n^2 \omega^2} \left[2\zeta\omega_n e^{-\zeta\omega_n t} \cos(\omega_n \sqrt{1-\zeta^2} t) \right.$$
$$\left. + \frac{2\zeta^2 \omega_n^2 - (\omega_n^2 - \omega^2)}{\omega_n \sqrt{1-\zeta^2}} e^{-\zeta\omega_n t} \sin(\omega_n \sqrt{1-\zeta^2} t) - 2\zeta\omega_n \cos(\omega t) + \frac{\omega_n^2 - \omega^2}{\omega} \sin(\omega t) \right].$$

The first two terms affect the transient response, not the steady-state regime, which is determined by the last two terms. In steady state

$$y(t) = kU_0 \left(\left(1 - \frac{\omega^2}{\omega_n^2}\right)^2 + \frac{4\zeta^2 \omega^2}{\omega_n^2} \right)^{-\frac{1}{2}} \sin\left(\omega t - \arctan\left(\frac{\frac{2\zeta\omega}{\omega_n}}{\left(1 - \frac{\omega^2}{\omega_n^2}\right)} \right) \right).$$

Notice that this is the expected response from the application of the frequency response concept (Eq. 4.5). If the amplitude of the response is made equal to $kU_0/\sqrt{2}$, the bandwidth (resulting frequency) is given by

$$BW = \omega_n \sqrt{1 - 2\zeta^2 + \sqrt{2 - 4\zeta^2 + 4\zeta^4}}, \tag{4.14}$$

where the cutoff frequency matches the bandwidth but not the corner frequency in this case. When ζ varies between 0 and 1, the BW is directly proportional to ω_n and varies between $1.55 \, \omega_n$ and $0.64 \, \omega_n$. Note that for $\zeta = 0.707$, $BW \approx \omega_n$, [2] (Fig. 4.2).

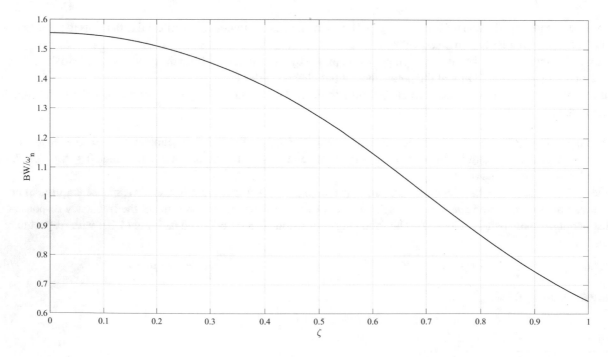

Fig. 4.2 Relationship between ζ and BW/ω_n (normalized bandwidth)

In the case of overdamped systems, the Bode diagram is constructed from the two first-order systems that form it, taking advantage of the properties of the logarithmic scales (see Sect. 4.3).

4.4.1.3 References Related to this Concept

- [2] Bavafa-Toosi, Y. (2017). *Introduction to linear control systems*. Academic Press-Elsevier. ISBN: 978-0-12-812748-3. Chapter 4, Sect. 4.6.2, pp. 276–279.
- [4] Golnaraghi, F., & Kuo, B. C. (2017). *Automatic control systems* (10th ed.). McGraw Hill Education. ISBN: 978-1-25-964384-2. Chapter 10, Sect. 1.2 and 2.
- [5] Ogata, K. (2010). *Modern control engineering* (5th ed.). Prentice Hall ISBN: 978-0-13-615673-4. Chapter 7, Sects. 2-4, pp. 403–445.
- [6] Bolzern, P., Scattolini, R., & Schiavoni, N. (2009). *Fundamentos de control automático (Fundamentals of automatic control, in Spanish)*. ISBN: 978-84-481-6640-3. Chapter 6, Sect. 6, pp. 148–151.
- [9] Franklin, G. F., Powell, J. D., & Emani-Naeni, A. (2010). *Feedback control of dynamic systems* (6th ed.). Pearson. ISBN: 978-0-13-500150-9. Chapter 6, Sect. 1, pp. 319–321.
- [11] Shahian, B., & Hassul, M. (1993). *Control system design using MATLAB®*. Prentice Hall. ISBN: 0-13-174061-X. Chapter 1, Sect. 5, paragraph 2, pp. 11–16.

Application   Interactive tool: f_second_order

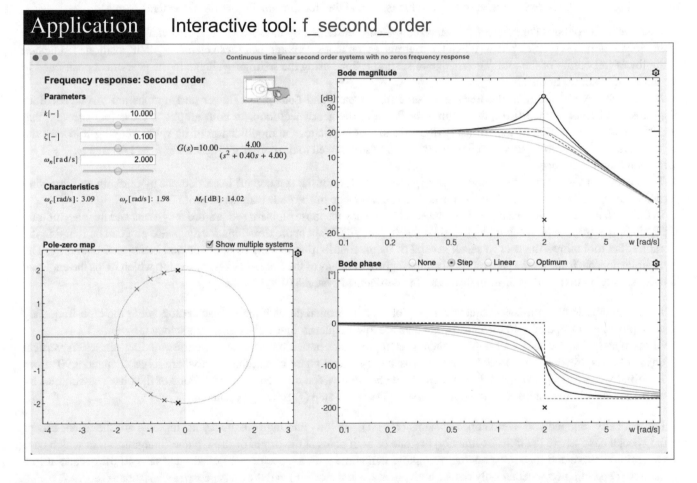

Frequency Response: Second Order

The main objective of this card is to analyze the frequency response of second-order LTI systems according to the values of the parameters of their descriptive transfer functions. Representative frequencies can be determined, together with the main values of magnitude and phase at those frequencies. Bandwidth and corner frequencies can also be determined.

The layout of the tool has four main areas:

Parameters: The upper-left side shows the numerical parameters and the transfer function that define the system under study in a standardized format:

$$G(s) = \frac{k\omega_n^2}{s^2 + 2\zeta\omega_n s + \omega_n^2},$$

together with indices describing the frequency response of the selected system (**Characteristics**), which in this case are the cutoff frequency ω_c [rad/s] (bandwidth), the resonant frequency ω_r [rad/s], and the peak magnitude M_r [dB]. The value of the static gain k, the relative damping factor ζ and the undamped natural frequency ω_n can be modified through their respective textboxes or sliders. To enter negative values of the parameters, the user has to change their value in the corresponding textbox (and the scroll bar limits will automatically take into account the new sign of the parameter). When the values of k, ζ, and ω_n are changed, the symbolic representation of the transfer function and the frequency characteristics are automatically updated in this area, as well as the data in the rest of graphic representations. When the two poles are real, textboxes and sliders corresponding to ω_n and ζ are transformed into τ_1 and τ_2, which are the time constants associated with those real poles, thus representing in the **Characteristics** area only the value of ω_c and the transfer function in time constants format.

Pole–zero map: It contains the poles of the analyzed systems, described by the symbol $\times$, which can be clicked and dragged to any location in the complex plane. When the user places the mouse pointer over these elements, their position is indicated in the lower-left corner of the tool. The change of scaling is done using the settings available in the gearwheel positioned in the upper right corner of the chart.

In the right area of the tool, the **Bode magnitude** (upper part) and **Bode phase** (lower part) graphs are displayed. Both diagrams can be also scaled using the settings available in the gearwheel icon. In both graphs, a cross (**x**) appears, that corresponds to the location of the corner frequency and that facilitates the modification of its value, moving it toward the right or left. If the two poles are real, there are two symbols **x**, linked to the corner frequencies $\omega_{cf_1} = 1/\tau_1$ and $\omega_{cf_2} = 1/\tau_2$ corresponding to each one.

The change of the magnitude at low frequency (and therefore of the static gain) is carried out by clicking and dragging vertically the thick black colored line (--) located on the right side of the **Bode magnitude** plot.

In this tool, it is possible to visualize both the exact frequency response plots for second-order systems and their asymptotic approximations (represented using black dashed lines). Since different approximations for the phase plot can be found in the literature, the tool allows the user to select several of them, contributing to their comparison and better understanding. The asymptotic approach allows the generation of simple approximations of the frequency response plot, which helps the engineer to build simple mental models of it. In the case of underdamped systems ($0 \leq \zeta < 1$):

- In the magnitude plot, the low-frequency asymptote is a horizontal dashed line in black color, while the high-frequency asymptote has a slope of -40 dB/decade that cuts the low-frequency one at the corner frequency ($\omega_{cf} = \omega_n$).
- In the phase plot, the low- and high-frequency asymptotes are horizontal lines at the corresponding phases. A vertical dashed line is drawn at the corner frequency, crossing the actual phase plot at the point where it has changed $\pm 90°$ from its initial value[11] (step shape). It is also possible to choose between different representations of the phase asymptotes by activating the circular buttons on the phase curve: $\odot$ None $\odot$ Step $\odot$ Linear $\odot$ Optimum.

For critically damped and overdamped systems ($\zeta \geq 1$), the low- and high-frequency asymptotes are the same, except in the special case where the transfer function has one or both poles at the origin of the s-plane (integrators), in which case the magnitude curve and its corresponding asymptote will come with a slope of -20 dB/decade or -40 dB/decade at low frequency (depending on whether only one or both poles are at the origin) and the phase curve at $-90°$ or $-180°$.

The option Show multiple systems introduces five different transfer functions, always highlighting in bold the system selected through its representative poles in the **Pole–zero map** graph or over any point of its frequency response, reflecting the values of the frequency, magnitude and phase corresponding as a label in that point selected in the graphs. When Show multiple systems is activated, in the Options menu several alternatives appear to initialize the location of the different poles and to be able to compare the response of various representations of systems according to their parameters:

- Damping factor effect: It initializes stable systems with different values of the relative damping factor ζ maintaining k constant and ω_n constant ($\omega_n = 2$ rad/s).

---

[11] Depending on the sign of the static gain and if the poles are in the LHP or RHP.

- Natural frequency effect: It initializes several stable systems with a set of values of the undamped natural frequency ω_n maintaining ζ and k constant ($\zeta = 0.5$, $k = 1$).
- Constant imaginary part: It initializes several underdamped systems with different values of their real parts and with a same constant imaginary part. Changing the scale settings is required in this case.
- Constant real part: In this case, the real part is constant and the same for all systems representations ($s = -5$), while the imaginary part is initialized at different values.

Notice that the Nyquist and Nichols diagrams of the system treated in this section can be analyzed using the f_concept interactive tool.

4.4.1.4 Homework

1. Select a second-order underdamped system with $\zeta = 0.5$, $\omega_n = 3$ rad/s and $k = 1$. Analyzing the Bode plots, determine the value of ω_n, ω_r, ω_{cf} and $\omega_c = BW$. Are all the same? Justify the answer.
2. Determine, using the tool, a transfer function of a second-order system having a static gain equal 2, a peak magnitude of $M_r = 5$ dB and a resonant frequency $\omega_r = 1$ rad/s. Compare the results with those expected from theory.
3. Check the Show multiple systems option. For the displayed systems, calculate the cutoff frequency ω_c (matching the bandwidth BW), the resonant frequency ω_r and the peak magnitude M_r using the tool. Which of the displayed systems will not have overshoot when a step signal is introduced at the input?
4. For a second-order system with $k = 1$ and $\omega_n = 2$ rad/s, analyze the frequency response obtained for values of $\zeta = 0.1, 0.2, 0.3, 0.5, 0.7, 1.0$ (use the option Damping factor effect). Calculate (using the tool) the cutoff frequency ω_c, the resonant frequency ω_r, the peak magnitude M_r, and the magnitude of $|G(j\omega_n)|$, as well as the difference between the real and the asymptotic plots, both in magnitude and in phase (use all possible approximations in this case).
5. Using the f_concept interactive tool, repeat the previous exercise and check that in the polar diagram, the frequency point whose distance from the origin is maximum corresponds to the resonant frequency. Also, check in the Nichols diagram that the vertical distance between the points represented by $\omega = 0$ and $\omega = \omega_r$ is the peak value of $G(j\omega)$ in dB.
6. Using the f_concept interactive tool, build with the editor of poles and zeros (repository) a transfer function

$$G(s) = \frac{1}{s(\tau s + 1)}.$$

Draw the polar diagram of this transfer function, indicating where the zero frequency and infinite frequency points are located.
7. Using again the f_second_order interactive tool, check the Show multiple systems option and select in the Options menu the Natural frequency effect item. Then select sequentially all the systems in the **Bode magnitude** plot and analyze the value of the peak magnitude M_r in each case. Why does it have the same value in all cases? Justify the answer. Use the tool to calculate the values of the bandwidth and the resonant frequency ω_r of all represented systems. Which system will be faster when a step signal is introduced at its input? Which one will be slower? Justify the answer.
8. Now choose the menu option Constant imaginary part. For each of the represented systems, indicate the value of k, ζ, ω_n, ω_c, ω_r, and M_r. Determine which system will have the maximum overshoot when a step input is entered and which will have the slowest response. Justify the answer. Place the two poles closest to the imaginary axis on the j-axis. Justify what is happening with the magnitude and phase plots. What is the value of the relative damping factor ζ? What kind of time response can be expected?
9. Select the option Constant real part. For each of the systems represented, indicate the value of k, ζ, ω_n, ω_c, ω_r, and M_r. Move some complex conjugated pair of poles away from the real axis vertically. Explain what happens to the magnitude and phase plots in the Bode diagram. Indicate which position of the poles produces a relative damping factor $\zeta = 0.707$ and the value of the associated undamped natural frequency ω_n. In that case, calculate the value of ω_c, ω_r and M_r. Why $\omega_n = \omega_c$? Why $\omega_r < \omega_c$? Justify the answers.
10. Using the interactive tool t_second_order (card 3.3), simulate examples 1, 2, 7, 8, and 9 and check the time responses obtained and whether the expected behavior matches the justifications you have given in these sections.
11. Start the tool or press Reset in the Options menu. For the default parameter settings, indicate whether there can be a frequency value so that when a sine wave of that frequency is entered at the input, the output will grow indefinitely. Which value is obtained in this case in the peak magnitude M_r?
12. Analyze the frequency response of an overdamped second-order system

$$G(s) = \frac{k}{(\tau_1 s + 1)(\tau_2 s + 1)},$$

using the tool and considering $k = 1$ and the following cases:

a. $\tau_2 = 1$ s, $\tau_1 = 0.1$ s.
b. $\tau_2 = 2$ s, $\tau_1 = 1$ s.
c. $\tau_2 = \tau_1 = 1$ s.
d. $\tau_2 = 1$ s, $\tau_1 = -1$ s.
e. $\tau_2 = -1$ s, $\tau_1 = -0.1$ s.

13. Select a transfer function $G(s) = 1/((0.02s + 1)(0.2s + 1))$. Notice that ω_c (BW) in this case is 4.95 (the BW of a first-order transfer function given by $G(s) = 1/(0.2s + 1)$ is 5). Click and drag on the pole in $s = -50$ approaching it to the one in $s = -5$. Notice that when the pole is in $s = -10$, $\omega_c = 4.18$ and when both poles are in $s = -5$, the bandwidth is approximately 3.22 rad (Ref. [2], Example 4.7).

14. What is the bandwidth of the system represented by the transfer function $4/(s^2 + s + 1)$? (note that the gain does not play any role). Compare the value obtained analytically with that provided by the tool. What is the bandwidth of $G(s) = 1/(s + 10)^2$ and $3/(s^2 + 9)$? (Ref. [2], Examples 4.8 and 4.9).

4.5   Effect of a Zero on the Frequency Response of Continuous-Time First-Order Linear Systems

4.5.1   Interactive Tool: f_first_order_zero

4.5.1.1 Concepts Analyzed in the Card and Learning Outcomes
- Frequency response of a first-order system with a zero. Bode diagram. Asymptotic approximations.
- Relationship between relative pole and zero locations with frequency response. Minimum and maximum gain. Minimum and maximum phase lag. Phase lead and phase lag.

4.5.1.2 Summary of Fundamental Theory
In this card, the frequency response of systems represented by the following transfer function are studied:

$$G(s) = k\frac{\beta s + 1}{\tau s + 1}, \tag{4.15}$$

where τ is the time constant associated with the pole, β is the time constant associated with the zero, and k is the static gain.

As in the case of first-order systems, first of all, the special case of a zero at the origin of the s-plane (differentiator) will be analyzed

$$G(s) = s \rightarrow G(j\omega) = j\omega \rightarrow |G(j\omega)| = |\omega|; \lfloor G(j\omega) = \arctan\left(\frac{\omega}{0}\right) = 90°. \tag{4.16}$$

The logarithmic magnitude of $j\omega$ in dB is $20\log(|\omega|)$ and its phase angle is constant and equal to $90°$. In a Bode diagram, the magnitude $20\log(|\omega|)$ is a line of slope 20 dB/decade passing through the point (0 dB, $\omega = 1$ rad/s).[12]

For a zero (factor $j\beta\omega + 1$), the Bode diagram can be obtained as for a pole (factor $1/(j\tau\omega + 1)$). The low-frequency magnitude asymptote is a straight line with a zero slope at 0 dB, which at the corner frequency $\omega_{cf} = 1/\beta$ will cut off the high-frequency asymptote which has a slope of 20 dB/decade. The phase plot goes from $0°$ (low frequency) to $90°$ (high frequency).

The shape of the frequency response of first-order systems with a zero depends on the relative position of the pole and zero and the values of the characteristic time constants. Note that, in this case, the system is not strictly causal (since the polynomial of the numerator and that of the denominator have the same degree). In the scope of closed-loop systems studied in Chap. 7, an MP transfer function with one pole and one zero in the LHP in which $\beta < \tau$ (zero to the left of the pole) is called *lag controller* (*compensator, network*), while in the case $\beta > \tau$ (zero to the right of the pole in the LHP), it is called *lead controller* (*compensator, network*), due to the characteristics that the phase plot has in each of these situations (negative or positive phase contribution). Figure 4.3 shows the Bode and Nyquist diagrams of these kinds of systems [5] where the pole and the zero are separated one decade in this example.

---

[12] If the transfer function contains the factor $(j\omega)^n$ the log magnitude becomes $20n\log(|\omega|)$.

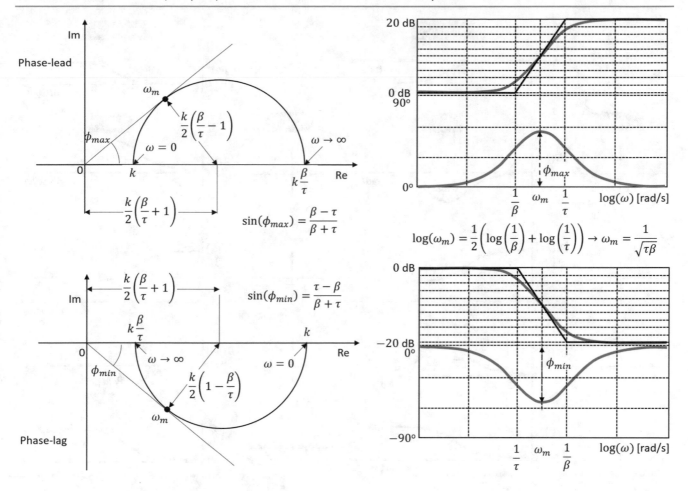

Fig. 4.3 Bode and Nyquist diagrams of a first-order system with a zero

Phase-lead controllers act as high-pass filters, while phase-lag controllers have low-pass filter characteristics. In these cases (pole and zero in the LHP, including the possibility of an integrator or a differentiator), the magnitude $\rho = |G(j\omega)|$ and phase $\phi = \lfloor G(j\omega)$ asymptotes will have the characteristics shown in Table 4.2.

If there is no integrator or differentiator in the transfer function, between the low-frequency asymptote ($\rho_{\omega\to 0}$) and the high-frequency asymptote ($\rho_{\omega\to\infty}$), the magnitude in the Bode diagram will evolve:

- **Phase lag:** An asymptote with a slope -20 dB/decade between $\omega_{cf_1} = 1/\tau$ and $\omega_{cf_2} = 1/\beta$, ω_{cf_1} being the lower frequency corner frequency and ω_{cf_2} the higher frequency one.
- **Phase lead:** An asymptote with a slope 20 dB/decade between $\omega_{cf_1} = 1/\beta$ and $\omega_{cf_2} = 1/\tau$.

If there is an integrator or a differentiator, there is only one corner frequency at $\omega_{cf} = 1/\beta$ or $\omega_{cf} = 1/\tau$ respectively. The phase angle will be determined in both cases by (see Fig. 4.3): $\phi = \arctan(\beta\omega) - \arctan(\tau\omega)$, reaching the minimum (maximum) phase lag for $\omega_m = 1/\sqrt{\tau\beta}$, where

Table 4.2 Characterization of asymptotes in a first-order linear system with a zero

Asymptote	Pole and zero in $s \neq 0$	Integrator	Differentiator
$\rho_{\omega\to 0}$	0 dB/decade slope	-20 dB/decade slope	20 dB/decade slope
$\phi_{\omega\to 0}$	0° line	$-90°$ line	90° line
$\rho_{\omega\to\infty}$	0 dB/decade slope	0 dB/decade slope	0 dB/decade slope
$\phi_{\omega\to\infty}$	0° line	0° line	0° line

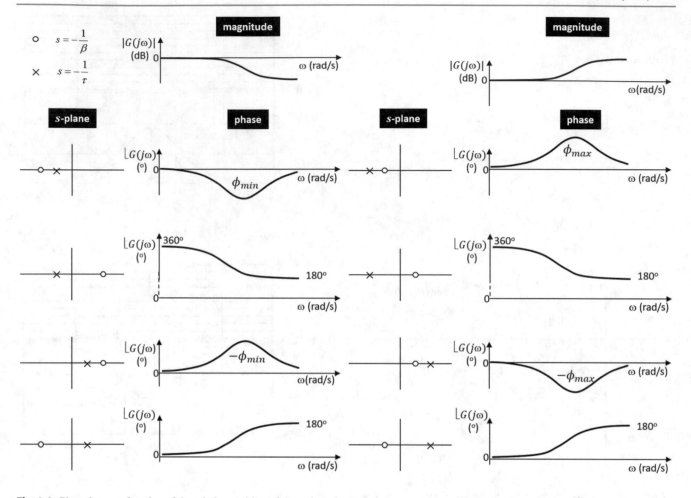

Fig. 4.4 Phase lag as a function of the relative position of the pole and zero

$$\sin \phi_{min} = -\frac{1 - (\tau/\beta)}{1 + (\tau/\beta)} \ , \ \sin \phi_{max} = \frac{1 - (\tau/\beta)}{1 + (\tau/\beta)}. \tag{4.17}$$

For systems with zeros or poles in the open RHP (NMP systems), the magnitude plot is similar to the case explained in the previous paragraph (corresponding to an MP system), but the phase plot changes as a function of the location of the pole and/or zero in the right half plane of the complex plane (and of its relative position), and may cause phase shifts of up to 360° depending on the individual contributions (see Fig. 4.4). These characteristics will be analyzed through exercises and will be deepened in Sect. 4.8.

4.5.1.3 References Related to this Concept

- [5] Ogata, K. (2010). *Modern control engineering* (5th ed.). Prentice Hall. ISBN: 978-0-13-615673-4. Chapter 7, pp. 403–409 and 493–511.
- [9] Franklin, G. F., Powell, J. D., & Emani-Naeni, A. (2010). *Feedback control of dynamic systems* (6th ed.). Pearson. ISBN: 978-0-13-500150-9. Chapter 6, Sect. 7, pp. 366–383.

Application   Interactive tool: f_first_order_zero

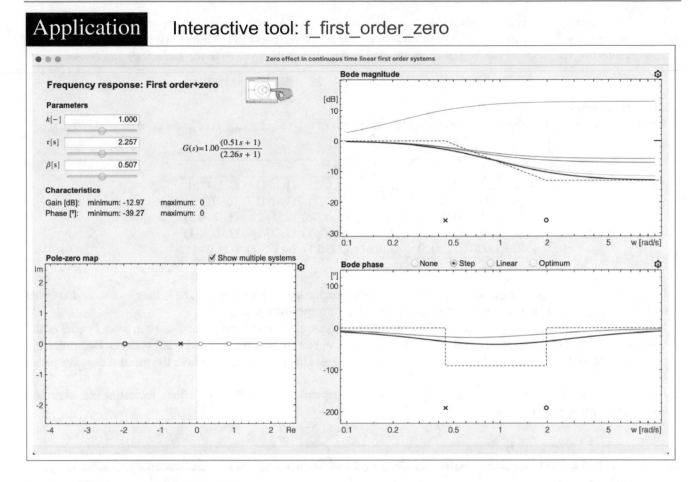

Frequency Response: First Order + Zero

> This card analyzes the effect of a zero in the frequency response of first-order LTI systems. The concepts learned here are very useful also for designing lead-lag compensators in the frequency domain, as will be shown in Chap. 7.

Section **Parameters** (upper-left area) offers the possibility of entering and modifying the values of the parameters k, τ, and β that define the transfer function of a first-order system with a zero (through textboxes and sliders). Negative values must be entered through the textboxes or modifying the location of the pole and or zero in the **Pole–zero map** area. On the right is the symbolic representation of the transfer function determined by these parameters. The maximum and minimum gain and phase values corresponding to that parameters configuration are presented in the **Characteristics** area.

In the **Pole–zero map** area, the pole and the zero of the system can be moved interactively along the real axis, their position displayed in the lower-left corner of the tool. The scale can be modified using the settings available in the gearwheel icon. Next to the title of the graph, the option Show multiple systems can be enabled, which takes the system under analysis as the nominal one and includes four new zeros in different positions on the real axis (keeping fixed pole), facilitating comparisons between MP and NMP systems.

The right area of the tool is dedicated to the analysis of the frequency response of the system, using **Bode magnitude** and **Bode phase** plots. By default, the diagrams are shown for one system only (the selected system) and with the classical asymptotic representations (see Sect. 4.3). Notice that the Nyquist and Nichols diagrams of the system treated in this section can be analyzed using the f_concept interactive tool.

When placing the mouse over any point of the real magnitude or phase plots, the values of the frequency (w) and magnitude (m, dB) or phase (p, °) corresponding to that frequency are displayed, being very useful when measuring certain characteristics on the graph. That information is also displayed in a more complete way in the lower-left corner of the tool. The change of the magnitude at low frequency (and therefore, of the static gain) is done in these tabs by clicking and dragging up/down the black thick line (--) located in the right part of the **Bode magnitude** plot.

In the two graphs, there is a cross symbol **x** and a circle **o** corresponding to the location of the two corner frequencies (ω_{cf_1} and ω_{cf_2}) associated with the pole and the zero of the system. These two symbols can be shifted to the right or left, so that, in addition to affecting the shape of the frequency response plot, they update their values in the **Parameters** and **Pole–zero map** areas.

4.5.1.4 Homework

1. For the following combinations of k, τ y β, calculate the values of ω_{cf_1}, ω_{cf_2}, $\rho_{\omega \to 0}$, $\rho_{\omega \to \infty}$, $\phi_{\omega \to 0}$, $\phi_{\omega \to \infty}$, ω_m, $\phi_{min(max)}$.

			(k, τ, β)		
(1,1,0.5)	(1,1,2)	(1,1,∞)	(1,1,-2)	(1,1,-1)	(1,1,-0.5)
(1,0.5,1)	(1,2,1)	(1,∞,1)	(1,-2,1)	(1,-1,1)	(1,-0.5,1)
(1,-1,0.5)	(1,-1,1)	(1,-1,2)	(1,-1,∞)	(1,-1,-2)	(1,-1,-0.5)
(1,0.5,-1)	(1,1,-1)	(1,2,-1)	(1,∞,-1)	(1,-2,-1)	(1,-0.5,-1)
(0.5,1,0.5)	(2,1,0.5)	(0.5,0.5,1)	(2,0.5,1)	(-1,0.5,1)	(-1, 1, -1)

In the interactive tools, it is logically not possible to enter an infinite value to a parameter, so in the cases indicated in the table, what has to be done is to enter a very high value in the corresponding parameter.

2. Using the Show multiple systems option, comparatively analyze some of the combinations indicated in rows 1 and 3 of the table from the previous exercise. Discuss the differences that can be seen between MP and NMP systems. Determine the transfer function which produces the largest positive phase lag and the system that produces the greatest negative phase lag.

3. For the above examples, compare the accuracy of the different asymptotic phase approximations, indicating the maximum error that occurs in each case.

4.6 Effect of a Zero on the Frequency Response of Continuous-Time Second-Order Linear Systems

4.6.1 Interactive Tool: f_second_order_zero

4.6.1.1 Concepts Analyzed in the Card and Learning Outcomes
- Frequency response of a system represented by a second-order transfer function with a zero.
- Asymptotic approximations to the magnitude and phase plots of the Bode diagram of a second-order transfer function with a zero.
- Relationship between the location of the poles and the zero of the second-order system with the frequency response. Minimum and maximum gain. Minimum and maximum phase lag.

4.6.1.2 Summary of Fundamental Theory
In the previous card, the theoretical concepts related to the effect of a zero in the frequency response of a first-order system have been analyzed. For an MP second-order system (the two poles and the zero are in the left half plane of the complex plane), the following hold:

- If the system is overdamped, each real pole contributes in magnitude a slope of -20 dB/decade from its corner frequency $\omega_{cf_i} = 1/\tau_i$ and a phase ranging from 0 to $-90°$ (with a total contribution of $-180°$).
- If the system is underdamped, the two complex poles directly provide a slope of -40 dB/decade from the corner frequency $\omega_{cf} = \omega_n$ and a phase ranging from $0°$ to $-180°$.

The zero provides a slope of 20 dB/decade from its corner frequency $\omega_{cf} = 1/\beta$ and a phase ranging from 0 to $90°$. Therefore, due to the additive character of Bode's logarithmic plots, the frequency response will change its shape depending on the relative position between the zero and the poles of the system, as analyzed for the first-order case in Fig. 4.4. The same visual analysis can be done in this case using the interactive tool.

4.6.1.3 References Related to this Concept

- [4] Golnaraghi, F., & Kuo, B. C. (2017). *Automatic control systems* (10th ed.). McGraw Hill Education. ISBN: 978-1-25-964384-2. Chapter 10, Sect. 3.1.
- [5] Ogata, K. (2010). *Modern control engineering* (5th ed.). Prentice Hall. ISBN: 978-0-13-615673-4. Chapter 7, pp. 403–412.
- [11] Shahian, B., & Hassul, M. (1993). *Control system design using MATLAB®*. Prentice Hall, ISBN: 0-13-174061-X. Chapter 1, Sect. 5, paragraph 3, pp. 16–19.

Application   Interactive tool: f_second_order_zero

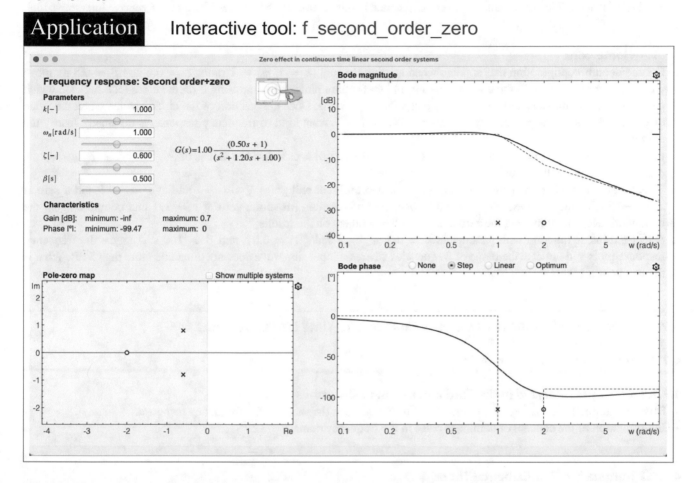

Frequency Response: Second Order + Zero

> This interactive tool helps to analyze the Bode diagram of a second-order transfer function with a zero, both in the underdamped and overdamped cases. It is quite interesting to analyze how the frequency response is influenced by the relative position between the poles and the zero in the MP case and also how a zero in the RHP makes the system NMP.

On the upper-left, the **Parameters** area offers in this case four sliders and their associated textboxes to modify the parameters k, β, ζ, and ω_n (or alternatively τ_1 and τ_2 in case of selecting two real poles), which define the transfer function of a second-order system with a zero, also represented on the right hand side. The **Characteristics** area shows the values of maximum and minimum gain and phase corresponding to that configuration of the parameters.

The **Pole–zero map** includes an interactive pole and two interactive zeros. When placing the mouse over them, their position is displayed in the lower-left corner of the tool. The complex conjugated poles can be dragged to the real axis, becoming two real poles that change ω_n and ζ in **Parameters** are into τ_1 and τ_2. The graph includes a gearwheel for scaling purposes and on the same graph the option Show multiple systems can be enabled, which adds four additional zeros to the actual selected configuration.

The right area of the tool has the **Bode magnitude** and **Bode phase** plots, representing both the frequency response curves and the asymptotic ones. When passing the mouse over the magnitude or phase curve, the frequency values (w) and magnitude (m) or phase (p) are displayed over them, as well as in the lower-left corner of the tool. The low-frequency gain is modified using the thick dashed line (--) located to the right of the **Bode magnitude** plot.

In the two graphs, there is a circle o linked to the system's zero corner frequency. In the case of two complex conjugated poles, a cross symbol ✕ is drawn to indicate the corner frequency associated with the two complex conjugated poles (ω_n). If the poles are real, two crosses ✕ are drawn in the diagrams, corresponding to the two corner frequencies ($\omega_{cf_1} = 1/\tau_1$ and $\omega_{cf_2} = 1/\tau_2$) associated with those poles. These two symbols can be moved to the right or left, so that, in addition to affecting the shape of the frequency response plot, their values are automatically refreshed in the **Parameters** and **Pole–zero map** areas.

The Nyquist and Nichols diagrams of the system treated in this section can be analyzed using the f_concept interactive tool.

4.6.1.4 Homework

1. For the default configuration that appears when starting the tool ($k = 1$, $\omega_n = 1$ rad/s, $\zeta = 0.6$ and $\beta = 0.5$ s), move the circle (o) represented in the **Bode magnitude** and **Bode phase** plots, superimposing it on the cross (✕). Indicate which parameter of the transfer function is changed and its actual value. Does the position of the circle change when the value of ζ increases or decreases? Justify the answer. When $\zeta = 1$, what kind of frequency response is obtained? Justify the answer.

2. For $k = 1$, $\zeta = 0.1$, $\omega_n = 1$ rad/s, and $\beta = \{1.0, 0.1, -0.1, -1.0\}$ s, calculate the values of ω_r, M_r, $\rho_{\omega \to 0}$, $\rho_{\omega \to \infty}$, $\phi_{\omega \to 0}$, $\phi_{\omega \to \infty}$. Comment on the results.

3. Select a system with static gain equal to one, with two different real poles at $s = -1$ and at $s = -0.5$ and a zero at $s = -0.3$. Calculate the system bandwidth for that configuration. Move the zero at $s = -1$ and compute again the bandwidth. Repeat by moving the zero at $s = -0.5$. Comment on the results.

4. Select a transfer function with parameters $k = 0.3$, $\omega_n = 1$ rad/s, $\zeta = 0.01$, and $\beta = 100$ s. Analyze the frequency response plots and indicate the range of frequencies where an input sine wave does not attenuate more than 3 dB. What is the bandwidth of this system?

4.7   Frequency Response of Generic Continuous-Time Linear Systems

4.7.1   Interactive Tool: f_generic

4.7.1.1 Concepts Analyzed in the Card and Learning Outcomes

- Effect of adding poles, zeros, integrators, and differentiators in the shape of the frequency response.
- Interpretation of the concept of dominant poles in the frequency domain.

4.7.1.2 Summary of Fundamental Theory

The creation of Bode diagrams of generic systems (of any order) is done by representing the transfer function as a product of basic factors. The main advantage of the logarithmic graphical representation is the conversion of multiplicative factors into additive factors, by the definition of logarithmic gain. The different types of factors that can be found in a transfer function are as follows: Static gain (k), integrators ($1/(j\omega)$), differentiators ($j\omega$), first-order terms ($(j\omega\tau + 1)^{\pm 1}$), second-order terms ($(j\omega/\omega_n)^2 + (j\omega/\omega_n)2\zeta + 1)^{\pm 1}$, and time delay $e^{-j\omega t_d}$. Table 4.3 shows the main logarithmic and phase gain characteristics associated with these factors.[13] A generic transfer function can thus be represented as

$$G(j\omega) = \frac{k \prod_{\ell=1}^{q}(j\omega\beta_\ell + 1) \prod_{\ell=1}^{r}\left(\left(\dfrac{j\omega}{\omega_{n_{z_\ell}}}\right)^2 + j\omega\left(\dfrac{2\zeta_{z_\ell}}{\omega_{n_{z_\ell}}}\right) + 1\right)}{(j\omega)^{\aleph} \prod_{i=1}^{p}(j\omega\tau_i + 1) \prod_{i=1}^{h}\left(\left(\dfrac{j\omega}{\omega_{n_i}}\right)^2 + j\omega\left(\dfrac{2\zeta_i}{\omega_{n_i}}\right) + 1\right)} e^{-j\omega t_d}.$$

This transfer function includes q real zeros, $2r$ complex conjugated zeros (r pairs of complex conjugated zeros), $\aleph$ poles at the origin of the s-plane, p poles on the real axis, $2h$ complex conjugated poles (h pairs of complex conjugated poles), and time delay t_d. The logarithmic magnitude of $G(j\omega)$ is

---

[13] Note that in the case of real zeros, their associated time constant is called β_ℓ in this text, instead of also using τ_ℓ as for the poles.

Table 4.3 Characteristics of basic factors of logarithmic magnitude and phase ($\nu = \omega/\omega_n$ for poles, $\nu = \omega/\omega_{n_z}$ for zeros)

Factor	Logarithmic gain [dB]	Phase [°]
Positive static gain	$20 \log(k)$	0
Negative static gain	$20 \log(|k|)$	180
Integrator	$-20 \log(|\omega|)$	-90
Multiple integrator	$-20\aleph \log(|\omega|)$	$-90\aleph$
Differentiator	$20 \log(|\omega|)$	90
Multiple differentiator	$20\aleph \log(|\omega|)$	$90\aleph$
Real pole	$-10 \log(\tau^2\omega^2 + 1)$	$-\arctan(\tau\omega)$
Real zero	$10 \log(\tau^2\omega^2 + 1)$	$\arctan(\tau\omega)$
Complex conjugated poles	$-10 \log((1-\nu^2)^2 + 4\zeta^2\nu^2)$	$-\arctan\left(\dfrac{2\zeta\nu}{1-\nu^2}\right)$
Complex conjugated zeros	$10 \log((1-\nu^2)^2 + 4\zeta^2\nu^2)$	$\arctan\left(\dfrac{2\zeta\nu}{1-\nu^2}\right)$
Time delay	0	$-\omega t_d \dfrac{180°}{\pi}$

$$20 \log(|G(j\omega)|) = 20 \log(|k|) + 20 \sum_{\ell=1}^{q} \log(|j\omega\beta_\ell + 1|) + 20 \sum_{\ell=1}^{r} \log\left(\left|\left(\frac{j\omega}{\omega_{n_{z\ell}}}\right)^2 + j\omega\left(\frac{2\zeta_{z\ell}}{\omega_{n_{z\ell}}}\right) + 1\right|\right) -$$

$$-20\aleph \log(\omega) - 20 \sum_{i=1}^{p} \log(|j\omega\tau_i + 1|) - 20 \sum_{i=1}^{h} \log\left(\left|\left(\frac{j\omega}{\omega_{n_i}}\right)^2 + j\omega\left(\frac{2\zeta_i}{\omega_{n_i}}\right) + 1\right|\right),$$

and the Bode diagram can be obtained by adding up the graph due to each individual factor. Note that the time delay does not contribute to the magnitude curve because $|e^{-j\omega t_d}| = 1$, and therefore, $|e^{-j\omega t_d}|_{dB} = 0$.

The phase plot in degrees is obtained as

$$\phi(j\omega) = \lfloor G(j\omega) = \lfloor k + \sum_{\ell=1}^{q} \arctan(\beta_\ell\omega) + \sum_{\ell=1}^{r} \arctan\left(\frac{2\zeta_{z\ell}\omega_{n_{z\ell}}\omega}{\omega_{n_{z\ell}}^2} - \omega^2\right) -$$

$$-90° \aleph - \sum_{i=1}^{p} \arctan(\tau_i\omega) - \sum_{i=1}^{h} \arctan\left(\frac{2\zeta_i\omega_{n_i}\omega}{\omega_{n_i}^2} - \omega^2\right) - \omega t_d \frac{180°}{\pi},$$

which is simply the sum of the phase angles due to each individual factor of the transfer function. As can be seen, the phase due to the time delay has a linear evolution in ω (although it is not represented in the Bode diagram using a line due to the logarithmic scale on the frequency axis).

If the system is MP with positive gain, there is an univocal correspondence between the Bode magnitude and phase diagrams, so that the transfer function of a system can be estimated from a Bode plot. From the magnitude diagram, the low-frequency slope is given by $(-20\aleph)$, while the high-frequency slope is $-20(n - m)$, provided by the relative degree of the transfer function. From low to high-frequency, each ±20 dB/decade change in slope provides the corner frequency of one zero/pole. The same applies to ±40 dB/decade change in slope for a pair of (possibly conjugate) zeros/poles. If all poles and zeros are in the LHP and the phase diagram does not fulfill the expected one from the composition of the set of poles, zeros, integrators, and differentiators obtained from the magnitude diagram, it will indicate the presence of a time delay which can be computed from the difference between the expected phase $\lfloor G(j\omega)$ and that shown in the Bode phase diagram $\phi(\omega)$ at a selected high frequency as $(180°/\pi)\omega t_d = \lfloor G(j\omega) - \phi(\omega)$. The experimental determination of the transfer function from data is treated in Sect. 4.9.

On the other hand, when the concept of *dominant poles* has been discussed in Section 3.6, it has been found that an approach to a high-order transfer function $G(s)$ containing its dominant poles tends to reproduce the slow dynamics of the system, overlooking the fast ones. It is, therefore, expected that the frequency response of the dominant pole approach will not differ too much from that of the original system for low values of the frequency ω. As often said, a dominant pole approach represents a low-frequency model of the system. In fact, when plotting asymptotic diagrams, poles and zeros with small time constants or high natural frequencies produce very unrepresentative contributions to the shape of the diagrams for low-frequency values. Regarding bandwidth, in the case of real poles, the smallest rightmost pole in magnitude is larger than the bandwidth. The further apart the other poles are from this pole, the better this pole represents the bandwidth (in Ref. [2] different examples of the analysis of bandwidth can be found, which can be reproduced with the interactive tool introduced in this section).

4.7.1.3 References Related to this Concept

- [1] Åström, K. J. & Murray, R. M. (2014). *Feedback systems: An introduction for scientists and engineers* (2nd ed.). Princeton University Press. ISBN: 9780691193984. Chapter 9, Exercise 9.9, pp. 9–26.
- [2] Bavafa-Toosi, Y. (2017). *Introduction to linear control systems*. Academic Press-Elsevier. ISBN: 978-0-12-812748-3. Chapter 2, Remark 2.19, pp. 118–121; Sect. 4.6.4, p. 281.
- [5] Ogata, K. (2010). *Modern control engineering* (5th ed.). Prentice Hall. ISBN: 978-0-13-615673-4. Chapter 7, pp. 403–445.
- [6] Bolzern, P., Scattolini, R., & Schiavoni, N. (2009). *Fundamentos de control automático (Fundamentals of automatic control, in Spanish)*. McGraw-Hill. ISBN: 978-84-481-6640-3. Chapter 6, Sect. 6, paragraph 6, page 155, Sect. 9, pp. 167–168.
- [8] Dorf, R. C., & Bishop, R. H. (2011). *Modern control systems* (12th ed.). Prentice Hall. ISBN: 978-0-13-602458-3. Chapter 8, pp. 556–601.
- [9] Franklin, G. F., Powell, J. D., & Emani-Naeni, A. (2010). *Feedback control of dynamic systems* (6th ed.). Pearson. ISBN: 978-0-13-500150-9. Chapter 3, Sect. 4, pp. 134–137; Chapter 6, Sect. 1, pp. 314–33.
- [12] Barrientos, A., Sanz, R., Matía, F., & Gambao, E. (1996). *Control de sistemas continuos. Problemas resueltos (Control of continuous systems. Problems solved)*. McGraw-Hill. ISBN: 84-481-0605-9. Chapter 7, Sect. 2, pp. 233–235.
- [13] Truxal, J. G. (1955). *Automatic feedback control system synthesis*. McGraw-Hill. ISBN: 978-00-7065-310-8. pp. 212–219.

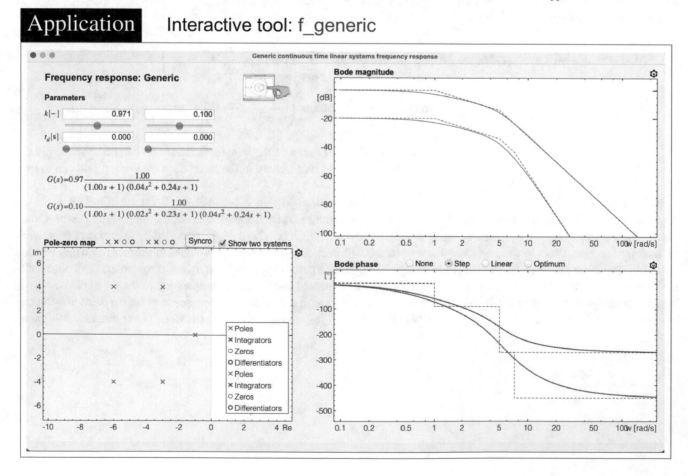

Frequency Response: Generic Systems

This tool facilitates the analysis of the frequency response of generic systems, with an arbitrary number of poles and zeros, including integrators and differentiators and dead-time. Comparisons among different configurations can be easily performed.

Note to the reader: Almost all software packages (like Sysquake or MATLAB®) implement primitives for the computation of frequency-domain diagrams that include inconsistent definition of angles (see [2] for details), providing differences in the phase plots of frequency diagrams of high-order systems. In this text, the authors have tried to be consistent with the definition of angles and related conventions to obtain Bode phase plots also consistent with the used definitions (mostly in the case of NMP systems treated in the next section and with definitions of relative stability margins in Chap. 6). However, as the primitives are embedded in the software code used, it is possible that not all user-tested examples will in all cases produce consistent results, as [2] points out.

The upper-left area of the tool contains the **Parameters** section, where a symbolic representation of the transfer function and two sliders are displayed, making it easy to change the values of the static gain k and the time delay t_d. The values of both parameters can also be set via textboxes (for instance to enter static gains of different sign). By activating Show two systems over the **Pole–zero map**: plot (lower-left area of the tool) that will be explained later, the parameters of the original system are shown in red color, while those corresponding to the second (new) system are shown in blue (modifiable through associated sliders and textboxes). The two symbolic representations of the corresponding transfer functions also become visible. No textboxes or associated sliders are included in this area to other dynamic parameters (poles and zeros).

Pole–zero map: It includes the poles and zeros editor and repository introduced in the tool **Time response: Generic** to configure any kind of transfer function by adding or removing poles ($\times$), zeros ($\circ$), integrators ($\mathbf{x}$), and differentiators ($\mathbf{o}$). System causality is guaranteed at all times and the total number of poles is limited to 4 to avoid high computational load. The system configuration is done by clicking and dragging the graphic elements that appear in the repository (elements $\times$, $\circ$, $\mathbf{x}$, $\mathbf{o}$) over the complex plane (or by double-clicking on them or by taking them out to that repository in case the user wants to remove items). When the Show two systems option is active, the synchronization button (Synchro) makes the second system (represented in blue) exactly the same as the first one (red) in order to facilitate analyzing the effect of modifying the parameters of one of the two systems in relation to the initial configuration.

The right area of the tool displays the **Bode magnitude** and **Bode phase** graphs, including both the frequency response and the asymptotic curves (explained in detail in the card 4.2) and the frequency (w) and magnitude (m) or phase (p) values when the mouse is placed over them (detailed information is also displayed in the lower-left corner of the tool). When time delay is included, the phase of the original system is represented using a dotted line, while the phase of the delayed system is represented in solid line. Different asymptotic approximations of the phase plot can be selected over the phase graph.

Three examples have been included by default in the Options menu:

- Example 1 uses as reference the transfer function $P_{11}(s)$ (see Chap. 1), with two complex conjugated poles and a real pole.
- Example 2 ($P_5(s)$): It shows two systems, represented by the following transfer functions, to allow comparison of the Bode plots at low, medium, and high frequencies and the effect of the relative degree between the denominator and numerator polynomials of the transfer function:

$$G(s) = \frac{1}{s(s+1)}, \quad G(s) = \frac{s(0.4s+0.4)}{s^2+0.7s+1.75}.$$

- Example 3 ($P_7(s)$): It includes two second-order systems with real poles at the LHP in the same location and a zero of equal absolute value but located in one case at the RHP and at the LHP in the other. This is a very useful example to analyze the concept of minimum phase and nonminimum phase in second-order systems with a zero. Scaling is required using the settings available in the gearbox placed at the **Pole–zero map** plot.
- In addition, an arbitrary transfer function can be entered using MATLAB® formats ((NUM,DEN) and ZPK) explained in Table 1.2 in Chap. 1.

The Nyquist and Nichols diagrams of generic systems can be analyzed using the f_concept interactive tool.

4.7.1.4 Homework

1. Consider the following transfer function:

$$G(s) = \frac{5(0.1s + 1)}{s(0.5s + 1)((\frac{s}{50})^2 + \frac{0.6s}{50} + 1)}.$$

Draw Bode's diagram in two different ways. First, convert $G(s)$ to a MATLAB®-type format((NUM,DEN) or ZPK) and enter it through the Options menu. Secondly, and using the Show two systems option, incrementally add element by element of the transfer function (from low to high frequency) to analyze how the full diagram can be built by adding the constitutive elements of the transfer function. Indicate the value of the different corner frequencies. Calculate the value of the magnitude and phase at the frequencies $\omega = 1$, 5 and 10 rad/s.

2. Now select the Example 1 ($P_{11}(s)$) from the Options menu. By activating the Show two systems box, move the two less-dominant complex conjugated poles (in blue) away from the imaginary axis and comment on what happens to the magnitude and phase plots. Would it be possible to obtain similar magnitude and phase plots in both cases? Make an analysis using the tool and justify the conclusions.

3. Select the Example 2 ($P_5(s)$) from the Options menu of the tool. By activating the Show two systems box, add an integrator to the system represented in blue. Compare the frequency responses of the two systems and justify the differences and similarities found.

4. Select theExample 3 ($P_7(s)$) from theOptions menu. By activating the Show two systems box, analyze how the frequency response of the system shown in red changes as the zero is dragged toward the left half plane, until it coincides with the zero of the system shown in blue (changing the scale of the **Pole–zero map** plot is required.

5. Using the interactive tool, draw the Bode diagrams for the following transfer function (or an approximate one—it is recommended to enter the complex conjugated poles first or enter the transfer function directly from the Options menu):

$$G(s) = \frac{10(s + 3)}{s(s + 2)(s^2 + s + 2)} = \frac{10(s + 3)}{s^4 + 3s^3 + 4s^2 + 4s}.$$

Indicate the corner frequencies on the magnitude curve and the associated value. Calculate the maximum error obtained with the asymptotic curves at the corner frequencies.

6. Using the interactive tool, build with the editor of poles and zeros a transfer function

$$G(s) = \frac{e^{-t_d s}}{\tau s + 1}.$$

Analyze the Bode diagram of that transfer function and describe what happens to the magnitude and phase plots when the time delay increases. Using the interactive tool f_concept (card 4.2), build with the editor of poles and zeros the same transfer function. Draw the polar diagram of that transfer function, indicating where the points of zero frequency and infinite frequency are located. Analyze how, in the presence of time delay, the polar diagram has a spiral shape. Justify this result.

7. Using the interactive tool, consider the electric motor model modified from the one described in [6]

$$G(s) = \frac{3.98}{(8.56\,s + 1)(0.22\,s + 1)} = \frac{2.1134}{(s + 0.1168)(s + 4.5455)} = \frac{2.1134}{s^2 + 4.6623\,ss + 0.5309},$$

which describes the relationship between armature voltage and rotation speed. This transfer function can be reasonably approximated by the following transfer function:

$$G_a(s) = \frac{3.98}{(8.56\,s + 1)},$$

which preserves the dominant pole. Compare the Bode diagrams of the two transfer functions and indicate the range of frequencies for which the approximation given by $G_a(s)$ can be considered valid (a "good approach").

8. For the next transfer function, described in [6]

$$G(s) = \frac{(0.5\,s + 1)}{(0.1\,s + 1)(0.002s^2 + 0.02s + 1)(s^2 + 0.1s + 1)} = \frac{(0.5\,s + 1)}{0.0004s^5 + 0.00604\,s^4 + 0.221\,s^3 + 1.028s^2 + 0.32s + 1},$$

propose a second-order dominant pole without zeros approach to that transfer function. Represent the Bode diagram of both and indicate the range of frequencies over which such an approximation could be considered valid.

9. Simultaneously, draw the Bode diagrams of the systems with transfer function [12]:

$$G(s) = \frac{10(s - 5)}{(s + 1)(s + 7)} = \frac{10(s - 5)}{s^2 + 8s + 7}, \quad G(s) = \frac{10(s + 5)}{(s + 1)(s + 7)} = \frac{10(s + 5)}{s^2 + 8s + 7}.$$

Compare both and comment on the results.

10. When a rational approximation of a time delay is required, the so-called *Padé approximation* is frequently used

$$G(s) = e^{-t_d s} \approx \frac{1 - s\frac{t_d}{2}}{1 + s\frac{t_d}{2}} = -\left(\frac{s - \frac{2}{t_d}}{s + \frac{2}{t_d}}\right).$$

Thus, a time delay is approximated by an RHP zero. Check by using the tool (e.g. by choosing $t_d = 1$ s) that the two transfer functions are similar when $\omega < 1/t_d$ [1]. The time delay transfer function can be introduced by using the (NUM,DEN) format just setting the numerator and the denominator as constants equal to 1, or by double-clicking on or by the default poles and zeros to the repository. Note that, by the angle criterion chosen in the tools, a difference of 360° can be seen in the phase, being, therefore, equivalent. This problem can also be analyzed using the f_concept tool.

11. Activating the Show two systems box, consider $G_1(s) = (1 - s)/(1 + s)$ and $G_2(s) = e^{-s}$ (this last system has to be included using the (NUM,DEN) format as in the previous exercise, or by double-clicking on or by dragging the default poles and zeros back to the repository). Verify that both systems have the same magnitude and similar phase lags (or multiple of 360°) in a low-frequency range ([2], pp. 119–121). Notice that $G_1(s)$ constitutes an approximation of a time delay. This example is a good introduction to next section. what if $G_1(s)$ is written as $G_1(s) = (-1 + s)/(1 + s)$?

12. Using the interactive tool f_concept (card 4.2), analyze the Bode diagram, the Nyquist diagram and the Nichols diagram of the following generic systems whose transfer functions have the following structure (the value of the parameters can be freely chosen by the user). Notice that these frequency response plots already appeared in Truxal's classic book [13]:

$$G_1(s) = \frac{k}{\tau s + 1}, \qquad G_2(s) = \frac{k}{(\tau_1 s + 1)(\tau_2 s + 1)}, \qquad G_3(s) = \frac{k}{s},$$

$$G_4(s) = \frac{k}{s(\tau s + 1)}, \qquad G_5(s) = \frac{k}{s(\tau_1 s + 1)(\tau_2 s + 1)}, \qquad G_6(s) = \frac{k(\beta s + 1)}{s(\tau_1 s + 1)(\tau_2 s + 1)},$$

$$G_7(s) = \frac{k}{s^2}, \qquad G_8(s) = \frac{k}{s^2(\tau s + 1)}, \qquad G_9(s) = \frac{k(\beta s + 1)}{s^2(\tau s + 1)},$$

$$G_{10}(s) = \frac{k}{s^3}, \qquad G_{11}(s) = \frac{k(\beta s + 1)}{s^3}, \qquad G_{12}(s) = \frac{k(\beta_1 s + 1)(\beta_2 s + 1)}{s^3},$$

$$G_{13}(s) = \frac{k(\beta s + 1)}{s^2(\tau_1 s + 1)(\tau_2 s + 1)}, \qquad G_{14}(s) = \frac{k}{(\tau_1 s + 1)(\tau_2 s + 1)(\tau_3 s + 1)},$$

$$G_{15}(s) = \frac{k(\beta_1 s + 1)(\beta_2 s + 1)}{s(\tau_1 s + 1)(\tau_2 s + 1)(\tau_3 s + 1)(\tau_4 s + 1)}.$$

It is very important to analyze the shapes that are obtained and draw conclusions that will help to construct any diagram of a generic system. These representations will be very useful to understand the stability criteria in the frequency domain of feedback systems.

Check using those examples that in the case of Nyquist (polar) diagrams, if a generic representation of the system is considered with $k > 0$

$$G(j\omega) = \frac{k(j\omega\beta_1 + 1)(j\omega\beta_2 + 1)\dots(j\omega\beta_m + 1)}{(j\omega)^N(j\omega\tau_1 + 1)(j\omega\tau_2 + 1)\dots(j\omega\tau_n + 1)}, \tag{4.18}$$

(where $n > m$), depending on the number of integrators (terms $(j\omega)$ in the denominator):

- For $\aleph = 0$, the starting point (zero frequency) of the polar diagram will be on the positive real axis, the tangent in the polar diagram in $\omega = 0$ being perpendicular to the real axis. The end point corresponding to $\omega = \infty$ is at the origin, and the curve is tangent to one of the axes.
- For $\aleph = 1$, at low frequencies, the polar diagram is asymptotic to a line parallel to the negative imaginary axis. In $\omega = \infty$, the magnitude returns to zero and the curve converges toward the origin being tangent to one of the axes.
- If $\aleph = 2$, at low frequencies the polar diagram is asymptotic to a line parallel to the real negative axis, while for $\omega = \infty$, the magnitude is zero, the curve being tangent to one of the axes.

As mentioned above, for $\omega = \infty$, the geometric places are tangent to one of the axes (negative imaginary axis for $n - m = 1$, negative real axis for $n - m = 2$ and positive imaginary axis for $n - m = 3$). The complicated shapes of the polar diagram plots are often due to the dynamics of the numerator.

13. Consider the example in [9], which deals with the frequency response representing a mechanical system with two equal masses coupled by a spring. The transfer function that models the system is

$$G(s) = \frac{0.01(s^2 + 0.01s + 1)}{s^2((s^2/4) + 0.02(s/2) + 1)}.$$

Use the tool to calculate the Bode diagram. Discuss the results obtained with the asymptotic magnitude and phase approximations. Which asymptotic phase approximation provides the best results in this example?

4.8   Nonminimum Phase Systems

4.8.1   Interactive Tool: f_nonminimum_phase

4.8.1.1 Concepts Analyzed in the Card and Learning Outcomes
- Nonminimum phase concept. Analysis of the frequency response of nonminimum phase systems.
- Evaluation of the phase lag in nonminimum phase systems from the components of the transfer function.

4.8.1.2 Summary of Fundamental Theory
In the literature, different definitions of nonminimum phase (NMP) systems can be found. Following [5], transfer functions having neither poles nor zeros in the open right half s-plane (RHP) are *minimum phase* (MP) transfer functions, whereas those having poles and/or zeros in the open RHP are *nonminimum phase* (NMP) transfer functions. Systems with MP transfer functions are called MP systems, whereas those with NMP transfer functions are called NMP systems. Other texts only consider NMP to systems that are causal and stable whose inverses are causal and unstable, but from our perspective RHP poles and zeros have to be taken into account, and also time delays due to the interpretation of what "nonminimum phase" means. The difference between MP and NMP systems in the frequency domain is that for systems with the same magnitude characteristic, the range of the phase angle of the MP transfer function is minimal among all systems with the same magnitude diagram (the transfer function is uniquely determined from the magnitude plot), while the range of the phase angle for any NMP transfer function is greater than that of the equivalent MP system and its transfer function cannot be obtained by analyzing the magnitude plot alone. Therefore, a given NMP system will have a greater phase contribution than the MP system with the equivalent magnitude response [8]. This implies that transfer functions with zeros in the open RHP are classified as NMP transfer functions [8]. But this characteristic of having greater phase contribution than the MP system also happens if the system has poles on the open RHP (the system is unstable) or time delay (as the phase lag increases with frequency) [1]. Moreover, although this case is not considered in textbooks as an NMP feature (only in some of them, as [10]), having a negative static gain implies and addition of $\pm 180°$ (depending on the criterion, $+180°$ in this book), so that this fact also fits the feature of having a greater phase contribution than the MP system with the equivalent magnitude response. Many examples of NMP systems can be found in [2], p. 119.

The time response of an NMP system with zeros in the RHP is characterized by an inverse response, where the reaction to a step change in the input is initially in one direction, but after a transient, it evolves in the opposite direction.

One way to tackle the problem of NMP transfer functions is to use a graph of the pattern of poles and zeros in s-plane. If a zero (the same applies for poles) is located at $s = z$ and an arbitrary point s is selected on the plane, the vector that starts at $s = z$ and ends at point s represents the length and angle of vector $(s - z)$, as shown in Fig. 4.5. When plotting the frequency response, the points of interest are those that meet $s = j\omega$ (the points on the positive imaginary axis, from $\omega = 0$ to $\omega = \infty$).

If the transfer function is factored as follows:

$$G(s) = \frac{\kappa \prod\limits_{\ell=1}^{q} (s - z_\ell) \prod\limits_{\ell=1}^{r} (s^2 + 2\zeta_{z_\ell} \omega_{n_{z_\ell}} s + \omega_{n_{z_\ell}}^2)}{s^{\aleph} \prod\limits_{i=1}^{p} (s - p_i) \prod\limits_{i=1}^{h} (s^2 + 2\zeta_i \omega_{n_i} s + \omega_{n_i}^2)},$$

whose frequency response is given by

$$G(j\omega) = \frac{\kappa \prod\limits_{\ell=1}^{q} (j\omega - z_\ell) \prod\limits_{\ell=1}^{r} (-\omega^2 + j2\zeta_{z_\ell} \omega_{n_{z_\ell}} \omega + \omega_{n_{z_\ell}}^2)}{(j\omega)^{\aleph} \prod\limits_{i=1}^{p} (j\omega - p_i) \prod\limits_{i=1}^{h} (-\omega^2 + j2\zeta_i \omega_{n_i} \omega + \omega_{n_i}^2)}.$$

This transfer function can be expressed in polar form $G(j\omega) = \rho e^{j\phi}$, where

$$\rho = \frac{|\kappa| \prod\limits_{\ell=1}^{q} \rho_{z_\ell} \prod\limits_{\ell=1}^{2r} \rho_{z_\ell}}{\rho_0^{\aleph} \prod\limits_{i=1}^{p} \rho_{p_i} \prod\limits_{i=1}^{2h} \rho_{p_i}} \; ; \; \phi = \phi_\kappa + \sum\limits_{\ell=1}^{q} \phi_{z_\ell} + \sum\limits_{\ell=1}^{2r} \phi_{z_\ell} - \aleph\phi_0 - \sum\limits_{i=1}^{p} \phi_{p_i} - \sum\limits_{i=1}^{2h} \phi_{p_i}. \tag{4.19}$$

In those formulas ρ_{z_ℓ}, ρ_{z_ℓ}, ρ_0, ρ_{p_i}, and ρ_{p_i} are the modulus and ϕ_{z_ℓ}, ϕ_{z_ℓ}, ϕ_0, ϕ_{p_i} and ϕ_{p_i} are the phases of the vectors with origin in the corresponding singularities of $G(s)$ (poles and zeros) and with the end at $j\omega$ (see an example Fig. 4.6), while ϕ_κ is the argument of the gain term κ. By analyzing the modulus and phase of individual vectors when increasing ω from zero to infinity, the evolution of $G(j\omega)$ can be determined. In the example shown in Fig. 4.6, $\phi = \phi_\kappa + \phi_{z_1} + \phi_{z_2} - \phi_0 - \phi_{p_1}$.

Therefore, the phase of the transfer function will be the sum of the phase angles of the zeros minus those of the poles. For each frequency ω, the phase can be calculated knowing that the positive angles are counted from the positive real axis in a counter clockwise direction, which is the standard convention for measuring angles and also the criterion followed to state that the phase of a negative static gain is 180°. The authors fully agree with [2] in the sense that the lack of general consensus on this convention is a source of confusion for students. The previous reasoning can be applied to analyze the phase of NMP systems also using the convention that a negative gain term κ gives a phase shift of 180°, or analyze the main elements present in the transfer function individually taking into account the value of the associated static gain k, as done in [2]:

1. *Case 1—NMP zero with negative static gain (but positive[14] κ), $(s - z)$, $z > 0$:* As ω increases from 0 to ∞, the angle goes from 180° to 90°, which is the angle that should be added to that of the complete transfer function ϕ.

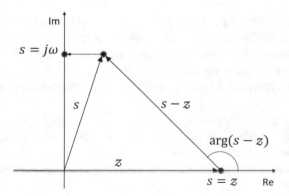

Fig. 4.5 Evaluation of the magnitude and phase angle of a vector $(s - z)$

---

[14] κ is associated with the complete transfer function, but here its sign is considered in the form in which poles and zeros are represented.

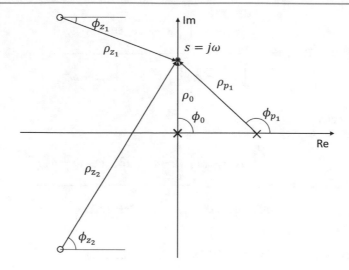

Fig. 4.6 Calculation of $G(j\omega)$ in graphic form

Table 4.4 Contributions in phase of the different elements of the transfer function

$z > 0, p > 0, \sigma > 0, \omega > 0$	Phase for $\omega = 0$	Phase for $\omega = \infty$
Zero at $s = 0$	$90°$ $\forall\omega$	$90°$
Zero at $s = z$; $(s - z)$	$180°$	$90°$
Zero at $s = z$; $(z - s) = -(s - z)$	$0°$ $(360°)$	$-90°$ $(270°)$
Zero at $s = -z$; $(s + z)$	$0°$	$90°$
Zero at $s = -z$; $-(z + s) = -(s + z)$ (negative gain)	$180°$	$270°$
Zeros at $s = \sigma \pm j\omega$; $(s - \sigma)^2 + \omega^2$	$360°$	$180°$
Zeros at $s = -\sigma \pm j\omega$; $(s + \sigma)^2 + \omega^2$	$0°$	$180°$
Pole at $s = 0$	$-90°$ $\forall\omega$	$-90°$
Pole at $s = p$; $1/(s - p)$	$-180°$	$-90°$
Pole at $s = p$; $1/(p - s) = -1/(s - p)$	$0°$	$90°$
Pole at $s = p$; $1/(s + p)$	$0°$	$-90°$
Pole at $s = p$; $-1/(p + s) = -1/(s + p)$ (negative gain)	$180°$	$90°$
Poles at $s = \sigma \pm j\omega$; $1/((s - \sigma)^2 + \omega^2)$	$-360°$	$-180°$
Poles at $s = -\sigma \pm j\omega$; $1/((s + \sigma)^2 + \omega^2)$	$0°$	$-180°$

2. *Case 2—NMP zero with positive gain (but negative κ)*, $(z - s) = -(s - z)$, $z > 0$: The analysis in [2] states that the phase of this term is the counter clockwise that angle the vector $z - s$ has, as ω increases from 0 to ∞, and thus the angle varies from $0°$ to $-90°$. It could also be obtained by adding $180°$ (as in this case κ is negative) to the previous case, so that the angle goes from $360°$ to $270°$, which is the same result obtained using MATLAB®. Notice that both results are equivalent.
3. *Case 3—NMP pole with negative static gain (but positive κ)*, $1/(s - p) = -1/(p - s)$, ; $p > 0$: This is the same than case 1, but for a pole (so that its phase contribution is negative, following (4.19)), and the angle varies from $-180°$ to $-90°$ as ω increases from 0 to ∞.
4. *Case 4—NMP pole with positive static gain (but negative κ)*, $1/(p - s) = -1/(s - p)$, ; $p > 0$: Similar to Case 2 (but with negative phase contribution) or adding $180°$ to Case 3 (due to negative κ), the angle varies from $0°$ to $90°$.
5. *Note:* in the case of complex conjugate poles or zeros (see Fig. 4.6), for $\omega = 0$ the angle contribution is zero (there is a cancellation of the same positive and negative angle). In the case of complex conjugate zeros, when $\omega \to \infty$, the contribution will be $2 \cdot 90° = 180°$, while a pair of conjugate poles contribute with $2 \cdot (-90°) = -180°$.

Table 4.4 summarizes the contributions of the different elements to the phase of a generic transfer function.

Section 7.4.5 in [2] introduces some guidelines for the determination of NMP systems from the Bode diagram. The magnitude diagrams of both MP and NMP systems are the same. From this amplitude diagram:

- The type of the system can be obtained from the low-frequency slope ($-20\aleph$).
- The high-frequency slope is $-20(n-m)$.

From the phase diagram:

- If the correspondences $\omega = 0 \leftrightarrow \phi = -90\aleph$ and $\omega \to \infty \leftrightarrow \phi = -90(n-m)$ hold, then either the system is MP or NMP with and equal number of NMP poles and zeros with the same signs.
- If the correspondence $\omega = 0 \leftrightarrow \phi = -90\aleph$ does not hold, the system is not MP with positive gain; it is either MP with negative gain or NMP.

4.8.1.3 References Related to this Concept

- [2] Bavafa-Toosi, Y. (2017). *Introduction to linear control systems*. Academic Press-Elsevier. ISBN: 978-0-12-812748-3. Chapter 7, Sect. 7.4, pp. 547–552.
- [5] Ogata, K. (2010). *Modern control engineering* (5th ed.). Prentice Hall. ISBN: 978-0-13-615673-4. Chapter 7, pp. 415–418.
- [6] Bolzern, P., Scattolini, R., & Schiavoni, N. (2009). *Fundamentos de control automático (Fundamentals of automatic control, in Spanish)*. McGraw-Hill. ISBN: 978-84-481-6640-3. Chapter 6, Sect. 7, pp. 157–158.
- [10] Rohrs, C. E., Melsa, J. L., & Schultz, D. G. (1994). *Sistemas de control lineal (Linear control systems)*. McGraw-Hill. ISBN: 970-10-0411-6. Chapter 5, pp. 298–302.

Application — Interactive tool: f_nonminimum_phase

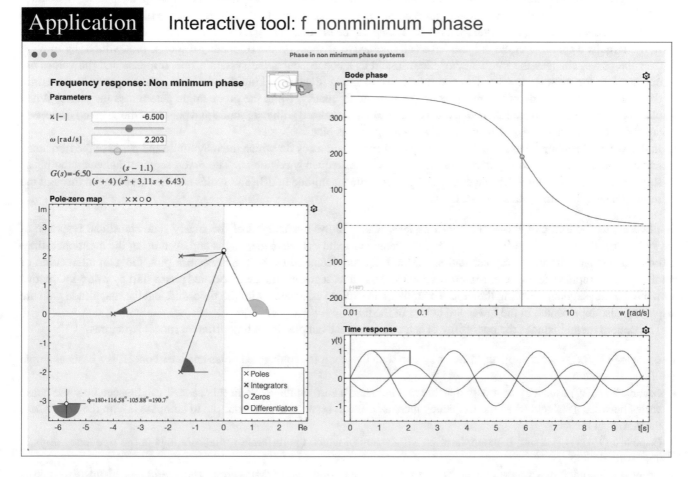

Frequency Response: Nonminimum Phase

This tool is devoted to analyzing in the frequency domain how to obtain the phase of NMP systems in a visual way.

The **Parameters** area of this tool only includes the possibility of modifying the gain term κ (a ZPK format is used to represent the transfer function) and the specific frequency for which the phase shift is being calculated, through textboxes (used to change the sign of κ) and sliders. Below these parameter values, a symbolic representation of the transfer function is displayed.

The main novelty of this tool is that in the **Pole–zero map** plot, which includes the repository of poles and zeros to build arbitrary transfer functions (used as in the other tools, for example, those explained in Sects. 3.6 and 4.7), the phase angles associated with each dynamic element (poles, zeros, integrators, and differentiators) are shown, so that the corresponding phase angle of the transfer function can be analyzed for a particular frequency. This frequency can be selected in the **Parameters** area or by means of a white circle located on the positive imaginary axis that can be shifted from $j0$ to $j\infty$.

In the lower-left part of the plot, the overall phase contribution is shown. As with other interactive tools, if the mouse pointer is placed over any dynamic element, its coordinates in the complex plane are displayed in the lower-left corner of the tool. The scale can be modified using the settings available in the gearwheel icon.

On the right side of the tool, different representations of the frequency response are displayed, which can be selected through the drop-down menu of Options located in the upper frame of the application. By default, the **Bode diagram (phase)** appears, since it is the representation where the concepts associated with the individual phase angle contribution by each of the dynamic elements that make up the transfer function are best displayed. The other selectable options are as in the f_concept tool: Bode diagram (phase and magnitude), Nyquist diagram, Nichols diagram, and Complete view. The latter includes all representations, which have the following interactive elements:

- In the **Bode magnitude** and **Bode phase**, a black vertical line allows the selected frequency to be changed interactively. On the **Bode magnitude**, a magenta circle is drawn, indicating the logarithmic magnitude for the given frequency. In the same way, a green circle is represented on the **Bode phase** plot marking the lag at the selected frequency. By placing the mouse cursor over the magnitude curve (at any point or on the magenta circle), the magnitude of the diagram can be modified (affecting only the static gain of the system). The scale can be modified using the settings available in the gearwheel icon.
- In the **Nyquist diagram**, a circle is included as an interactive element on the frequency response plot, which, by moving over the curve, changes the frequency of interest, and as a consequence, the system's **Time response** and the values are displayed in the **Parameters** area. The non-interactive magenta segment linking the circle over the response to the origin determines the magnitude of the response at the selected frequency, while the green angle determines the phase. When placing the mouse over the circle above the plot, the values associated with magnitude and phase for the selected frequency are showed. The scale can also be modified as in the previous plot.
- In the **Nichols diagram**, the user can interactively modify the frequency for which the magnitude and phase of the frequency response are being studied (by accessing the circle on the frequency response). The green segment indicates the phase shift in this diagram, while the magenta one represents the magnitude in dB. The scales can also be changed through the settings available in the gearwheel icon.

In all cases, the bottom part shows the **Time response**, which allows the analysis of the steady-state sinusoidal response of the system for the value of the frequency set in the **Parameters** and **Pole–zero map** areas and even using the mentioned lines and circles available in all the representations in the frequency domain. In the **Time response** plot, the gain attenuation or amplification is represented by a non-interactive vertical magenta segment and the associated phase shift by a non-interactive horizontal green segment. Placing the mouse over them, the values associated with the time shift and the magnitude gain are displayed, also represented in the lower-left corner of the tool.

The Options menu includes the possibility of selecting several examples including different model structures:

- Example 1 ($P_{11}(s)$) uses a stable third-order system with two complex conjugated poles and a real one. It is a basic example of an MP system.
- Example 2 ($P_4(s)$) uses a system with a stable real pole and a zero on the RHP. In this case, it can be seen how the phase varies between $180°$ and $0°$ as the frequency increases. It can serve as a base example to compare it with the case where the zero is at the LHP.
- Example 3 ($P_{11}(s)$) is a simple example of an MP system in which the different representations in the frequency domain appear by default.
- Example 4 includes two stable real poles and two complex conjugated NMP zeros. The usefulness of this example is to analyze the phase contribution of those zeros.

4.8.1.4 Homework

1. Test all possible combinations of MP and NMP zeros and poles with negative and positive gains for $G(s) = \pm(s \pm 1)/((s \pm 10)(s \pm 100))$. Note that, in all cases, the magnitude diagram has the slope $-20\aleph$ dB/decade, where $\aleph$ is the type of the system, in this case 0 at low frequencies and $-20(n-m)$ dB/decade at high frequencies, where $n-m$ is the relative degree of the system, difference between the order of the polynomial of the denominator of the transfer function and that of the numerator, in this case $-20(2-1) = -20$ dB/decade (Ref. [2], Sect. 7.4, pp. 547–552).

2. For the following sets of transfer functions, compare the magnitude and phase plots obtained for each of them. Comment on the results and indicate the phase shift values for $\omega = 0$, $\omega \to \infty$ and $\omega = 1$ rad/s.

Combinations of (κ, p, z):	$(1, -5, -1), (1, -5, 1), (-1, -5, -1), (-1, -5, 1),$
Group 1: $G(s) = \dfrac{\kappa(s-z)}{(s-p)}$	$(1, 5, -1), (-1, 5, -1), (1, 5, 1), (-1, 5, 1)$
Combinations of (κ, p, z):	$(1, -5, -1), (1, -5, 1), (-1, -5, -1), (-1, -5, 1),$
Group 2: $G(s) = \dfrac{\kappa(s-z)}{s(s-p)}$	$(1, 5, -1), (-1, 5, -1), (1, 5, 1), (-1, 5, 1)$
Combinations of (κ, p_1, p_2):	$(1, -5, -1), (1, -5, 1), (-1, -5, -1), (-1, -5, 1),$
Group 3: $G(s) = \dfrac{\kappa s}{(s-p_1)(s-p_2)}$	$(1, 5, -1), (-1, 5, -1), (1, 5, 1), (-1, 5, 1)$
Combinations of $(\zeta_z, \omega_{n_z}, \zeta, \omega_n)$:	$(0.1, 1, 0.5, 1), (-0.1, 1, 0.5, 1), (0.1, 1, -0.5, 1),$
Group 4: $G(s) = \dfrac{s^2 + 2\zeta_z\omega_{n_z}s + \omega_{n_z}^2}{s^2 + 2\zeta\omega_n s + \omega_n^2}$	$(0.1, 1, 0.5, 2), (-0.1, 1, 0.5, 2), (0.1, 1, -0.5, 2)$

3. Using Example 2 ($P_4(s)$) from the Options menu, analyze the difference in the various frequency-domain and time-domain representations between the default setting and the one resulting from locating the zero at $s = -1$. What are the most noticeable differences you see in the Bode phase plot? And in the steady-state regime sinusoidal response of the system?

4. Using Example 3 ($P_{11}(s)$) included in the Options menu, repeat the previous exercise placing the real pole at $s = 0.72$.

4.9 Model Fitting in the Frequency Domain

4.9.1 Interactive Tool: f_model_fitting

4.9.1.1 Concepts Analyzed in the Card and Learning Outcomes

- Fitting models to experimental data in the frequency domain.

4.9.1.2 Summary of Fundamental Theory

The frequency response concept, when dealing with plants that can be described by stable linear models, can be used to determine a transfer function model from experimental data, which can be incorporated in frequency response diagrams, mainly the Bode plot. These data are obtained in the following way: the system is brought to an operating point and a sinusoidal signal $U_0 \sin(\omega_0 t)$ is introduced at the input, waiting the necessary time until the system reaches a steady-state regime. In this situation, the amplitude and time shift (directly related to the phase through the selected frequency) of the output sine wave signal is measured. Both the input and output signals, therefore, represent increments or deviation variables from the operating point, which allows the linear model to represent the dynamics of the system around that operating point.

The magnitude of the frequency response in $\omega = \omega_0$ is equal to the ratio between the amplitudes of the output sine wave and that of the input sine wave, its phase being equal to the difference between their phase shifts. If this operation is repeated using sinusoidal signals of the same amplitude and different frequencies $\omega = \omega_1, \omega_2, ..., \omega_i$, points of the frequency response can be obtained for the frequencies used in the generation of the input sines. These points can be placed in a Bode diagram (previously converting the magnitude to decibels and the time shifts to phase in degrees) and interpolate these points with an appropriate approximation. Then, using asymptotic approximations, a transfer function can be postulated to model that frequency response, as already commented in Sect. 4.7.

There are three factors that make asymptotic magnitude and phase diagrams ideal for the solution of this problem. First, it is relatively easy to approximate the graph of the magnitude determined experimentally by the graph of the asymptotic magnitude. Since the graph only contains slope lines that are integer multiples of ± 20 dB/decade, the asymptotic approximation can be obtained in a considerably simple manner. Secondly, the shape of the transfer function can be obtained by inspection from the asymptotic approximation. This characteristic stems from the fact that the corner frequencies of the asymptotic approximation correspond to the location of the poles and zeros of the transfer function, with only a slight difficulty in considering complex conjugated poles or zeros. Thirdly, the joint consideration of the experimentally determined magnitude and phase plots uniquely specifies the transfer function. Using only the magnitude graph, the shape of the transfer function (and its corresponding corner frequencies) can be determined and then the phase in MP systems. The procedure is more complex and there is no guarantee of success if the system is NMP, as there is not a direct correlation between magnitude and phase (see Sect. 4.7).

4.9.1.3 References Related to this Concept

- [6] Bolzern, P., Scattolini, R., & Schiavoni, N. (2009). *Fundamentos de control automático (Fundamentals of automatic control, in Spanish)*. McGraw-Hill. ISBN: 978-84-481-6640-3. Chapter 6, Sect. 5, pp. 145–146.
- [10] Rohrs, C. E., Melsa, J. L., & Schultz, D. G. (1994). *Sistemas de control lineal (Linear control systems)*. McGraw-Hill. ISBN: 970-10-0411-6. Chapter 5, pp. 309–320, 1994.

Application   Interactive tool: f_model_fitting

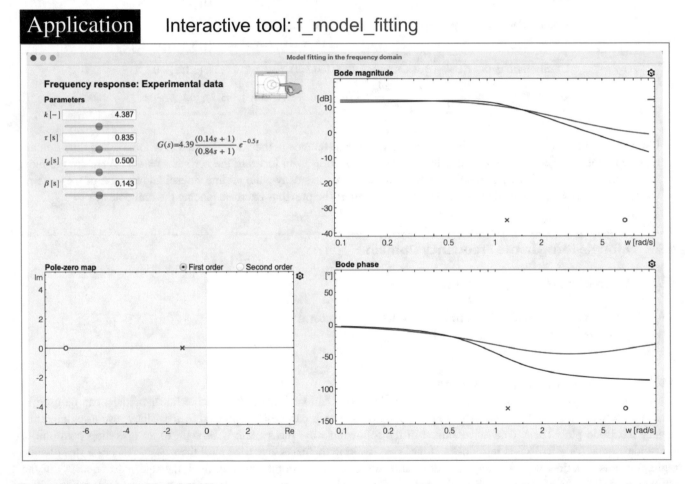

Frequency Response: Model Fitting

The frequency response of a system can readily be obtained experimentally by exciting the system with sinusoidal input signals about a particular operating point. Corresponding points on the Bode diagram can be drawn from the amplitude and time shift of the output response in steady state (converting the quotient between the output and input magnitude to dB and the time shift to phase shift through the corresponding frequency). Then, asymptotic approximations help to interpolate these points and to identify system transfer functions from the experimental Bode diagrams. The interactive tool presented in this section helps to estimate these transfer functions using in this case first or second-order structures (which may include time delay and a zero).

Parameters: The upper-left side shows the numerical parameters and the transfer function that is modeled on the system under study. In this tool, the possible structures of transfer functions that model the experimental data have been fixed:

- First-order transfer function with a zero and a delay, represented by:

$$G(s) = k \frac{(\beta s + 1)}{(\tau s + 1)} e^{-t_d s}.$$

- Second-order transfer function with a zero and delay, represented by:

$$G(s) = k\omega_n^2 \frac{(\beta s + 1)}{(s^2 + 2\zeta\omega_n s + \omega_n^2)} e^{-t_d s} \text{ or } G(s) = k \frac{(\beta s + 1)}{(\tau_1 s + 1)(\tau_2 s + 1)} e^{-t_d s},$$

whose parameters can be set via textboxes or sliders in the **Parameters** area, which includes the symbolic representation of the chosen transfer function. The choice between the possible models is made by marking a circular selection button (*radio button*) located over the **Pole–zero map** plot (lower-left area of the tool), which shows the poles and zeros of the selected system, their location being displayed in the lower-left corner of the tool when the mouse is placed over them.

In the right area of the tool, the **Bode magnitude** and **Bode phase** diagrams are shown, including fine black color lines corresponding to the frequency response obtained from experimental data and others in red thick lines representing the frequency response of the system that is postulated as the most suitable model to reproduce the frequency response obtained from experimental data. Both diagrams can be enlarged or reduced using the settings available in the gearwheel icon. In both graphs, red crosses ✕ are shown, which correspond to the location of the corner frequencies to be modified by dragging them to the right or left, so that their values are automatically refreshed in the **Parameters** and **Pole–zero map** areas. In the case of first-order systems, the red cross represents the cutoff frequency associated with the pole (and therefore the time constant), while in the case of second-order systems, the red cross corresponds to the natural frequency if the system is underdamped. If the system is overdamped, two crosses are drawn representing the corner frequencies associated with each real pole. A red circle ○ can also be moved affecting the location of the zero, this action being updated in the other areas of the tool. The change of the low-frequency magnitude (and therefore, of the static gain) is done in this tool by clicking and dragging up/down the red thick line that appears on the right side of the **Bode magnitude** plot.

The lower-left area of the tool is dedicated to the representation in the complex plane (**Pole–zero map**). In this graph, it is possible to select as a model a first or second-order system with a zero, appearing in each case the poles and the associated zero (the time delay has no representation in this diagram, as it is not a rational term). The position of the poles and the zero of the selected system can be modified by dragging them to another location within this diagram, their position being indicated in the lower-left corner of the interactive tool.

4.9.1.4 Homework

1. Select from the Options menu Example 1, 2, 3 and 4. Fit the two possible transfer functions to the experimental data as well as possible. Indicate the range of frequencies for which the models are valid.

References

1. Åström, K. J., & Murray, R. M. (2014). *Feedback systems: An introduction for scientists and engineers* (2nd ed.). Princeton University Press.
2. Bavafa-Toosi, Y. (2017). *Introduction to linear control systems*. Academic Press-Elsevier.
3. Horowitz, I. (1993). *Quantitative feedback design theory (QFT)*. QFT Publications.
4. Golnaraghi, F., & Kuo, B. C. (2017). *Automatic control systems* (10th ed.). McGraw Hill Education.
5. Ogata, K. (2010). *Modern control engineering* (5th ed.). Prentice Hall.
6. Bolzern, P., Scattolini, R., & Schiavoni, N. (2009). *Fundamentos de control automático (Fundamentals of automatic control)*. McGraw-Hill.
7. D'Azzo, J. J., Houpis, C. H., & Sheldon, S. N. (2003). *Linear control system analysis and design with MATLAB®* (5th ed.). Marcel Dekker Inc.
8. Dorf, R. C., & Bishop, R. H. (2011). *Modern control systems* (12th ed.). Prentice Hall.
9. Franklin, G. F., Powell, J. D., & Emani-Naeni, A. (2010). *Feedback control of dynamic systems* (6th ed.). Pearson.
10. Rohrs, C. E., Melsa, J. L., & Schultz, D. G. (1994). *Sistemas de control lineal (Linear control systems)*. McGraw-Hill.
11. Shahian, B., & Hassul, M. (1993). *Control system design using MATLAB®*. Prentice Hall.
12. Barrientos, A., Sanz, R., Matía, F., & Gambao, E. (1996). *Control de sistemas continuos. Problemas resueltos (Control of continuous systems. Problems solved)*. McGraw-Hill.
13. Truxal, J. G. (1955). *Automatic feedback control system synthesis*. McGraw-Hill.

Relationship Between Model Parameters with Physical Models

5.1 Introduction

In Chap. 2, different representative physical processes are analyzed as illustrative examples to explain the concepts related to obtaining linear models from nonlinear differential equations. In Chaps. 3 and 4, interactive tools have been developed to introduce basic concepts of time and frequency response of LTI systems represented by transfer functions.

This chapter is devoted to exploring the relationships between the parameters of the physical models of the systems studied in Chap. 2 with their representative transfer functions (in nonlinear systems obtained by linearizing the system about an operating point). The interactive tools help to analyze how the physical parameters influence the dynamic behavior of systems.

5.2 The Tank Level System Transfer Function

5.2.1 Interactive Tool: tank_level_tf

5.2.1.1 Concepts Analyzed in the Card and Learning outcomes
- Obtaining the transfer function of a dynamical system linearized around an operating point.
- Study the relationship between the parameters of a transfer function with those of the differential equation representing the linearized system model.
- Analysis of the influence on the unit step response of changing parameters in the transfer function.

5.2.1.2 Summary of Fundamental Theory
Most dynamical systems have a behavior that can be represented by physical laws of nonlinear nature. Using a Taylor series expansion in the nonlinear terms about an operating point, a linear model is obtained in the form of an ODE, from which, applying the Laplace transform with zero initial conditions, it is easy to get the external description of the system in the form of a transfer function. In this section, the tank level system [1, 2] is again used to analyze the relationship between the nominal parameters of the transfer function with those of the original physical model.

As explained in Sect. 2.2, the process consists of a tank with an outlet orifice or discharge section, into which a variable fluid flow rate enters by the action of a pump. It is described by the following nonlinear differential equation [2]:

$$A\frac{dh(t)}{dt} = q(t) - a\sqrt{2gh(t)}, \tag{5.1}$$

where $h(t)$ represents the height of the tank, A its section, a the area of the discharge section, g the acceleration of gravity at sea level, and $q(t)$ the inlet volumetric flow.

If the system is linearized around a steady-state operating point, a first-order linear model is obtained

Supplementary Information The online version contains supplementary material available at https://doi.org/10.1007/978-3-031-09920-5_5.

$$\tau \frac{d\tilde{h}(t)}{dt} + \tilde{h}(t) = k\tilde{q}(t), \tag{5.2}$$

where $\tau = \frac{A}{a}\sqrt{\frac{2\overline{h}}{g}}$, $k = \frac{\tau}{A}$, $\tilde{h}(t)$ is the deviation variable (height of fluid) with respect to the operating point $(\overline{q}, \overline{h})$.

Applying the Laplace transform with zero initial conditions in the deviation variables ($\tilde{h}(0) = 0$, $\tilde{q}(0) = 0$), the transfer function is obtained as the quotient between the Laplace transform of the output $\tilde{h}(t)$ ($H(s)$) and that of the input $\tilde{q}(t)$ ($Q(s)$):

$$G(s) = \frac{H(s)}{Q(s)} = \frac{k}{\tau s + 1} = \frac{\frac{1}{a}\sqrt{\frac{2\overline{h}}{g}}}{\frac{A}{a}\sqrt{\frac{2\overline{h}}{g}}s + 1}. \tag{5.3}$$

From equation (3.13), the time evolution of the level inside the tank for an inlet flow step input Δq is given by

$$\tilde{h}(t) = k\Delta q(1 - e^{-\frac{1}{\tau}t}), \quad t \geq 0, \tag{5.4}$$

with k and τ previously determined.

Note: As pointed out in Sect. 2.2, if the output flow is selected as the output of the process, it is easy to check that the transfer function in that case should be

$$G_a(s) = \frac{1}{\tau_a s + 1}$$

with $\tau_a = \frac{A}{a}\sqrt{\frac{2\overline{h}}{g}}$.

5.2.1.3 References Related to this Concept

- [1] Guzmán, J. L., Vargas, H., Sánchez-Moreno, J., Rodríguez, F., Berenguel, M., & Dormido, S. (2007). *Distance Education Research Trends. Chapter 6: Education research in engineering studies: Interactivity, virtual and remote labs*. Nova Science Publishers Inc., pp. 131–167.
- [2] Johansson, K. H. (2000). The quadruple-tank process: A multivariable laboratory process with an adjustable zero. *IEEE Transactions on Control Systems Technology, 8*(3), 456–465.
- [3] Åström, K. J., & Murray, R. M. (2014). *Feedback systems: An introduction for scientists and engineers* (2nd ed.). Princeton University Press. ISBN: 9780691193984.
- [4] Dorf, R. C., & Bishop, R. H. (2011). *Modern control systems* (12th ed.). Prentice Hall. ISBN: 978-0-13-602458-3. Example 2.13, pp. 94–104.
- [5] Golnaraghi, F. and B.C. Kuo. Automatic control systems. Tenth edition. McGraw Hill Education, ISBN: 978-1-25-964384-2, 2017.
- [6] Ogata, K. (2010). *Modern control engineering* (12th ed.). Prentice Hall. ISBN: 978-0-13-615673-4.

Table 2.1 in Chap. 2 contains references on how this process is analyzed in different textbooks as further reading.

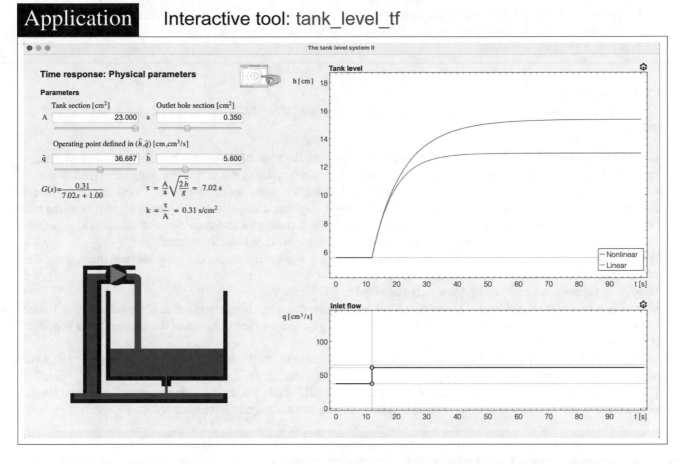

Tank Level: Transfer Function

In the application described in the interactive tool, it is possible to compare the time response of the open-loop system obtained with the nonlinear model with that from the linearized system. The first-order transfer function that describes the linearized system is also shown, as well as the relationship between the parameters of the physical system and those of the transfer function (static gain and time constant). The tool also allows analysis at different operating points and with different tank geometries. This provides quantitative and qualitative information on the validity interval of these models. System parameters can be changed by interacting with the geometric description of the tank.

In the upper-left part of the tool, it is possible to modify the values of the fundamental variables that define the dynamics of the system: the area of the tank (A), the area of the discharge hole (a), and the variables that define the operating point ($\overline{q}, \overline{h}$). A symbolic representation of the transfer function is also displayed, which changes as the values of the indicated parameters are manipulated. The formula relating these parameters to the static gain (k) and the time constant (τ) is also shown.

If the sliders of $\overline{q}$ or $\overline{h}$ are accessed, the value of the linked variable is modified according to Eq. (2.12) and an internal scaling of the intervals in which these variables are modified takes place. There is also a link between the intervals of A and a so that the simulations shown in the upper-right hand image make physical sense.

The images on the right area of the tool compare the time response of the linear model and that of the nonlinear system, as a result of a change in the fluid input flow rate represented in the bottom right plot. Note that the base value of the change is defined by $\overline{q}$, which can be modified in this graph or in the scroll bars on the upper-left area. When the pointer is placed over any point in the time evolution of the height or the flow, a label is then visible providing the coordinates (t, h) or (t, q) that define that point in the corresponding graph.

The tool presents in the lower-left part an outline of the process to be modeled. The figure is interactive, in the sense that its attributes can be modified by dragging over the geometry. The area A can be changed by clicking on the tank perimeter and dragging in the horizontal direction; the same applies to the area of the discharge orifice a (note that there is a limitation of minimum and maximum values so that the simulation has a physical sense and the graph can be represented in the space left for this purpose). The user can also change the operating point by clicking on the liquid surface and dragging it in a vertical direction. The equilibrium flow rate $\overline{q}$ can be changed by moving the liquid level in the pipe just before the elbow to the left of the pump in the vertical direction.

5.2.1.4 Homework

For a steady state defined by the default setting when starting the tool ($\overline{q} = 112.87$ cm^3/s, $\overline{h} = 8$ cm, $A = 17.32$ cm^2 and $a = 0.9$ cm^2):

1. Obtain a linear model of the open-loop system by measuring on the time response graph of the nonlinear system the gain and time constant that define an approximate linear model of the system valid at that operating point (reaction curve method). Compare the results with the linearized theoretical model, analyzing the discrepancies. Comment on the results. Repeat the process again changing the amplitude of the input step that is displayed in the lower-right part of the tool, bringing the final value of the flow to 140 cm^3/s. What are the new static gain and time constant values? Do they differ much from the first you have calculated? If the answer is yes, please indicate possible reasons.
2. If the steady-state flow is set to $\overline{q} = 130$ cm^3/s, indicate how this new value influences the static gain and the time constant of the system. What is the new steady-state level?
3. Repeat the previous exercise setting the steady-state flow to $\overline{q} = 90$ cm^3/s.
4. Using the values of the variables and parameters from the previous exercise, change the section of the tank to $A = 10$ cm^2. Indicate how this decrease in the tank section affects the static gain and the time constant of the system. What happens if it is increased to $A = 20$ cm^2?
5. Using the settings of the previous exercise, indicate how an increase in the discharge section to $a = 0.3$ cm^2 affects the static gain and time constant of the system.
6. If a sinusoidal input $\tilde{q}(t) = \sin(\omega t)$ excites the system, analytically obtain the steady-state regime and represent it using the interactive tool f_concept (card 4.2), reasoning about the maximum change in level achieved.

5.3 Variable Section Tank Level System Transfer Function

5.3.1 Interactive Tool: spherical_tank_level_tf

5.3.1.1 Concepts Analyzed in the Card and Learning outcomes
- Obtaining the transfer function of a dynamical system linearized around an operating point.
- Study the relationship between the parameters of a transfer function with those of the differential equation representing the linearized system model.
- Analysis of the influence on the unit step response of changing parameters in the transfer function.

5.3.1.2 Summary of Fundamental Theory
The nonlinear differential equation describing the dynamics of the fluid level inside a spherical tank with a discharge hole was obtained in Eq. (2.23). The linearized model is given by

$$\frac{d\tilde{h}(t)}{dt} = \frac{1}{\pi\overline{h}(2r - \overline{h})}\left(\tilde{q}(t) - a\sqrt{\frac{g}{2\overline{h}}}\tilde{h}(t)\right). \tag{5.5}$$

Rearranging the terms in the nominal form $\tau\frac{d\tilde{h}(t)}{dt} + \tilde{h}(t) = k\tilde{q}(t)$, it is obtained:

$$\tau = \frac{1}{a}\sqrt{\frac{2\overline{h}}{g}}\pi\overline{h}(2r - \overline{h}) = k(\pi\overline{h}(2r - \overline{h})),$$

$$k = \frac{1}{a}\sqrt{\frac{2\overline{h}}{g}} = \frac{\tau}{\pi\overline{h}(2r - \overline{h})}. \tag{5.6}$$

By applying the Laplace transform to the linear differential equation with zero initial conditions in the deviation variables, the typical first-order transfer function is obtained

$$G(s) = \frac{H(s)}{Q(s)} = \frac{k}{\tau s + 1},$$

where $H(s)$ is the Laplace transform of $\tilde{h}$ and $Q(s)$ is the Laplace transform of $\tilde{q}$. Notice that most textbooks remove the $(\tilde{\cdot})$ from the variables once the system has been linearized (in a linear system the behavior does not depend on any operating point). From Eq. (3.13), the time evolution of the level inside the tank for an inlet flow step input Δq is given by

$$\tilde{h}(t) = k\Delta q(1 - e^{-\frac{1}{\tau}t}), \quad t \geq 0, \tag{5.7}$$

with k and τ previously determined.

5.3.1.3 Reference Related to this Concept

- [7] Tavakolpour-Saleh, A., Setoodeh, A., & Ansari, E. (2016). Iterative learning control of two coupled nonlinear spherical tanks. *International Journal of Mechanical and Mechatronics Engineering, World Academy of Science, Engineering and Technology, 10*(11), 1862–1869.

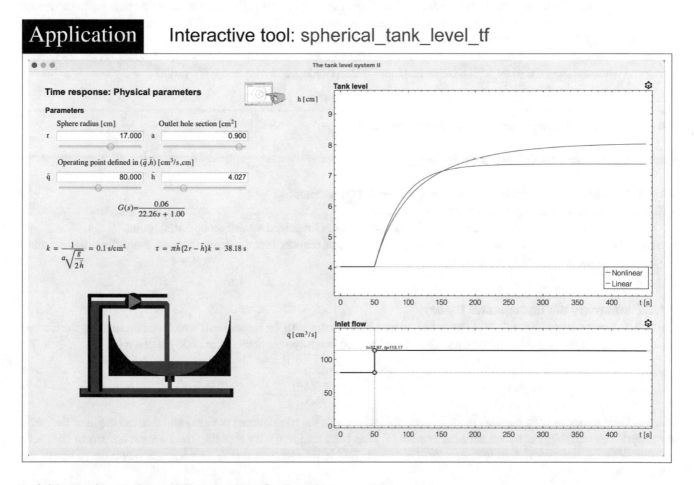

Variable Section Tank Level System: Transfer Function

This interactive tool is the same as that presented in Sect. 2.3, but includes a symbolic representation of the transfer function so that, the relationship of the static gain and time constant can be directly linked to the model parameters and selected operating point. Steps responses are also shown.

The explanation of this interactive tool is the same that can be found in card 5.2, with the only difference that the radius of the sphere r is shown instead of the section A, as, in this case, it depends on the level of fluid inside the tank.

5.3.1.4 Homework

For a steady state defined by the default setting when starting the tool ($\bar{h} = 8$ cm, $r = 17.32$ cm^2 and $a = 0.9$ cm^2) and the given input step:

1. Obtain a linear model of the open-loop system by measuring on the time response graph of the nonlinear system the gain and time constant that define an approximate linear model of the system valid at that operating point (reaction curve method). Compare the results with the linearized theoretical model, analyzing the discrepancies. Comment on the results. Repeat the process again changing the amplitude of the input step that is displayed in the lower-right part of the tool, bringing the final value of the flow to $150 \text{ cm}^3/\text{s}$. What are the new static gain and time constant values? Do they differ much from the first calculated? If the answer is yes, please indicate possible reasons. Repeat the procedure again bringing the final value of the low to $80 \text{ cm}^3/\text{s}$.
2. Using the default settings, change the radius of the sphere to $r = 12$ cm^2. Indicate how this decrease affects the static gain and the time constant of the system. What happens if it is increased to $r = 22$ cm^2?
3. Indicate how an increase in the discharge section to $a = 0.3$ cm^2 affects the static gain and time constant of the system.
4. Reset the tool in the Session menu. If the steady-state flow is set to $\bar{q} = 50 \text{ cm}^3/\text{s}$, indicate how this new value influences the static gain and the time constant of the system with respect to the previous exercise. What is the new steady-state level? Repeat setting the steady-state flow to $\bar{q} = 140 \text{ cm}^3/\text{s}$.
5. Using the interactive tool f_concept (card 4.2), represent the frequency diagrams of the tank level. What is the steady-state regime when exciting the plant with a sinusoidal input of amplitude 1 and frequency $1/\tau$ rad/s?

5.4   Ball and Beam System Transfer Function

5.4.1   Interactive Tool: ball_and_beam_control

5.4.1.1 Concepts Analyzed in the Card and Learning Outcomes

- Obtaining a double integrator transfer function model.
- Obtaining the transfer function of an unstable dynamical system linearized around an operating point.
- Study the relationship between the gain of a double integrator transfer function with those of the differential equation representing the linearized system model.

5.4.1.2 Summary of Fundamental Theory

The ball and beam system (Fig. 2.7) is characterized as a beam coupled to the motor shaft which can be tilted to the desired angle. The linear approximation of the dynamics of the ball and beam system obtained in (2.33) is given by

$$\left(\frac{I_b}{r^2} + m\right)\ddot{x} = mg\theta, \tag{5.8}$$

which is valid under the assumptions made in Sect. 2.4, where x is the translational position and θ is the angle of the beam shaft (deviation variables, although here the symbol ($\tilde{\cdot}$) has been removed). By applying the Laplace transform with zero initial conditions, the typical double integrator is obtained. Taking into account that $I_b = \frac{2}{5}mR^2$, R being the radius of the ball (see Fig. 2.7)

$$G(s) = \frac{X(s)}{\Theta(s)} = \frac{mg}{\left(\frac{I_b}{r^2} + m\right)}\frac{1}{s^2} = \frac{g}{1 + \frac{2}{5}\left(\frac{R}{r}\right)^2}\frac{1}{s^2} = \frac{k}{s^2}. \tag{5.9}$$

It can be seen how the gain multiplying the double integrator depends on the structural parameters of the system R and r. This transfer function is also obtained if the beam dynamics are taken into account, when they are sufficiently fast to be negligible in comparison to those of the ball. This simplified model allows to explain and design basic controllers, as will be analyzed in Sect. 9.4. The process transfer function in a real experiment can be determined by applying a known voltage to the actuator and recording the time taken for the ball to roll the length of the beam. The gain of the transfer function is then obtained from the standard equations of motion [8].

This system is unstable as its representative transfer function has two poles at the origin (double integrator). If a step input of amplitude U_0 (bounded input) is introduced, it is obtained

$$Y(s) = G(s)U(s) = \frac{k}{s^2}\frac{U_0}{s} = \frac{kU_0}{s^3}.$$

From line 4 of Table 3.2, it is easy to see that $y(t) = \frac{kU_0}{2}t^2$, the position growing parabolically in time (unbounded output), so the system is unstable in open loop, requiring a feedback controller to stabilize it. The physical meaning is clear, if the angle of the beam shaft is changed from zero to U_0, the ball rolls with constant acceleration.

5.4.1.3 References Related to this Concept

- [8] Wellstead, P. E., Chrimes, V., Fletcher, P. R., Moody, R., & Robins, A. J. (1989). Ball and beam control experiment. *The International Journal of Electrical Engineering & Education, 16*, 21–39.
- [9] Acharya, M., Bhattarai, M., & Poudel, B. (2014). Real time motion assessment for positioning in time and space critical systems. *International Journal of Applied Research and Studies (iJARS), 3*(7), 1–11.
- [10] Shahian, B., & Hassul, M. (1993). *Control system design using MATLAB®*. Prentice Hall. ISBN: 0-13-174061-X.

Ball and Beam System: Transfer Function

In this case, no explicit tool has been developed for the analysis of the open-loop time response of the nonlinear and linearized models because when excited with a step input, its output grows indefinitely. It represents an unstable system from the BIBO point of view, and therefore, its analysis only makes sense in Chap. 9 where it will be stabilized through feedback (see Sect. 9.4).

5.4.1.4 Homework

1. What is the natural equilibrium of this system? Can other operating points be defined (and reachable through feedback, as treated in Chap. 9)?
2. Using integrators, draw a block diagram relating Θ with X and reason about what represents the output of each block.
3. What kind of response is expected when entering a unit impulse to the system? Reason about the physical meaning.
4. What kind of response is achieved when exciting it with a unit amplitude sine? Reason about the physical meaning.
5. Draw the pole-zero diagram of this system and reason about stability.

5.5 Inverted Pendulum on a Cart System Transfer Function

5.5.1 Interactive Tool: inverted_pendulum_control

5.5.1.1 Concepts Analyzed in the Card and Learning outcomes
- Obtaining the transfer function of a dynamical system linearized around an operating point.
- Stability analysis of a second-order system and poles location.

5.5.1.2 Summary of Fundamental Theory

The equations of motion of the inverted pendulum on a cart system (Fig. 2.8) have been analyzed in Sect. 2.5. Applying Laplace transform to Eq. (2.41), the transfer function relating the angle of the pendulum to the applied force can be obtained as

$$\frac{\Theta(s)}{U(s)} = \frac{-1}{Mls^2 - (M+m)g} = \frac{-1}{Ml\left(s + \sqrt{\frac{M+m}{Ml}g}\right)\left(s - \sqrt{\frac{M+m}{Ml}g}\right)} = \frac{\frac{-1}{(M+m)g}}{\frac{Ml}{(M+m)g}s^2 - 1}. \tag{5.10}$$

The inverted pendulum simplified linear model has two symmetric poles, one on the negative real axis and another on the positive real axis

$$\text{Stable pole } \quad p_1 = -(\sqrt{g(M+m)}/\sqrt{Ml}),$$
$$\text{Unstable pole } \quad p_2 = (\sqrt{g(M+m)}/\sqrt{Ml}).$$

Hence, it is an open-loop unstable plant, as one natural mode of the system will follow a positive exponential profile.

It is important to point out that the initial set of differential equations are of fourth-order (second derivative in x and second derivative in θ), but due to the assumptions and simplifications made, a second-order system is obtained. In fact, if Eqs. (2.45) and (2.46) are used instead of using Eq. (2.41), a transfer function is obtained

$$(I + ml^2)s^2 \Theta(s) = mgl\Theta(s) - mls^2 X(s), \tag{5.11}$$
$$(M + m)s^2 X(s) = U(s) - bsX(s) - mls^2 \Theta(s), \tag{5.12}$$

where $U(s)$, $X(s)$ and $\Theta(s)$ are the Laplace transforms of the deviation variables $\tilde{u}(t)$, $\tilde{x}(t)$ and $\tilde{\theta}(t)$ respectively. Applying Laplace transform with zero initial conditions (as it corresponds to the deviation variables) to the first equation, it results in

$$X(s) = \frac{1}{mls^2}\left(-\frac{(I+ml^2)}{ml} + \frac{g}{s^2}\right)\Theta(s).$$

Substituting in the second one and denoting $q = (ml)^2 - (M+m)(I+ml^2)$, the following transfer function is obtained:

$$G(s) = \frac{\Theta(s)}{U(s)} = \frac{\frac{ml}{q}s}{s^3 - \frac{b(I+ml^2)}{q}s^2 + \frac{mgl(M+m)}{q}s + \frac{mglb}{q}}. \tag{5.13}$$

5.5.1.3 References Related to this Concept

- [6] Ogata, K. (2010). *Modern control engineering* (5th ed.). Prentice Hall. ISBN: 978-0-13-615673-4. Chapter 3, example 3–5, pp. 68–72.
- [10] Shahian, B., & Hassul, M. (1993). *Control system design using MATLAB®*. Prentice Hall. ISBN: 0-13-174061-X. Chapter 7, Sect. 9, pp. 217–219; Appendix A, Sect. 4, pp. 476–488.

Inverted Pendulum on a Cart: Transfer Function

In this case, no explicit tool has been developed for the analysis of the time response of the nonlinear and linearized models because when excited with a step input, its output grows indefinitely. It represents an unstable system from the BIBO point of view and therefore its analysis only makes sense in Chap. 9 where it will be stabilized through feedback (see Sect. 9.4).

5.5.1.4 Homework

1. What is the natural equilibrium of this system? Can other operating points being defined (and reachable through feedback, as treated in Chap. 9)?
2. Using integrators, draw a block diagram relating u with θ and reason about what represents the output of each block.
3. What kind of response is expected when entering a unit impulse to the system? Reason about the physical meaning.
4. What kind of response is achieved when exciting it with a unit amplitude sine? Reason about the physical meaning.
5. Draw the pole-zero diagram of this system and reason about stability.

5.6   DC Motor System Transfer Function

5.6.1   Interactive Tool: DC_motor_control

5.6.1.1 Concepts Analyzed in the Card and Learning outcomes
- Transfer function of a first-order plus integrator linear system.

5.6.1.2 Summary of Fundamental Theory
The fundamental ODEs defining the dynamics of the DC motor were obtained in Sect. 2.6 (Eqs. (2.53) and (2.54))

$$L_a \frac{di_a(t)}{dt} + R_a i_a(t) + K_b \frac{d\theta(t)}{dt} = e_a(t), \tag{5.14}$$

$$J_l \frac{d^2\theta(t)}{dt^2} + b_l \frac{d\theta(t)}{dt} = K_t i_a(t). \tag{5.15}$$

Notice that $\dot{\theta}(t)$ represents the angular velocity of the motor. Recalling that the nominal ODE of a first-order system can be represented as

$$\tau \frac{dy(t)}{dt} + y(t) = ku(t),$$

it is easy to check that in Eq. (5.14) the so called *electrical time constant* of the armature is given by $\tau_a = L_a/R_a$, while from (5.15), the *mechanical time constant* is $\tau_l = J_l/b_l$. As pointed out in Sect. 3.6, τ_l is several times larger than τ_a, because the inductance L_a is quite small [11].

As the previous differential equations are linear, the Laplace transform can be applied assuming zero initial conditions and that there is no load torque disturbance T_l. The following equations are obtained:

$$L_a s I_a(s) + R_a I_a(s) + K_b s \Theta(s) = E_a(s), \tag{5.16}$$

$$J_l s^2 \Theta(s) + b_l s \Theta(s) = K_t I_a(s), \tag{5.17}$$

where $I_a(s)$, $E_a(s)$ and $\Theta(s)$ are the Laplace transforms of $i_a(t)$, $e_a(t)$ and $\theta(t)$ respectively.

By eliminating $I_a(s)$ from Eqs. (5.16) and (5.17), the transfer function relating the input voltage to the angular position can be obtained

$$G(s) = \frac{\Theta(s)}{E_a(s)} = \frac{K_t}{s(L_a J_l s^2 + (L_a b_l + R_a J_l)s + R_a b_l + K_t K_b)}, \tag{5.18}$$

which is a third-order transfer function including an integrator. That integrator comes from the relationship between angular position with angular velocity.

The armature inductance L_a is generally small and can be neglected. In this case, the transfer function is reduced to

$$G(s) = \frac{\Theta(s)}{E_a(s)} = \frac{k}{s(\tau s + 1)}, \tag{5.19}$$

with

$$k = \frac{K_t}{(R_a b_l + K_t K_b)} \quad \text{and} \quad \tau = \frac{R_a J_l}{(R_a b_l + K_t K_b)}.$$

Due to the presence of an integrator (pole at the origin of the s-plane), the system is marginally stable in open loop. If the time evolution to a step input is computed

$$Y(s) = G(s)U(s) = \frac{k}{s(\tau s + 1)} \frac{U_0}{s} = \frac{c_1}{\tau s + 1} + \frac{c_2}{s^2} + \frac{c_3}{s},$$

with c_1, c_2 and c_3 being the residues, which can be computed as

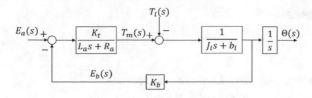

Fig. 5.1 Block diagram of the DC motor model

$$c_1 = Y(s)(\tau s + 1)|_{s=-\frac{1}{\tau}} = kU_0\tau^2, c_2 = Y(s)s^2|_{s=0} = kU_0, c_3 = \frac{d}{ds}[Y(s)s^2]|_{s=0} = -kU_0\tau,$$

so that, from lines 2, 4 and 5 of Table 3.2, the step response is obtained as

$$y(t) = kU_0\tau \left(e^{-\frac{t}{\tau}} - 1 - \frac{t}{\tau} \right), \tag{5.20}$$

so that as t increases, the output follows a ramp response (unbounded response, as the angular position is increasing with a constant slope after the transient).

The transfer function relating the angular velocity with the input voltage is given by

$$G(s) = \frac{\Omega(s)}{E_a(s)} = \frac{k}{\tau s + 1}, \tag{5.21}$$

with $\Omega(t) = \dot{\theta}(t)$ and thus $\Omega(s) = s\Theta(s)$.

An equivalent block diagram of the DC motor model is given in Fig. 5.1. The back EMF is represented by a negative feedback loop as it opposes the applied voltage. The blocks can be obtained by applying the Laplace transform to Eqs. (2.48)–(2.52). This block diagram inherently expresses one of the drawbacks associated with the use of the transfer function, as it "hides" significant internal variables by only focusing on the input–output relationship. This is both the advantage and the weakness of the transfer function representation. All plants with the same input–output characteristics, i.e. the same descriptive differential equations, have the same transfer function. This may represent an advantage when designing control schemes (the same scheme can be used to control systems of different nature), but at the same time it is possible to speculate about the physical origin of the problem [12].

5.6.1.3 References Related to this Concept

- [4] Dorf, R. C., & Bishop, R. H. (2011). *Modern control systems* (12th ed.). Prentice Hall. ISBN: 978-0-13-602458-3. Chapter 2, Example 2.5, pp. 70–74.
- [6] Ogata, K. (2010). *Modern control engineering* (5th ed.). Prentice Hall. ISBN: 978-0-13-615673-4. Chapter 3, Example A-3–9, pp. 95–97.

DC Motor: Transfer Function

In this case, no explicit tool has been developed for the analysis of the time response because when excited with a step input, its output grows indefinitely. It represents an unstable system from the BIBO point of view, and therefore, its analysis only makes sense in Chap. 9 where it will be stabilized through feedback (see Sect. 9.4).

5.6.1.4 Homework

1. What is the mechanical time constant of the motor? And the electrical one? How are they affected by an increase in J_l?
2. What kind of response is expected when entering a unit impulse to the system? Reason about the physical meaning.
3. What kind of response is achieved when exciting it with a unit amplitude sine? Reason about the physical meaning.
4. Add a torsional union between the motor and the load. Develop the model and the block diagram, comparing the result with the nominal one ([12], p. 79).

5. Following the diagram in Fig. 5.1, obtain the transfer function from $T_l(s)$ to $\Theta(s)$.

6. Estimate the parameters of the first-order differential equation describing the angular velocity dynamics of a DC motor that achieves a steady-state angular velocity of 5000 rpm when $e_a = 12$ V and has a time constant of 0.1 s. Can you also estimate the physical parameters J_l, b_l, K_t, K_b, and R_a with this information? Reference [13], Question 2.43.

7. The tool uses the simplified transfer function (5.19) obtained assuming that the armature inductance L_a is generally small and can be neglected. If this is not the case, the transfer function is given by a third-order system with an additional pole (Eq. (5.18)). What is the electrical time constant and the mechanical time constant?

8. Using the interactive tool f_concept (card 4.2), represent the frequency diagrams of the DC motor considering that its output is the angular velocity. Using those diagrams, determine the steady-state regime when the motor is excited with a sinusoidal input of amplitude 1 and frequency $1/\tau$ rad/s.

9. Example from [14], Exercise 3.18. Following the explanation in the introduction of Chap. 3, and considering that the output is the angular velocity of the motor, assume that the applied voltage is zero and that the initial velocity is not zero. Notice that applying a zero voltage at motor terminals is equivalent to putting a short circuit at the armature motor terminals and that a nonzero-induced voltage is present if the velocity is not zero. Depict the free response under these conditions. What can be done to force the free response to vanish faster? What happens with the electric current? Can the free response vanish faster if the armature resistance is closer to zero? Try to relate to the term "dynamic braking"? Why does the time constant depend on motor inertia? What does this mean from the point of view of Newton's Laws of Mechanics? Assume that a voltage different from zero is applied when the initial velocity is zero. Depict the system response. Why does the final velocity not depend on the inertia J_l? Give an interpretation from the point of view of physics.

References

1. Guzmán, J. L., Vargas, H., Sánchez-Moreno, J., Rodríguez, F., Berenguel, M., & Dormido, S. (2007). Education research in engineering studies: Interactivity, virtual and remote labs. In *Distance Education Research Trends* (pp. 131–167). Nova Science Publishers Inc.

2. Johansson, K. H. (2000). The quadruple-tank process: A multivariable laboratory process with an adjustable zero. *IEEE Transactions on Control Systems Technology, 8*(3), 456–465.

3. Åström, K. J., & Murray, R. M. (2014). *Feedback systems: An introduction for scientists and engineers* (2nd ed.). Princeton University Press.

4. Dorf, R. C., & Bishop, R. H. (2011). *Modern control systems* (12th ed.). Prentice Hall.

5. Golnaraghi, F., & Kuo, B. C. (2017). *Automatic control systems* (10th ed.). McGraw Hill Education.

6. Ogata, K. (2010). *Modern control engineering* (5th ed.). Prentice Hall.

7. Tavakolpour-Saleh, A., Setoodeh, A., & Ansari, E. (2016). Iterative learning control of two coupled nonlinear spherical tanks. *International Journal of Mechanical and Mechatronics Engineering, World Academy of Science, Engineering and Technology, 10*(11), 1862–1869.

8. Wellstead, P. E., Chrimes, V., Fletcher, P. R., Moody, R., & Robins, A. J. (1989). Ball and beam control experiment. *The International Journal of Electrical Engineering & Education, 15*, 21–39.

9. Acharya, M., Bhattarai, M., & Poudel, B. (2014). Real time motion assessment for positioning in time and space critical systems. *International Journal of Applied Research and Studies (IJARS), 3*(7), 1–11.

10. Shahian, B., & Hassul, M. (1993). *Control system design using MATLAB®*. Prentice Hall.

11. de Silva, C. W. (2009). *Modeling and control of enginnering systems*. Taylor & Francis Group, CRC Press.

12. Rohrs, C. E., Melsa, J. L., & Schultz, D. G. (1994). *Sistemas de control lineal (Linear control systems)*. McGraw-Hill.

13. de Oliveira, M. C. (2017). *Fundamentals of linear control*. A concise approach: Cambridge University Press.

14. Hernández-Guzmán, V. M., & Silva-Ortigoza, R. (2019). *Automatic control with experiments*. Springer.

6.1   Problem 1. First-Order System

Consider a dynamical system described by the following differential equation:

$$u(t)\frac{dy(t)}{dt} = y^2(t) + y(t)u(t) + u(t).$$

1. Obtain the transfer function describing the behavior of the system around the operating point (also equilibrium point) defined by $(\overline{u}, \overline{y})$.
2. Indicate the range of values of $\overline{u}$ for which the linearized system is stable (local stability).
3. For a value of $\overline{u}$ within the range that guarantees local stability, analytically calculate the time response $y(t)$ of the linearized system around that operating point when a step of amplitude $+0.5$ is introduced at the input. Compare the theoretical response with that obtained using the interactive tool t_first_order (card 3.2).
4. If the system is excited with a sinusoidal signal in the form $\tilde{u}(t) = \sin(t)$, obtain the equation of the output of the system $\tilde{y}(t)$ around the operating point defined by $\overline{u}$ chosen in question 3. Check the obtained result in steady-state regime with the interactive tool f_concept (card 4.2).
5. For $\overline{u}$ selected in question 3, obtain the Bode and Nyquist diagrams of the linearized system. Check the results using the interactive tools f_first_order (card 4.3) and f_concept.

6.1.1   Solution

1. It is a first-order nonlinear system, where the steady-state conditions are obtained by making the time derivatives equal to zero (equilibrium point):

$$0 = \overline{y}^2 + \overline{y}\,\overline{u} + \overline{u} \to \overline{u}(1 + \overline{y}) = -\overline{y}^2 \to \overline{u} = -\frac{\overline{y}^2}{1 + \overline{y}}.$$

Linearizing the differential equation (the time dependence is omitted for the sake of simplicity):

$$f(\dot{y}, y, u) = 0 \to f(\dot{y}, y, u) = u\dot{y} - y^2 - yu - u = 0,$$

$$f(\dot{y}, y, u) \approx \left.\frac{\partial f}{\partial \dot{y}}\right|_{(\overline{u},\overline{y},\overline{\dot{y}})}\dot{\tilde{y}} + \left.\frac{\partial f}{\partial y}\right|_{(\overline{u},\overline{y},\overline{\dot{y}})}\tilde{y} + \left.\frac{\partial f}{\partial u}\right|_{(\overline{u},\overline{y},\overline{\dot{y}})}\tilde{u} = \overline{u}\dot{\tilde{y}} - (2\overline{y} + \overline{u})\tilde{y} + (-\overline{y} - 1)\tilde{u},$$

$$\overline{u}\dot{\tilde{y}} = (2\overline{y} + \overline{u})\tilde{y} + (\overline{y} + 1)\tilde{u}.$$

Once the linearized ODE is obtained, the Laplace transform can be applied with zero initial conditions to obtain the transfer function $G(s)$, as the quotient of the Laplace transform of the output $Y(s)$ and that of the input $U(s)$:

$$\overline{u}sY(s) = (2\overline{y} + \overline{u})Y(s) + (\overline{y} + 1)U(s),$$

Supplementary Information The online version contains supplementary material available at https://doi.org/10.1007/978-3-031-09920-5_6.

$$Y(s)\,(\overline{u}s - (2\overline{y} + \overline{u})) = U(s)(\overline{y} + 1),$$

$$G(s) = \frac{Y(s)}{U(s)} = \frac{\overline{y} + 1}{\overline{u}s - (2\overline{y} + \overline{u})} = \frac{-\frac{\overline{y}+1}{2\overline{y}+\overline{u}}}{-\frac{\overline{u}}{2\overline{y}+\overline{u}}s + 1}.$$

2. For a linear system to be stable, its poles (roots of the characteristic equation) must lie in the LHP, that is, their real part must be negative. So, in this example $s = (2\overline{y} + \overline{u})/\overline{u}$ and using the steady-state relationship $\overline{u} = -\overline{y}^2/(1 + \overline{y})$, it is obtained:

$$s = \frac{-\sqrt{\overline{u}^2 - 4\overline{u}}}{\overline{u}} < 0, \ \overline{u}(\overline{u} - 4) > 0.$$

The intervals ensuring stability (depending on the selected sign of the square root) are $\overline{u} \in (-\infty, 0) \cup (4, \infty)$.

3. Using the transfer function in time constants format $G(s) = \frac{k}{\tau s + 1}$, the response $y(t)$ can be obtained, with $k = -\frac{\overline{y}+1}{2\overline{y}+\overline{u}}$ and $\tau = -\frac{\overline{u}}{2\overline{y}+\overline{u}}$:

$$Y(s) = G(s)U(s) \rightarrow Y(s) = \frac{k/\tau}{s + 1/\tau} \cdot \frac{0.5}{s} \rightarrow Y(s) = \frac{0.5k}{s} + \frac{-0.5k}{s + 1/\tau}.$$

Using the inverse Laplace transform:

$$Y(s) = \frac{0.5k}{s} - \frac{0.5k}{s + 1/\tau} \rightarrow \tilde{y}(t) = 0.5k(1 - e^{-\frac{t}{\tau}}).$$

As $\tilde{y}(t)$ is a deviation variable, the real translation to the operating point defined by $(\overline{u} = 4.5, \overline{y} = -3)$, for which $k = -4/3$, and $\tau = 3$ is given by:

$$\tilde{y}(t) = -0.6667 + 0.6667e^{-\frac{t}{3}} \rightarrow y(t) = -3.6667 + 0.6667e^{-\frac{t}{3}}.$$

Notice that the interactive tool t_first_order only works on the deviation variable describing the linear part of the response (Fig. 6.1). Some representative values can be verified using the tool: $\tilde{y}(0) = 0$, $\tilde{y}(3) = -0.421$, $\tilde{y}(\infty) = -0.6667$.

4. For a sinusoidal input, using decomposition into single fractions as follows:

$$Y(s) = G(s)U(s) = \frac{-4/3}{3s + 1} \cdot \frac{1}{s^2 + 1^2} = \frac{-0.4}{s + 1/3} + \frac{0.4s - 0.1333}{s^2 + 1}.$$

The time response using the inverse Laplace transform can be obtained as follows:

$$\tilde{y}(t) = -1.2\,e^{-\frac{t}{3}} + 0.4\cos(t) - 0.1333\sin(t).$$

When reaching the steady-state regime, the exponential term vanishes and the response is that provided by the cosine and sine functions. From the frequency response concept, it is known that in steady-state regime, the output signal must be a sinusoidal one $y(t \rightarrow \infty) = B\sin(\omega t + \phi)$, with the same frequency that the input signal ($\omega = 1$ rad/s), amplitude equal to $B = |G(j)| = 0.4217$ (the amplitude of the input sine is one) and phase lag $\phi = \lfloor G(j) = 108.43° = 1.8925$ rad (see next exercise). Using the properties of trigonometric functions (sum of angles), it can be seen that $y(t) = 0.4\cos(t) - 0.1333\sin(t) = 0.4217\sin(t + 1.8925)$. This can be analyzed in the steady-state sinusoidal responses provided by the interactive tool f_concept (card 4.2), as shown in Fig. 6.2, where the transfer function is represented in ZPK format.

5. The frequency response is first analytically computed using: $G(s) = -4/3/(3s + 1)$.

$$\text{Gain:}\ \ k = -4/3 \rightarrow \begin{cases} |G(j\omega)|_{dB} = 20\log(|-4/3|) = 2.493 \text{ dB}, \\ \lfloor G(j\omega) = \arctan\left(\frac{0}{-4/3}\right) = 180°. \end{cases}$$

$$\text{Pole:}\ \ \frac{1}{3s + 1} \xrightarrow{s=j\omega} \frac{1}{1 + j3\omega} \cdot \frac{1 - j3\omega}{1 - j3\omega} = \frac{1 - j3\omega}{1 + 9\omega^2} \rightarrow \begin{cases} \text{Re}(G(j\omega)) = \frac{1}{1+9\omega^2}, \\ \text{Im}(G(j\omega)) = \frac{-3\omega}{1+9\omega^2}, \end{cases}$$

$$|G(j\omega)|_{dB} = -10\log(1 + 9\omega^2), \ \lfloor G(j\omega) = \arctan(-3\omega)/1.$$

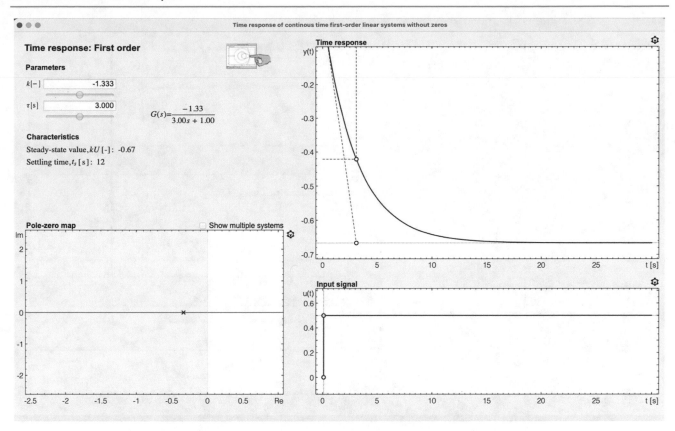

Time response: First order

Parameters

$k[-]$ -1.333

$\tau[s]$ 3.000

$$G(s) = \frac{-1.33}{3.00s + 1.00}$$

Characteristics

Steady-state value, kU [-]: -0.67

Settling time, t_s [s]: 12

Pole-zero map   □ Show multiple systems

Time response

Input signal

Fig. 6.1 Problem 1: Response $\tilde{y}(t)$ to a step of amplitude 0.5 using the interactive tool t_first_order

Table 6.1 includes some representative values for the pole contribution.

The magnitude plot has a contribution given by the static gain (constant equal to 2.493 dB) and that provided by the real pole, with 0 dB at low frequency, -3 dB at $\omega = 1/3$ rad/s, which is the corner and cutoff frequency from which the magnitude asymptote decreases linearly with a slope of -20 dB/decade. At $\omega = 1$ rad/s the magnitude is 0.4217 (-7.5 dB), as commented in the previous exercise. The phase has the contribution of the gain (180° by convention, as it is negative) and that of the real pole following the arctangent profile, starting in 0° at low frequencies, $-45°$ at $\omega = 1/3$ rad/s and $-90°$ when $\omega \to \infty$. At $\omega = 1$ rad/s the phase is 108.43°, as has been also mentioned in the previous exercise. Figure 6.3 shows the Bode diagram.

Regarding the Nyquist diagram, using the real and imaginary parts of $G(j\omega)$, some representative values are obtained in Table 6.2. Figure 6.4 shows the Nyquist diagram.

6.2 Problem 2. Second-Order System

A dynamical system is described by the following differential equation:

$$\frac{d^2 y(t)}{dt^2} + u(t)\frac{dy(t)}{dt} + y^2(t) = u^2(t).$$

1. Obtain the transfer functions describing the behavior of the system around the operating points defined by $\bar{u} = 1.667$ (the reader can select a different value). Select the transfer function that represents the locally-stable linear system.
2. Calculate analytically the unit step response, and check it with the interactive tool t_second_order (card 3.3). Compute the static gain, peak overshoot, peak time, rise time, settling time, and compare them with those shown in the interactive tool.

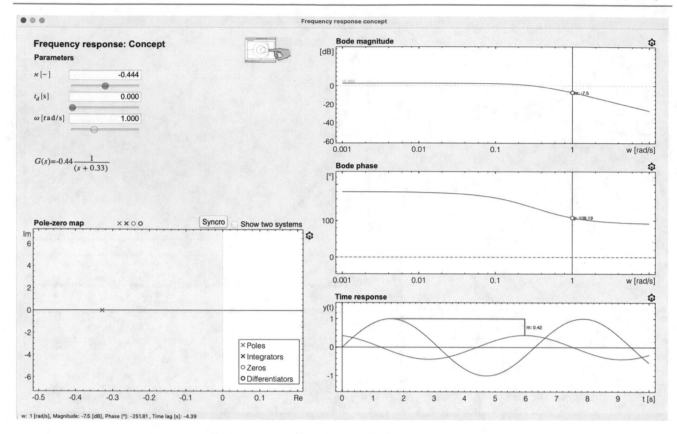

Fig. 6.2 Problem 1: Response to sinusoidal input using the interactive tool f_concept

Table 6.1 Problem 1: Representative values of the pole contribution in the Bode diagram

$\omega = 0$ rad/s	$\omega = 1/3$ rad/s	$\omega = 1$ rad/s	$\omega \to \infty$ rad/s
$\lvert G(j0) \rvert$ dB $= 0$,	$\lvert G(j\frac{1}{3}) \rvert$ dB $= -3$,	$\lvert G(j1) \rvert$ dB $= -10$,	$\lvert G(j\infty) \rvert$ dB $= -\infty$,
$\lfloor G(j0) = 0°$,	$\lfloor G(j\frac{1}{3}) = -45°$,	$\lfloor G(j1) = -71.57°$,	$\lfloor G(j\infty) = -90°$.

3. Calculate the range of values of $\overline{u}$ providing different kinds of responses: overdamped, critically damped, underdamped, critically stable and unstable. Check the results with the interactive tool t_second_order.

4. For $\overline{u} = 1.667$, calculate the associated peak magnitude M_r [dB] and the resonant frequency ω_r [rad/s]. Compare the theoretical results with those provided by the interactive tool f_second_order (card 4.4) and comment on the results.

5. Plot the Nyquist diagram of the same transfer function used in the previous exercise. Using the interactive tool f_concept (card 4.2), build the same transfer function using the pole-zero editor and repository. Draw the polar diagram, indicating the location of the points of zero and infinite frequencies, and those corresponding to peak magnitude, resonant frequency and natural frequency.

6. Analyze the effect of adding a time delay to the frequency response of the system, both in Bode and Nyquist diagrams.

6.2.1 Solution

1. This second-order nonlinear system must be linearized to obtain its transfer function. First, the steady-state analysis is carried out using $\overline{u} = 1.667$ and making the time derivatives equal to 0: $\overline{y}^2 = \overline{u}^2 \to \overline{y} = \pm\overline{u}$. Linearizing the differential equation, taking into account that $\overline{\dot{y}} = 0$ (derivative of a constant):

$$f(\ddot{y}, \dot{y}, y, u) = \ddot{y} + u\dot{y} + y^2 - u^2 = 0,$$

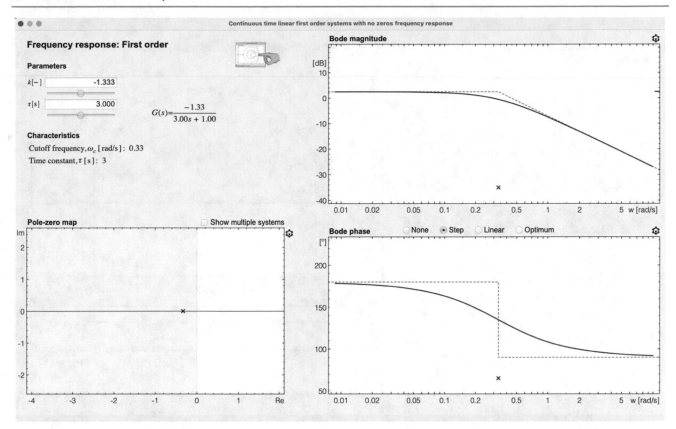

Fig. 6.3 Problem 1: Bode diagram using the interactive tool f_first_order

Table 6.2 Problem 1: Representative values of the pole contribution in the Nyquist diagram

$\omega = 0$ rad/s	$\omega = 1/3$ rad/s	$\omega = 1$ rad/s	$\omega \to \infty$ rad/s
$\mathrm{Re}(G(j0)) = -\frac{4}{3}$,	$\mathrm{Re}(G(j\frac{1}{3})) = -\frac{4}{16}$,	$\mathrm{Re}(G(j1)) = -\frac{4}{30}$,	$\mathrm{Re}(G(j\infty)) = 0$,
$\mathrm{Im}(G(j0)) = 0$,	$\mathrm{Im}(G(j\frac{1}{3})) = \frac{4}{6}$,	$\mathrm{Im}(G(j1)) = \frac{4}{10}$,	$\mathrm{Im}(G(j\infty)) = 0$.

$$f \approx \left.\frac{\partial f}{\partial \ddot{y}}\right|_{(\overline{u},\overline{y})} \ddot{\tilde{y}} + \left.\frac{\partial f}{\partial \dot{y}}\right|_{(\overline{u},\overline{y})} \dot{\tilde{y}} + \left.\frac{\partial f}{\partial y}\right|_{(\overline{u},\overline{y})} \tilde{y} + \left.\frac{\partial f}{\partial u}\right|_{(\overline{u},\overline{y})} \tilde{u},$$

$$f \approx \ddot{\tilde{y}} + \overline{u}\dot{\tilde{y}} + 2\overline{y}\tilde{y} - 2\overline{u}\tilde{u},$$

$$\frac{d^2\tilde{y}(t)}{dt} + \overline{u}\frac{d\tilde{y}(t)}{dt} + 2\overline{y}\tilde{y}(t) = 2\overline{u}\tilde{u}(t),$$

$$\downarrow \mathscr{L}$$

$$s^2 Y(s) + \overline{u}s Y(s) + 2\overline{y}Y(s) = 2\overline{u}U(s),$$

$$G(s) = \frac{Y(s)}{U(s)} = \frac{2\overline{u}}{s^2 + \overline{u}s + 2\overline{y}}.$$

It can be seen that it is a strictly causal system, since the order of the polynomial of the numerator is smaller than that of the denominator. This could also be deduced from the differential equation, since the order of the derivatives of the input is lower than the order of the derivatives of the output. The steady-state values are given by $(\overline{u}, \overline{y}) = (1.667, \pm 1.667)$. Due to the square root there are two possible values of $\overline{y}$, only one of them should make sense, which is that providing a stable system ($\overline{y} = 1.667$):

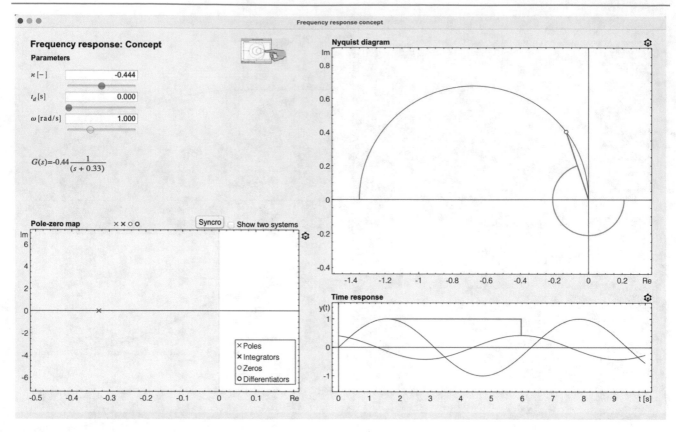

Fig. 6.4 Problem 1: Nyquist diagram using the interactive tool f_concept

$$G(s) = \frac{3.3334}{s^2 + 1.6667\, s + 3.3334}\,.$$

The roots of the characteristic polynomial are two complex conjugates: $p_1 = -0.83 + 1.6j$ and $p_1^* = -0.83 - 1.6j$, with negative real part (these poles lie in the LHP), resulting in a stable system. It can be easily checked that using $\bar{y} = -1.667$ results in an unstable system with a pole in the RHP.

2. The unit step response is analyzed by making $U(s) = 1/s$ and decomposing into simple fractions:

$$G(s) = \frac{Y(s)}{U(s)} \rightarrow Y(s) = G(s)U(s) \rightarrow Y(s) = \frac{2\bar{u}}{(s^2 + \bar{u}s + 2\bar{y})}\frac{1}{s}\,,$$

$$Y(s) = \frac{c}{s} + \frac{ds + f}{s^2 + \bar{u}s + 2\bar{y}} = \frac{cs^2 + c\bar{u}s + c2\bar{y} + ds^2 + fs}{(s^2 + \bar{u}s + 2\bar{y})s}\,.$$

By combining both equations the values of c, d and f (residues) can be obtained as follows:
- $s^2 \rightarrow c + d = 0$,
- $s^1 \rightarrow c\bar{u} + f = 0$,
- $s^0 \rightarrow c2\bar{y} = 2\bar{u}$.

Therefore, solving the system, the following values are obtained: $c = \bar{u}/\bar{y}$, $d = -\bar{u}/\bar{y}$ and $f = -\bar{u}^2/\bar{y}$. As $\bar{u} = \bar{y} = 1.667 \rightarrow c = 1, d = -1, f = -\bar{u}$ and $Y(s)$ is given by:

$$Y(s) = \frac{\bar{u}/\bar{y}}{s} + \frac{-\bar{u}/\bar{y}s - \bar{u}^2/\bar{y}}{s^2 - \bar{u}s + 2\bar{y}} = \frac{1}{s} - \frac{s + \bar{u}}{s^2 + \bar{u}s + 2\bar{u}}\,.$$

The inverse Laplace transform of $1/s$ is 1 (see Table 3.2), but the second term must be operated on to arrive at some known expression. First, it is written in factored form, knowing that its roots are $-0.83 \pm 1.6j$:

$$\frac{s+\overline{u}}{s^2+\overline{u}s+2\overline{u}} = \frac{s+1.6667}{(s+0.83+1.62\,j)(s+0.83-1.62\,j)} = \frac{s+1.6667}{(s+0.83)^2+1.62^2}\ .$$

It can be observed that the equation has the form $\dfrac{s+2a}{(s+a)^2+\omega^2}$, where $a = 0.83$ and $\omega = 1.62$. Therefore, it is split into two terms (one for $(s+a)$ and another for a), and the second is multiplied and divided by ω, as shown below. Thus, one expression will be obtained corresponding to the damped sine and another to the damped cosine:

$$\frac{s+a}{(s+a)^2+\omega^2} + \frac{a}{\omega}\frac{\omega}{(s+a)^2+\omega^2},$$

and using Table 3.2, the following expression is obtained:

$$Y(s) = \frac{1}{s} - \frac{s+0.83}{(s+0.83)^2+1.62^2} - \frac{0.83}{1.62}\frac{1.62}{(s+0.83)^2+1.62^2},$$

$$\downarrow \mathscr{L}^{-1}$$

$$\tilde{y}(t) = 1 - e^{-0.83t}\cos(1.62t) - \left(\frac{0.83}{1.62}\right)e^{-0.83t}\sin(1.62t).$$

Figure 6.5 shows the evolution of $\tilde{y}(t)$ using the interactive tool t_second_order. Selecting representative points in the time response (for different values of t) it is possible to verify that the obtained expression is correct. Notice that for $\overline{y} = -1.667$, the time response would growth indefinitely as the system is unstable.

3. By comparing the obtained transfer function with that of a standard or normalized second-order system, the parameters k, ζ and ω_n can be easily obtained.

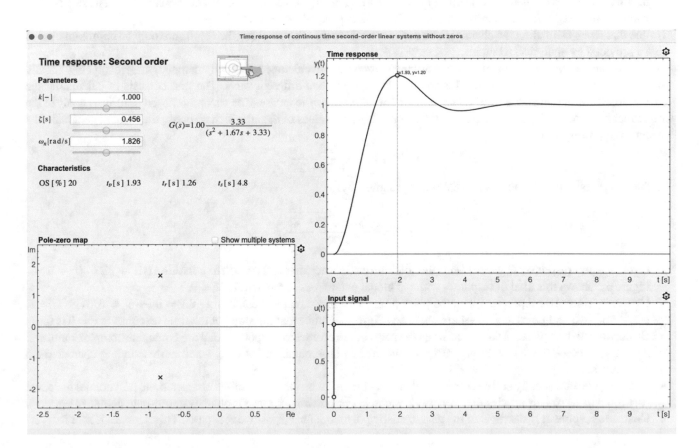

Fig. 6.5 Problem 2: Step response $\tilde{y}(t)$ using the interactive tool t_second_order

$$G(s) = \frac{3.3334}{s^2 + 1.6667\,s + 3.3334} = \frac{k\omega_n^2}{s^2 + 2\zeta\omega_n s + \omega_n^2},$$

so that $\omega_n = \sqrt{3.3334} = 1.826$ rad/s, $\zeta = \frac{1.6667}{2\omega_n} = 0.456$, and $k = 1$.

- **Maximum percentage overshoot:** $OS\,[\%] = 100\exp\left(-\frac{\zeta\pi}{\sqrt{1-\zeta^2}}\right) \rightarrow OS\,[\%] = 20\%$.

- **Peak time:** $t_p = \dfrac{\pi}{\omega_n\sqrt{1-\zeta^2}} \rightarrow t_p = 1.933$ s.

- **Rise time:** $t_r = \dfrac{\pi - \phi}{\omega_n\sqrt{1-\zeta^2}} = \dfrac{\pi - \arccos(\zeta)}{\omega_n\sqrt{1-\zeta^2}} \rightarrow t_r = 1.258$ s.

- **Settling time:** $t_s = \dfrac{4}{\zeta\omega_n} \rightarrow t_s = 4.804$ s.

These values can also be estimated from the time response plots of the interactive tool t_second_order shown before, as follows:

- **Maximum overshoot:** For its estimation, the maximum output y_{max} and steady-state y_∞ values must be observed. As the system has unit static gain, $y_\infty = 1.0$ and $y_{max} = 1.2$, in the figure above. The peak percentage overshoot is calculated as follows: $OS\,[\%] = 100 \cdot \frac{y_{max}-y_\infty}{y_\infty} = 20\%$.

- **Peak time:** It is defined as the time it takes for the system output to reach its maximum value, measured from the instant when the output starts to evolve (all characteristic times are calculated from this time instant). Analyzing the plot, $t_p \approx 1.93$ s (remember that when placing the mouse over the time response in the interactive tool, a label showing both the time and magnitude of the signal is displayed).

- **Rise time:** In this case computed as the time the plant output takes for reaching the steady-state value for the first time, $t_r \approx 1.25$ s.

- **Settling time:** It is computed as the time after which the plant output is confined in a band around $\pm 2\%$ of its steady-state value (in this case the time elapsed until the value 0.98 is reached), $t_s \approx 4.6$ s.

Notice that when estimating the characteristic parameters from the time response the results are only an approximation to those provided by analytical formulas.

4. Next, the range of values of $\bar{u}$ that produce different responses (overdamped, critically damped, underdamped, critically stable, and unstable) are calculated. This can be developed in two different ways. The first consists of calculating the expression for ζ and analyzing the values it takes for each type of response. The second (chosen one) is to analyze the value that the roots of the denominator would take for different values of $\bar{u}$, and which response they would generate. The system is represented by:

$$G(s) = \frac{2\bar{u}}{s^2 + \bar{u}s + 2\bar{u}},$$

so that the roots of the characteristic polynomial are given by:

$$p_{1,2} = \frac{-\bar{u} \pm \sqrt{\bar{u}^2 - 8\bar{u}}}{2}.$$

- **Overdamped response:** Obtained when the poles are distinct negative real roots. The condition is $\bar{u}^2 - 8\bar{u} > 0 \rightarrow \bar{u} > 8$. Figure 6.6 shows this kind of response, with the model parameters computed for $\bar{u} = 9$.

- **Critically damped response:** When the poles of the system are real and equal $\bar{u}^2 - 8\bar{u} = 0 \rightarrow \bar{u} = 8$. With $\bar{u} = 8$, $\zeta = 1$ and two real poles at $s = -4$ are obtained (Fig. 6.7). Notice that the slope of the time response at $t = 0$ is 0.

- **Underdamped response:** Is obtained when the roots of the characteristic polynomial are conjugate complex numbers: $\bar{u}^2 - 8\bar{u} < 0 \rightarrow 0 < \bar{u} < 8$. Figure 6.8 shows the response obtained for $\bar{u} = 1$ and corresponding characteristic parameters.

- **Critically or marginally stable response:** It is that obtained when the roots of the characteristic polynomial have real part equal to zero. This implies $\bar{u} = 0$, but in this case a system without sense is obtained (zero-gain double integrator).

- **Unstable response:** If the roots of the characteristic polynomial have positive real parts. This happens if $\bar{u} < 0$.

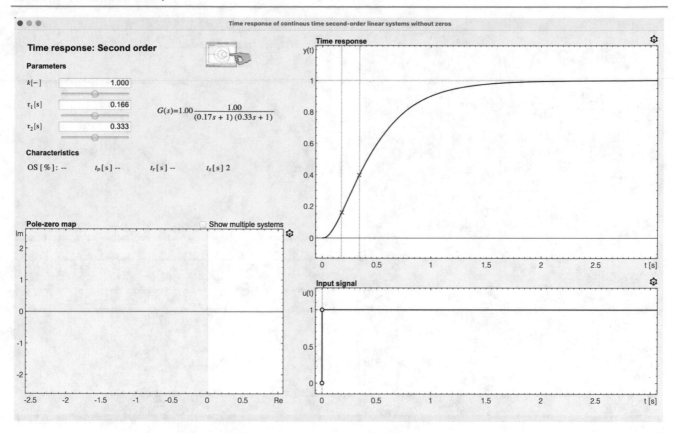

Fig. 6.6 Problem 2: Overdamped response using the interactive tool t_second_order

5. In order to calculate the peak magnitude M_r [dB] and the resonant frequency ω_r [rad/s] associated with the stable system, the parameters computed in the previous question are used ($k = 1$, $\omega_n = 1.826$ rad/s and $\zeta = 0.456$). The peak magnitude is obtained from the following equation:

$$M_r = \frac{1}{2\zeta\sqrt{1-\zeta^2}} = \frac{1}{2\cdot 0.456\sqrt{1-0.456^2}} = 1.232,$$

and converted into dB: M_r [dB] $= 20\log(1.232) = 1.8125$.
The resonant frequency is computed at the peak magnitude:

$$\omega_r = \omega_n\sqrt{1 - 2\zeta^2} = 1.8258\sqrt{1 - 2(0.456^2)} = 1.395 \text{ rad/s}.$$

These results can be verified using the interactive tool f_second_order (card 4.4), where it can be seen that the characteristic parameters are the same obtained analytically (Fig. 6.9). It is important to notice here that if the system would not have static gain equal to one, M_r should be measured as the difference between the peak magnitude in the Bode diagram and the low-frequency value of the magnitude. The reader can verify this just by changing $k = 2$ and measuring this difference defining M_r [dB] in the Bode diagram.

6. The Nyquist diagram of the linearized system is analyzed both analytically and using the interactive tool f_concept. By substituting $s = j\omega$ in the transfer function:

$$G(s) = \frac{3.3334}{s^2 + 1.6667\,s + 3.3334} \rightarrow G(j\omega) = \frac{1}{\left(\frac{j\omega}{\omega_n}\right)^2 + \left(\frac{2\zeta\omega}{\omega_n}\right)j + 1},$$

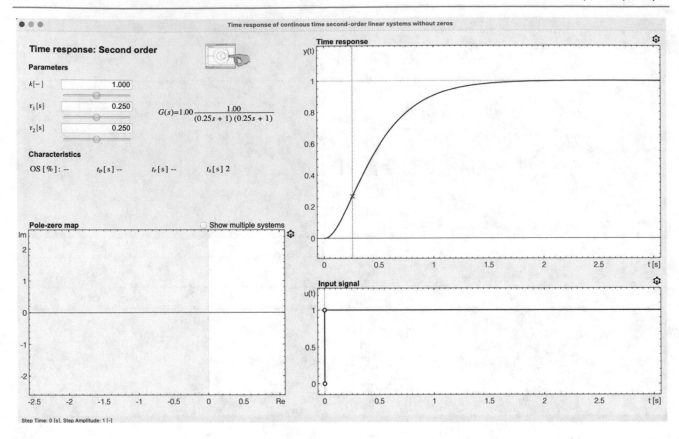

Fig. 6.7 Problem 2: Critically damped response using the interactive tool t_second_order

$$G(j\omega) = \frac{1}{1 - \left(\frac{\omega}{\omega_n}\right)^2 + j\frac{2\zeta\omega}{\omega_n}} \cdot \frac{1 - \left(\frac{\omega}{\omega_n}\right)^2 - j\frac{2\zeta\omega}{\omega_n}}{1 - \left(\frac{\omega}{\omega_n}\right)^2 - j\frac{2\zeta\omega}{\omega_n}} = \frac{1 - \left(\frac{\omega}{\omega_n}\right)^2 - j\frac{2\zeta\omega}{\omega_n}}{\left[1 - \left(\frac{\omega}{\omega_n}\right)^2\right]^2 + \frac{4\zeta^2\omega^2}{\omega_n^2}}.$$

The real and imaginary parts of $G(j\omega)$ are given by:

$$\text{Re}(G(j\omega)) = \frac{1 - \frac{\omega^2}{\omega_n^2}}{\left[1 - \left(\frac{\omega}{\omega_n}\right)^2\right]^2 + \frac{4\zeta^2\omega^2}{\omega_n^2}}; \quad \text{Im}(G(j\omega)) = \frac{-\frac{2\zeta\omega}{\omega_n}}{\left[1 - \left(\frac{\omega}{\omega_n}\right)^2\right]^2 + \frac{4\zeta^2\omega^2}{\omega_n^2}}.$$

Computing $\text{Re}(G(j\omega))$ and $\text{Im}(G(j\omega))$ for frequency values of interest, it is obtained:

- $\omega = 0$ rad/s; Re $= 1$, Im $= 0$.
- $\omega \to \infty$; Re $= 0$, Im $= 0$.
- $\omega = \omega_r = 1.395$ rad/s; Re $= 0.6312$; Im $= 1.059$.
- $\omega = \omega_n = 1.826$ rad/s; Re $= 0$; Im $= -1.096$.

To define the corresponding transfer function in the interactive tool f_concept, the user has to drag the poles and zeros to their corresponding location in the s-plane and select the Nyquist plot as representative diagram, so that the previous calculations can be verified in the tool (Fig. 6.10).

7. A time delay is translated into a term of the form $e^{-t_d s}$ multiplying the transfer function:

$$G(j\omega) = \frac{e^{-jt_d\omega}}{1 - \frac{\omega^2}{\omega_n^2} + \frac{2\zeta\omega}{\omega_n}j} = \frac{1 - \left(\frac{\omega}{\omega_n}\right)^2 - \frac{2\zeta\omega}{\omega_n}j}{\left[1 - \left(\frac{\omega}{\omega_n}\right)^2\right]^2 + \frac{4\zeta^2\omega^2}{\omega_n^2}}e^{-jt_d\omega}.$$

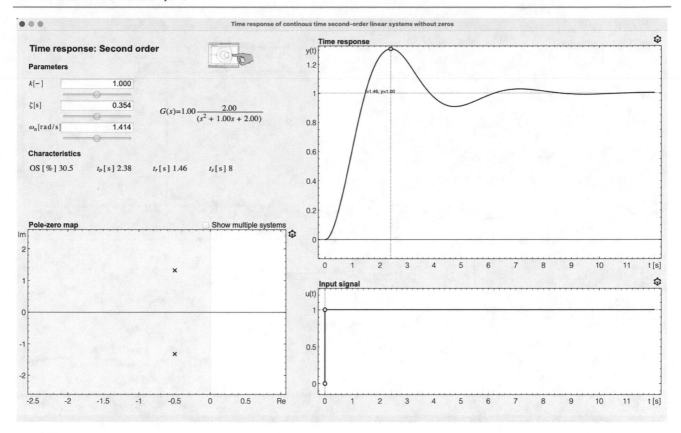

Fig. 6.8 Problem 2: Underdamped response using the interactive tool t_second_order

In the Bode magnitude plot, using the properties of logarithms,

$$|G(j\omega)|_{\text{dB}}^{\text{delay}} = |G(j\omega)|_{\text{dB}} + 20\log(|e^{-jt_d\omega}|),$$

it is easy to see that the term $e^{-jt_d\omega}$ is a complex number of modulus 1 and phase $-t_d\omega$, so that the second term on the right is zero and the time delay does not affect the magnitude curve of the Bode diagram. Regarding the phase plot, it is formed by the addition of the phases of each term, so that the phase due to the time delay must be added to that of the system without delay:

$$e^{-jt_d\omega} = \cos(t_d\omega) - j\sin(t_d\omega) \rightarrow \lfloor e^{-jt_d\omega} = \arctan\left(\frac{-\sin(t_d\omega)}{\cos(t_d\omega)}\right) = -t_d\omega.$$

The time delay contributes $-t_d\omega$ rad to all the points of the Bode diagram, leading it towards more negative phases. It should be clarified that the calculated delay is in radians, so to convert it to degrees, it must be multiplied by $\frac{180}{\pi}$. If the frequency scale were not logarithmic, the effect of the delay would be a straight line with slope $-t_d$, but its effect in the Bode plot is a logarithmic decrease, as shown in Fig. 6.11, where both the frequency response of the system with and without delay are shown using the interactive tool f_generic (card 4.7). It can be seen that, while the magnitude is unaffected, the phase shift considerably increases as the frequency does. This will have serious implications in closed-loop systems (in open loop, the time delay does not affect stability, it only causes a time displacement in the output signal with respect to the input signal).

In the Nyquist diagram, the time delay generates a logarithmic spiral. This is because, as frequency increases, the magnitude decreases but the phase lag increases (towards more negative values). Therefore, a point that previously had a certain magnitude and a certain phase will now have the same magnitude but a larger (more negative) phase lag, so this point will have shifted (rotated). Therefore, the Nyquist diagram of a system with time delay is in the form of a logarithmic spiral, which will end at the origin when the magnitude is zero. Figure 6.12 shows the effect produced by the delay in a

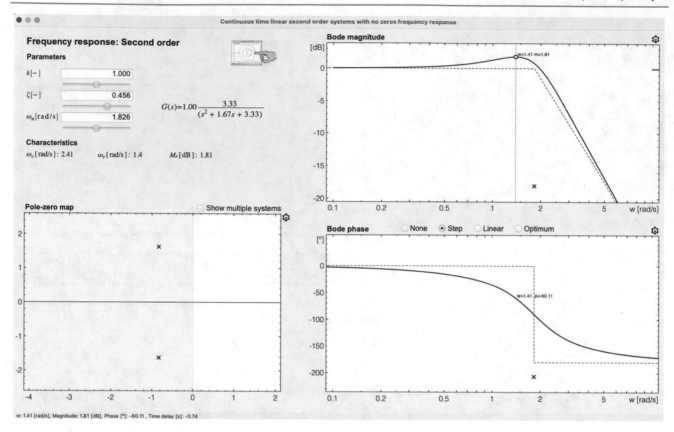

Fig. 6.9 Problem 2: Peak magnitude and resonant frequency using the interactive tool f_second_order

Nyquist diagram, which helps understanding by visualization. It can be seen, in addition to the logarithmic spiral shape, that for the same frequency, the corresponding point is rotated clockwise but with the same magnitude.

6.3   Problem 3. First-Order System with a Zero

Consider de dynamical system described by:

$$u(t)\frac{dy(t)}{dt} + y(t) = -0.5y(t)\frac{du(t)}{dt} + 2u(t).$$

1. Obtain the transfer function describing the behavior of the system about an operating point defined by constant values of the input and the output $(\overline{u}, \overline{y})$, expressing the parameters as a function of $\overline{u}$.
2. Indicate the range of values of $\overline{u}$ for which the linearized system, represented by the previous transfer function, is stable.
3. Provide the range of values of $\overline{u}$ for which the linearized process presents inverse response (NMP).
4. For $\overline{u} = 1.667$ (other values can be selected), obtain the analytical expression of the time response $\tilde{y}(t)$ of the linearized system when it is excited with a step input of amplitude 2. Compare the theoretical response with that given by the interactive tool t_first_order_zero (card 3.4).
5. For the same value of $\overline{u}$, calculate the Bode and Nyquist diagrams, and check them with those obtained using the interactive tools f_first_order_zero (card 4.5), f_generic (card 4.7) and f_nonminimum_phase (card 4.8).

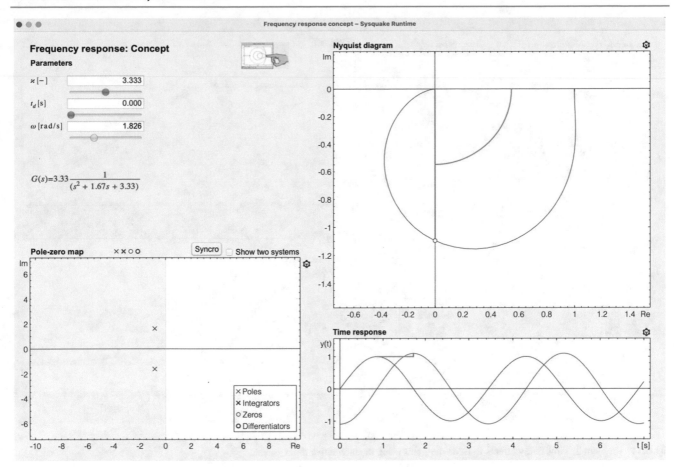

Fig. 6.10 Problem 2: Nyquist diagram using the interactive tool f_concept

6.3.1 Solution

1. For this nonlinear system the steady-state analysis provides $\overline{y} = 2\overline{u}$. Linearizing applying Taylor series expansion about $(\overline{u}, \overline{y})$ and taking into account that $\dot{\overline{u}} = \dot{\overline{y}} = 0$ (as $\overline{u}$ and $\overline{y}$ are constant):

$$f(\dot{y}, y, \dot{u}, u) = u\dot{y} + y + 0.5y\dot{u} - 2u = 0, \quad f \approx \frac{\partial f}{\partial \dot{y}}\bigg|_{(\overline{u},\overline{y})} \dot{\tilde{y}} + \frac{\partial f}{\partial y}\bigg|_{(\overline{u},\overline{y})} \tilde{y} + \frac{\partial f}{\partial \dot{u}}\bigg|_{(\overline{u},\overline{y})} \dot{\tilde{u}} + \frac{\partial f}{\partial u}\bigg|_{(\overline{u},\overline{y})} \tilde{u},$$

$$f \approx \overline{u}\dot{\tilde{y}} + (1 + 0.5\dot{\overline{u}})\tilde{y} + (\dot{\overline{y}} - 2)\tilde{u} + 0.5\overline{y}\dot{\tilde{u}}, \quad f \approx \overline{u}\dot{\tilde{y}} + \tilde{y} - 2\tilde{u} + \overline{u}\dot{\tilde{u}}, \quad \overline{u}\frac{d\tilde{y}(t)}{dt} + \tilde{y}(t) = 2\tilde{u}(t) - \overline{u}\frac{d\tilde{u}(t)}{dt},$$

$$\downarrow \mathscr{L}$$

$$\overline{u}sY(s) + Y(s) = 2U(s) - \overline{u}sU(s) \rightarrow G(s) = \frac{Y(s)}{U(s)} = \frac{-\overline{u}s + 2}{\overline{u}s + 1}.$$

It can be seen that it is a causal system, since the order of the numerator and the denominator are the same. This conclusion can also be reached by analyzing the initial differential equation, since the orders of the derivatives of the input and the output are the same. So, there will be a direct transmission of the input signal to the output signal.

2. For the system to be stable, its pole must lie in the LHP (negative real part). It can be easily observed from the transfer function that the pole is placed at $s = -1/\overline{u}$ and therefore the condition is $\overline{u} \in (0; \infty)$.

3. A system is NMP when it presents a zero or a pole in the RHP or when it has a time delay (in this case, no delay is present). In order to determine the conditions for having an inverse response, knowing that the zeros are the roots of the numerator of the transfer function:

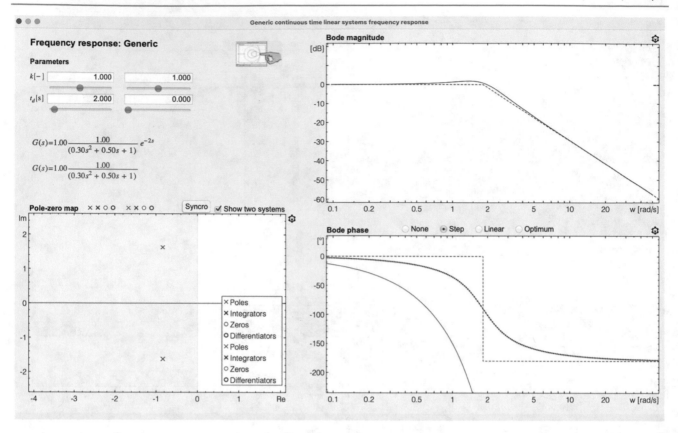

Fig. 6.11   Problem 2: Time delay effects in Bode diagrams using the interactive tool f_generic

$$-\overline{u}s + 2 = 0 \rightarrow s = \frac{2}{\overline{u}} > 0 \rightarrow \overline{u} > 0.$$

Therefore, the system will have inverse response (thus being NMP) as long as the steady-state value of the input defining the operating point is positive. Notice that if $\overline{u} < 0$, the zero will be located in the LHP, but then the pole will lie in the RHP providing an unstable system (and thus NMP). For $\overline{u} = 0$ the system is static (the transfer function is a gain).

4. A step input signal of amplitude 2 is represented by $2/s$ in Laplace transform, so that:

$$Y(s) = G(s)U(s) = \frac{-\overline{u}s + 2}{\overline{u}s + 1}\frac{2}{s} = \frac{c}{s} + \frac{d}{\overline{u}s + 1} = \frac{c\overline{u}s + ds + c}{s(\overline{u}s + 1)},$$

$Y(s)$ and $U(s)$ being the Laplace transforms of $\tilde{y}$ and $\tilde{u}$ respectively. Equating the terms in s^0 and s^1, it follows that $c = 4$ and $d = -6\overline{u}$. So the final expression of $Y(s)$ results as follows:

$$Y(s) = \frac{4}{s} - \frac{6\overline{u}}{\overline{u}s + 1} = \frac{4}{s} - 6\frac{1}{s + 1/\overline{u}} \xrightarrow{\mathscr{L}^{-1}} \tilde{y}(t) = 4 - 6e^{-\frac{t}{\overline{u}}}.$$

Particularizing this response for $\overline{u} = 1.667$, $\tilde{y}(t) = 4 - 6e^{-\frac{t}{1.667}}$. Three characteristic points of the equation are:

- $t = 0 \rightarrow \tilde{y}(0) = -2$,
- $t \rightarrow \infty \rightarrow \tilde{y}(\infty) \rightarrow 4$,
- $\tilde{y} = 0 \rightarrow t = -1.667 \ln(4/6) = 0.6758$.

The response obtained using the interactive tool t_first_order_zero (card 3.4) is shown in Fig. 6.13. It can be seen that the calculated points coincide. Furthermore, the zero of the system is in the RHP and there is a direct transmission of the input to the output so that at time $t = 0$ the output is not 0, as expected.

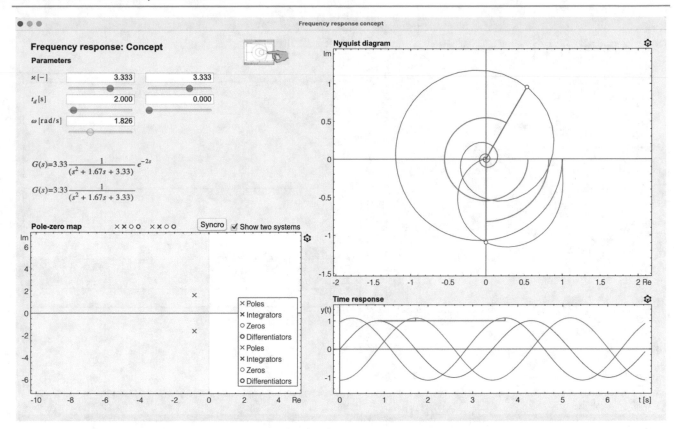

Fig. 6.12 Problem 2: Time delay effects in Nyquist diagram using the interactive tool f_concept

5. The Bode diagram can be determined by analyzing the effect of poles and zeros. The transfer function can be represented in time-constants format (where the static gain k is explicitly shown) or in pole-zero format (where the transfer constant or gain term κ is explicit).

$$G(s) = \frac{-1.667\,s + 2}{1.667\,s + 1} \rightarrow \begin{cases} G(s) = 2\frac{-0.833\,s+1}{1.667\,s+1} \rightarrow k = 2, \\ G(s) = -1\frac{(s-1.2)}{(s+0.6)} \rightarrow \kappa = -1. \end{cases}$$

Analyzing the contribution of the terms to the magnitude plot:

- $k = 2 \rightarrow |k|_{\,\mathrm{dB}} = 20\log(2) = 6$ dB.
- $(1.667s + 1) \rightarrow$ the transfer function has a pole at $s = -0.6$, with 0 dB gain at low frequencies and decrease following the -20 dB/decade asymptote from $\omega = 0.6$ rad/s, where the gain of this term is -3 dB (corner frequency).
- $(-0.8333s + 1) \rightarrow$ the transfer function has a zero at $s = 1.2$. Therefore, this term provides an asymptotic gain of 0 dB at low frequencies, and from $\omega = 1.2$ rad/s the magnitude will grow following the asymptote straight line of 20 dB/decade slope, with a gain of 3 dB at $\omega = 1.2$ rad/s (corner frequency).

By adding the effects of each term, the magnitude plot can be obtained. It starts from 6 dB, then starts to drop with a slope of -20 dB/decade due to the influence of the pole (at $\omega = 0.6$ rad/s, where the magnitude of the Bode plot should be decreased by -3 dB, but the influence of the zero location makes this magnitude being approximately 4 dB instead of 3 dB). Before the location of the zero at $\omega = 1.2$ rad/s, the magnitude plot starts to track a zero slope asymptote near 0 dB. For the analysis of the phase plot and taking into account that the system is NMP, the pole-zero representation is selected as explained in the interactive tool f_nonminimum_phase (card 4.8).

Following the established convention, the angle provided by κ is $+180°$ (κ is negative). The points of interest for the analysis of the phase (Fig. 6.14) are:

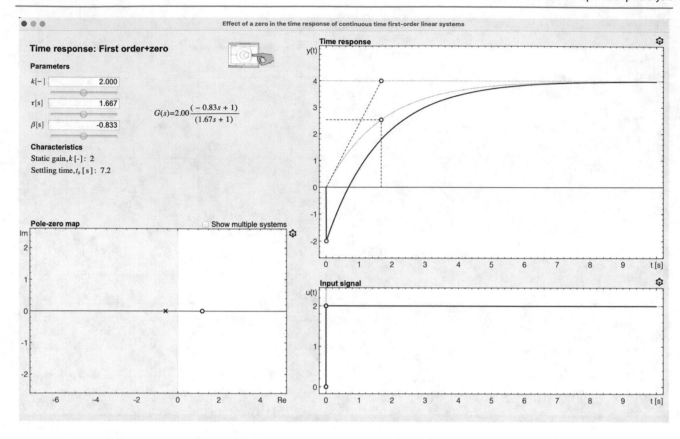

Fig. 6.13 Problem 3: Step response using the interactive tool t_first_order_zero

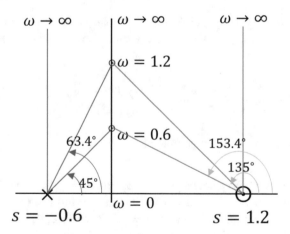

Fig. 6.14 Problem 3: Phase analysis in NMP example

$$\lfloor G(j\omega) = \lfloor \kappa + \lfloor \text{zero} - \lfloor \text{pole},$$
$$\omega = 0 \text{ rad/s} \rightarrow \lfloor G(j0) = 180° + 180° - 0° = 360°,$$
$$\omega \rightarrow \infty \text{ rad/s} \rightarrow \lfloor G(j\infty) = 180° + 90° - 90° = 180°,$$
$$\omega = 0.6 \text{ rad/s} \rightarrow \lfloor G(j0.6) = 180° + 153.4° - 45° = 288.4°,$$
$$\omega = 1.2 \text{ rad/s} \rightarrow \lfloor G(j1.2) = 180° + 135° - 63.4° = 251.6°.$$

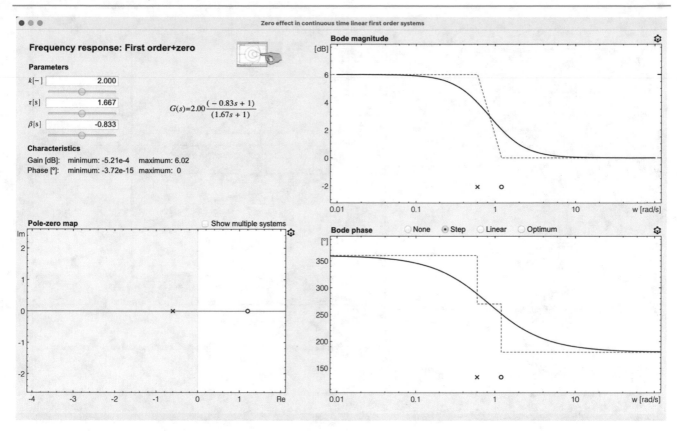

Fig. 6.15 Problem 3: Bode diagram using the interactive tool f_first_order_zero

The phase plot starts at 360° at low frequencies, has values of 288.4° and 251.6° at frequencies 0.6 and 1.2 rad/s respectively, and reaches 180° at high frequencies. This is corroborated with the tools f_second_order_zero (card 4.6) and f_nonminimum_phase, as shown in Figs. 6.15 and 6.16.

For the Nyquist diagram, representative points can be computed:

$$G(j\omega) = \frac{-j1.667\omega + 2}{j1.667\omega + 1} \cdot \frac{1 - j1.667\omega}{1 - j1.667\omega} = \frac{-j1.667\omega - 1.667^2\omega^2 + 2 - j3.333\omega}{1.667^2\omega^2 + 1}$$

$$\mathrm{Re}(G(j\omega)) = \frac{2 - 2.778\omega^2}{1 + 2.778\omega^2}; \quad \mathrm{Im}(G(j\omega)) = \frac{-5\omega}{1 + 2.778\omega^2}$$

- $\omega = 0$; Re $= 2$, Im $= 0$.
- $\omega \to \infty$; Re $= -1$, Im $= 0$.
- Re $= 0 \to \omega = \sqrt{\frac{2}{2.778}} = 0.848$ rad/s $\to$ Im $= -1.26$.
- $\omega = 0.59$ rad/s; Re $= 0.525$, Im $= -1.499$.
- $\omega = 1.199$ rad/s; Re $= -0.399$, Im $= -1.2$.

These values can be checked with the interactive tool f_nonminimum_phase (card 4.8), as shown in Fig. 6.17.

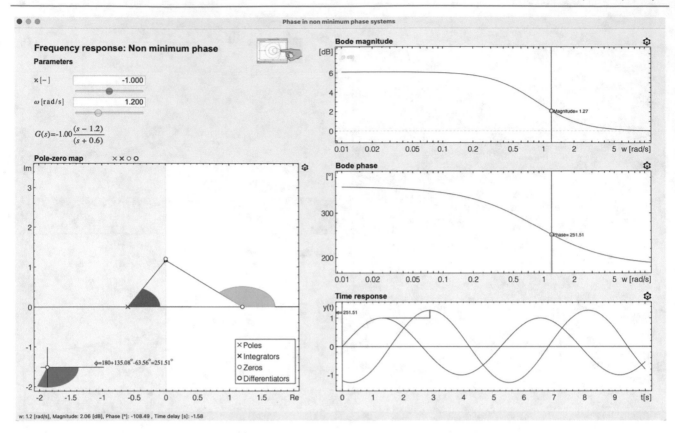

Fig. 6.16 Problem 3: Bode diagram using the interactive tool f_nonminimum_phase

6.4　Problem 4. Second-Order System with a Zero

Consider the very simplified model structure of a water-steam heat exchanger,[1] having no physical sense, but reproducing the structure of the real equations, to facilitate analysis (modified from [1], example 2.16, pp. 59–61).

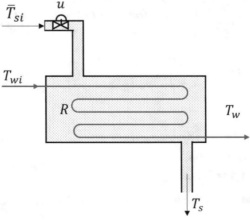

Problem 4: Simplified heat exchanger diagram

It is described by two differential equations:

---

[1] Steam enters the chamber at the top, passing through the control valve, and cooler steam leaves at the bottom. There is a constant flow of water through the pipe extracting heat from the steam and thus increasing its temperature.

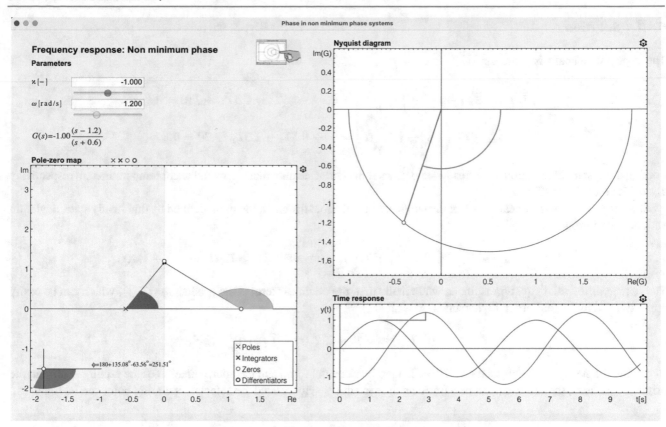

Fig. 6.17 Problem 3: Nyquist diagram using the interactive tool f_nonminimum_phase

$$\frac{dT_s(t)}{dt} = u(t)(\overline{T}_{si} - T_s(t)) - \frac{1}{R}(T_s(t) - T_w(t)),$$

$$\frac{dT_w(t)}{dt} = T_{wi}(t) - T_w(t) + \frac{1}{R}(T_s(t) - T_w(t)),$$

where $T_s(t)$ is the outlet steam temperature, $\overline{T}_{si}$ is the inlet steam temperature, which is considered constant in this exercise, $T_w(t)$ is the outlet water temperature, $T_{wi}(t)$ is the inlet water temperature (disturbance), $u(t)$ is the control signal related to the valve aperture, and R is a heat transfer coefficient in adequate units. The objective is to obtain the linearized model and the transfer functions relating:

1. The outlet steam temperature with the control signal $G_{su}(s) = T_s(s)/U(s)$.
2. The outlet steam temperature with the inlet water temperature $G_{swi}(s) = T_s(s)/T_{wi}(s)$.
3. The outlet water temperature with the control signal $G_{wu}(s) = T_w(s)/U(s)$.
4. The outlet water temperature with the inlet water temperature $G_{wwi}(s) = T_w(s)/T_{wi}(s)$.

There may be pole-zero cancellation in the derivation of these transfer functions. In this problem, $R = 2$ and the values of inputs (manipulable and disturbances) defining the operating point are: $\overline{u} = 1, \overline{T}_{si} = 110$ and $\overline{T}_{wi} = 20$. The reader can select different values. For the analysis of the nonlinear model, its implementation and simulation (for instance in Simulink®) are recommended. Once linearized, the outputs provided by the linear models can be compared with those of the nonlinear model against step changes in valve opening and water inlet temperature around the operating point. Using the interactive tools t_second_order (card 3.3), t_second_order_zero (card 3.5), and t_generic (card 3.5), analyze the time responses. The frequency responses of the four transfer functions above can be studied using the interactive tools f_concept (card 4.2), f_second_order (card 4.4), f_second_order_zero (card 4.6), and f_generic (card 4.7).

6.4.1   Solution

The steady-state analysis provides:

$$\overline{U}\,\overline{T}_{si} + \overline{T}_s\left(-\overline{u} - \frac{1}{R}\right) + \frac{1}{R}\overline{T}_w = 0 \rightarrow -1.5\overline{T}_s + 0.5\overline{T}_w + 110 = 0,$$

$$\overline{T}_{wi} - \overline{T}_w\left(1 + \frac{1}{R}\right) + \frac{1}{R}\overline{T}_s = 0 \rightarrow 0.5\overline{T}_s - 1.5\overline{T}_w + 20 = 0.$$

Solving the system of equations, the steady-state value of the outlet steam temperature and water temperature are respectively $\overline{T}_s = 87.5$ and $\overline{T}_w = 42.514$.

In order to perform the linearization of the nonlinear model about the operating point defined by this steady state, deviation variables are introduced:

$$T_s(t) = \overline{T}_s + \tilde{T}_s(t); \;\; T_w(t) = \overline{T}_w + \tilde{T}_w(t); \;\; u(t) = \overline{u} + \tilde{u}(t); \;\; T_{wi}(t) = \overline{T}_{wi} + \tilde{T}_{wi}(t).$$

The second differential equation is linear, while in the first, the nonlinear term is the product $u(t)T_s(t)$, which can be easily approximated by its first-order Taylor series expansion:

$$u(t)\,T_s(t) \approx \overline{u}\,\overline{T}_s + \overline{u}\,\tilde{T}_s(t) + \overline{T}_s\,\tilde{u}(t).$$

Replacing the steady-state values, the variables in their incremental form, and the approximation of the nonlinear term in the original equations, the linear equations describing the system behavior around the selected operating point can be found as:

$$\frac{d\tilde{T}_s(t)}{dt} = \tilde{u}(t)(\overline{T}_{si} - \overline{T}_s) + \tilde{T}_s(t)\left(-\overline{u} - \frac{1}{R}\right) + \frac{1}{R}\tilde{T}_w(t),$$

$$\frac{d\tilde{T}_w(t)}{dt} = \tilde{T}_{wi}(t) - \left(1 + \frac{1}{R}\right)\tilde{T}_w(t) + \frac{1}{R}\tilde{T}_s(t),$$

to which the Laplace transform with zero initial conditions in the deviation variables can be applied:

$$sT_s(s) = (\overline{T}_{si} - \overline{T}_s)U(s) - \left(\overline{u} + \frac{1}{R}\right)T_s(s) + \frac{1}{R}T_w(s),$$

$$sT_w(s) = T_{wi}(s) - \left(1 + \frac{1}{R}\right)T_w(s) + \frac{1}{R}T_s(s),$$

where, as usual, $T_s(s)$, $U(s)$, $T_w(s)$ and $T_{wi}(s)$ are the Laplace transforms of $\tilde{T}_s(t)$, $\tilde{u}(t)$, $\tilde{T}_w(t)$ and $\tilde{T}_{wi}(t)$ respectively. Now, the expression of the outputs $T_s(s)$ and $T_w(s)$ as a function of the inputs $U(s)$ and $T_{wi}(s)$ (as $T_{si} = \overline{T}_{si}$ is considered constant through the exercise) can be obtained. The complete development is not complex but takes some time. From the first equation:

$$T_s(s) = \frac{\overline{T}_{si} - \overline{T}_s}{s + \overline{u} + \frac{1}{R}}U(s) + \frac{1}{Rs + \overline{u}R + 1}T_w(s).$$

Now operating on the second equation and substituting in the first one:

$$T_w(s)\left[s + 1 + \frac{1}{R}\right] = T_{wi}(s) + \frac{1}{R}T_s(s),$$

$$T_w(s)\left[s + 1 + \frac{1}{R}\right] = T_{wi}(s) + \frac{1}{R}\left[\frac{\overline{T}_{si} - \overline{T}_s}{s + \overline{u} + \frac{1}{R}}U(s) + \frac{1}{Rs + \overline{u}R + 1}T_w(s)\right],$$

$$T_w(s)\left[\frac{(s + 1 + \frac{1}{R})(R^2s + R^2\overline{u} + R) - 1}{R^2s + R^2\overline{u} + R}\right] = T_{wi}(s) + \frac{\overline{T}_{si} - \overline{T}_s}{Rs + R\overline{u} + 1}U(s).$$

From the above equation, $G_{wwi}(s)$ and $G_{wu}(s)$ can be obtained.

$$G_{wwi}(s) = \frac{T_w(s)}{T_{wi}(s)} = \frac{R^2 s + R^2 \overline{u} + R}{R^2 s^2 + (R^2 \overline{u} + 2R + R^2)s + (R^2 \overline{u} + R\overline{u} + R)}.$$

Dividing by R^2 to get a normalized form and replacing the values of the parameters:

$$G_{wwi}(s) = \frac{T_w(s)}{T_{wi}(s)} = \frac{s + (\overline{u} + \frac{1}{R})}{s^2 + (\overline{u} + \frac{2}{R} + 1)s + (\overline{u} + \frac{\overline{u}}{R} + \frac{1}{R})} = \frac{s + 1.5}{(s+1)(s+2)}.$$

$$G_{wu}(s) = \frac{T_w(s)}{U(s)} = \frac{\overline{T}_{si} - \overline{T}_s}{Rs + R\overline{u} + 1} \cdot \frac{R^2 s + R^2 \overline{u} + R}{R^2 s^2 + (R^2 \overline{u} + 2R + R^2)s + (R^2 \overline{u} + R + R\overline{u})}.$$

The values of the parameters are then replaced and the final expression of $G_{wu}(s)$ is:

$$G_{wu}(s) = \frac{T_w(s)}{U(s)} = \frac{22.5(4s + 6)}{8s^3 + 36 s^2 + 52 s + 24} = \frac{11.25(s + 1.5)}{(s + 1.5)(s + 1)(s + 2)} = \frac{11.25}{(s + 2)(s + 1)}.$$

The transfer functions that remain to be calculated are those of $T_s(s)$ with respect to $T_{wi}(s)$ and to $U(s)$. $T_w(s)$ is extracted from the second equation, and introduced in the first.

$$T_w(s) = \frac{1}{s + 1 + \frac{1}{R}} T_{wi}(s) + \frac{1}{Rs + R + 1} T_s(s),$$

$$T_s(s)\left[s + \overline{u} + \frac{1}{R}\right] = (\overline{T}_{si} - \overline{T}_s)U(s) + \frac{1}{Rs + R + 1} T_{wi}(s) + \frac{1}{R^2 s + R^2 + R} T_s(s),$$

so that $G_{swi}(s)$ and $G_{su}(s)$ can be obtained after replacing the values of the parameters (notice that there is an internal pole-zero cancellation):

$$G_{swi}(s) = \frac{T_s(s)}{T_{wi}(s)} = \frac{4s + 6}{(2s + 3)(4s^2 + 12s + 8)} = \frac{0.5}{(s + 1)(s + 2)},$$

$$G_{su}(s) = \frac{T_s(s)}{U(s)} = \frac{(\overline{T}_{si} - \overline{T}_s)(R^2 s + R^2 + R)}{R^2 s^2 + (R^2 \overline{u} + R^2 + 2R)s + (R^2 \overline{u} + R\overline{u} + R)} = \frac{22.5(s + 1.5)}{(s + 1)(s + 2)}.$$

Therefore, the transfer functions are:

$$G_{su}(s) = \frac{T_s(s)}{U(s)} = \frac{22.5(s + 1.5)}{(s + 1)(s + 2)} , \; G_{swi}(s) = \frac{T_s(s)}{T_{wi}(s)} = \frac{0.5}{(s + 1)(s + 2)},$$

$$G_{wu}(s) = \frac{T_w(s)}{U(s)} = \frac{11.25}{(s + 2)(s + 1)} , \; G_{wwi}(s) = \frac{T_w(s)}{T_{wi}(s)} = \frac{s + 1.5}{(s + 1)(s + 2)}.$$

$$T_s(s) = G_{su}(s)U(s) + G_{swi}(s)T_{wi}(s) = \frac{22.5(s + 1.5)}{(s + 1)(s + 2)}U(s) + \frac{0.5}{(s + 1)(s + 2)}T_{wi}(s),$$

$$T_w(s) = G_{wu}(s)U(s) + G_{wwi}(s)T_{wi}(s) = \frac{11.25}{(s + 2)(s + 1)}U(s) + \frac{s + 1.5}{(s + 1)(s + 2)}T_{wi}(s).$$

Figure 6.18 shows the comparison between the nonlinear and linear models subjected to a step in the control signal of 0.25 at $t = 25$ s (the abscissa axis is in seconds) and one of amplitude 1 in the inlet water temperature at $t = 50$ s. The fitting between both models depend on the kind of nonlinearity and the amplitude of change in the input signals (in such a way that the deviation variables can be considered as such). It can be seen that the linear model adjusts better to changes in the water outlet temperature, having a greater difference in the steam temperature.

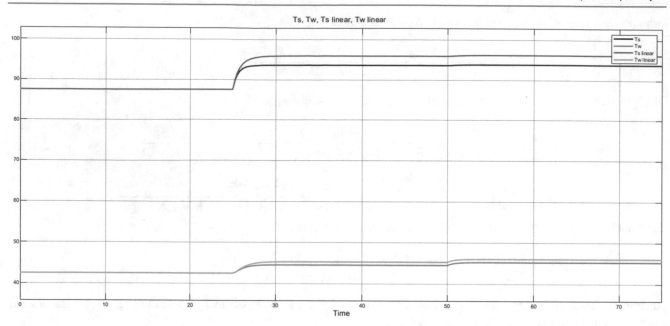

Fig. 6.18 Problem 4: Example of results comparing the nonlinear and linear models

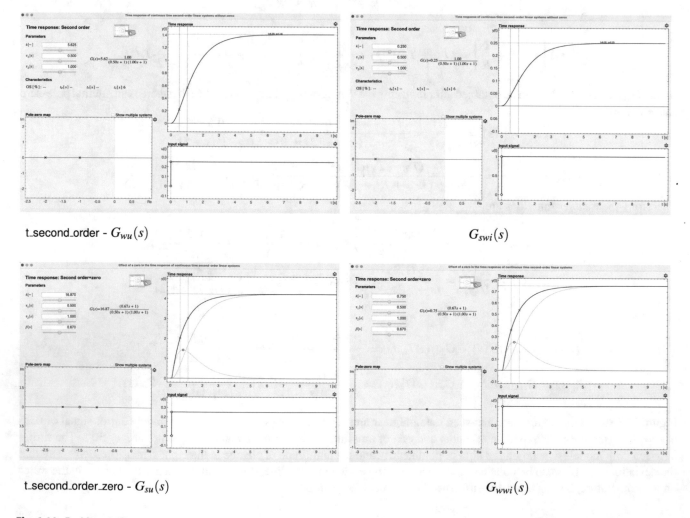

Fig. 6.19 Problem 4: Time response

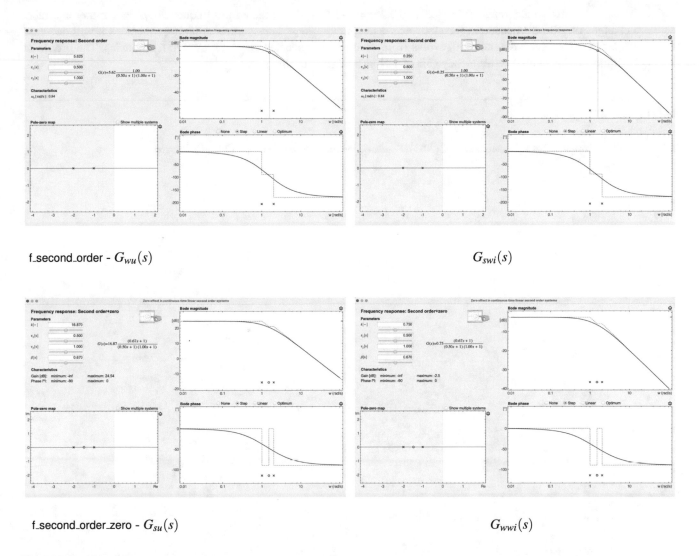

f_second_order - $G_{wu}(s)$ $G_{swi}(s)$

f_second_order_zero - $G_{su}(s)$ $G_{wwi}(s)$

Fig. 6.20 Problem 4: Frequency response

The time response can be analyzed using the interactive tools t_second_order and t_second_order_zero, considering a step of amplitude 0.25 in u and amplitude 1 in T_{wi}.

$$G_{wu}(s) = 5.625 \frac{1}{(0.5\,s + 1)(s + 1)}, \quad G_{swi}(s) = 0.25 \frac{1}{(0.5\,s + 1)(s + 1)},$$

$$G_{su}(s) = 16.87 \frac{(0.67\,s + 1)}{(0.5\,s + 1)(s + 1)}, \quad G_{wwi}(s) = 0.75 \frac{0.67s + 1}{(0.5\,s + 1)(s + 1)}.$$

The results are shown in Fig. 6.19.

In the same way, the frequency response can be analyzed in the Bode diagram using the tools f_second_order and f_second_order_zero, as shown in Fig. 6.20.

$G_{su}(s)$ has a magnitude of $20 \log(16.87) = 24.54$ dB at low frequencies. The corner frequencies are $\omega = 1$ rad/s (pole), $\omega = 1.5$ rad/s (zero) and $\omega = 2$ rad/s (pole). Each pole introduces an asymptote of -20 dB/decade and the zero one of 20 dB/decade slope. The phase plot starts at $0°$ at low frequencies (the system has positive static gain and no poles or zeros at the origin). Each pole contributes with an arctangent curve from 0 to $-90°$ and the zero from 0 to $90°$, so the phase at high frequencies is $-90°$ and, as the poles are close to the zero, contributions partially cancel each other. The profile of $G_{wwi}(s)$ is similar, but with the gain $20 \log(0.75) = -2.5$ dB.

$G_{swi}(s)$ has a low-frequency magnitude of $20\log(0.25) = -12.04$ dB. At the corner frequencies $\omega = 1$ and $\omega = 2$ rad/s there are changes of -20 dB/decade in the slope of the magnitude asymptotes so that at high frequencies it will decrease at -40 dB/decade. The phase diagram will evolve from $0°$ at low frequencies to $-180°$ at high frequencies. The profile of $G_{wu}(s)$ is similar to this, with low-frequency gain of $20\log(5.625) = 15$ dB.

Reference

1. Franklin, G. F., Powell, J. D., & Emani-Naeni, A. (2015). *Feedback control of dynamic systems* (7th ed.). Pearson.

Part III
Closed-Loop Analysis and Design

The second part of the book is devoted to analyzing and designing feedback control systems. Chapter 7 deals with methods for the analysis of stability of linear systems. After introducing the fundamentals of block diagrams and block diagram algebra, a basic feedback loop is adopted, in which a system represented by its transfer function is controlled using a simple proportional gain. This helps to focus on the properties of feedback. Three main methods are analyzed: root locus, Nyquist stability criterion, and classical relative stability margins (gain and phase margins). The influence of time delay in the stability of feedback loops is also analyzed using an example of high visual content. Moreover, in the introduction to Chap. 7, several conceptual tools are also treated, also very useful in Chap. 8, where basic control design methods are introduced, both in the time domain (mainly proportional–integral–derivative—PID—controllers) and in the frequency domain (phase-lead and phase-lag compensators).

The introduction to Chap. 8 summarizes basic relationships between time- an frequency-domain specifications and the role of sensitivity functions in feedback control. After the introduction, this chapter starts with the analysis of steady-state errors in feedback control systems, explains basic PID and phase-lead/lag control structures, and introduces several design techniques. The last card presented introduces two loop shaping interactive tools that do not follow the layout of the previous ones, but they summarize all the contents explained in the book; the reader will find them very useful for control analysis and design purposes, and to visually analyze the relationship between time-domain and frequency-domain specifications fulfillment. Inherent to the design in all these methods are the concepts of stability and sensitivity of feedback loops, that being the main reason for summarizing them in the introductions to Chaps. 7 and 8.

Chapter 9 applies all the concepts treated in the previous ones to the selected illustrative plants, which models and representative transfer functions have been introduced in Chaps. 2 and 5 respectively. Moreover, Chap. 10 includes solved exercises covering the most important aspects on closed-loop stability and control design, using in all cases the interactive tools explained in the text.

Notice that the time delay in the interactive tools devoted to closed-loop analysis and design is implemented using a discrete-time formulation in the simulations, so that in extreme cases (for instance, closed-loop systems with very high gains and small stability margins) the simulations may become unstable due to numerical integration issues.

Closed-Loop Systems and Stability

7.1 Introduction

This chapter covers basic concepts related to feedback and its effect on the behaviour and stability of closed-loop systems. As pointed out by [1], stability is crucial in the study of system dynamics and all controlled systems must be stable. Stability captures de idea that *small causes have small consequences forever*, what is a requirement to be able to predict the future behavior of a system with a degree of certainty and accuracy. Although stability can be seen as a qualitative property of a system, it is also important to quantify how stable a system is (for instance, when the system faces a change in its parameters).

Sensitivity measures how important signals (mainly the system input and output) are influenced by other external signals such as load disturbances or measurement noise.

Robustness indicates how changes in the parameters of a system or in the environment may affect its behavior. It is a quality of a control system that shows little sensitivity to effects that were not considered in the analysis and design phase, e.g. disturbances, unmodeled dynamics, noise, etc.

The ideal for a system is to have good stability properties, low sensitivity to load disturbances and noise and good robustness properties. This will be formalized in this and the following chapter. Four interactive tools are included here, where the *root locus method*, the *Nyquist stability criterion*, the concepts of *relative stability margins* and the *influence of time delay on stability* are studied. Basic background knowledge of the Laplace transform and transfer functions is assumed and can be reviewed in the introduction to Chap. 3.

7.1.1 Block Diagrams and Block Diagram Algebra

The properties of the Laplace transform and the definition of transfer functions allow engineers to work comfortably with *block diagrams* and perform algebraic operations on them. Figure 7.1 shows the main block diagram structures used in control, as well as the operations that can be performed on them using *block diagram algebra*.

Control systems are an interconnection of components that form a system configuration that will provide the desired response [2]. Figure 7.2 represents the block diagram of a two-degree-of-freedom (2-DoF) closed-loop system. As pointed out by [3], a general feedback loop is influenced by three external signals: the setpoint r, the load disturbance d, and the measurement noise n. The most important signals from the point of view of control are the process output y_p, the measured process output y and the controller output u_c. Let R, D, N, Y_p, Y and U_c be their respective Laplace transforms (Fig. 7.2). Control is based on the use of algorithms and feedback in engineering systems. In a control loop, the process output is passed through a transfer function $H(s)$ located in the feedback loop and the output of this block is brought to the system input. The transfer function $G(s)$ represents the *plant (system)*, $C(s)$ the *controller*, $H(s)$ is the feedback loop transfer function (usually linked to the *sensor* dynamics or analog filters for noise cancellation[1]) and $F(s)$ is a transfer function that allows the *reference* to be filtered[2] and gives the control system structure two degrees of freedom (generally "forgotten" in basic control

---

[1] Often unit feedback is used ($H(s) = 1$), because the sensor dynamics are much faster than the process dynamics and can be neglected.
[2] $F(s) = 1$ if there is no need for reference filtering (pure error feedback).

Supplementary Information The online version contains supplementary material available at https://doi.org/10.1007/978-3-031-09920-5_7.

© Springer Nature Switzerland AG 2023
J. L. Guzmán et al., *Automatic Control with Interactive Tools*,
https://doi.org/10.1007/978-3-031-09920-5_7

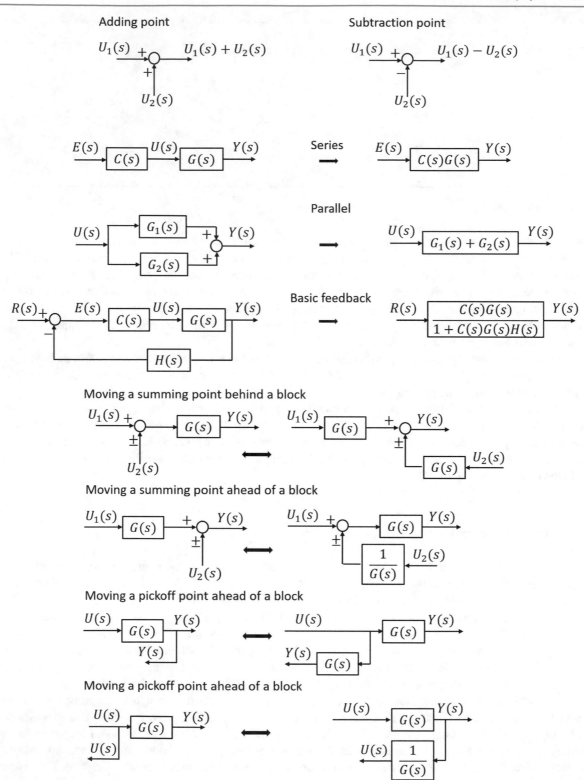

Fig. 7.1 Structures and operations in block diagram algebra

courses). It is considered that a disturbance enters the plant input (called load disturbance); this is a prototype problem, as the disturbances can appear in many other places in the system.

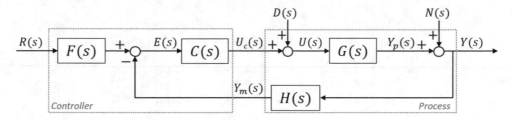

Fig. 7.2 Block diagram of a two-degree-of-freedom (2-DoF) closed-loop system

To summarize, the internal, input and output signals of the loop are:

- $Y(s)$: It is the *measured output* of the process described by $G(s)$ (variable used for control purposes).
- $Y_m(s)$: It is the measured process output when the sensor dynamics are taken into account ($H(s) \neq 1$). $Y_m = Y$ when unit feedback ($H(s) = 1$) is considered.
- $Y_p(s)$: It is the process output, which is the real variable to control.
- $U(s)$: It is the *input* to the system. It coincides with the control signal $U_c(s)$ in closed-loop systems without load disturbances ($D(s) = 0$), since the input to the process is provided by the controller $C(s)$.
- $U_c(s)$: It is the *output* of the controller $C(s)$.
- $R(s)$: It is the *reference* signal or *setpoint* that the closed-loop system must track.
- $E(s)$: It is the *error* signal, the result of comparing the reference with the measured process output. This signal is the input to the controller $C(s)$.
- $D(s)$: It is the *load disturbance* at the process input. It is an exogenous variable that drives the process away from its desired behavior and negatively affects the system output. It is always useful to understand disturbances, their source and where they enter the system [4].
- $N(s)$: It is the *measurement noise*, usually associated with measuring instruments (sensors), which often cause high-frequency random signals.

An *open-loop system* uses an actuator to control the process directly without using feedback, and thus the output has no effect upon the input signal. Using the concept of transfer function and the basic operations described in Fig. 7.1, it is easy to see that if $F(s) = 1$ (the reference is not filtered) in Fig. 7.2, the open loop (without feedback or $H(s) = 0$) can be represented by the product $C(s)G(s)$, so that $Y(s) = C(s)G(s)R(s)$. As the error signal $E(s) = R(s) - Y(s) = (1 - C(s)G(s))R(s)$, the tracking error is zero ("perfect control") if $C(s) = G(s)^{-1}$. Therefore, necessary (but not sufficient) conditions for the possibility of open-loop control is that $G(s)$ be both stable and MP and lack of model and controller[3] uncertainty [5]. Open-loop control systems behave properly only if the plant model is very accurate, if the plant parameters do not change in time, and if there are no external disturbances. In other cases, control systems must include feedback loops. Output sensitivity can be understood as the percent of change in the output due to a change in the plant parameters, so that in open loop a percent change in the plant produces the same percent change in the output. The same applies to disturbances.

A *closed-loop system* relies on *feedback*, consisting of redirecting a certain proportion of the output signal of a system towards the input. It implies that two or more dynamical systems are connected together, so that each system influences the other and their dynamics are strongly coupled [6]. A closed-loop system (Fig. 7.2 with $H(s) \neq 0$, usually $H(s) = 1$) uses a measurement of the *output signal* (y) and a comparison with the desired output (*reference* (r) or *setpoint*) to generate an *error signal* (e), which is used by the *controller* to generate a *control signal* (u_c) that is applied to the *actuator*, which is the device that provides the driving power (u) to the process [2]. As will be analyzed, feedback can increase robustness by reducing sensitivity to model uncertainty. In Chap. 8, the analysis of sensitivity to model uncertainties, disturbances and noise are dealt with. The basic feedback scheme based on measuring, calculating and acting is the central concept in automatic control systems, where negative feedback is normally used. The output signal is redirected to the input and subtracted from the reference signal. This is the case considered in this book. In the case of positive feedback, the output is added to the reference.

---

[3] The controller always exhibits a certain degree of uncertainty, due to precision errors in the control algorithms and hardware (related to actuators, amplifiers and controllers). Moreover, this realization of perfect control could lead to very large control signals that would not be realizable by the actuator.

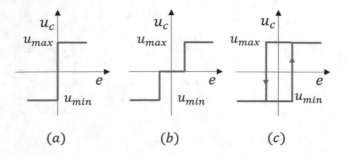

Fig. 7.3 Static characteristic (input–output) of on-off controllers: **a** Ideal, **b** with dead-zone, **c** with hysteresis

When the reference signal is constant, it is often said that it is a *regulation* problem, although in this text no distinction is made between the terms control and regulation. On the other hand, when the output signal must follow a changing reference value, it is known as the *tracking* problem.

7.1.2 Simplest Forms of Feedback

The simplest forms of feedback are on-off control, proportional control and proportional–integral–derivative control [6]. On-off control is a feedback mechanism which can be described by:

$$u_c(t) = \begin{cases} u_{max}, & \text{if } e(t) > 0 \\ u_{min}, & \text{if } e(t) < 0, \end{cases} \tag{7.1}$$

which is a nonlinear control law where the maximum corrective action (u_{max} or u_{min}) is always used, and the response is a steady oscillation (limit cycle). Modifications by introducing dead zone or hysteresis can be used [6], as shown in Fig. 7.3.

The analysis in this chapter is done using proportional control, where:

$$u_c(t) = Ke(t), \tag{7.2}$$

which is the simplest linear control law (small errors provide small control signals) and has one tuning parameter, the proportional gain K. The general case of PID control is dealt with in Chap. 8, where a term proportional to the integral of the error and another proportional to the error derivative are added to (7.2), with coefficients providing degrees of freedom to fulfilling desired closed-loop specifications. This book does not treat the case in which the control action is constrained by u_{min} and u_{max}, as in that case the closed-loop system becomes nonlinear, requiring nonlinear control algorithms or special mechanism such the *anti-windup* action in PID controllers [3]. Notice that if the actuator saturates, the control law (7.2) should include the conditions $u_c(t) = u_{max}$ if $e(t) \geq e_{max}$ and $u_c(t) = u_{min}$ if $e(t) \leq e_{min}$, where $e_{max} = u_{max}/K$ and $e_{min} = u_{min}/K$. The interval (e_{min}, e_{max}) is called the *proportional band*, because the behavior of the control law is linear when the error is in that interval [6]. Figure 7.4 shows typical time responses of on-off control and proportional control.

7.1.3 Basic Relations in Feedback Loops

The key aspects in the design of control systems are to ensure that the closed-loop system is *stable* (bounded inputs produce bounded outputs) and that it has a *desired behavior* (good disturbance rejection characteristics, fast response to changes in the setpoint, low sensitivity to unmodeled dynamics, etc.). These properties are established through a variety of modeling and analysis techniques that capture the essential dynamics of the system, and facilitate the exploration of possible behaviors in the presence of uncertainty, noise and even components failure. Uncertainty enters the control loop through the noise in sensors and actuators, external disturbances and uncertain dynamics that come from errors in the model parameters or unmodeled effects [6].

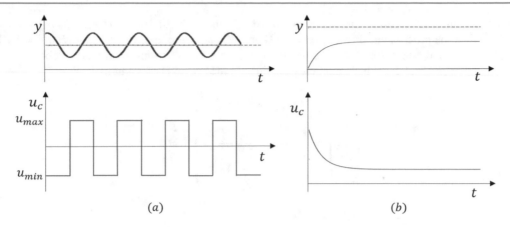

Fig. 7.4 Examples of time responses of: **a** On-off control, **b** Proportional control

Using the properties of the Laplace transform and the transfer function (block diagram algebra) and defining the *tracking error* as $E(s) = F(s)R(s) - Y(s)$ the following expressions can be obtained:

$$
\begin{aligned}
Y(s) &= \frac{F(s)C(s)G(s)}{1+C(s)G(s)H(s)}R(s) + \frac{G(s)}{1+C(s)G(s)H(s)}D(s) + \frac{1}{1+C(s)G(s)H(s)}N(s), \\
Y_p(s) &= \frac{F(s)C(s)G(s)}{1+C(s)G(s)H(s)}R(s) + \frac{G(s)}{1+C(s)G(s)H(s)}D(s) - \frac{C(s)G(s)H(s)}{1+C(s)G(s)H(s)}N(s), \\
U_c(s) &= \frac{C(s)F(s)}{1+C(s)G(s)H(s)}R(s) - \frac{C(s)G(s)H(s)}{1+C(s)G(s)H(s)}D(s) - \frac{C(s)H(s)}{1+C(s)G(s)H(s)}N(s).
\end{aligned}
\tag{7.3}
$$

Notice that the previous transfer functions, when $H(s) = 1$, are the so-called *Gang of Six* introduced by Professors Åström and Hägglund in [3]. The special case when $F(s) = 1$ is called a system with pure error feedback.

Assuming that $D(s) = 0$ and $N(s) = 0$, the closed-loop transfer function $G_{cl}(s)$ that relates the output to the reference is given by:

$$
G_{cl}(s) = \frac{Y(s)}{R(s)} = \frac{F(s)C(s)G(s)}{1+C(s)G(s)},
\tag{7.4}
$$

which is the basic feedback case described in the block diagrams in Fig. 7.1. If non-unit feedback were used, the previous expression would be:

$$
G_{cl}(s) = \frac{F(s)C(s)G(s)}{1+C(s)G(s)H(s)}.
\tag{7.5}
$$

Also notice that if $|C(j\omega)G(j\omega)| \gg 1$, from (7.4), tracking and disturbance rejection are improved, but this cannot be achieved in all frequencies, as happens with noise attenuation, as will be analyzed in this chapter. The main advantages of feedback are that it provides robustness against uncertainty and allows modifying the dynamic behavior of a system (feedback can provide good behavior from bad components), allowing to "create" systems with a linear input–output response. Feedback allows to reduce the *sensitivity* of a system, understood in this framework as the ratio of the change in the closed-loop transfer function with respect to the change in the process transfer function (or parameters) for a small incremental change [2]. The main drawbacks are that it can produce *instability* in a system if it is not used properly, it requires an increase in components (mainly sensors, which introduce noise and inaccuracies into the loop), increasing the complexity of the system, and inherently couples parts of a system. Another cost of feedback is loss of gain, since closing a simple loop around a system represented by the transfer function $L(s)$ results in $L(s)/(1 + L(s))$. The gain reduction of the loop is $1/(1 + L(s))$, which is the factor that reduces the sensitivity of the system to variations and disturbances of the parameters [2] and is therefore called the *sensitivity function*.

Using the loop transfer function $L(s) = C(s)G(s)$ (in this chapter unit feedback is considered and $C(s) = K$ is a proportional gain), the stability of the closed-loop system can be determined. The same applies to closed-loop design through adequately "shaping" $L(s)$, as is studied in Chap. 8.

Sensitivity functions are very useful for frequency-domain design methods [6]. From the analysis of the transfer functions in Fig. 7.2 and Eqs. (7.4), the process has three inputs (U_c, D and N) and one output ($Y = Y_m$, as $H(s) = 1$). The controller has two inputs (R and Y) and one output (U_c). As pointed out by [6], the following relations are obtained from the block diagram in Fig. 7.2, some of which have been explicitly obtained in the previous paragraphs:

$$
\begin{pmatrix} Y \\ Y_p \\ U_c \\ U \\ E \end{pmatrix} = \begin{pmatrix} \frac{FCG}{1+CG} & \frac{G}{1+CG} & \frac{1}{1+CG} \\ \frac{FCG}{1+CG} & \frac{G}{1+CG} & \frac{-CG}{1+CG} \\ \frac{FC}{1+CG} & \frac{-CG}{1+CG} & \frac{-C}{1+CG} \\ \frac{FC}{1+CG} & \frac{1}{1+CG} & \frac{-C}{1+CG} \\ \frac{F}{1+CG} & \frac{-G}{1+CG} & \frac{-1}{1+CG} \end{pmatrix} \begin{pmatrix} R \\ D \\ N \end{pmatrix}.
\tag{7.6}
$$

From here, the *Gang of Six* transfer functions [6] is obtained:

$$
TF = \frac{FCG}{1+CG}, \quad T = \frac{CG}{1+CG}, \quad S = \frac{1}{1+CG},
$$
$$
GS = \frac{G}{1+CG}, \quad CS = \frac{C}{1+CG}, \quad CFS = \frac{CF}{1+CG}.
\tag{7.7}
$$

In general frequency response design methods, these six transfer functions can be used to specify the expected closed-loop behavior (in Sect. 8.11, two basic loop shaping techniques are introduced). When pure error feedback is used ($F(s) = 1$), the system is completely characterized by four transfer functions (*Gang of Four*) [6]:

- S: Sensitivity function.
- T: Complementary sensitivity function.
- GS: Load sensitivity function (also called input sensitivity function).
- CS: Noise sensitivity function (also called output sensitivity function).

Notice that the controller $C(s)$ affects both load disturbances and measurement noise.

Stability is a measure of system behavior. In Chap. 3, through the analysis of the time step response of LTI systems, it was established that a system is stable if all the poles of the transfer function have negative real parts (called asymptotically stable systems, because the output tends to its steady-state value asymptotically). The concept of marginally stable systems was also introduced in Sect. 3.3 (which means that only some limited inputs make the output unbounded).

From this chapter onwards, the focus will be on the stability of closed-loop systems, which are the main objective of automatic control (an unstable closed-loop system has no practical value). Therefore, in the design of feedback control systems the poles of the characteristic equation ($J(s) = 1 + C(s)G(s) = 1 + L(s) = 0$), must lie in the LHP. A more general concept is that of bounded input-bounded output stability (external stability or BIBO stability), which means that a system will be stable if a bounded output is obtained when the corresponding input is bounded. Asymptotically stable systems also enjoy the property of BIBO stability [7]. Notice that these are the simplest definitions of stability. In Appendix D of reference [5], an excellent survey on stability concepts and tools can be found, where there are numerous definitions of stability, generally linked to the internal (state-space) representation of the system. Underlying all of them is the idea that as time tends to infinity, the output of the system cannot tend to infinity.

Absolute stability is a description of the system that reveals whether a closed-loop feedback system is stable or unstable, regardless of other attributes such as the degree of stability. An absolute stable system is called *stable* (the qualification of absolute is suppressed), and its degree of stability can be characterized. *Relative stability* provides a measure of how stable the system is. As a necessary and sufficient condition for a feedback system to be stable (in the absolute sense) is that all poles of its transfer function have negative real parts. Relative stability can be analyzed by examining the relative positions of the poles (when working with representations on the complex s-plane, Sect. 7.2) or with the so-called *relative stability margins* (phase margin and gain margin, Sect. 7.4), coming from a simplified interpretation of the Nyquist stability criterion (Sect. 7.3), when working with graphical representations of the frequency response. The concept of *conditional stability* is also dealt within the framework of systems with variable gain, if the system is stable (in the sense of BIBO stability) for certain gain values, but not for others. Using feedback, processes that are unstable in open loop can be stabilized, and in that case, as well as in the case of stable plants in open loop, feedback allows adjusting a certain closed-loop performance related to the design specifications.

This chapter introduces different analysis techniques to study the effect that feedback and variation of the gain of the system's loop function (static gain of $L(s) = C(s)G(s)$) have on closed-loop stability and performance.

The well-known *Routh-Hurwitz Stability Criterion* has not been introduced in this text, because the method is difficult to visualize and work with interactive tools, and also today computers can provide the roots of a high-order polynomial with precision. Moreover, this test gives only information regarding if the system is or not stable, without qualitative information. Readers are referred to classical textbooks that can be found in the References for its analysis.

7.2   Root Locus

7.2.1   Interactive Tool: root_locus

7.2.1.1 Concepts Analyzed in the Card and Learning Outcomes
- Stability and performance of basic feedback systems.
- Root locus concept.
- Construction of the root locus.
- Characteristic elements of the locus: Asymptotes, centroids and branches.
- Control structure selection using root locus.

7.2.1.2 Summary of Fundamental Theory
The first technique related to the concept of feedback studied in this chapter is the *root locus method*, which serves to examine the effect of feedback on the location of the poles and zeros of the closed-loop system, when varying from zero to infinity a characteristic parameter (usually the gain) of the open-loop system [2]. It is a technique in which, starting from information about the open-loop system, it is possible to deduce the location of the closed-loop poles, and therefore, the dynamic behavior of the feedback system and its asymptotic, relative and conditional stability. In this framework, *relative stability* is related to the relative damping of each root of the characteristic equation. The relative stability of a system is a property that can be related to the relative position of the real part of each root or pair of roots. It can also be defined as a function of the damping factors σ of each complex root pair and thus, as a function of the response speed and the overshoot. *Conditionally stable systems* are those stable for some intervals of the loop gain and unstable outside these intervals. The concept of *marginal stability* was introduced in Sect. 3.3: If the system has simple roots on the imaginary axis, with all other roots in the LHP, steady oscillations are obtained in steady-state regime for a bounded input, unless the input is a sinusoid (which is bounded) whose frequency is equal to the magnitude of the $j\omega$-axis roots. For this case, the output becomes unbounded. Such a system is called marginally stable, since only certain bounded inputs (sinusoids of the frequency of the poles) will cause the output to become unbounded [2].

Traditionally, a unit feedback system with a configuration as that shown in Fig. 7.5 is selected to introduce the root locus method, where K represents the proportional gain of a controller and $G(s)$ the transfer function of the open-loop system ($L(s) = KG(s)$ is the loop transfer function[4]). It is also considered that there are no disturbances neither noise affecting the system, so that in the analysis it is considered that $U_c(s) = U(s)$.

As previously mentioned, in the root locus method, the roots of the characteristic equation of the closed-loop system are represented for all values of a system parameter, usually the controller gain K. To apply the method, the transfer function is considered factored into poles and zeros (ZPK):

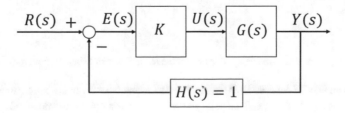

Fig. 7.5 Block diagram used for analyzing the root locus

---

[4] If $H(s) \neq 1$ the root locus is built with $L(s) = C(s)G(s)H(s)$.

$$G_\kappa(s) = \kappa \frac{(s - z_1)(s - z_2) \cdots (s - z_m)}{(s - p_1)(s - p_2) \cdots (s - p_n)} = \kappa G(s),$$

where κ is the *transfer constant* or *gain term*. To facilitate analysis, κ is embedded in K, so that the analysis is done using the following loop transfer function:

$$L(s) = KG(s) = K \frac{(s - z_1)(s - z_2) \cdots (s - z_m)}{(s - p_1)(s - p_2), \cdots (s - p_n)}.$$

The method is based on expressing the closed-loop transfer function so that the characteristic equation is given by:

$$`1 + L(s) = 1 + KG(s) = 0, \tag{7.8}$$

and it is assumed that there is no time delay.[5]

The root locus of a system is defined by the set of points of the complex plane for which there is a value of the parameter K (or in general another parameter of the loop transfer function) that makes that point a pole of the closed-loop system. By using this method, the effect of varying the gain, or adding open-loop poles or zeros, on the location of closed-loop poles can be analyzed. If the characteristic equation is expressed as follows:

$$1 + KG(s) = 1 + K\frac{b(s)}{a(s)} \rightarrow a(s) + Kb(s) = 0.$$

For small K ($K \rightarrow 0$), $a(s) = 0$ must be met, while for very large K ($K \rightarrow \infty$), $Kb(s) = 0$, so it can be deduced that the closed-loop roots location starts at the poles of the open-loop system and ends at the zeros of the open-loop transfer function (including zeros at infinity[6]). It can also be understood that when negatively feeding back a stable NMP system, for high gains it becomes unstable, because closed-loop poles approach NMP zeros. For simple examples like those presented in Fig. 7.6, the location of closed-loop poles can be easily analyzed by solving the closed-loop characteristic equation for values of K ranging from zero to infinity and plotting the roots (closed-loop poles). In more general cases, the root locus methods allows to figure out where the closed-loop poles are going to be located when changing a parameter, without having to solve the characteristic equation, being very useful also for controller structure selection.

In what follows, it is assumed that $K \geq 0$ (the case of negative gains will be commented on at the end of the section). From Eq. (7.8), $KG(s) = -1 = |-1|e^{\pm j\pi(2l+1)}$, providing the so-called:

- **Magnitude condition:** $|KG(s)| = 1$.
- **Angle condition:** $\lfloor KG(s) = \pm 180°(2l + 1)$, $l = 0, 1, 2, \ldots$.

The values of s that satisfy both magnitude and angle conditions are the roots of the characteristic equation (closed-loop poles), and the rules for the construction of the root locus are obtained by applying these conditions to test points in the complex plane.[7]

A series of steps to build the root locus are summarized below and their proof can be found in classical textbooks, as those included in the references section. More important than the construction procedure,[8] is to know the theoretical foundations and implications that the construction of the locus entails. The most relevant *rules* for building the root locus are summarized here:

---

[5] There are variants of the method for systems with time delays that are not treated in this text. Some of them make use of polynomial approximations of the time delay.

[6] Any rational fraction has the same number of poles as zeros if one takes into account that it can have poles or zeros at infinity.

[7] Recall that the angles of the segments joining the open-loop poles and zeros with a test point s are measured counterclockwise from the positive real axis.

[8] The location of the closed-loop poles can be computed through a polynomial roots numerical solver for different values of K or using interactive tools like the one explained in this card. The advent of interactive control system design software tools, which easily draw the root locus, allows teachers to focus on explaining how the introduction of new poles/zeros in the open-loop transfer function affects the root locus, rather than on explaining its construction rules.

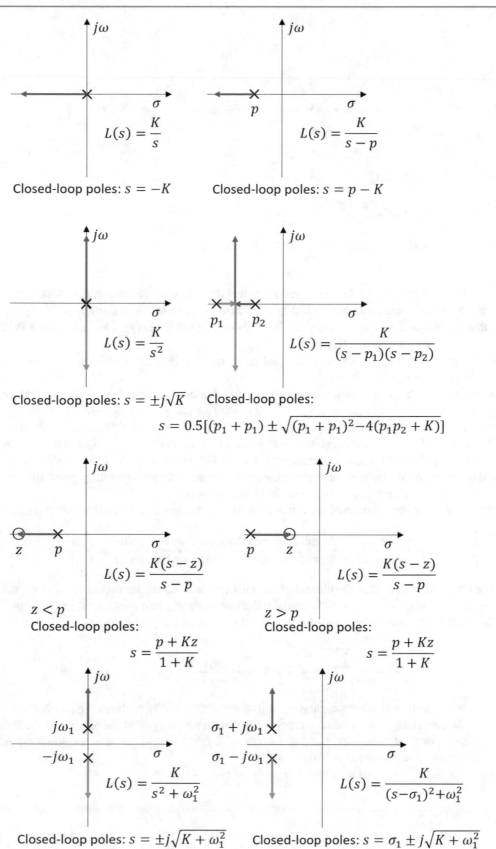

Fig. 7.6 Basic examples of the location of closed-loop poles using open-loop information and by varying the characteristic gain K from 0 to ∞

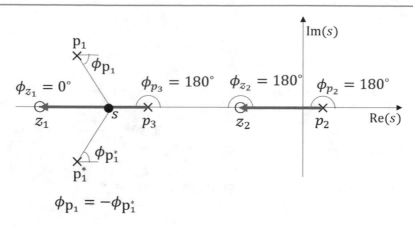

Fig. 7.7 Root locus on the real axis from the angle condition for $K \geq 0$

1. Write the characteristic equation in pole-zero notation, so that the parameter of interest K (which is modified to analyze its effect on the location of the closed-loop poles) appears clear in the form $1 + KG(s) = 0$. The system is supposed to be causal so that its relative degree $(n - m) \geq 0$, n being the number of poles (p_i, $i = 1...n$) and m the number of zeros (z_ℓ, $\ell = 1...m$) of the transfer function $G(s)$.

2. Place on the pole-zero map (s-plane) the poles ($\times$) and zeros (o) of $G(s)$. The locus starts at the poles of $G(s)$ and ends at the zeros of $G(s)$, including zeros at infinity.[9]

3. Draw the *geometric loci on the real axis*, which are those that leave an odd number of poles and zeros on the right.[10] By applying the angle condition $\lfloor L(s) = \sum \lfloor z_\ell - \sum \lfloor p_i = \pm 180°(2l + 1)$, for a point s on the real axis it holds that the contribution in angle of complex poles (ϕ_{p_i}) and zeros (ϕ_{z_ℓ}) on $\lfloor L(s)$ is zero (see Fig. 7.7 and the analysis done in Sect. 4.8, Figs. 4.5 and 4.6). The same applies to real poles and zeros left of s (ϕ_{p_i}, ϕ_{z_ℓ} left of s). Therefore, the angle condition applied to a point on the real axis becomes $\lfloor L(s) = \pm 180°(N_{z_r} - N_{p_r})$, N_{z_r} being the number of real zeros right of s and N_{p_r} the number of real poles right of s, so that the total number of real open-loop poles and zeros to the right of a point s on the real axis which belongs to the root locus must be odd.

   Notice that all points belonging to the root locus must also satisfy the magnitude condition evaluated in those roots:

   $$K = \frac{\text{product of the lengths between the point } s \text{ and the poles}}{\text{product of the lengths between the point } s \text{ and the zeros}} = \frac{\prod |s - p_i|}{\prod |s - z_\ell|}.$$

4. The root locus has n branches, since the characteristic equation has n roots, starting at open-loop poles and ending at open-loop zeros. There are $n - m$ *asymptotes* (straight lines representing root locations for very large values of s). The asymptotes intersect in the *centroid* η (located on the real axis) and will depart with angles γ, being:

   $$\eta = \frac{\sum_{i=1}^{n} p_i - \sum_{\ell=1}^{m} z_\ell}{n - m}; \quad \gamma = \frac{180°(2l + 1)}{n - m}; \; l = 0, 1, 2, ..., n - m - 1,$$

   where p_i and z_ℓ enter with their corresponding signs in the equation of the centroid. The derivation can be found for instance in [5]. The underlying idea is that a test point s is chosen far away from the origin and the limit of $G(s)$ when $s \to \infty$ is applied. For the asymptotes, $\lim_{s \to \infty} KG(s) = K/s^{n-m} = -1$, so that for positive gain $(n - m)\lfloor s = \pm 180°(2l + 1) \to \lfloor s = \gamma = \pm 180°(2l + 1)/(n - m)$.

   For the centroid, the derivation is not so trivial. For $K \to \infty$ and $s \to \infty$ then:

---

[9] If the transfer function considered is not causal, there will be poles at infinity.

[10] This rule (and others) does not apply when $K < 0$, positive feedback and cases of NMP systems. Some comments are included at the end of the section.

$$\lim_{K\to\infty,s\to\infty}(1+KG(s)) = \lim_{K\to\infty,s\to\infty}\left(1+K\frac{b(s)}{a(s)}\right) =$$

$$= \lim_{K\to\infty,s\to\infty}1+\frac{K}{s^{n-m}+(\sum p_i-\sum z_\ell)s^{n-m-1}+\ldots} \approx 1+K\frac{1}{(s-\eta)^{n-m}} = 0.$$

Notice that if $n=m$ the root locus has no asymptotes.

From this condition it is also easy to visualize that the root locus is symmetric with respect to the real axis. This is a logical condition, as the root locus is the locus of closed-loop poles, which are always symmetric with respect to the real axis, because they are the roots of the characteristic equation which has real coefficients. The real axis is always the horizontal axis of symmetry of the open-loop pole/zero configuration [5].

5. Calculate the *points of departure and arrival on the real axis* of the locus (*break-away*, *break-in*, *breakpoints*, *bifurcations* or *saddle points*), which correspond to points on the s-plane in which there are multiple roots of the characteristic equation (and therefore cancel the characteristic polynomial and its derivative). These breakpoints, which are points of confluence and dispersion of branches, are determined from the roots of the polynomial equation:

$$1+KG(s) = 1+K\frac{b(s)}{a(s)} = 0 \to K = -\frac{a(s)}{b(s)},$$

$$\frac{d\left(\frac{-1}{G(s)}\right)}{ds} = \frac{dK}{ds} = 0 = -\left(\frac{da(s)}{ds}b(s)-a(s)\frac{db(s)}{ds}\right),$$

and it must be verified that they belong to the root locus and, therefore, in them the value of K is positive.

It is easy to see that if there is root locus between two poles or two zeros, there is an odd number of breakpoints between them. If the root locus exists between a pole and a zero, there is not a breakpoint between them (or an even number of breakpoints). There may exist complex breakpoints lying out of the real axis. Moreover, if at a point of the root locus $dK/ds \neq 0$, there is one and only one branch of the root locus passing through this point [5].

6. Determine the *departure angles* (*arrival*) of the root locus from the complex poles (towards the complex zeros). If a test point is selected and moved in the precise vicinity of the complex pole (or complex zero), it is considered that the sum of the angular contributions of the other poles and zeros does not substantially change. Therefore, the angle of arrival (or departure) from the root locus of a complex pole (or complex zero) is found by applying the angle condition $\sum_{\ell=1}^{m}\phi_{z_\ell}-\sum_{i=1}^{n}\phi_{p_i} = \pm180°$, that is, subtracting from $\pm180°$ the sum of all vector angles from all other poles and zeros to the complex pole (or complex zero) in question, including appropriate signs:

   - *Departure angle from a complex pole* $\phi_{p_k} = 180°(2l+1)-\sum_{i=1}^{n}\phi_{p_i}+\sum_{\ell=1}^{m}\phi_{z_\ell} = 180°(2l+1)-\sum_{i=1}^{n}\angle(p_k-p_i)+\sum_{\ell=1}^{m}\angle(p_k-z_\ell); \ i\neq k.$
   - *Arrival angle to a complex zero* $\phi_{z_k} = 180°(2l+1)-\sum_{\ell=1}^{m}\phi_{z_\ell}+\sum_{i=1}^{n}\phi_{p_i} = 180°(2l+1)-\sum_{\ell=1}^{m}\angle(z_k-z_\ell)+\sum_{i=1}^{n}\angle(z_k-p_i); \ \ell\neq k.$

   where it is easy to see that ϕ_{p_i} is the angle that forms with the real axis the vector that joins the pole p_i with the complex pole (zero), and ϕ_{z_ℓ} is the angle that forms with the real axis the vector that joins the zero z_ℓ with the complex pole (zero), see an example in Fig. 7.8.

7. Determine the *crossing points with the imaginary axis*, for example by making $s=j\omega$ in the characteristic equation and equating both the real part and the imaginary part to zero. This gives the values of K (critical gain) and ω for which there are closed-loop roots on the imaginary axis. Note that the Routh-Hurwitz test can be used at this point (as commented, it is not treated in this book, but can be found in classical books, [2, 5, 8]). As the gain term κ is embedded in K in the explanations, it must be taken into account here that the gain of the controller producing j-axis crossings, is obtained by dividing the obtained one by κ (the corresponding frequency does not change).

8. The number of *separate loci* of the geometric root locus is equal to the number of poles of the transfer function, assuming that $(n-m)\geq0$.

9. Take some test points and trace the final root locus. It is recommended to take some test points near the origin of the s plane and near the $j\omega$ axis, given their importance in the relative stability of the closed-loop system.

It is important to realize that the poles, zeros, and the gain of the open-loop system $KG(s)$ are used to build the root locus, not those of the closed-loop transfer function. This is one of the aspects of classical control, which uses the open-loop transfer function to predict closed-loop behavior.

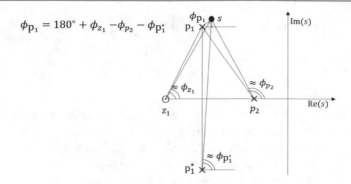

Fig. 7.8 Example of departure angle from a complex pole

What happens if $K < 0$ or $\kappa < 0$?

There may be occasions when there is interest in analyzing the case of the location of the closed-loop poles when the possible values of the parameter are negative. An equivalent situation occurs when there are positive feedback loops, where the characteristic equation is given by $1 - KG(s) = 0$, which produces $KG(s) = 1$. This can also happen in NMP systems, depending on the sign of the resulting gain κ. The variations brought about are given by the interpretation of the basic equation that provides the root locus. In this case, the angle of $G(s)$ becomes $0°$ or a multiple of $360°$. This changes the following rules:

3. The locus on the real axis now exists in those locations that leave an even number of poles plus zeros on the right.
4. The angle of the asymptotes is now $\gamma = \frac{360°(2l+1)}{n-m}$; $l = 0, 1, 2, ..., n - m - 1$.
6. Departure and arrival angles from/to complex poles/zeros:
   Departure angle: $\phi_{p_k} = 360°(2l + 1) - \sum_{i=1}^{n} \phi_{p_i} + \sum_{\ell=1}^{m} \phi_{z_\ell}$; $i \neq k$.
   Arrival angle: $\phi_{z_k} = 360°(2l + 1) - \sum_{\ell=1}^{m} \phi_{z_\ell} + \sum_{i=1}^{n} \phi_{p_i}$; $\ell \neq k$.

What happens if the effect of a parameter different from the gain is relevant?

The root locus technique can be applied with respect to any parameter μ that appears linearly in the characteristic equation, so that it can be transformed to the format $1 + \mu G'(s) = 0$. The same rules apply to it as to the basic root locus.[11] Rather than giving a theoretical development, the idea is summarized for a simple example. Consider the following transfer function and algebraic manipulations, where μ is the uncertain parameter instead of the gain K:

$$G(s) = \frac{s + 1}{s^2(s + \mu)} \rightarrow 1 + G(s) = 0 \rightarrow s^2(s + \mu) + s + 1 = 0 \rightarrow$$

$$\rightarrow \mu = \frac{-s^3 - s - 1}{s^2} = -\frac{1}{G'(s)} \rightarrow G'(s) = \frac{s^2}{s^3 + s + 1}.$$

Representing the root locus of $G'(s)$ provides the location of the closed-loop poles when μ varies from 0 to ∞. In this example, it is easy to check that for $\mu \leq 1$ the closed-loop system is unstable. It can be also checked using the characteristic equation $s^3 + \mu s^2 + s + 1 = 0$, $s = j\omega \rightarrow \mu = 1$.

What happens in the presence of time delays?

There are also variants of the method for systems with time delays. In such cases, the phase-angle condition is directly applied by adding the phase-lag introduced by the time delay, or either a polynomial approximation to the exponential term representing the time delay in the Laplace domain is used [9]. The best approximation (discussed in Chap. 4) is that of Padé:

$$e^{-t_d s} \approx \frac{1 - s(t_d/2)}{1 + s(t_d/2)}.$$

---

[11] When more than one parameter changes, the locus of the roots is called the *root contour* [5].

Issues on control design

It is important to remark that the previous explanation is the classical approach to the root locus method, but thanks to tools like the one introduced in this section, the method is not only useful to analyze the effect a variation in a parameter plays in the location of closed-loop poles, but also for control design purposes. In fact, once determined the region in the LHP where the dominant closed-loop poles must be placed to guarantee the given specifications, the root locus method can be used to find a controller $C(s)$ to fulfill those specifications (including both controller structure and parameters selection). If a dynamic controller structure is proposed, the root locus can be drawn to assess whether the desired set of closed-loop poles can (potentially) be achieved with such controller structure. If not, the dynamic part of the controller can be adapted by introducing more zeros and/or poles that modify the root locus in the desired way. As will be analyzed in the next chapter, a general procedure to design a controller using the root locus method implies [10]:

1. Determine the desired positions of the closed-loop poles that satisfy the given specifications. This is often achieved by assuming that the closed-loop system has a certain configuration of dominant poles.
2. Choose $C(s)$ so that all branches of the root locus:
   a. Are in the LHP for at least some value of the gain K (this is a necessary condition for a stabilizing controller to exist).
   b. Traverse the respective zones of the desired pole locations (this is a necessary condition to achieve the desired performance).
3. Choose K so that the closed-loop poles take on the desired values at the root locus.

One of the problems identified, which makes its application difficult, is how best to explain how to use the root locus technique to achieve control system designs that are relatively easy to implement and also meet specifications with good performance, including high disturbance rejection and good stability margins.

For control design purposes, the three main rules designers use to apply are number 3, 4 and 8 of those enumerated before. If $n - m > 2$ necessarily some branches of the root locus will penetrate the RHP and consequently the system will be unstable for sufficiently large values of K. There is also a *relative degree rule* ([5], Remark 5.14), which is a conservation law for systems satisfying $n \geq m + 2$, which states that the sum of the closed-loop poles is equal to the sum of the open-loop poles, is constant, and independent of K. Thus, the total distance some closed-loop poles go to the left is equal to the total distance other poles go to the right.

It is important for students to visualize the root locus of some characteristic systems, so that they can internalize how the addition of poles and zeros influences the shape of the roots location. For this purpose, interactive tools such as the one presented in this section are very useful. Figure 7.9 shows the root locus of basic systems, which allows general conclusions to be drawn on how the inclusion of poles and/or zeros of the open-loop transfer function $L(s)$ affects the root locus structure itself. The central idea of designing a controller $C(s)$ using the root locus is to try to curve the branches of the root locus of the uncompensated system towards the LHP, in order to ensure that the system is stable. From Fig. 7.9 some useful insights can be made:

- Two poles located on the real axis of the LHP, and a zero positioned to the left of these poles, is sufficient to make the root locus curve to the left.
- To achieve a circular pattern that pulls the root locus to the left, the number of zeros required in the open-loop transfer function must be equal to the number of poles minus one.
- The placement of the leftmost zero has a substantial bearing on how far the root locus can be bent to the left, which in turn has a direct influence on the potential bandwidth of the closed-loop system.

7.2.1.3 References Related to this Concept

- [2] Dorf, R. C., & Bishop, R. H. (2011). *Modern control systems* (12th ed.). Prentice Hall. ISBN: 978-0-13-602458-3.
- [5] Bavafa-Toosi, Y. (2017). *Introduction to linear control systems.* Academic Press-Elsevier. ISBN: 978-0-12-812748-3. Chapter 5.
- [6] Åström, K. J., & Murray, R. M. (2014). *Feedback systems: An introduction for scientists and engineers* (2nd ed.). Princeton University Press. ISBN: 9780691193984.
- [7] Bolzern, P., Scattolini, R., & Schiavoni, N. (2009). *Fundamentos de control automático (Fundamentals of automatic control).* McGraw-Hill, ISBN: 978-84-481-6640-3.
- [8] Ogata, K. (2010). *Modern control engineering* (5th ed.). Prentice Hall, ISBN: 978-0-13-615673-4.
- [10] Díaz, J. M., Costa-Castelló, R., & Dormido, S. (2012). An interactive approach to control systems analysis and design by the root locus technique. *Revista Iberoamericana de Automática e Informática Industrial, 18*(2), 172–188. DOI:https://doi.org/10.4995/riai.2020.13811.

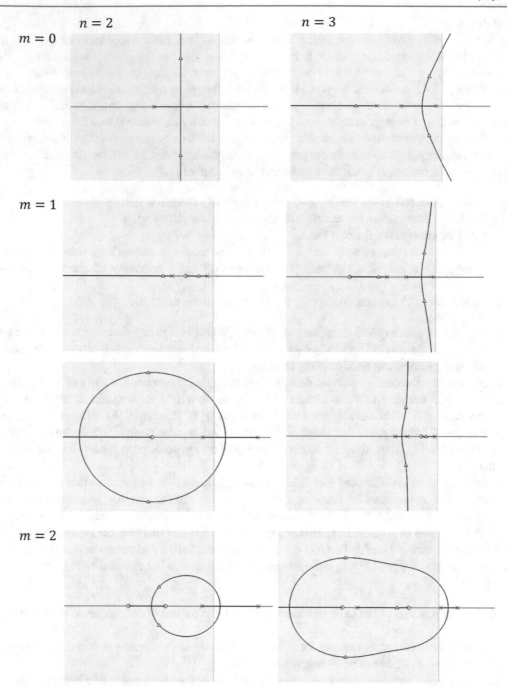

Fig. 7.9 Representative root loci. Open-loop poles (n) and zeros (m) are represented by the symbols $\times$ and $\circ$ respectively. Closed-loop poles are represented by the symbol $\triangle$

- [11] Barrientos, A., Sanz, R., Matía, F., & Gambao, E. (1996). *Control de sistemas continuos. Problemas resueltos (Control of continuous systems. Problems solved)*. Mc Graw-Hill, ISBN: 84-481-0605-9.
- [12] D'Azzo, J. J., Houpis, C. H., & Sheldon, S. N. (2003). *Linear control system analysis and design with MATLAB*® (5th ed.). Marcel Dekker Inc., ISBN: 0-8247-4038-6.
- [13] Franklin, G. F., Powell, J. D., & Emani-Naeni, A. (2010). *Feedback control of dynamic systems* (6th ed.). Pearson.
- [14] Shahian, B., & Hassul, M. (1993). *Control system design using MATLAB*®. Prentice Hall, ISBN: 0-13-174061-X.
- [15] Truxal, J. G. (1955). *Automatic feedback control system synthesis*. McGraw-Hill.

Application   Interactive tool: root_locus

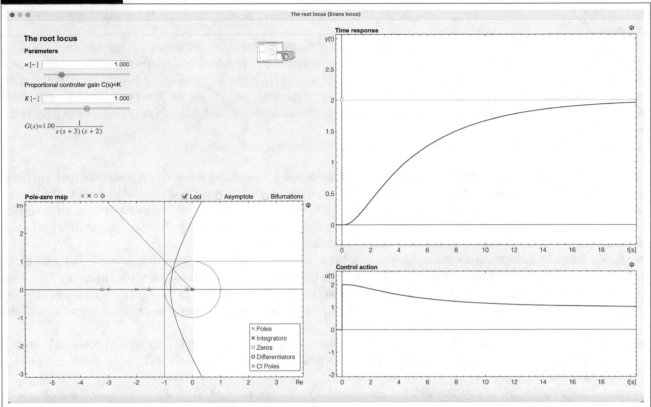

Root Locus

The main objective of this tool is to analyze the root locus of negative feedback linear systems for $K \geq 0$, but also for systems with $\kappa < 0$ providing the equivalent to a positive feedback (negative gain) root locus. The interactive tool allows the user to understand the main underlying concepts, and to visualize the elements that usually help the user to construct the root locus: asymptotes, centroids, departure and arrival angles, and others. Moreover, it is very useful to analyze and visualize the effect of including open-loop poles and zeros on the location of closed-loop poles, being a powerful tool for controllers' structure selection.

The tool is divided into three areas, although, in this case, the fundamental one is that corresponding to the **Pole-zero map**, which is very rich in visual contents.

The **Parameters** area is reserved only for modifying the controller gain K (whose values have been limited so that it is always positive) through a textbox and a slider. A symbolic representation of the resulting transfer function is included. The system gain κ can also be modified here (note that the transfer function is represented in ZPK format, allowing both for positive and negative values of κ), although the analysis is usually performed based on the changes in the proportional gain of the K controller.

The **Pole-zero map** includes the editor of poles and zeros (repository from which poles, zeros, integrators and differentiators can be dragged into/out (or double-clicking) the s-plane) and the possibility of activating or deactivating the representation of Asymptotes (and corresponding centroids) and Bifurcations (arrival and departure points on the real axis) of the root locus. The label shown in the plot in this case, in addition to the poles, zeros, integrators and differentiators (with their associated symbols), includes the centroid symbol ($\diamond$) and that of the bifurcation points ($\square$). The poles of the closed-loop system (solutions of the characteristic equation) are represented on the root locus (represented in blue color) using a triangle ($\triangle$) as a symbol. The Loci option activates the geometric loci of points in the s-plane with constant ζ, ω_n, σ and ω_d, to facilitate the analysis of the fulfillment of closed-loop specifications. Using the poles and zeros editor, any system can be configured (MP, NMP). By placing the mouse over these elements in the plot, the corresponding representative values are shown in the lower-left corner of the tool. If the poles or zeros are complex conjugated, the departure or arrival angles of the root locus are added to this basic information. In the case of the closed-loop poles, in addition to their location, the value of the associated gain K is also displayed. The same applies to the position of the bifurcation points and centroids (as well as the angle of the asymptotes).

The upper-right area of the tool represents the **Time response** of the closed-loop system for the poles and zeros configuration shown in the **Pole-zero map** (blue color is always used to represent the closed-loop system response). The reference is represented in a dashed green line. The lower graph shows the Input signal to the system (generated from the error between the reference and the closed-loop system output). In both graphs, if the mouse is placed over the time plots, the value associated with the time (t) and the corresponding signal (y,u) is displayed. The scale can be modified using the settings available in the gearwheel icon.

The Options menu includes by default five examples with typical initial configurations:

- Example 1: It represents a third-order system with two real poles and an integrator. It is a very typical case found in many textbooks to analyze the values of the asymptotes and loop gain from which the roots of the closed loop cut the imaginary axis and the system becomes unstable (stability conditioned to the value of the gain). It is also very useful for analyzing the concept of dominant poles, because by increasing the gain, one of the poles of the closed-loop system moves away from the origin and has less and less influence on the dynamic response.
- Example 2: It has a similar structure to the previous one, but a zero is included in the numerator, so one of the roots of the closed loop will reach the location of that zero for very high loop gains. The inclusion of the zero also causes two of the asymptotes to be located on the RHP in this case. By moving the location of the zero to the right from its initial location, dragging it along the axis, it is interesting to analyze how the root locus changes its profile and how the zeros "attract" the locus, making the closed-loop system stable.
- Example 3: It represents a high-order system with an integrator, a real pole, two complex conjugated poles and two real zeros. For positive gains it can be verified that the system will always be stable and there will be two dominant complex conjugated poles.
- Example 4: It shows the case of a system with two conjugated complex zeros in the LHP, and two conjugated complex poles in the RHP (unstable open-loop system). It is an interesting example to analyze how an unstable open-loop system can be stabilized by increasing the loop gain (with positive gains and negative feedback). It is also useful to analyze the concept of *conditional stability*.
- Example 5: It represents a high-order system with two real poles, two conjugated complex poles and two conjugated complex zeros. If the less dominant real pole moves away from the imaginary axis, it changes the configuration of separated loci. If one of the poles is moved into the RHP, the system becomes conditionally stable.

In addition, the Options menu includes the possibility of entering generic transfer functions in polynomial and zero-pole formats: Introduce plant (NUM,DEN) and Introduce plant ZPK.

7.2.1.4 Homework

1. For $L(s) = \dfrac{K}{s(s+1)(s+2)}$, represent the root locus for $K > 0$ and for $K < 0$ (notice that in this case what has to be done is to select $\kappa = -1$ in the interactive tool, as K is limited to be positive). Find all the elements described in the classical *rules* to build the root locus.
2. Select the Example 1 ($P_8(s)$ in Table 1.2) from the Options menu. Indicate if there is a gain value that makes the closed-loop system unstable (consider only the case of positive gain) and, if so, provide the value of that gain. Where is the centroid of the asymptotes and what angles do these asymptotes form with the real positive axis? Add a real zero from the repository to the left of the two real poles and indicate if the position of the bifurcation point changes. What are the angles of the asymptotes in this case? In which range of gain K values are there no complex conjugated poles in the closed loop?
3. Exercise 2.2 from reference [16]. A servomotor described by transfer function (5.18) is included in a feedback system in series with a gain K for positioning of a tool in a tooling machine. The following figure shows the poles of the closed-loop system for different values of the gain K. Find (without calculations), the value of K corresponding to the step responses shown below. Can the position of the poles be inferred from this information? Use also the root_locus interactive tool to try solving the exercise.

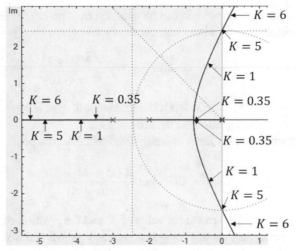

Location of closed-loop poles for different values of K

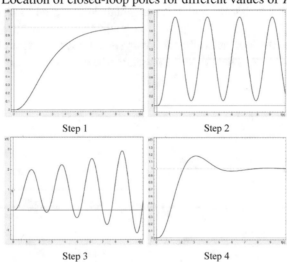

Step 1   Step 2

Step 3   Step 4

4. Select the Example 2 from the Options menu, representing the product of transfer functions $P_2(s)P_6(s)$ in Table 1.2. Indicate the value of the loop gain that makes the system unstable. Determine the value of that gain that causes a 50% percentage peak overshoot in the closed-loop response. What is the value of the peak time? What order is the closed-loop system? Is it appropriate to use peak overshoot and peak time as time characteristics of the closed-loop system response? Calculate the location of the centroid, asymptote angles, and bifurcation points.

5. Select the Example 3 from the Options menu (product $P_6(s)P_{12}(s)$ in Table 1.2). Answer in this case to the same questions formulated in the previous exercise.

6. Select the Example 4 from the Options menu. Analyze if there is any possible location of the zeros (in the real axis) providing an overdamped response in the closed-loop system. Justify the answer. Also indicate a possible location of the zero closest to the imaginary axis that makes the asymptote the imaginary axis itself.

7. Select the Example 5 from the Options menu. Indicate if there is any value of the loop gain that makes the closed-loop system unstable. Is it possible to obtain an overdamped response for any value of the gain? Is there a centroid or a bifurcation in this case? Move the two complex conjugated zeros so that the polynomial of the transfer function's numerator is approximately $s^2 + 0.8s + 8$. Calculate the range of values of K that makes the closed-loop system stable. Return to the original settings. Move the pole that is initially at $s = -2.05$ to $s = -4$. How many separate loci does the root locus have? Where are the centroid and the bifurcation point located?

8. Select the Example 5 from the Options menu. Using the initial configuration of poles and zeros, indicate the gain values that, from your point of view, cause the closed-loop system to have less relative stability.

9. Select a system with a pole at $s = -1$ and a zero at $s = 1$. Is there any value of K that stabilizes the closed-loop system? What kind of time response do you observe?

10. The takeoff of an aircraft is inherently unstable and can be modeled by a transfer function $G(s)$. To control the vehicle, an adjustable thruster control system is used whose transfer function is given by $C(s)$ [2]:

$$G(s) = \frac{1}{s(s-1)}, \quad C(s) = \frac{K(s+2)}{(s+10)}.$$

The loop transfer function is given by $L(s) = C(s)G(s)$. Determine the K gain interval for which the system is marginally stable and the roots of the characteristic equation for the critical gain.

11. The control system of a car suspension tester has a unit negative feedback on a process with transfer function [2]:

$$L(s) = KG(s) = \frac{K(s^2 + 4s + 8)}{s^2(s+4)}.$$

The dominant roots must have $\zeta = 0.5$. Determine the value of K and the position of the dominant roots.

12. Select a system that has an integrator, a pole at $s = -1$, a pole at $s = -2$ and a real zero (its initial position can be $s = -4$). Analyze the different shapes of the root locus for $\kappa = 1$ and $K = 1$ depending on the zero position. Justify if there is a location of the zero for which three real poles are obtained in closed loop, and if an overdamped response is achieved.

13. Using the interactive tool, analyze the root locus of the following generic systems from Truxal's classical book [15], which loop transfer function $L(s) = KG(s)$ has the following structure (the user can choose any value of the parameters), when the gain K is modified:

$$L_1(s) = \frac{K}{\tau s + 1}, \quad L_2(s) = \frac{K}{(\tau_1 s + 1)(\tau_2 s + 1)}, \quad L_3(s) = \frac{K}{s},$$

$$L_4(s) = \frac{K}{s(\tau s + 1)}, \quad L_5(s) = \frac{K}{s(\tau_1 s + 1)(\tau_2 s + 1)}, \quad L_6(s) = \frac{K(\beta s + 1)}{s(\tau_1 s + 1)(\tau_2 s + 1)},$$

$$L_7(s) = \frac{K}{s^2}, \quad L_8(s) = \frac{K}{s^2(\tau s + 1)}, \quad L_9(s) = \frac{K(\beta s' + 1)}{s^2(\tau s + 1)},$$

$$L_{10}(s) = \frac{K}{s^3}, \quad L_{11}(s) = \frac{K(\beta s + 1)}{s^3}, \quad L_{12}(s) = \frac{K(\beta_1 s + 1)(\beta_2 s + 1)}{s^3},$$

$$L_{13}(s) = \frac{K(\beta s + 1)}{s^2(\tau_1 s + 1)(\tau_2 s + 1)}, \quad L_{14}(s) = \frac{K}{(\tau_1 s + 1)(\tau_2 s + 1)(\tau_3 s + 1)},$$

$$L_{15}(s) = \frac{K(\beta_1 s + 1)(\beta_2 s + 1)}{s(\tau_1 s + 1)(\tau_2 s + 1)(\tau_3 s + 1)(\tau_4 s + 1)}.$$

14. Study the stability of the systems in the following figure when K varies between 0 and ∞ [11]:

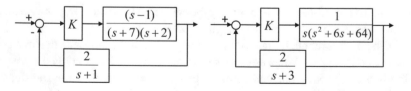

15. Based on the root locus for $K > 0$, analyze the stability of the following systems described in [6] and [11]:

$$L_1(s) = \frac{K(s+6)^2}{s(s+1)^2}, \quad L_2(s) = \frac{2K(s+4)}{(s^2 + 2s + 2)(s+2)(s+3)},$$

$$L_3(s) = \frac{K(s+3)(s+2)}{(s^2 + 2s + 2)(s+1)}.$$

16. Consider the system $L(s) = K \dfrac{s + 0.5}{(s^2 - 10)(s^2 + s + 10)}$ (reference [5], Example 5.12). Find the value of gain and frequency where the system has two poles on the j-axis.

17. Represent the root locus of the systems described by the following transfer functions (reference [10]):

$$L_1(s) = KG_1(s) = K \frac{1 - s}{(1 + s)(1 + 0.5s)}, \quad L_2(s) = KG_2(s) = K \frac{s - 1}{(s + 1)(s + 2)}.$$

As can be observed, both have the same location of poles and zeros, but the root locus is different. What is the reason?

18. A pathological case in root locus drawing (reference [5], Example 5.16) if given by the transfer function:

$$L(s) = \frac{K(bs + 64)}{s(s + 1)(s^2 + 7s + 25)} = \frac{K(bs + 64)}{s^4 + 8s^3 + 32s^2 + 25s}.$$

Using the (NUM,DEN) format in the Options menu, enter that transfer function ($[b, 64], [1, 8, 32, 25, 0]$) for values of $b = 38, 39$ and 40. This is a case in which small perturbations in one parameter of the transfer function result in a different root locus, requiring software tools instead of hand drawing for the analysis. The user should have to select adequate values of K for stabilizing the closed loop, as well as applying variable scaling in the graphs.

19. Consider the following plant, controller and sensor dynamics:

$$G(s) = \frac{1700}{s(s^2 + 80s + 1700)}; \quad C(s) = K; \quad H(s) = \frac{1}{0.02s + 1},$$

where it is desired that the rightmost poles have $\zeta = 0.6$. Using the (NUM,DEN) format in the Options menu of the interactive tool root_locus and introducing the transfer function $1700, [0.02, 2.6, 114, 1700, 0]$, find the controller C to achieve the specification on ζ and analyze the achieved time response for a reference step input. Are the rightmost poles dominant? In reference [5] (Example 5.17) the complete development of this example can be found so that it can be compared with the results achieved using the interactive tool.

20. The root locus analysis of the following transfer function is proposed in [5] (Example 5.18).

$$L(s) = \frac{K(2 - s)}{s(s + 1)(s^2 + 6s + 12)}.$$

Draw the root locus using the interactive tool entering the transfer function through the (NUM,DEN) option in the form $[-1, 2], [1, 7, 18, 12, 0]$. Check if the value of κ is correct and reason about the gain of the loop transfer function and the implication in the rules applied to build the locus. Check if the system is stable for any value of K and analyze other features of the root locus, as the angle of departure of the complex poles.

21. Considering the following loop transfer function:

$$L(s) = \frac{K(\beta s + 1)}{s^2(\tau s + 1)},$$

select any positive value of τ and analyze the stability of the closed-loop system when K varies between 0 and ∞ considering three cases: $\beta = 1.5\tau$, $\beta = \tau$ and $\beta = 0.5\tau$. Indicate the value of the gain at which the system becomes unstable in each case. In the unstable case, propose a modification to the function $L(s)$ (by adding poles or zeros) that stabilizes the system for any value of K.

22. Exercise 3.6 from reference [16]. Using the interactive tool, plot the root locus for the following transfer functions. For which values of K are the systems stable? What conclusions on the principal shape of the step response can be drawn from the root locus?

A Ferris wheel: $\quad$ A Mars rover: $\quad$ A magnetic floater:

$$L_1(s) = \frac{K(s + 2)}{s(s + 1)(s + 3)}, \quad L_2(s) = \frac{K}{s(s^2 + 2s + 2)}, \quad L_3(s) = \frac{K(s + 1)}{s(s - 1)(s + 6)}.$$

23. Modification of Exercise 3.9 from reference [16]. The transfer function of a chemical reactor is given by:

$$G(s) = \frac{1}{(s+1)(s-1)(s+5)}.$$

   a. Use a proportional controller and draw a root locus with respect to K. Calculate the value of K that stabilizes the system.
   b. Add a zero in $s = -2$ and draw the root locus with respect to K. For which values of K does the controller $C(s) = K(s+2)$ stabilize the system?

7.3   The Nyquist Stability Criterion

7.3.1   Interactive Tool: Nyquist_criterion

7.3.1.1 Concepts Analyzed in the Card and Learning Outcomes
- Frequency-domain stability analysis.
- Understand the Nyquist stability criterion and the role of the Nyquist plot.
- Stability analysis through visualization.

7.3.1.2 Summary of Fundamental Theory
The Nyquist stability criterion determines the stability of a closed-loop system from the frequency response of the open-loop system and the open-loop poles and zeros. It uses the polar diagram as the basic representation, where the imaginary part of the loop transfer function is shown on the ordinate axis ($\mathrm{Im}(L(j\omega))$) and the real part on the abscissa axis ($\mathrm{Re}(L(j\omega))$). If the characteristic equation is analyzed with unit feedback:

$$J(s) = 1 + C(s)G(s) = 1 + L(s) = 0. \tag{7.9}$$

For the system to be stable, all the roots of the characteristic equation must lie in the open LHP. The Nyquist stability criterion helps to graphically determine the stability of a closed-loop system from the open-loop frequency response plot, without having to determine the closed-loop poles. The criterion is based on the complex plane contour transformation theorem (Cauchy theorem of the complex variable theory, known as the *principle of argument*). Specifically, since to guarantee stability, it must be ensured that all the zeros (roots) of $J(s)$ are on the LHP, a transformation of the RHP in the $J(s)$ plane is performed (a closed contour in the s-plane results in a closed contour in the $J(s)$ plane). Notice that this transformation is used to explain the Cauchy theorem, but the Nyquist stability criterion is applied using $L(s)$ instead of $J(s)$, as explained below.

Cauchy theorem (principle of argument): If a contour Γ_s in the s-plane surrounds Z zeros and P poles of $J(s)$ (considering a multiplicity of poles and zeros; and that $J(s)$ and thus $L(s)$ are analytic, nonzero on Γ_s, and with a finite number of zeros and poles inside Γ_s) and it does not pass through any pole or zero of $J(s)$ when the path is traversed in the clockwise direction along the contour, the corresponding contour Γ_J in the $J(s)$-plane surrounds the origin of such plane N = Z − P times in the same direction. Note that a positive number N indicates that there are more zeros than poles in the function $J(s)$ and a negative number N indicates that there are more poles than zeros. The number P is easily determined for $J(s)$ from the function $L(s)$ (it is the number of unstable poles of $L(s)$). Therefore, if N is determined from the graph of $J(s)$, it will be easy to determine the number of zeros of $J(s)$ (poles of the closed-loop transfer function) in the closed contour in the s-plane.

   Cauchy's theorem can be better understood by considering $J(s)$ in terms of the angle due to each pole and zero when the contour Γ_s is traversed clockwise [2]. Let's consider a simple example:

$$J(s) = \frac{(s-z_1)(s-z_1^*)}{(s-p_1)(s-p_2)} = \underbrace{\frac{|s-z_1||s-z_1^*|}{|s-p_1||s-p_2|}}_{|J(s)|} \underbrace{(\phi_{z_1} + \phi_{z_1^*} - \phi_{p_1} - \phi_{p_2})}_{\lfloor J(s)}.$$

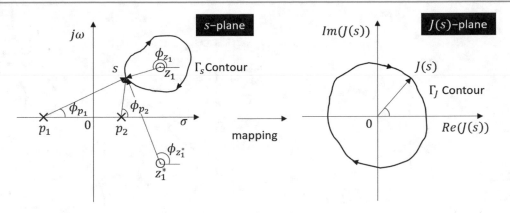

Fig. 7.10 Interpretation of the principle of argument

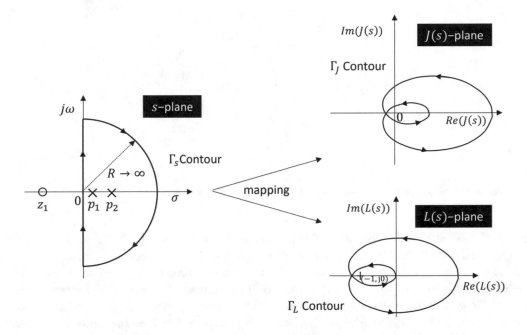

Fig. 7.11 Nyquist contour covering the whole open RHP and example of mapping in the $J(s)$ and $L(s)$ planes

By looking at Fig. 7.10, when the closed contour is traversed in the s-plane, the total angle change for $\phi_{z_1^*}$, ϕ_{p_1} and ϕ_{p_2} is $0°$, while ϕ_{z_1} traverses $360°$. This reasoning can be extended to the rounding of P poles and Z zeros.

The application of the Nyquist criterion is that, for a system to be stable, all zeros of $J(s)$ (closed-loop poles) must be in the LHP. Therefore, the roots of a stable system (the zeros of $J(s)$) must be on the left side of the $j\omega$ axis in the s-plane. A contour Γ_s is chosen in the s-plane which encloses the whole RHP and, using Cauchy's theorem, it is determined if any of the zeros in $J(s)$ is located inside of Γ_s. The contour Γ_s is formed by the complete $j\omega$ axis from $\omega = -\infty$ to ∞ and a semi-circular path of infinite radius in the RHP (Fig. 7.11). The application of Cauchy's theorem involves drawing Γ_J in the $J(s)$-plane and determining the number N of encirclements around the origin. Then the number of zeros of $J(s)$ inside the contour Γ_s (zeros of $J(s)$ in the RHP) is Z = P + N. Therefore, if P = 0, as is usually the case (if it is non-zero it implies unstable open-loop poles), it follows that the number of unstable closed-loop poles is equal to N, the number of encirclements around the origin of the $J(s)$-plane.

As $L(s)$ is generally obtained in a factorized form, and $L(s) = J(s) - 1$, the number of encirclements around the origin of the $J(s)$-plane in the clockwise direction will be equal to the number of encirclements in the same direction around the critical point $-1 + j0$ in the $L(s)$-plane. Therefore, the simplified *Nyquist stability criterion* is established as [2]: "A feedback system is stable if, and only if, the contour Γ_L in the $L(s)$-plane does not circle the point $(-1, 0)$ when the number of poles of $L(s)$ in the RHP is zero (P = 0)". When the number of poles of $L(s)$ in the RHP is different from zero, the Nyquist

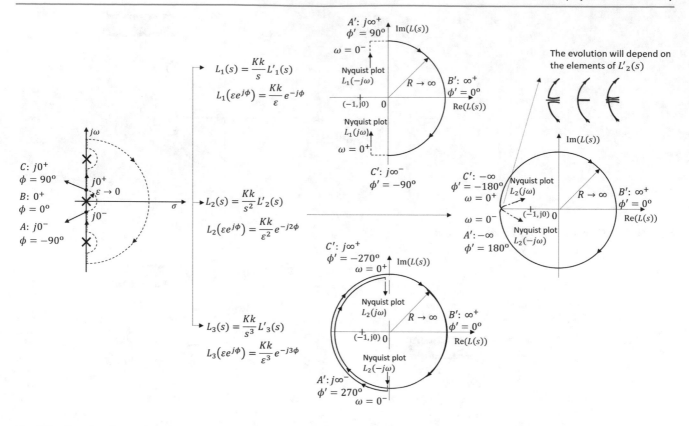

Fig. 7.12 Nyquist contour with singularities in the j-axis, with $L_i(s)$ in times constants format

criterion becomes: "A feedback control system is stable if, and only if, for the contour Γ_L, the number of counterclockwise encirclements around the point $(-1, 0)$ is equal to the number of poles of $L(s)$ with real positive parts".

The basis of the two previous concepts is the fact that, for the transformation of $L(s)$, the number of roots (or zeros) of $1 + L(s)$ in the RHP is represented by Z = N + P. Obviously, if the number of poles of $L(s)$ in the RHP is zero (P = 0), for a closed-loop stable system it is necessary that N = 0 and the contour Γ_L should not surround the point $(-1, 0)$. Also, if P is different from zero, and for a closed-loop stable system, it is required that Z = 0, then N = −P, which involves P counterclockwise encirclements.

In practice, the basic approach consists in drawing the Nyquist diagram both for positive and negative frequencies (the negative frequency part is the mirror image of the positive one with respect to the real axis) and then count the number of encirclements. This works if the system does not have zeros/poles on the j-axis. If the open-loop system has integrators or poles on the imaginary axis (the same applies for differentiators and zeros), the Γ_s contour is drawn in such a way as to avoid passing through these singularities, by introducing a small semicircle of radius $\epsilon \to 0$ that surrounds the singularity in a counterclockwise direction (the semicircle is included in the RHP). In that contour $s = \epsilon e^{j\phi}$, where $\epsilon \to 0$ and ϕ varies from $-90°$ in $\omega = 0^-$ to $+90°$ in $\omega = 0^+$ in the case of a zero or pole at the origin. The points corresponding to $s = j0^+$ and $s = j0^-$ of the geometric locus of $L(s)$ in the $L(j\omega)$ plane are $-j\infty$ and $j\infty$ respectively. When $L(s) = K/s$, it becomes

$$L(\epsilon e^{j\phi}) = \frac{K}{\epsilon e^{j\phi}} = \frac{K}{\epsilon} e^{-j\phi},$$

where $K/\epsilon \to \infty$ when $\epsilon \to 0$ and $-\phi$ varies from 90° to −90° when the representative point s moves along the semicircle. Therefore, the points $L(j0^-) = j\infty$ and $L(j0^+) = -j\infty$ are joined by a semicircle of infinite radius in the right semi-plane of the $L(j\omega)$ plane. When $L(s)$ has a factor $1/s^\aleph$, where $\aleph = 2, 3, ...,$ the plot of $L(s)$ has $\aleph$ clockwise semicircles of infinite radius from the origin [8]. These considerations also apply to poles and zeros on the $j\omega$ axis. For instance, if $L(s)$ has two poles at the origin (type-2, $\aleph = 2$), $\lim_{s \to \epsilon e^{j\phi}} \frac{1}{s^2} = \frac{1}{\epsilon^2} e^{-j2\phi}$, so that when ϕ varies between $-90°$ to $+90°$ in the semicircle of radius ϵ in the s-plane, those points move from $+180°$ to $-180°$ in the $L(s)$-plane (Fig. 7.12). An alternative procedure can be analyzed in [5].

7.3.1.3 References Related to this Concept

- [2] Dorf, R. C., & Bishop, R. H. (2011). *Modern control systems* (12th ed.). Prentice Hall. ISBN: 978-0-13-602458-3.
- [5] Bavafa-Toosi, Y. (2017). *Introduction to linear control systems*. Academic Press-Elsevier. ISBN: 978-0-12-812748-3.
- [6] Åström, K. J., & Murray, R. M. (2014). *Feedback systems: An introduction for scientists and engineers* (2nd ed.). Princeton University Press, ISBN: 9780691193984.
- [8] Ogata, K. (2010). *Modern control engineering* (5th ed.). Prentice Hall, ISBN: 978-0-13-615673-4.
- [11] Barrientos, A., Sanz, R., Matía, F., & Gambao, E. (1996). *Control de sistemas continuos. Problemas resueltos (Control of continuous systems. Problems solved)*. Mc Graw-Hill, ISBN: 84-481-0605-9.

Application   Interactive tool: Nyquist_criterion

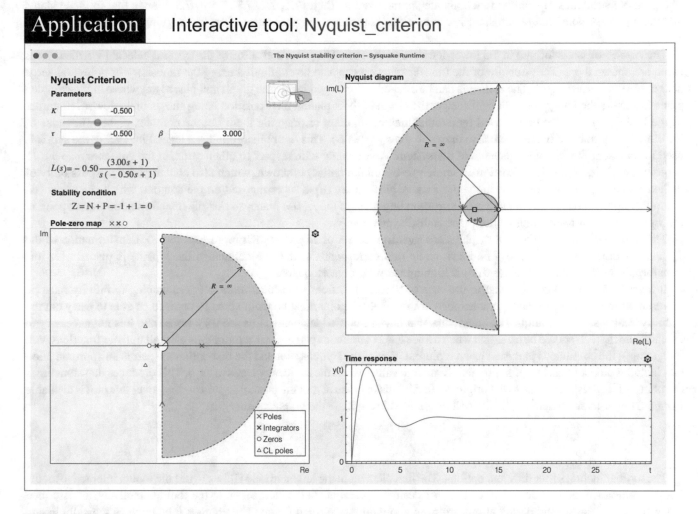

Nyquist Criterion

This interactive tool is devoted to graphically understanding the Nyquist criterion, and how the stability of the closed-loop system can be analyzed based on the open-loop transfer function and by the application of Cauchy's theorem. In this tool, an attempt has been made to find a tradeoff between visualization and conceptualization, trying to preserve the spirit of classic books [17], by selecting appropriate use cases and a representation of the contours at infinity. Due to the fact that it is quite difficult to represent contours at infinity, the internal parameters have been hidden to the user, while constraining the possible cases for analysis to twenty-nine (which cover mostly all reasonable cases).

This tool has taken a great deal of effort to develop and has resulted in a solution that is reasonably robust and has been tested in a number of scenarios, but as there are countless situations that can arise, it may need to be reset on occasion.

The upper-left area is dedicated to setting the loop transfer function gain K through a textbox (useful for changing the sign of the gain) or a slider. The structure of the transfer function $L(s)$ has to be entered using the structures of plants in the Options menu (from $P1(s)$ to $P29(s)$, all of them with unit static gain so that the open-loop gain is represented by K) or through the editor of poles and zeros (repository). The first way has the advantage that it starts from pre-set initial conditions for the transfer function which has been selected to be a case study of interest. The second way has the advantage that with the interactive addition/deletion of poles/zeros/integrators, one has a very visual perception of the effect of these modifications on the shape of the Nyquist plot. All cases of proper transfer functions have been implemented up to order three. Twenty-nine possible structures (including all in Table 1.2) are therefore available, with arbitrary parameters that, once selected, can be changed in this area of the tool using the corresponding textboxes and sliders. The position of poles and zeros can be modified in the **Pole-zero map** by dragging them to desired locations. A symbolic representation of the transfer function $L(s)$ is also shown, as well as the formula indicating compliance with the Nyquist criterion ($Z = P + N$), which is evaluated automatically. The possible structures of transfer functions are summarized in Tables 7.1, 7.2, 7.3, 7.4 to 7.5. Notice that in those plants including pairs of poles and/or zeros, these can be either complex conjugates or real, just by dragging them out/into the real axis.

The **Pole-zero map** is located in the lower-left area of the tool. The poles and zeros of the systems selected via the Options menu are shown there. The structure of the transfer function can also be configured using the repository of poles and zeros next to the chart title ($\times \times \circ$). The elements can be clicked and dragged inside the graph and placed anywhere in the complex plane (logically the integrators will be placed at the origin of the s-plane), their position being shown in the lower-left corner of the tool. They can also be removed by double-clicking on them or dragging them out the complex plane. The poles of the closed-loop transfer function are also drawn in the s plane ($\triangle$). This figure incorporates a legend indicating the symbols used to represent the dynamic elements of the system. As a novelty with respect to other cards, in the **Pole-zero map** the Γ_s contour that includes the RHP (shadowed semicircle of infinite radius) is shown, which also contemplates the possibility of the existence of poles and/or zeros in the $j\omega$-axis. A black circle ($\circ$) is superimposed on the contour, which can be moved around using the mouse and which has its corresponding point in the $L(j\omega)$ plane, so that the transformation from a point in the s plane to its corresponding $L(j\omega)$ plane is displayed directly.

The upper-right area of the tool displays the **Nyquist diagram** of $L(j\omega)$, which represents the Γ_L transformation of the Γ_s contour drawn in the s-plane. The black circle that corresponds to the one defined in the s-plane is placed over this transformation. The critical point $(-1, 0)$ is highlighted with a black square.

It must be taken into account that in the case of transfer functions with integrators, a curve tending to infinity must be represented in a finite space and the coherence of the $-1 + j0$ point must be maintained in order to be able to carry out the stability analysis. The techniques and heuristics that have been used to achieve this are very varied and it is not necessary to explain them here. It should be clear that when a trace has a continuous part and a discontinuous part (all transfer functions with integrators), the continuous part is an exact representation of the Nyquist plot and the discontinuous part is an approximation that includes adjustments by splines to the asymptotic values when the frequency tends to zero. This produces that sometimes the point $-1 + j0$ is very close to the origin, so that in this interactive tool the possibility of zooming over this plot is available from the upper menu, where a specific toolbar is added to facilitate the graphical analysis:

The toolbar includes four different options. The first element in the left hand side allows to use the tool in interactive mode. So, this option should be selected when the user desires to manipulate the elements into the tool interactively, as described along this text. The second option allows to make zoom on the selected figure. The zoom can be done as typically in any software, by using the mouse to select an area, and afterwards the selected area is made larger. On the other hand, it can be used just only by clicking on any part of the figure. To make zoom out, the user should click with the mouse together with shift key at the same time on any part of the figure. The third option can be used to make zoom only in the x-axis, by following the same procedure than for the previous zoom option. Finally, the last option allows to change the scale of the figure by dragging the mouse in any direction. It is important to remark that the first option should be selected to use the tool with the interactive capabilities.

There are also some situations in which the Nyquist curve seems to behave badly because the trajectory becomes very large. This is normal in some transfer functions when their elements approach the imaginary axis, since there is a change of curvature in the trace when passing from one half-plane to another. This cannot be avoided and what has to be done is to avoid placing the poles/zeros in these areas. This situation also occurs when there are poles and zeros whose relative positions are changing, since the contribution of the phases of these elements is also modified (as in the case analyzed of first order systems with a zero).

Under the Nyquist graph of $L(j\omega)$ the **Time response** produced by a unit step in the reference to the closed-loop system is shown.

Table 7.1 Plants 1–6 considered in the interactive tool Nyquist_criterion and stability conditions (SC) or necessary stability conditions (NSC)
List of plant structures

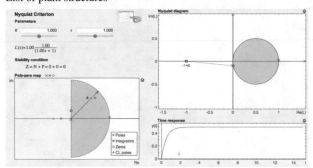

$$P_1(s) = \frac{K}{(\tau s + 1)}$$
SC: $\tau(K+1) > 0$

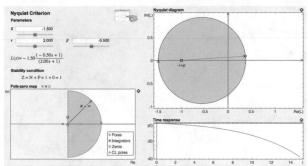

$$P_2(s) = \frac{K(\beta s + 1)}{(\tau s + 1)}$$
SC: $(\tau + K\beta)(K+1) > 0$

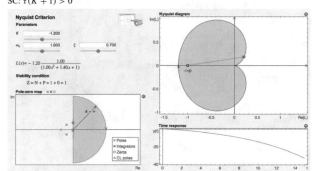

$$P_3(s) = \frac{K\omega_n^2}{(s^2 + 2\zeta\omega_n s + \omega_n^2)}$$
SC: $\zeta > 0, K > -1$

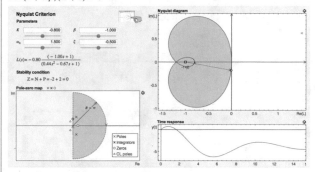

$$P_4(s) = \frac{K\omega_n^2(\beta s + 1)}{(s^2 + 2\zeta\omega_n s + \omega_n^2)}$$
SC: $K > \max\left(-1, -\frac{2\zeta}{\beta\omega_n}\right)$

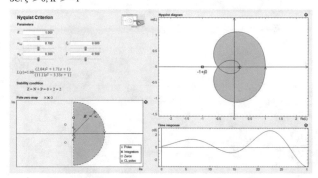

$$P_5(s) = \frac{K\omega_n^2}{\omega_{n_z}^2}\frac{(s^2 + 2\zeta_z\omega_{n_z}s + \omega_{n_z}^2)}{(s^2 + 2\zeta\omega_n s + \omega_n^2)}$$
SC: $K > \max(C5)$ or $K < \min(C5)$
$$C5 = \left(-1, -\frac{\omega_{n_z}^2}{\omega_n^2}, -\left(\frac{\omega_{n_z}}{\omega_n}\right)\left(\frac{\zeta}{\zeta_z}\right)\right)$$

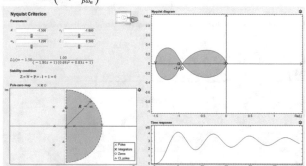

$$P_6(s) = \frac{K\omega_n^2}{(\tau s + 1)(s^2 + 2\zeta\omega_n s + \omega_n^2)}$$
SC: $\tau > 0, \zeta > 0$, and $-1 < K < \frac{2\zeta(1 + 2\tau\zeta\omega_n + \tau^2\omega_n^2)}{\tau\omega_n}$
or $\tau < 0, \zeta > 0, (1 + 2\tau\zeta\omega_n) < 0$ and
$$-\frac{2\omega_n\zeta\tau^2 + 2\zeta + |\tau| - |\tau|\omega_n - 4|\tau|\omega_n\zeta^2}{|\tau|\omega_n} < K < -1$$

Table 7.2 Plants 7–12 considered in the interactive tool Nyquist_criterion and stability conditions (SC) or necessary stability conditions (NSC)

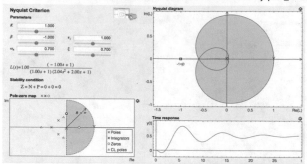

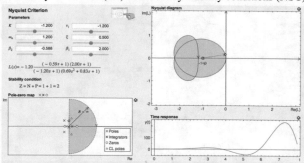

$$P_7(s) = \frac{K\omega_n^2(\beta s + 1)}{(\tau s + 1)(s^2 + 2\zeta\omega_n s + \omega_n^2)}$$

NSC: $\tau > 0$, $\zeta > -\frac{1}{2\tau\omega_n}$, and $K > \max{(C7)}$ or

$\tau > 0$, $\zeta < -\frac{1}{2\tau\omega_n}$, and $K < \min{(C7)}$

$$C7 = \left(-1, -\frac{(1+2\tau\zeta\omega_n)(2\zeta+\tau\omega_n)-\tau\omega_n}{\omega_n\beta(1+2\tau\zeta\omega_n)-\tau\omega_n}\right)$$

$$P_8(s) = \frac{K\omega_n^2}{\omega_{n_z}^2}\frac{(s^2 + 2\zeta_z\omega_{n_z}s + \omega_{n_z}^2)}{(\tau s + 1)(s^2 + 2\zeta\omega_n s + \omega_n^2)}$$

NSC: $\tau > 0$ and $K > \max{(C8)}$ or

$\tau < 0$ and $K < \min{(C8)}$

$$C8 = \left(-1, -\frac{\omega_{n_z}^2(1+2\tau\zeta\omega_n)}{\omega_n^2}, -\frac{\omega_{n_z}(2\zeta\omega_n+\tau\omega_n^2)}{2\zeta_z\omega_n^2}\right)$$

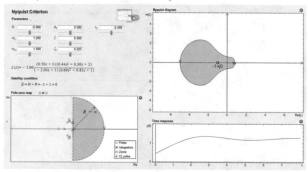

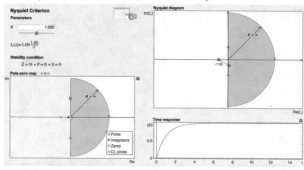

$$P_9(s) = \frac{K\omega_n^2}{\omega_{n_z}^2}\frac{(\beta s + 1)(s^2 + 2\zeta_z\omega_{n_z}s + \omega_{n_z}^2)}{(\tau s + 1)(s^2 + 2\zeta\omega_n s + \omega_n^2)}$$

NSC: $K > \max{(C9)}$ or $K < \min{(C9)}$

$$C9 = \left(-1, -\frac{\omega_{n_z}^2\tau}{\omega_n^2\beta}, \frac{\omega_{n_z}^2(1+2\tau\zeta\omega_n)}{\omega_n^2(1+2\beta\zeta\omega_{n_z})}, -\frac{\omega_{n_z}(2\zeta+\tau\omega_n)}{\omega_n(2\zeta+\beta\omega_{n_z})}\right)$$

$$P_{10}(s) = \frac{K}{s}$$

SC: $K > 0$

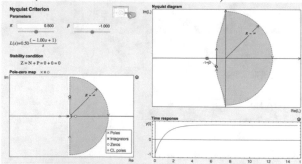

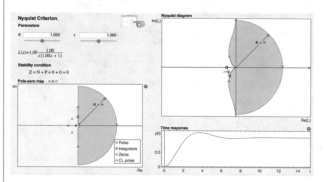

$$P_{11}(s) = \frac{K(\beta s + 1)}{s}$$

SC: $K(1 + K\beta) > 0$

$$P_{12}(s) = \frac{K}{s(\tau s + 1)}$$

SC: $\tau > 0$ and $K > 0$

Table 7.3 Plants 13–18 considered in the interactive tool Nyquist_criterion and stability conditions (SC) or necessary stability conditions (NSC)

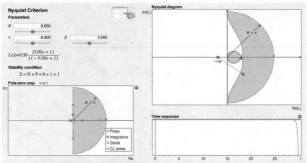

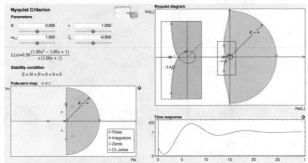

$$P_{13}(s) = \frac{K(\beta s + 1)}{s(\tau s + 1)}$$

SC: $\tau > 0$ and $K > \max(\text{C13})$ or

$\tau < 0$ and $K < \min(\text{C13})$

$$\text{C13} = \left(0, -\frac{1}{\beta}\right)$$

$$P_{14}(s) = \frac{K(s^2 + 2\zeta_z \omega_{n_z} s + \omega_{n_z}^2)}{\omega_{n_z} s(\tau s + 1)}$$

SC: $K > \max(\text{C14})$ or $K < \min(\text{C14})$

$$\text{C14} = \left(0, -\tau\omega_{n_z}^2, -\frac{\omega_{n_z}}{2\zeta_z}\right)$$

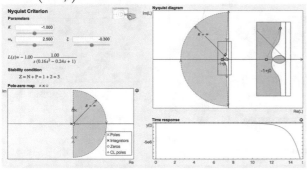

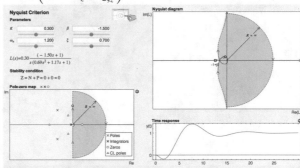

$$P_{15}(s) = \frac{K\omega_n^2}{s(s^2 + 2\zeta\omega_n s + \omega_n^2)}$$

SC: $\zeta > 0$ and $0 < K < 2\zeta\omega_n$

$$P_{16}(s) = \frac{K\omega_n^2(\beta s + 1)}{s(s^2 + 2\zeta\omega_n s + \omega_n^2)}$$

SC: $\zeta > 0, 0 < K < \frac{2\zeta\omega_n}{1 - 2\zeta\beta\omega_n}$

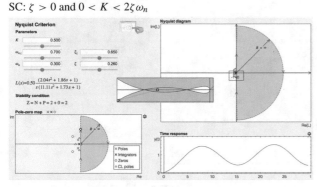

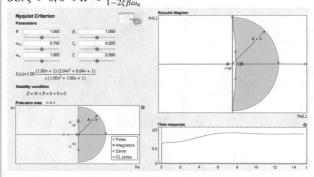

$$P_{17}(s) = \frac{K\omega_n^2}{\omega_{n_z}^2} \frac{(s^2 + 2\zeta_z\omega_{n_z}s + \omega_{n_z}^2)}{s(s^2 + 2\zeta\omega_n s + \omega_n^2)}$$

SC: C17a[a], C17b[b], C17c[c],
C17d[d]

$$P_{18}(s) = \frac{K\omega_n^2}{\omega_{n_z}^2} \frac{(s+1)(s^2 + 2\zeta_z\omega_{n_z}s + \omega_{n_z}^2)}{s(s^2 + 2\zeta\omega_n s + \omega_n^2)},$$

with $\omega_{n_z} = 0.7$ and $\omega_n = 1$

NSC: $K > \max(\text{C18})$ or $K < \min(\text{C18})$

$$\text{C18} = \left(0, -\frac{0.49}{\beta}, -(0.98\zeta + 1.4\zeta_z), -\frac{0.49}{1.4\zeta_z + 0.49\beta}\right)$$

[a]C17a: $\zeta > 0, \zeta_z > 0$; if $\zeta_z\zeta > \frac{(\omega_n - \omega_{n_z})^2}{4\omega_n\omega_{n_z}}$, stable $\forall K > 0$; if $\zeta_z\zeta < \frac{(\omega_n - \omega_{n_z})^2}{4\omega_n\omega_{n_z}}$, stable $\forall K > 0 \in [0, r_1] \cup [r_2, \infty], r_1 > 0, r_2 > 0$ are the roots of $AK^2 + BK + C = 0$; $A = 2\zeta_z\omega_n^2$, $B = (4\zeta_z\zeta\omega_n\omega_{n_z}^2 + \omega_n^2\omega_{n_z} - \omega_{n_z}^3)$, $C = 2\zeta\omega_{n_z}^3\omega_n$,

[b]C17b: $\zeta > 0, \zeta_z < 0$; stable $\forall K \in [0, r_1], r_1 > 0$ is the positive root of $AK^2 + BK + C = 0$ in C17a.

[c]C17c: $\zeta < 0, \zeta_z > 0$; stable $\forall K \in \left[\max\left(r_1, -\frac{2\zeta\omega_{n_z}^2}{\omega_n}\right), \infty\right]$; $r_1 > 0$ is the positive root of $AK^2 + BK + C = 0$ in C17a.

[d]C17d: $\zeta < 0, \zeta_z < 0$; if $\zeta_z\zeta > \frac{(\omega_n - \omega_{n_z})^2}{4\omega_n\omega_{n_z}}$, unstable $\forall K > 0$; if $\zeta_z\zeta < \frac{(\omega_n - \omega_{n_z})^2}{4\omega_n\omega_{n_z}}$, stable $\forall K \in \left[\max\left(r_1, -\frac{2\zeta\omega_{n_z}^2}{\omega_n}\right), r_2\right], r_1 > 0, r_2 > 0$ are the roots of $AK^2 + BK + C = 0$ in C17a.

Table 7.4 Plants 19–24 considered in the interactive tool Nyquist_criterion and stability conditions (SC) or necessary stability conditions (NSC)

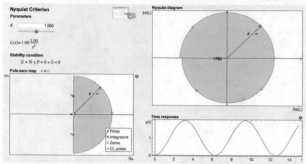

$$P_{19}(s) = \frac{K}{s^2}$$

SC: Marginally stable $\forall K$

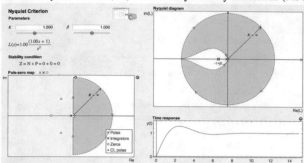

$$P_{20}(s) = \frac{K(\beta s + 1)}{s^2}$$

SC: $K > 0$ and $\beta > 0$

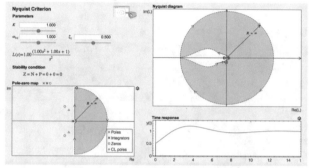

$$P_{21}(s) = \frac{K(s^2 + 2\zeta_z \omega_{n_z} s + \omega_{n_z}^2)}{\omega_{n_z}^2 s^2}$$

SC: $K > 0$ and $\zeta_z > 0$ or $K < -\omega_{n_z}^2$ and $\zeta_z > 0$

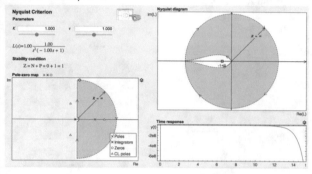

$$P_{22}(s) = \frac{K}{s^2(\tau s + 1)}$$

SC: Unstable $\forall K$ and τ

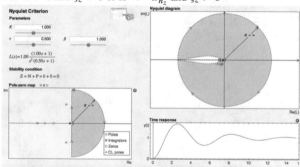

$$P_{23}(s) = \frac{K(\beta s + 1))}{s^2(\tau s + 1)}$$

SC: $\tau > 0$, $K > 0$ and $\beta > \tau$

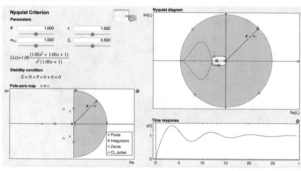

$$P_{24}(s) = \frac{K}{\omega_{n_z}^2} \frac{(s^2 + 2\zeta_z \omega_{n_z} s + \omega_{n_z}^2)}{s^2(\tau s + 1)}$$

SC: $\tau > 0$ and $K > \max\left(0, \frac{\omega_{n_z}^2(\tau \omega_{n_z} - 2\zeta_z)}{2\zeta_z}\right)$ or

$\tau < 0$ and $K < \min\left(-\omega_{n_z}^2, \frac{\omega_{n_z}^2(\tau \omega_{n_z} - 2\zeta_z)}{2\zeta_z}\right)$

Table 7.5 Plants 25–29 considered in the interactive tool Nyquist_criterion and stability conditions (SC) or necessary stability conditions (NSC)

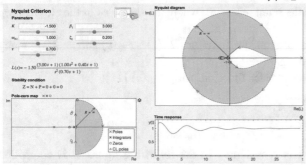

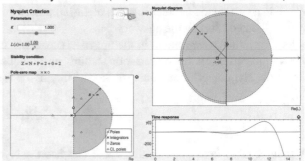

$$P_{25}(s) = \frac{K}{\omega_{n_z}^2} \frac{(\beta s + 1)(s^2 + 2\zeta_z \omega_{n_z} s + \omega_{n_z}^2)}{s^2(\tau s + 1)}$$

with $\tau = 0.7$ and $\omega_{n_z} = 1$

NSC: $2\zeta_z + \beta \omega_{n_z} > 0$ and

$(K > \max(C25)$ or $K < \min(C25))$

$$C25 = \left(0, -\frac{\tau \omega_{n_z}^2}{\beta}\right)$$

$$P_{26}(s) = \frac{K}{s^3}$$

SC: Unstable $\forall K$

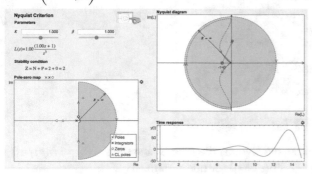

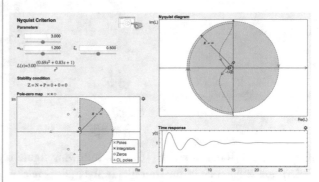

$$P_{27}(s) = \frac{K(\beta s + 1)}{s^3}$$

SC: Unstable $\forall K$ and β

$$P_{28}(s) = \frac{K}{\omega_{n_z}^2} \frac{(s^2 + 2\zeta_z \omega_{n_z} s + \omega_{n_z}^2)}{s^3}$$

SC: $\zeta_z > 0$ and $K < \frac{\omega_{n_z}^3}{2\zeta_z}$

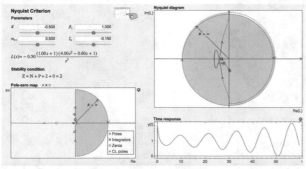

$$P_{29}(s) = \frac{K(\beta s + 1)(s^2 + 2\zeta_z \omega_{n_z} s + \omega_{n_z}^2)}{s^3}$$

NSC: $1 + 2\beta \zeta_z \omega_{n_z} > 0$, $\beta \omega_{n_z} + 2\zeta_z > 0$, and

$(K > \max(C29)$ or $K < \min(C29))$

$$C29 = \left(0, -\frac{\omega_{n_z}^2}{\beta}\right)$$

7.3.1.4 Homework

1. When in a system the geometric locus of $L(j\omega)$ passes through the point $-1 + j0$, where are the poles of the closed-loop system located? Justify the answer.
2. Select from the Options menu in the upper-left corner of the tool the transfer functions $P_1(s)...P_{29}(s)$ and analyze, using the Nyquist criterion, the closed-loop stability as a function of the gain K.
3. Using the Nyquist criterion, analyze the stability of the following transfer function using $P_{13}(s)$ [11]:

$$L(s) = K\frac{(s + 4)}{s(s - 2)}.$$

4. A system $G(s)$ is asymptotically stable and has the Nyquist plot represented in the following figure. It is controlled using feedback and a proportional gain so that $L(s) = KG(s)$. For what values of $K > 0$ is the closed-loop system asymptotically stable?

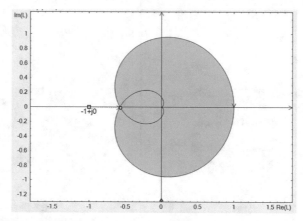

5. Exercise 3.20 from reference [16]. The following figure represents the Nyquist diagram of a system $G(s) = b(s)/a(s)$. Determine which one of the root loci in the figure matches $a(s) + Kb(s) = 0$ for this system.

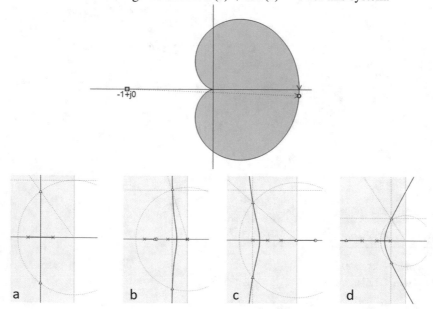

6. Use the tool to analyze the closed-loop stability with unit feedback of the following transfer functions (the subindex indicates the plant structure to be chosen from the Options menu), calculating the range of K values for which the closed-loop system is stable:

$$L_1(s) = \frac{K}{s-1}, \qquad L_{12}(s) = \frac{K}{s(s-2)}, \qquad L_{13}(s) = \frac{K(s+1)}{s(s-1)},$$

$$L_{13}(s) = \frac{K(s-1)}{s(s+1)}, \qquad L_{15}(s) = \frac{K}{s(s+1)^2}, \qquad L_{15}(s) = \frac{K}{s(s^2+s+4)},$$

$$L_4(s) = \frac{K(s-1)}{(s+2)^2}, \qquad L_{23}(s) = \frac{K(s+2)}{s^2(s+4)}, \qquad L_4(s) = \frac{K(s+1)}{(s-1)(s+2)},$$

$$L_{17}(s) = \frac{K(s+3)^2}{s(s+0.5)^2}, \qquad L_6(s) = \frac{K}{(s+1)^3}, \qquad L_{20}(s) = \frac{K(s+1)}{s^2},$$

$$L_{27}(s) = \frac{K(s+1)}{s^3}, \qquad L_{28}(s) = \frac{K(s+1)^2}{s^3}, \qquad L_6(s) = \frac{K}{(s-1)(s+2)(s-3)}.$$

7. Consider a feedback system with a transfer function (structure $P_{17}(s)$) given by:

$$L(s) = \frac{K(s+7)^2}{s(s+1)^2},$$

which is a slightly modified example from [6]. Calculate the values of K which make the closed-loop system stable. Is the system conditionally stable? How many times does the curve intersect the negative real axis?

8. As studied in Sect. 5.5, the linearized dynamics of an inverted pendulum can be represented by the transfer function $G(s) = 1/(s^2 - 1)$, where the input is the torque at the base and the output is the angle of the pendulum to the vertical (example from [6], page 9-10). The pendulum is to be stabilized with a controller with transfer function $C(s) = K(s+2)$ which, as will be analyzed in Chap. 8, corresponds to a proportional–derivative control. The loop transfer function with unit feedback is:

$$L(s) = \frac{K(s+2)}{s^2 - 1}.$$

Using the structure $P_4(s)$, apply the Nyquist stability criterion to determine the range of proportional controller gain that ensures a stable closed-loop system.

9. Analyze the stability of the system of the figure when K varies between 0 and ∞, applying the Nyquist criterion [11] (select $P_7(s)$ as plant structure). Compare the results with those obtained using the root locus method in Sect. 7.2 for the same system.

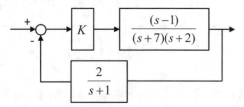

10. Considering the following loop transfer function (already studied in the root locus method and corresponding to the plant structure $P_{23}(s)$):

$$L(s) = \frac{K(\beta s + 1)}{s^2(\tau s + 1)},$$

select any positive value of τ and analyze the stability of the closed-loop system using the Nyquist_criterion interactive tool, when K varies between 0 and ∞, considering three cases: $\beta = 1.5\tau$, $\beta = \tau$ and $\beta = 0.5\tau$. Indicate the value of the gain for which the system becomes unstable in each case. In the unstable case, propose a modification to the function $L(s)$ (by adding poles or zeros) that stabilizes the system for any value of K.

7.4　Phase and Gain Margins

7.4.1　Interactive Tool: stability_margins

7.4.1.1　Concepts Analyzed in the Card and Learning Outcomes
- Understand the concept of relative stability through the basic definition of gain and phase margins.
- Gain and phase crossover frequencies and their role in closed-loop stability and performance.

7.4.1.2　Summary of Fundamental Theory
In the previous cards, the stability of the closed-loop system has been analyzed using information of the open-loop transfer function, both in the s-plane (root locus) and in the Nyquist plot (Nyquist stability criterion). In the first case, the relative stability is related to the damping factor σ of the closed-loop poles. In the second case, stability is related to the surroundings of the Nyquist contour of the point $-1 + j0$.

In practice, it is not enough that a system is stable [6]. There must also be some stability margins that describe how stable the system is and its robustness to perturbations and modeling errors (mainly due to uncertainty and unmodeled dynamics). This section introduces so-called classical *stability margins*, which quantify to some extent the sensitivity of a closed-loop system to these *modeling uncertainties*, which can be considered as the difference that always exists between the dynamics of the process to be controlled and that of the model that describes it. Therefore, *stability margins*, can be understood as a relative measure of the variation in the system's gain or phase that can cause the closed-loop system to become unstable. Stable systems with low stability margins only work in theory. When implemented in reality they are often unstable. The way in which modeling uncertainty has been quantified in classical control theory is to assume that due to modeling errors, there may be changes in system gain or phase (from those considered in the model). Typically, systems become unstable when their gains exceed certain limits or there is a large phase lag (i.e. negative phase associated with unmodeled poles and time delays). These tolerances on gain and phase uncertainty are called *gain margin* and *phase margin*.

The classical definition of stability margins works in general for MP systems with monotonically decreasing magnitude and phase (single gain and phase margins). In other kinds of systems, the definition involves a possible increase or decrease in the gain and/or phase that produce a change in the *stability condition* (if the closed-loop system is stable it becomes unstable or viceversa, if the closed-loop system is unstable it becomes stable). The objective of this card is to introduce the classical definition and examples where it works adequately, although further analysis will be done in the case of NMP systems.

This section considers the transfer function of the system in standardized time constants format. To perform the analysis in the frequency domain, s is substituted by $j\omega$ in the generic transfer function:

$$G(j\omega) = \frac{k \prod_{\ell=1}^{q}(j\omega\beta_\ell + 1) \prod_{\ell=1}^{r}\left(\left(\frac{j\omega}{\omega_{n_{z_\ell}}}\right)^2 + j\omega\left(\frac{2\zeta_{z_\ell}}{\omega_{n_{z_\ell}}}\right) + 1\right)}{(j\omega)^{\aleph} \prod_{i=1}^{p}(j\omega\tau_i + 1) \prod_{i=1}^{h}\left(\left(\frac{j\omega}{\omega_{n_i}}\right)^2 + j\omega\left(\frac{2\zeta_i}{\omega_{n_i}}\right) + 1\right)} e^{-j\omega t_d}. \tag{7.10}$$

This transfer function includes q real zeros, $2r$ complex conjugated zeros (r pairs of complex conjugated zeros), $\aleph$ poles at the origin of the s-plane, p poles on the real axis, $2h$ complex conjugated poles (h pairs of complex conjugated poles) and time delay t_d.

It is also assumed in this section that the system has unit feedback and that the loop transfer function is $L(s) = C(s)G(s) = KG(s)$, that is, to explain the concepts it will be assumed that the system represented by $G(s)$ has in series a proportional controller K, as in previous sections (see Fig. 7.5).

Recall the Nyquist stability criterion can be formulated using the following two rules:

1. A feedback control system is stable if, and only if, the Nyquist contour Γ_s mapped in the $L(s)$-plane (Γ_L), does not encircle the $(-1, 0)$ point ($N = 0$) when the number of poles of $L(s)$ in the RHP is zero ($P = 0$).
2. A feedback control system is stable if and only if, for the Nyquist contour Γ_s mapped in the $L(s)$-plane (Γ_L), the number of counterclockwise encirclements of the $(-1, 0)$ point is equal to the number of poles of $L(s)$ with positive real parts ($N = P$).

The Nyquist stability criterion is defined in terms of the *critical point* $(-1, 0)$ in the polar graph (which is the 0 dB and $-180°$ point in the Bode and Nichols diagrams). The proximity of the geometric location of $L(j\omega)$ to this point is a measure of the relative stability of a system, although other issues have to be taken into account, such as the number of encirclements around that point (even in the case the system is stable in open loop).

The classical stability margins are based on the assumption that the system is stable in open loop (P = 0), magnitude and phase are monotonous decreasing, and there is only one intersection with the critical lines (0 dB, $-180°$). Under these hypotheses, formally, the *gain margin* (GM) of a system is defined as the minimum value that the open-loop gain can be increased before the closed-loop system becomes unstable. For systems whose phase decreases monotonically with frequency, the gain margin can be calculated based on the frequency in which the phase of $L(j\omega)$ is $-180°$ (*phase crossover frequency*, ω_{pc}):

$$GM = \frac{1}{|L(j\omega_{pc})|}. \tag{7.11}$$

In decibels: $GM\ [dB] = -20 \log(|L(j\omega_{pc})|)$.

In the case of MP first- or second-order systems, the gain margin is infinite because the polar diagrams of such systems do not cut to the negative real axis (the $-180°$ line in Bode diagram). Notice that the conditions stated before are to ensure there is only one intersection of the curve with the critical line ($-180°$). In the Nyquist diagram, the GM can thus be defined as the amount by which gain can be increased before the Nyquist plot of the system crosses the critical point $(-1, 0)$.

Under the same hypotheses used for the basic definition of GM, the *phase margin* (PM) is defined as the amount of phase lag required for the stability limit defined by point $(-1, 0)$ to be reached (it is the smallest phase shift that causes $L(j\omega)$ to pass through point $(-1, 0)$). The *gain crossover frequency* (ω_{gc}) is defined as the frequency at which the Nyquist plot of the system intersects the unit circle (the loop transfer function $L(j\omega_{gc})$ has unit gain ($|L(j\omega_{gc})| = 1$, that is, 0 dB in Bode and Nichols charts). For a system whose gain decreases monotonically with frequency, the phase margin is given by:

$$PM = 180° + \lfloor L(j\omega_{gc}). \tag{7.12}$$

Notice that following this definition, the maximum time delay that can be introduced to an open-loop system before the closed loop becomes unstable is given by $t_d = (PM/\omega_{gc})(\pi/180°)$.

Under the simplifying hypotheses stated before, these margins have a simple graphic interpretation in the Nyquist diagram of the transfer function $L(j\omega)$ [6]. The *gain margin* is given by the inverse of the distance to the point where the transfer function $L(j\omega)$ cuts off the real negative axis. The *phase margin* is given by the smallest angle measured from the point $(-1, 0)$ to the point where $L(j\omega)$ cuts off the *unit circle* (circle centered on the origin passing through the point $(-1, 0)$), taken as a positive counterclockwise. This interpretation is correct as long as the gain or phase are monotonous. Figure 7.13 shows a graphical interpretation of the basic definitions for MP systems [8].

A geometrical interpretation of the phase and gain margins can also be made in the Nichols diagram, as they are defined in this case as distances to the critical point (0 dB, $-180°$). The gain margin is defined as the distance between the point at which $L(j\omega)$ cuts the vertical line of $-180°$ and the critical point (measured on that same vertical). The phase margin is the distance between the cutoff point of the $L(j\omega)$ curve with the horizontal one at 0 dB and the critical point, measured above that horizontal.

With the basic definitions, the system is considered to be stable in closed loop if the stability margins are positive, but this is not always true, as will be explained in next paragraphs. To obtain adequate closed-loop behavior taking into account possible modeling uncertainties, it is recommended [8] that the phase margin be between 30° and 60° and the gain margin be greater than 6 dB (between 6 and 14 dB, [6]). As briefly commented in this chapter, from Bode sensitivity integral considerations (Eq. (8.14)), the slope of the amplitude Bode plot at the gain crossover frequency should be around -30 dB/decade to obtain an adequate phase margin. These recommendations usually work, although there may be cases of systems with good phase and gain margins where the relative stability is poor, because there may be frequencies different to ω_{pc} and ω_{gc} for which the Nyquist diagram of the system is close to the critical point[12] (see reference [6], pages 9–14 and 9–15).

---

[12] In such cases, it is recommended to use another relative stability index called the *stability margin* [6], which is defined as the minimum distance to the critical point $(-1, 0)$ from any point on the frequency response curve.

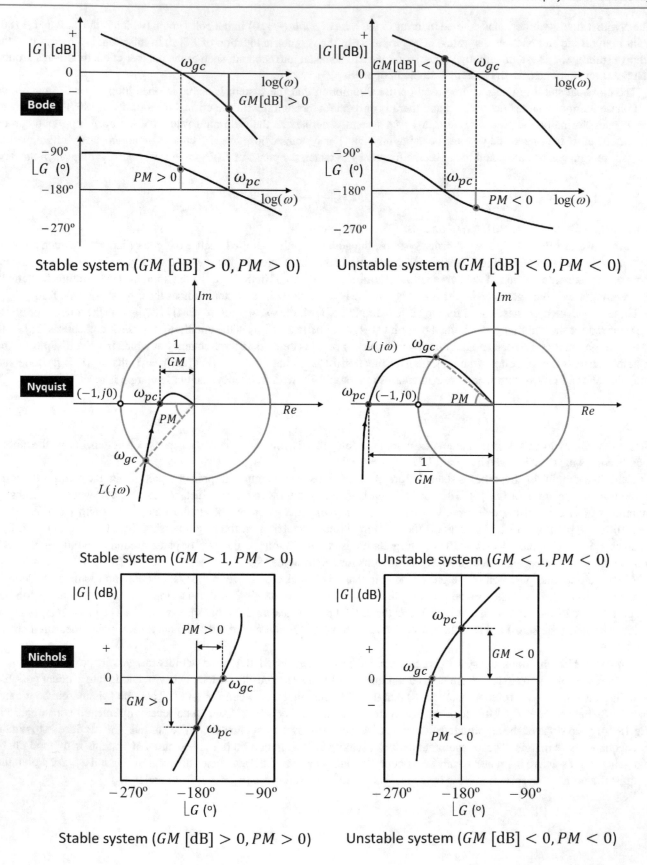

Fig. 7.13 Basic phase and gain margins for MP systems

A remark on complete definitions

This book deals with basic concepts to be easily understood by students, so the main contents students must know about PM and GM have been summarized above. Nevertheless, it has been considered necessary to include this remark to clarify those cases in which the classical definitions cannot be applied, as a complementary information both for teachers and students.

When the hypotheses used to introduce the basic definitions of gain and phase margins do not hold, it is recommended to analyze stability using the Nyquist criterion (for instance for NMP systems). Moreover, the previous basic definitions and interpretations provided in Fig. 7.13 may be incorrect if there are multiple frequencies where the 0 dB or $-180°$ lines are cut (conditionally stable systems).

In fact, it is easy to find examples (some of them simple ones) where the results provided by tools like the one presented in this section or MATLAB® seems to not agree with the previous definitions. For instance the transfer function:

$$L(s) = 30 \, K \frac{(0.3s + 1)(0.5s + 1)}{s^3(0.1s + 1)} = 45 \, K \frac{(s + 3.33)(s + 2)}{s^3(s + 10)}, \tag{7.13}$$

with $K = 1$ provides $GM = -4.51$ dB at $\omega_{pc} = 3.78$ rad/s and $PM = 8.52°$ at $\omega_{gc} = 5.13$ rad/s, while P $= 0$ and the system is stable in closed loop (as can be analyzed for instance using root locus). What the tools provide is correct, but in this case the phase does not decrease monotonically with frequency. Another example is the system given in Eq. (7.14):

$$L(s) = 1.2 \, K \frac{s(0.67s + 1)}{(0.2s + 1)(-s + 1)} = -4K \frac{s(s + 1.5)}{(s + 5)(s - 1)}, \tag{7.14}$$

with $K = 1$ provides $GM = -3.52$ dB at $\omega_{pc} = 1.73$ rad/s and $PM = -22.62°$ at $\omega_{gc} = 1$ rad/s, and the closed loop is stable. This is logical as in this case P $= 1$ and thus the natural way to perform stability analysis is through the Nyquist stability criterion. Many more examples can be found in [5]. The common mistakes in systems that do not fit the assumptions made in the classical definitions, are the determination of stability properties of MP or NMP systems in terms of the signs of their GM and PM, the GM and PM of systems which have multiple crossover frequencies, etc.

The correct definition should be that the *gain margin* is the amount by which gain can be increased or decreased before the system changes its *stability condition*, as there should be cases in which the gain must be decreased to stabilize a system. In the Nyquist diagram, the GM can thus be defined as the amount by which gain can be increased or decreased before the Nyquist plot of the system crosses the critical point $(-1, 0)$ and changes its *stability condition*.

As with the GM, the correct definition of *phase margin* is the amount the phase of a system can be increased or decreased, without changing the gain of the system, so that the stability condition of the system changes [5].

As pointed out in [18], the general stability condition stated for the Nyquist criterion Z $=$ P $+$ N $= 0$ can be reformulated in the Bode diagram with the following rules:

1. A feedback control system is stable if and only if the Bode diagram of $L(s)$ does not cross $-180°$ with $|L(j\omega)|_{dB} > 0$ (N $= 0$) when the number of poles of $L(s)$ in the RHP is zero (P $= 0$).
2. A feedback control system is stable if and only if the Bode plot of $L(s)$ does not intersect the point $(0$ dB, $-180°)$, and the net sum of its crossing of the ray $-180°$ with the modulus $|L(j\omega)|_{dB} > 0$ is equal to P. The crossings and their corresponding signs in the Bode diagram are illustrated in Fig. 7.14.

The number of crossings is equivalent to the number of encirclements of the critical point $(-1, 0)$ by the Nyquist contour. A single crossing in the Bode plot, for the frequency range $(0 < \omega < +\infty)$, is equivalent to two crossings for the full plot, for the frequency range $(-\infty < \omega < +\infty)$. For this reason it is sufficient to draw the Bode plot and then multiply the number of crossings by two, in order to calculate N. These rules can be considered as an extension of the procedures for the verification of stability which are based on phase margin or gain margin, and which are applied when there is only one intersection of the curve with the critical line $(-180°)$.

As known from the Nyquist stability criterion, for an NMP system where $L(s)$ is unstable, the stability condition is not satisfied unless the $L(j\omega)$ graph surrounds the point $(-1, 0)$, as Z $=$ P $+$ N and $P \neq 0$ in this case. Another interesting case already mentioned is that of conditionally stable systems, having two or more phase and/or gain crossover frequencies, as shown in Fig. 7.15 and in exercise 3 in the Homework subsection. In several texts, such as [8], it is stated that for stable systems having two or more gain crossover frequencies, the phase margin is measured at the highest gain crossover frequency. For these kind of systems, the classical definition of GM and PM does not provide adequate results regarding relative stability and software packages use to provide wrong results.

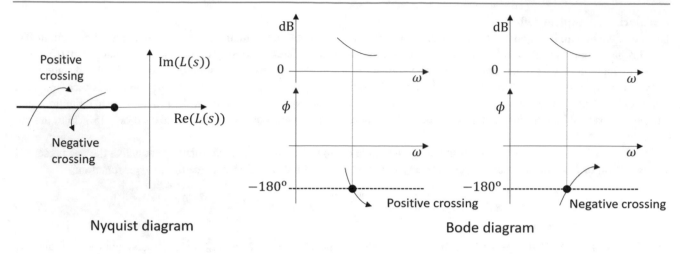

Fig. 7.14 Nyquist and Bode stability criteria

Bavafa-Toosi [5] defines two GMs for a system: one positive (in the increasing direction) and one negative (in the decreasing direction). For NMP systems the stability/instability properties have to be studied in terms of their GM/PM signs. The interested reader is encouraged to read [5], where several selected examples are developed. Here, a summary of the main concepts is included.

The definition of GM provided by Bavafa-Toosi is: "The amount by which the gain can be increased or decreased before the system becomes unstable if stable, or stable if unstable. Associated with a system, either MP or NMP, are one positive (GM^+) and one negative (GM^-) gain margin. For stable systems these are the destabilizing gain margins, whereas for unstable systems these are the stabilizing gain margins." If K is the current value of the gain of $L(s)$, it should be increased to $K^+ > K$ or decreased to $K^- < K$ so that the stability condition of the system changes. The quantities $K^+/K > 1$ and $K^-/K < 1$ are the GMs of the system (in dB: $GM^+ = 20 \log (K^+/K) > 0$ and $GM^- = 20 \log (K^-/K) < 0$). In practice, perturbations appear both in increasing and decreasing directions of gain. In the case GM^+ and/or GM^- are infinite, it means that the objective of changing the stability condition is not achievable [5]. For a given K, the most relevant GM is the smaller in absolute value (higher sensitivity). For unstable systems, GM^- is indicative of how much the gain should be decreased to make the system stable, while GM^+ indicates how much the gain must be increased to make the system stable.

Recalling the example in Eq. (7.13), for $K = 1$ the closed-loop system is stable, therefore $GM = -4.51 = GM^-$ indicates that the closed-loop system becomes unstable if the gain is decreased to $K = 10^{-4.51/20} = 0.595$ (affecting ω_{gc} and thus the PM). For $K = 0.1$, $GM = 15.49$ dB at $\omega_{pc} = 3.78$ rad/s and $PM = -34.33°$ at $\omega_{gc} = 1.62$ rad/s, providing an unstable

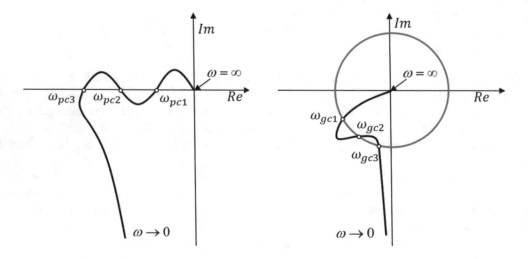

Fig. 7.15 Polar diagrams of systems with various phase or gain crossover frequencies

closed loop. If the gain is increased by a factor greater than $10^{15.49/20} = 5.95$ the closed-loop system becomes stable. Note than not all the real axis crossings result in a change in the stability condition [5].

In the case of the system represented by (7.14), for $K = 1$ the closed-loop system is stable while $GM = -3.52$ dB. This means that the closed-loop system will become unstable if the gain is decreased by $K = 10^{-3.52/20} = 0.668$.

Regarding the PM, there are also two PM, although the explanation is more difficult (see Example 6.25 in [5]) and has to be done in the Nyquist diagram.

- If ω_{gc} is in the third or fourth quadrants, then $PM > 0$. If $\lfloor L(s)$ is increased by PM (without changing the magnitude) and the stability condition changes, then $PM > 0$ is the PM of the system and it is denoted by PM^+ and it is said the system has a positive PM.
- If ω_{gc} is in the first or second quadrant, $PM < 0$. If $\lfloor L(s)$ is increased by PM (decrease it by $|PM|$, without changing the magnitude) and the stability condition changes, then $PM < 0$ is the PM of the system (denoted by PM^-) and it is said the system has a negative PM.

In a previous paragraph, it has been explained that the time delay of stable system can be increased by $t_d > (PM/\omega_{gc})(\pi/180°)$ seconds delay (in its forward path) before becoming unstable. Similarly, if the system is unstable then it means that with a delay $t_d > (PM/\omega_{gc})(\pi/180°)$ it will become stable. This time delay is sometimes called the *delay margin* of the system[13] [5]. Similar to the case of the GM, the smallest one in absolute value has to be selected. Depending on the system it may be either positive or negative.

The PM is measured as the angle the crossover point makes with the negative real axis. This angle is $\pm180°(2l + 1)$, so it is not prefixed to $-180°$ as the classical approach uses. The phase angle to compute the PM has to be measured with respect to either of the lines $\pm180°(2l + 1)$ which are nearer to it. Therefore, the computed phase margin is in the range $-180° \le PM \le 180°$. In negative angles, if the absolute value of the phase of the system is smaller than the absolute value of $\pm180°(2l + 1)$, the phase margin is positive, and otherwise it is negative. In positive angles, the relation is the opposite. That is, if the value of the phase of the system is smaller than the value of $180°(2l + 1) > 0$, the phase margin is negative, and otherwise it is positive [5].

In the example given by Eq. (7.14), $PM = -22.62°$ with $\omega_{gc} = 1$ rad/s in the second quadrant. The system is stable in closed loop with a negative PM. Notice that the phase curve evolves from 90° to 180° as frequency increases. In this case, the PM is measured with respect to the line $+180°$, as at $\omega_{gc} = 1$ rad/s $\lfloor L(j\omega_{gc}) = 157.4°$. This means that a positive increase of the phase of $L(s)$ around ω_{gc} can make the closed-loop unstable. For instance, if the zero in the numerator is placed at lower frequencies:

$$L(s) = -4K \frac{s(s + 1)}{(s + 5)(s - 1)},$$

then $\lfloor L(j\omega_{gc}) \approx 180.0°$ at $\omega_{gc} = 1.29$ rad/s, both GM and PM are almost zero and the closed-loop system starts to become unstable. For:

$$L(s) = -4K \frac{s(s + 0.9)}{(s + 5)(s - 1)},$$

the closed loop is unstable having $GM = 0.915$ dB at $\omega_{pc} = 1.2$ rad/s and $PM = 4.39°$ at $\omega_{gc} = 1.34$ rad/s.

Summarizing, it is not possible in general to determine the stability of a system by merely considering the signs of the GM and PM, as there exist stable and unstable MP and NMP systems with positive and negative PM, and even unacceptable computed PMs [5]. When the conditions for applying the simplified definition do not hold, it is better to apply the Nyquist stability criterion or to analyze the system as done with the examples in Eqs. (7.13) and (7.14). The reader can find much deeper explanations for systems with multiple gain and phase crossing frequencies in [5].

7.4.1.3 References Related to this Concept

- [2] Dorf, R. C. & Bishop, R. H. (2011). *Modern control systems.* (12th ed.). Prentice Hall. ISBN: 978-0-13-602458-3.
- [5] Bavafa-Toosi, Y. (2017). *Introduction to linear control systems.* Academic Press-Elsevier. ISBN: 978-0-12-812748-3. Chapter 6, Sect. 6.3, pages 472-503.
- [6] Åström, K. J., & Murray, R. M. (2014). *Feedback systems: An introduction for scientists and engineers* (2nd ed.). Princeton University Press, ISBN: 9780691193984.
- [8] Ogata, K. (2010). *Modern control engineering.* (5th ed.). Prentice Hall, ISBN: 978-0-13-615673-4.
- [14] Shahian, B. & Hassul, M. *Control system design using MATLAB®.* Prentice Hall, ISBN: 0-13-174061-X.

---

[13] If $PM > 0$ then $t_d > 0$ and are both realizable. If $PM < 0$ then $t_d < 0$.

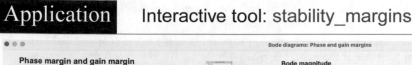

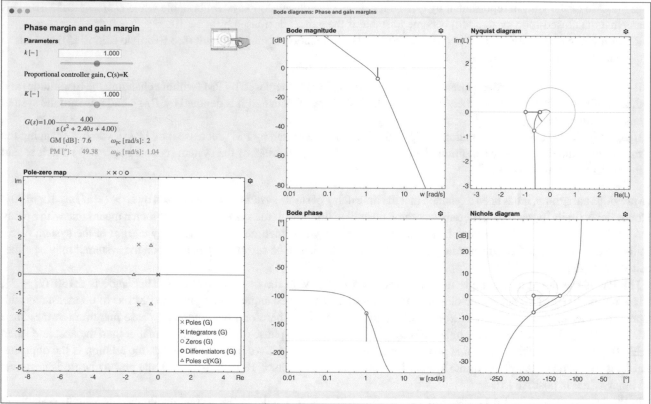

Stability Margins

In this tool the basic concept of phase margin and gain margin are analyzed through examples and using the different representations in the frequency domain: Bode, Nyquist and Nichols diagrams. Although the time delay has a fundamental influence on these relative stability margins, this tool does not incorporate the option to include time delay in the transfer function which describes the dynamics of the systems studied, since a specific tool has been developed to analyze the time delay limitations in closed-loop systems, which is explained in the next section.

The dynamic structure of the transfer function is established through the repository of poles and zeros ($\times$ $\times$ $\circ\circ$) over the **Pole-zero map**. The different dynamic elements can be clicked and dragged over the graph and placed at any point in the s-plane, although in this case the tool is limited to open-loop stable systems and thus poles on the RHP cannot be placed. Integrators and differentiators are directly placed at the origin. To extract items, simply double-click on them or drag them back to the repository. The Options menu allows to select different transfer functions structures. The graph also shows the poles of the closed-loop transfer function (Poles cl (KG)), using a blue triangle ($\triangle$). The zeros match those of the open-loop transfer function. A legend in the lower-right corner of the graph represents the symbols associated with each dynamic element.

The static gain k of $G(s)$ can be modified in the **Parameters** section, next to that of the proportional compensator K that is used to analyze the influence of a proportional term in the direct chain of the feedback loop on the closed-loop stability margins. In both cases, sliders and textboxes (useful to change the sign of the gains) are available. This section also shows a symbolic representation of the transfer function $G(s)$ and the values of:

- Gain margin (GM) and associated phase crossover frequency (ω_{pc}), in magenta color.
- Phase margin (PM) and associated gain crossover frequency (ω_{gc}), in green color.

The right area of the tool contains the system frequency response diagrams. In all of them, the frequency response of the loop transfer function $L(s) = KG(s)$ is represented using a blue line, identifying the segments or angle arcs that provide the value of the PM in green color and those that delimit the GM in magenta. As the time delay is not considered in this tool, interactivity is only associated with the GM, since a direct relationship has been established between it and the gain K of the proportional controller. Therefore, if in any of the frequency diagrams the magenta circle is dragged using the mouse over the frequency response, the gain K of the compensator and the values of the stability margins are automatically modified.

All the graphs incorporate gearwheels that are used to change the scales on the axes when changing their settings.

The Options menu includes a list with the twelve structures of transfer functions $P_1(s)...P_{12}(s)$ described in Table 1.2 in Chap. 1. There is also the option to include the transfer functions in (NUM,DEN) or ZPK formats. In this last case, unstable systems can be entered in the tool, but knowing that in that case there is no a relationship between the signs of the relative stability margins and closed-loop stability.

In addition to these functions, the above mentioned menu incorporates the option to choose between a Complete view of all frequency-domain plots (Bode diagram (magnitude and phase), Nyquist diagram and Nichols diagram), or to replace the Nichols diagram by the closed-loop response to unit step in the reference in $t = 0$ (Step response option). There are no interactive elements in this chart.

7.4.1.4 Homework

1. Select one by one the examples $P_1(s)$ up to $P_{12}(s)$ included in the Options menu of the tool. For the default parameter values, leaving $k = 1$ and $K = 1$, calculate the PM and GM that the unit feedback system has. Compute the value of K needed to make the system unstable. Compare the results with those obtained with the root_locus (card 7.2) and Nyquist_criterion (card 7.3) interactive tools.
2. Using the pole-zero editor, configure a system with $G(s) = k/s^n$. For $n = 1...4$, indicate the values of PM and GM. Do those values depend on K? Justify the answer.
3. Given the system described by:
$$L(s) = \frac{K(s^2 + 3s + 22)}{s(s + 1)(s + 2)},$$
compute the intervals of K for which the system is stable. Compare the results with those obtained with the root_locus interactive tool and reason about the concept of conditional stability.
4. Calculate the value of the PM and GM of the following systems. Which closed-loop system has the highest overshoot? Indicate the value of K from which said systems would become unstable. Compare the results with those obtained in the cards 7.2 and 7.3.

$$L_1(s) = \frac{K}{(s + 2)^2(s + 1)}, \quad L_2(s) = \frac{K}{(s + 1)^4}, \quad L_3(s) = \frac{K}{(s + 1)(s + 2)(s + 3)(s + 4)},$$

$$L_4(s) = \frac{K(s + 1)}{s(s + 2)(s + 3)}, \quad L_5(s) = \frac{K}{s(s + 1)^2}, \quad L_6(s) = \frac{K(s + 1)^2}{s(s + 2)},$$

$$L_7(s) = \frac{K(s + 2)}{s^2(s + 4)}, \quad L_8(s) = \frac{K(s + 1)}{s^3}, \quad L_9(s) = \frac{K}{s(s^2 + s + 4)},$$

$$L_{10}(s) = \frac{K(s + 3)^2}{s(s + 0.5)^2}, \quad L_{11}(s) = \frac{K}{(s + 1)^3}, \quad L_{12}(s) = \frac{K(s + 1)}{s^2}.$$

5. Enter a system that has the transfer function (reference [6], page 9–14):

$$L(s) = \frac{0.38(s^2 + 0.1s + 0.55)}{s(s + 1)(s^2 + 0.06s + 0.5)} = \frac{0.38(s^2 + 0.1s + 0.55)}{(s^4 + 1.06s^3 + 0.56s^2 + 0.5s)}.$$

Calculate the PM and GM and indicate whether you consider the system to have good relative stability. Show the step response of the closed-loop system. This is a clear example of a system in which good values of the phase and gain margins are obtained but a low relative stability, because the Nyquist diagram curve passes very close to the point (-1.0). Can you determine this "low" relative stability in the Bode diagram? Indicate the value of the K gain that produces the lowest value of relative stability.
6. Given the plant $L(s) = K/(s(s + 1))$, find the stability margins and the bandwidth of the closed-loop system for $K = 1$. In which frequency range is the input amplified and in which frequency range is it attenuated? Find K such that

$r(t) = \sin(0.8t)$ appears in the output with the same amplitude (Reference [5], Problem 4.27). It is recommended to use the stability_margins and f_concept (card 4.2) interactive tools.

7. Considering the following loop transfer function (already studied in the root locus method and Nyquist stability criterion):

$$L(s) = \frac{K(\beta s + 1)}{s^2(\tau s + 1)},$$

select any positive value of τ considering three cases: $\beta = 1.5\tau$, $\beta = \tau$ and $\beta = 0.5\tau$. Using the tool, indicate the range of values of K that guarantees a stable closed loop. In the unstable case, propose a modification to the function $L(s)$ (by adding poles or zeros) that stabilizes the system for any value of K.

8. Using all the interactive tools in this chapter (cards 7.2 to 7.4) analyze the stability of the example given by Eq. (7.13).

9. Using all the interactive tools in this chapter (cards 7.2 to 7.4) analyze the stability of the example given by Eq. (7.14).

10. This example exposes an incorrect interpretation of the classical concepts of stability margins when the loop transfer function is unstable. If the following transfer function is used (notice that it has to be included in the tool using the (NUM,DEN) or ZPK formats in the Options menu, as it is an unstable open-loop system):

$$L(s) = \frac{K}{s(1-s)},$$

stability_margins and MATLAB® provide: $GM = \infty$ and $PM = 128°$, while the closed-loop system is clearly unstable (it is easy to analyze for instance using root locus). So, positive stability margins do not guarantee closed-loop stability, as $P = 1$. This system cannot be stabilized only acting on the gain.

11. This example deals with open-loop NMP systems. Although the tool is oriented to explain the basic definitions for MP systems, the following examples can be analyzed using the (NUM,DEN) or ZPK formats in the Options menu to introduce the transfer functions. In each case, determine the values of K that provide stable closed-loop and reason about the signs of PM and GM following the explanations in the remark for NMP systems. The analysis can be complemented also reasoning with the root_locus (card 7.2) and Nyquist_criterion (card 7.3) interactive tools.

$$L_1(s) = \frac{K}{s-1}, \qquad L_2(s) = \frac{K}{s(s-2)}, \qquad L_3(s) = \frac{K(s+1)}{s(s-1)},$$

$$L_4(s) = \frac{K(1-s)}{s(s+1)}, \qquad L_5(s) = \frac{K(1-s)}{(s+2)^2}, \qquad L_6(s) = \frac{K(s+1)}{(s-1)(s+2)}.$$

7.5 Limitations Imposed by Time Delay in Closed-Loop Systems

7.5.1 Interactive Tool: limitations_delay

7.5.1.1 Concepts Analyzed in the Card and Learning Outcomes
- Effect of time delay on closed-loop performance.
- Influence of time delay on closed-loop stability.

7.5.1.2 Summary of Fundamental Theory
Time delays have a great influence on the stability of closed-loop systems (generally negative). Time delays are intrinsic to any physical system because signals are transmitted between components of the system with limited speed and measurements are done at the end point. Many examples can be found in [5] (Sect. 2.3.13).

Throughout this chapter, different techniques for analyzing the stability of feedback systems have been studied, using both the pole-zero map (root locus method) and the frequency diagrams (Nyquist stability criterion and relative stability margins) as a representation graph. It has been considered convenient to dedicate an interactive tool to analyze the limitations imposed by time delay in the stability of closed-loop systems, taking as main graphical representation the Nyquist diagram and the time response plots of the system input (control signal) and the system output.

A pure time delay is given by $y(t) = u(t - t_d)$ and in transfer function representation by $G(s) = e^{-t_d s}$. In the frequency domain $G(j\omega) = e^{-j\omega t_d}$, which has modulus 1 ($|e^{-j\omega t_d}| = 1$) and argument $\lfloor e^{-j\omega t_d} = -\omega t_d$. Therefore, pure time delays do not affect the magnitude curve in Bode diagrams, but introduce a linear lag with frequency that tends to destabilize the system (deteriorates the PM). In fact, for $\omega = 1/t_d$, a pure delay term introduces a lag of $-57.3°$ (-1 rad). This influence, which is often disregarded in design, can cause instability in a control system by producing negative phase margins. In the Nyquist diagram, the time delay shifts the point corresponding to ω_{gc} placed on the unit circle towards the critical point $(-1, 0)$, as it is analyzed in this card.

7.5.1.3 References Related to this Concept

- [2] Dorf, R. C., & Bishop, R. H. (2011). *Modern control systems*. (12th ed.). Prentice Hall. ISBN: 978-0-13-602458-3.
- [5] Bavafa-Toosi, Y. (2017). *Introduction to linear control systems*. Academic Press-Elsevier. ISBN: 978-0-12-812748-3. Section 2.3.13, pages 115–121.
- [8] Ogata, K. (2010). *Modern control engineering*. (5th ed.). Prentice Hall, ISBN: 978-0-13-615673-4.

Application   Interactive tool: limitations_delay

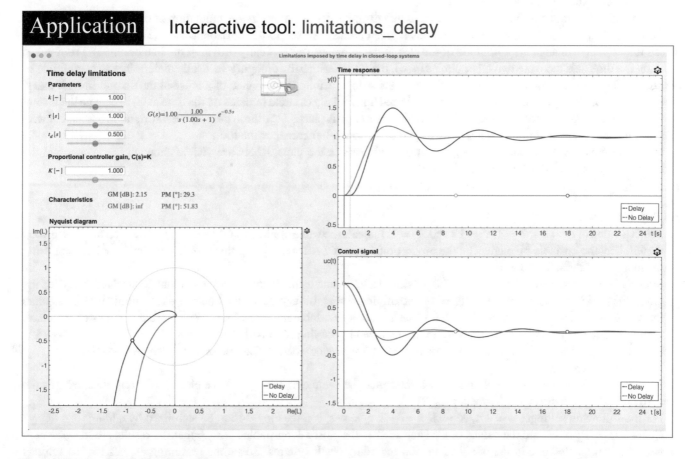

Time Delay and Stability Margins

This tool is devoted to analyzing the influence of the presence of time delay in the open-loop transfer function in the closed-loop stability and performance. The Nyquist diagram is used as main frequency-domain representation, also analyzing the effects on the closed-loop time response to a step reference and the influence of load disturbances and noise. In this interactive tool, a fixed transfer function structure of the system to be analyzed is used in the form $G(s) = \dfrac{k}{s(\tau s + 1)} e^{-t_d s}$, that is, a second-order system with an integrator and time delay. The reason for the choice is that it constitutes a system where it is very easy to visually analyze the influence of the time delay on the closed-loop stability and performance, that are the main concepts to explore.

The upper part of the tool includes a section where the process **Parameters** can be introduced, in this case the gain k, the time constant τ and the time delay t_d of $G(s)$. Next to the corresponding textboxes and sliders (red-coded), the symbolic representation of the transfer function $G(s)$ in red color is placed.

Under this section the **Proportional controller gain, C(s)=K** can be found, with associated textbox and slider ($L(s) = KG(s)$). The **Characteristics** section represents the numerical values of the phase margin (PM) and the gain margin (GM) of the system without delay (represented in magenta) and those corresponding to the system with time delay (in blue color).

The lower-left area of the tool is in this case devoted to the **Nyquist diagram** of $L(s) = KG(s)$, where the frequency responses of the system are displayed without delay and including delay, with the same colors as defined in the previous paragraph (the graph includes a descriptive legend). The only interactive element in this graph is a black circle (o) that when dragged modifies the value of the time delay of the blue system with respect to the original non-delayed one (magenta). Moving that point updates the values of t_d in the **Parameters** area and changes its value also in the closed-loop system time responses.

The right area of the tool is dedicated to the **Time response** of the two systems (with and without delay) in the upper part and that of the **Control signal** in the lower part, also incorporating legends with the color codes used, as well as a gearwheel icon to modify the scale when changing its settings. In the **Time response** plot a red dotted vertical line has been included that moving to the right increases the value of the time delay of the original system, causing the modification of this variable in the other areas of the tool already explained.

As a novelty, this tool has the possibility of including disturbances and noise in the simulation of the time response, being also used in Chap. 8. On the ordinate axis there is a green circle o, linked to the horizontal dashed line of the same color, whose function is to change the amplitude of the step in the reference (by vertically dragging the line or circle). From left to right, on the time axis there are two other circles (o, o), the first of which, by dragging vertically, determines the amplitude of the step load disturbance (at the system input), while the second defines the variance of a noise at the output. In both cases, when the circle is dragged vertically, it can also be shifted horizontally (a circle remains on the time axis) to change the instant the disturbance or noise is entered, respectively. When the mouse is placed over these circles, the information associated with them (signal amplitude and time instant) is displayed in the lower-left corner of the tool.

When the mouse is positioned over any point of the curves in the plots, labels are visible showing the instant and the variable (u or y) representing that point.

7.5.1.4 Homework

1. Select a value of $t_d = 0$ s, $k = 1$, $\tau = 1$ s and $K = 1$. Indicate the value of PM and GM for this setting. Can the closed-loop system become unstable by increasing the proportional gain K? After analyzing this problem with the tool, reason using the root_locus interactive tool (card 7.2).
2. Starting from the same configuration as the previous exercise and using the black circle (o) that determines the cut of the graph of $KG(j\omega)$ with the circle centered on the origin passing through the critical point $(-1, 0)$ (unit circle), calculate the value of the time delay from which the system becomes unstable in closed loop. Which are the values of the phase margin and the gain margin? Can the system be stabilized by varying K? Justify the answer.
3. Does t_d influence the PM? If t_d increases, does the PM improve or worsen? Can the system become unstable?
4. Does t_d influence the GM?
5. For a value of $\tau = 0.001$ s, $k = 1$ and $K = 1$ determine the value of t_d that makes the closed-loop unstable. What is the corresponding gain crossover frequency ω_{gc}?
6. For the following settings: $\tau = 1$ s, $K = 1$, $k = -2$, $t_d = 0$ s, can the closed-loop be stabilized by adding time delay? Change $t_d = 1$ s, what are the values of GM and PM? Is the closed-loop stable? Notice that the results are in agreement with the comments done in the previous section regarding NMP systems. The same reasoning applies for the settings $k = -1$, $\tau = -1$ s, $K = 1$ and $t_d = 5$ s, positive relative stability margins provide an unstable closed loop, as in this case the system is NMP.

References

1. Albertos, P., & Mareels, I. (2010). *Feedback and control for everyone*. Springer.
2. Dorf, R. C., & Bishop, R. H. (2011). *Modern control systems* (12th ed.). Prentice Hall.

3. Åström, K. J., & Hägglund, T. (2006). *Advanced PID control*. ISA—The Instrumentation Systems and Automation Society.

4. Åström, K. J. (2004). *Introduction to control*. Department of Automatic Control, Lund Institute of Technology, Lund University.

5. Bavafa-Toosi, Y. (2017). *Introduction to linear control systems*. Academic Press-Elsevier.

6. Åström, K. J., & Murray, R. M. (2014). *Feedback systems: An introduction for scientists and engineers* (2nd ed.). Princeton University Press.

7. Bolzern, P., Scattolini, R., & Schiavoni, N. (2009). *Fundamentos de control automático (Fundamentals of automatic control)*. McGraw-Hill.

8. Ogata, K. (2010). *Modern control engineering* (5th ed.). Prentice Hall.

9. Franklin, G. F., Powell, J. D., & Emani-Naeni, A. (2015). *Feedback control of dynamic systems* (7th ed.). Pearson.

10. Díaz, J. M., Costa-Castelló, R., & Dormido, S. (2021). An interactive approach to control systems analysis and design by the root locus technique. *Revista Iberoamericana de Automática e Informática Industrial, 18*(2), 172–188.

11. Barrientos, A., Sanz, R., Matía, F., & Gambao, E. (1996). *Control de sistemas continuos. Problemas resueltos (Control of continuous systems. Problems solved)*. McGraw-Hill.

12. D'Azzo, J. J., Houpis, C. H., & Sheldon, S. N. (2003). *Linear control system analysis and design with MATLAB®* (5th ed.). Marcel Dekker Inc.

13. Franklin, G. F., Powell, J. D., & Emani-Naeni, A. (2010). *Feedback control of dynamic systems* (6th ed.). Pearson.

14. Shahian, B., & Hassul, M. (1993). *Control system design using MATLAB®*. Prentice Hall.

15. Truxal, J. G. (1955). *Automatic feedback control system synthesis*. McGraw-Hill.

16. KTH - Royal Institute of Technology and Linköpings Universitet, Sweden. (2016). Reglerteknik ak med utvalda tentamenstal (automatic control exercises: Computer exercises, laboratory exercises). Retrieved July 01, 2021, from https://cutt.ly/jYkcFZV.

17. Thaler, R. G., & Brown, G. J. (1960). *Analysis and design of feedback control systems* (2nd ed.). McGraw Hill Book Co.

18. García-Sanz, M. (1999). Stability criteria in non-polar diagrams. *International Journal of Electrical Engineering Education, 36*, 65–72.

Control System Design

8.1 Introduction

Automatic control is essentially based on the concept of *feedback*. This concept is materialized in a control structure in which the *controller* can be understood as an operator, which, depending on the desired output of the plant (*reference*) and the actual *measured output*, provides the *control signal* (*action*) to be applied to the system. Feedback has not only good properties [1], as it attenuates the effects of disturbances, makes good systems from bad components, helps to reduce modeling uncertainties, compensates for process variations, enables tracking command signals, and stabilizes and shapes behavior, but also bad ones, as it may cause instability and feeds measurement noise into the system. So, any feedback control loop should take into account those considerations and should implement controllers that take advantage of the good properties, and attenuate or remove the bad ones. Moreover, it is also very important to take into account other important issues [1], such as robustness and limitations that can and cannot be achieved by feedback.

As commented in Chap. 7, in the framework of relative stability margins, robustness refers to the ability of a control system to withstand parameter variations in the process transfer function, while maintaining stability and desired performance. Robustness can be expressed by sensitivity to load disturbances and model uncertainty.

In Chap. 7, the basic concepts related to block diagrams, feedback loops, and closed-loop stability have been addressed, considering that only the transfer function representing the system and a proportional gain (controller) are placed in the loop transfer function ($L(s) = KG(s)$). This chapter deals with concepts related to control system design techniques based on both the *frequency domain* and the *time domain*, using as a basic block diagram described in Fig. 7.2 of Chap. 7. The feedback controller (C) should be designed to achieve low sensitivity to load disturbances (D), low injection of measurement noise (N), and high robustness to process variations and uncertainty, while the feedforward term F (reference filter) should help to achieve desired response to command signals (R). However, sometimes design for reference tracking, robustness, and disturbance attenuation cannot be separated (if only error can be measured) [1].

In general, the automatic control of feedback dynamical systems has the objective of ensuring that their time evolution follows a pattern predetermined by the designer, when a certain signal is introduced at their input. In control engineering, inputs in the form of impulse, step, ramp, and parabola are used when an analysis is made in the time domain, and sinusoids in the frequency domain (see Fig. 2.3 in Chap. 2). These inputs are also used to make comparisons between open-loop responses of dynamical systems, or to analyze compliance with certain performance specifications of a given control scheme in a closed loop. Moreover, as done in the interactive tools introduced in this chapter, response to a step load disturbance and output noise must always be considered to deal with the aforementioned control objectives.

In Sect. 1.6 of the introductory chapter of the book, the step-by-step procedure in control system design is summarized. This chapter deals with many of these steps, including the selection of the controller structure, controller parameter tuning, analysis of closed-loop properties, and simulation. The main design methods are included in the conceptual map of the book in Fig. 1.1.

8.1.1 Relation Between Time-Domain and Frequency-Domain Specifications

Time-domain techniques traditionally use *specifications* related to *closed-loop system* performance that can be assimilated to that of a first- or a second-order system. Although the correlation between time response and frequency response is approximate, time-domain specifications are usually adequately met in classical design methods in the frequency domain for

J. L. Guzmán et al., *Automatic Control with Interactive Tools*,
https://doi.org/10.1007/978-3-031-09920-5_8

step reference responses. As analyzed in Chap. 7, as an initial intuitive idea, a closed-loop system with small stability margins should provide time oscillatory responses.

The relationship between the tracking error and the reference is given by

$$\frac{E(s)}{R(s)} = \frac{1}{1 + L(s)}, \tag{8.1}$$

so that, in the frequency domain, it is important to analyze the shape of $L(j\omega)$.

As pointed out above, the relations of the specifications in the time and frequency domains are usually analyzed considering that the closed-loop system has one or two dominant poles and can be approximately modeled by standard first- (L_1) or second-order (L_2) transfer functions[1] given by

$$L_1(s) = \frac{K}{\tau s + 1}, \quad L_2(s) = \frac{\omega_n^2}{s(s + 2\zeta\omega_n)}. \tag{8.2}$$

Under unit feedback, the corresponding closed-loop transfer functions are given by

$$G_{cl_1}(s) = \frac{L_1(s)}{1 + L_1(s)} = \frac{K}{\tau s + (1 + K)}, \quad G_{cl_2}(s) = \frac{L_2(s)}{1 + L_2(s)} = \frac{\omega_n^2}{(s^2 + 2\zeta\omega_n s + \omega_n^2)}. \tag{8.3}$$

Notice that due to the structure of $L_2(s)$, ζ and ω_n here are descriptive parameters of the closed loop. In cards where both the open loop and the closed loop can be described by second-order transfer functions, the subscript $(\cdot)_{cl}$ shall be used to refer to the relative damping factor and the undamped natural frequency of the closed-loop transfer function.

Figure 8.1 shows the typical profile of the magnitude plot of $L_2(j\omega)$, with $\zeta = 0.5$ and $\omega_n = 1$ rad/s. Intuitively, as frequency is the inverse of time and vice versa, for a step reference, low frequencies are related to the steady-state response, high frequencies to the initial trend of the response, and frequencies around ω_{gc} to the transient response. From (8.1), if $|L(j\omega)|$ is large, small errors are obtained (as usually happens in steady state); if it is small (high frequencies), errors of the amplitude of the reference are obtained, what happens at the start of the time evolution after the reference is changed; and intermediate frequencies require obtaining relationships between ω_{gc}, which is the characteristic frequency of the open-loop transfer function, to closed-loop transient specifications.

Time- and frequency-domain specifications are related both to the transient and steady-state regimes.

The specifications related to the steady state are based on the analysis of *steady-state errors*, through the *static-error constants* K_p, K_v, and K_a, which interpretation, based on the *type* (number of integrators) of the open-loop transfer function $L(s)$, is provided in Sect. 8.2 of this chapter, both for time and frequency domains. Notice that G_{cl_1} in (8.3) has static gain $K/(1 + K)$, as there is a reduction in the closed-loop gain with respect to the open-loop gain as a consequence of feedback. If $L_1(s)$ is expressed in ZPK format, the closed-loop transfer function is given by

$$L_1(s) = \frac{K}{\tau s + 1} = \frac{\kappa}{s - p} \rightarrow G_{cl_1}(s) = \frac{K}{\tau s + (1 + K)} = \frac{\kappa}{s - p + K}, \tag{8.4}$$

with $p = -1/\tau$ and $\kappa = -Kp$. Notice that both in open loop and closed loop, in ZPK format, the numerator is κ, which is usually known as *gain-bandwidth product* GBW [3]. For first-order systems, the gain-bandwidth product is a constant.

The static gain of G_{cl_2} is one, because $L(s)$ is type-1, as it is explained in Sect. 8.2.

The specifications related to transient response and relative stability can be posed in the time and frequency domains, and for the simple cases considered in Eq. (8.2), several interesting relationships can be found, as discussed in what follows:

- Desired transient response, generally determined by the position of the *dominant closed-loop poles*. To achieve *overdamped* behavior, the specifications are usually expressed in terms of *closed-loop time constant* (τ_{cl}) or *settling time* (t_s), if the time response of the closed-loop system can be approximated by that of a first-order system (G_{cl_1} in (8.3)). From Sect. 3.2, it is known that $t_s \approx 4\tau_{cl}$. If the closed-loop system is modeled as an overdamped second-order system $G_{cl_{2o}}(s) = \frac{1}{(\tau_1 s + 1)(\tau_2 s + 1)}$, it is known from Sect. 3.3 that the settling time can be approximated by $t_s \approx 4(\tau_1 + \tau_2)$.

---

[1] Although the analysis is done considering a first- or a second-order closed-loop system, the obtained conclusions are use to work in higher-order systems. More details can be found in [2].

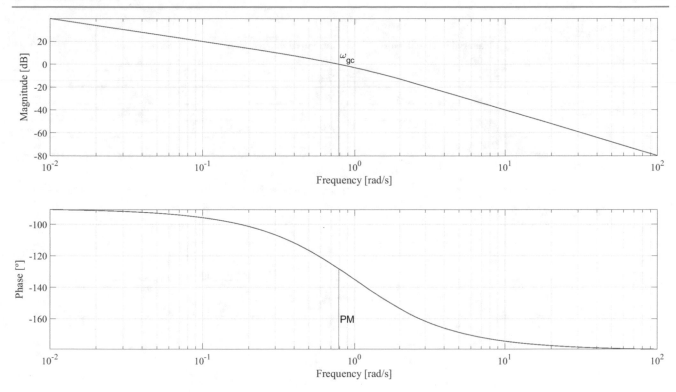

Fig. 8.1 Typical shape of $L(s)$ for obtaining a closed-loop second-order system

If the closed-loop system behavior is *underdamped* and can be approximated by a second-order transfer function (G_{cl_2} in (8.3)), the specifications are traditionally stated in terms of *damping* (maximum allowed *overshoot*, linked to the relative damping factor ζ), and characteristic response times (*rise time* t_r, *peak time* t_p, and *settling time* t_s, which allow obtaining the closed-loop *undamped natural frequency* ω_n, once ζ is known from the peak overshoot). In both cases, what is implicitly being done is imposing the desired location of the dominant poles of the closed-loop system. A filter in the reference can be added to cancel the effect of MP closed-loop zeros.

It is important to remark again that, although these characteristics of the dynamic response have been studied in Sects. 3.2 and 3.3 for standard first- and second-order systems, here they refer to closed-loop behavior.

There are also other control techniques (not treated in this book) in which the specifications are posed in terms of the integral of the error, like ISE, IAE, or ITAE treated in Sect. 3.8 in the framework of the identification of model parameters.

On the other hand, in classical frequency response-based design methods, the behavior of the closed-loop transient response is usually specified in an indirect way, through parameters which can be measured or described using the open- or closed-loop transfer function:

- Measured/described using the open-loop transfer function $L(s)$:
  - Affecting closed-loop damping and relative stability: *Phase margin* (PM).
  - Affecting closed-loop speed of response: *Gain crossover frequency* (ω_{gc}).
- Measured/described using the closed-loop transfer function G_{cl}:
  - Affecting closed-loop damping and relative stability: *Resonant peak* (M_r).
  - Affecting closed-loop speed of response: *Undumped natural frequency* (ω_n), *damped natural frequency* (ω_d), *resonant frequency* (ω_r), and *bandwidth* (BW).

Table 8.1 summarizes the previous statements. In Chaps. 3 and 4, different formulas were obtained for the previous parameters that are recalled here.

Table 8.1 Closed-loop specifications

	Measured in closed loop	Measured in open loop	Steady state
Time domain	$G_{cl}(s)$	$L(s)$	K_p, K_v, K_a
First order	τ_{cl}, t_s	$L_1(s)$	
Second order	OS, t_r, t_p, t_s	$L_2(s)$	
Frequency domain			K_p, K_v, K_a
First order	BW	PM, ω_{gc}	
Second order	$BW, M_r, \omega_n, \omega_d, \omega_r$	PM, ω_{gc}	

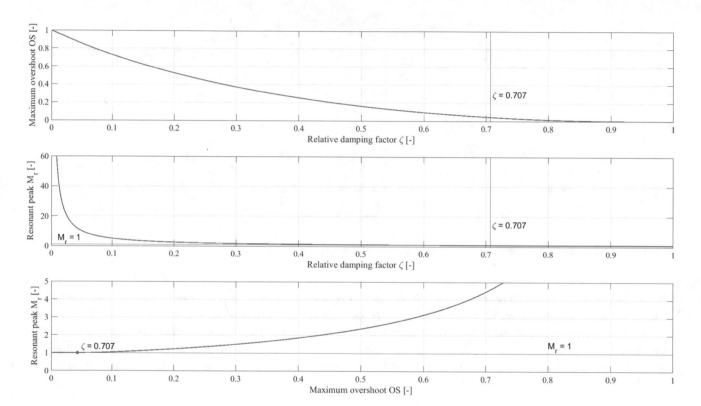

Fig. 8.2 Relations between the relative damping factor ζ, maximum overshoot (OS), and resonant peak (M_r)

In case the open-loop system is described by $L_2(s)$, providing a standard second-order closed-loop system $G_{cl_2}(s)$, the basic transient and stability specifications can be obtained and visualized. This is done in the following subsections.

8.1.1.1 Relation Between Closed-Loop Transient Response and Closed-Loop Frequency Response

First of all, the relation between the closed-loop parameters affecting damping (maximum overshoot OS and resonant peak M_r) is analyzed. Both parameters depend on the relative damping factor ζ:

$$OS = \exp\left(\frac{-\zeta\pi}{\sqrt{1-\zeta^2}}\right), \quad M_r = \frac{1}{2\zeta\sqrt{1-\zeta^2}}, \quad \forall \zeta \leq \frac{1}{\sqrt{2}}, \tag{8.5}$$

with M_r not in dB. These expressions were already analyzed in Sect. 3.3 (Eq. (3.26)) and Sect. 4.4 (Eq. (4.12)). Figure 8.2 graphically shows these relationships. Some interesting conclusions can be drawn:

- The smaller the ζ, the larger the OS and M_r.
- The resonant peak must be normally less than 1.5; the acceptable range of peak amplitude corresponds to damping rates of 0.4–0.7. Higher values of M_r mean large overshoots.

Regarding the speed of the closed-loop response, Fig. 8.3 shows the relationship of representative closed-loop normalized frequencies (all frequencies divided by ω_n) with the relative damping factor $\zeta \in (0, 1)$, based on the formulas:

$$\frac{\omega_d}{\omega_n} = \sqrt{1 - \zeta^2}, \quad \text{Sect. 3.3,}$$

$$\frac{\omega_r}{\omega_n} = \sqrt{1 - 2\zeta^2}, \quad \text{Sect. 4.4,}$$

$$\frac{BW}{\omega_n} = \sqrt{1 - 2\zeta^2 + \sqrt{2 - 4\zeta^2 + 4\zeta^4}}, \quad \text{Sect. 4.4.}$$

For the sake of space, the figure also shows the relation with ω_{gc}, that is, a frequency characteristic of the open-loop transfer function $L_2(s)$ and which role will be commented on in the next subsection. Interesting conclusions can be mentioned from this figure:

- For very small values of ζ, $\omega_n \approx \omega_d \approx \omega_r$. For higher values of ζ, $\omega_r < \omega_d < \omega_n$.
- As already known, ω_r only makes sense if $\zeta < 0.707$.
- As ζ varies from 0 to 1, BW varies from $1.55\omega_n$ to $0.64\omega_n$. For a value of $\zeta = 0.707$, $BW = \omega_n$. That is the reason why for most design considerations, it is assumed that the bandwidth of a second-order system can be approximated by ω_n.

The speed of response of the closed loop in the time domain is characterized by rise time t_r, peak time t_p, and settling time t_s, which are related to the relative damping factor ζ and undamped natural frequency ω_n through (3.27), (3.25), and (3.28), respectively:

$$t_r = \frac{\pi - \varphi}{\omega_n \sqrt{1 - \zeta^2}}, \quad \cos(\varphi) = \zeta, \quad t_p = \frac{\pi}{\omega_d} = \frac{\pi}{\omega_n \sqrt{1 - \zeta^2}}, \quad t_s \approx \frac{4}{\zeta \omega_n}. \tag{8.6}$$

Figure 8.4 shows the relations between closed-loop characteristic frequencies with rise time (the case $t_r \omega_{gc}$ is analyzed in the following subsection). From this figure, it can be seen that a rough approximation useful for design purposes is to consider

$$t_r BW \approx 3, \quad \text{for } \zeta < 0.8. \tag{8.7}$$

Figure 8.5 shows the relations between closed-loop characteristic times with bandwidth, which are useful for finding the values of closed-loop characteristic times knowing ζ and BW.

8.1.1.2 Relation Between Closed-Loop Transient Response and Open-Loop Frequency Response

As stated in Table 8.1, when considering a closed-loop second-order system represented by $G_{cl_2}(s)$, there are two characteristics that are measured using the open-loop transfer function $L_2(s)$ that have a determinant influence on closed-loop behavior. These characteristics are the PM (which influences damping and relative stability) and corresponding gain crossover frequency ω_{gc} (which influences the closed-loop speed of response).

From the application of the definition of PM to $L_2(s)$ in (8.2), it is obtained as

$$|L(j\omega_{gc})| = 1 \rightarrow \frac{\omega_n^2}{\omega_{gc} \left(\omega_{gc}^2 + 4\zeta^2 \omega_n^2\right)^{1/2}} = 1 \rightarrow \frac{\omega_{gc}^2}{\omega_n^2} = \sqrt{4\zeta^4 + 1} - 2\zeta^2, \tag{8.8}$$

$$PM = 180° - 90° - \arctan\left(\frac{\omega_{gc}}{2\zeta \omega_n}\right) = \arctan\left(\frac{2}{\sqrt{\left(4 + \frac{1}{\zeta^4}\right)^{1/2} - 2}}\right). \tag{8.9}$$

From (8.9), it is easy to see that there is a relationship between the PM and the relative damping factor ζ (Fig. 8.6), and thus with the peak overshoot OS (Sect. 3.3). Some interesting remarks can be done:

- In the range $0 \leq PM \leq 70°$, the relationship can be approximated by

$$\zeta \approx PM/100. \tag{8.10}$$

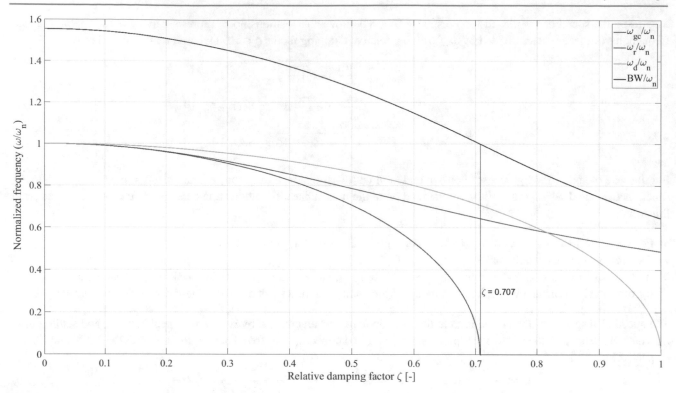

Fig. 8.3 Relations between the relative damping factor ζ and normalized frequencies

For $PM > 70°$, the closed-loop system does not oscillate.

- If PM increases, maximum overshoot decreases.
- The relation between rise time t_r and ω_{gc} has been plotted in Fig. 8.4. For low values of ζ, $\omega_{gc}t_r \approx 1.6$. From (8.6), and Figs. 8.3 and 8.4, for $\zeta \in (0, 0.3)$, $\omega_{gc}t_r \in (1.57, 1.80)$. For that reason, it is considered that

$$t_r \approx \frac{\pi}{2\omega_{gc}}, \text{ for low values of } \zeta. \tag{8.11}$$

For $\zeta \in (0.707, 0.9)$, $\omega_{gc}t_r \in (2.12, 3.2)$. This is the reason for doing the approximation:

$$t_r \approx \frac{\pi}{\omega_{gc}}, \text{ if } PM > 70°. \tag{8.12}$$

These relationships only work if $L_2(j\omega)$ cuts 0 dB with a slope ideally in the range $(-20, -40)$ dB/decade ("clean and smooth cutoff"). For instance, consider the transfer function of a plant and a controller given by

$$G(s) = \frac{0.1}{0.1s + 1}, \quad C(s) = \frac{10s + 1}{s}.$$

Figure 8.7 (up) shows the Bode diagram and stability margins of $L(s) = C(s)G(s)$, as well as the step response. As the $PM = 169°$ at $\omega_{gc} = 1$ rad/s, it is expected that the rise time of the closed-loop system is approximately $t_r = \pi/\omega_{gc} = 3.14$ s, but however, the rise time is higher than 50 s, because the crossing of $L(s)$ with 0 dB is almost flat (due to the presence of a zero in the controller). The same figure (down) shows the case in which the slope is sharper than -40 dB/decade, using

$$G(s) = \frac{1}{(1.5s + 1)^2}, \quad C(s) = \frac{0.5s + 1}{s}.$$

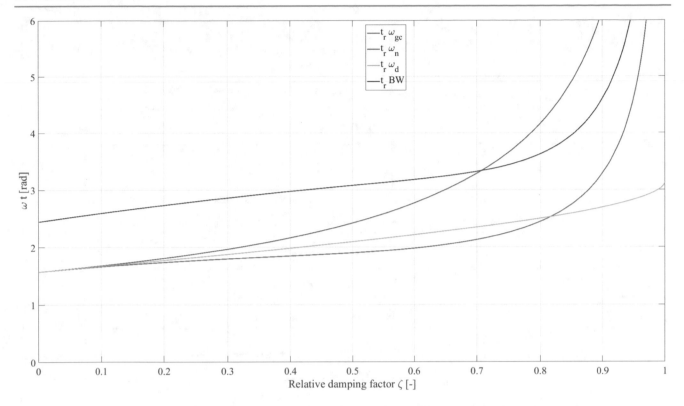

Fig. 8.4 Relations between closed-loop characteristic frequencies with rise time, as a function of the relative damping factor ζ

In this case, the relative stability margins are poor and such a choice must be avoided. The controllers used in the examples are proportional–integral (PI), defined in Sect. 8.4. These considerations are related to limitations imposed by Bode's sensitivity integral (8.14), discussed below.

8.1.1.3 Relation Between Closed-Loop Frequency Response and Open-Loop Frequency Response

From Eqs. (8.5) and (8.9), there is a relationship between PM and the resonant peak M_r, through the relative damping factor ζ (the natural frequency can be eliminated from the two equations). Figure 8.8 shows this relationship, from which the following conclusions are obtained:

- In the range $40° \leq PM \leq 60°$, the relationship can be approximated by

$$PM[\text{rad}] = \frac{2.3 - M_r}{1.25}, \tag{8.13}$$

  with M_r not in dB.
- If PM increases, the resonant peak decreases.

The characteristic open-loop design frequency ω_{gc} can be related to the closed-loop natural frequency through (8.8), and therefore to other closed-loop representative frequencies, as shown in Fig. 8.3, so that

- For very small values of ζ, $\omega_{gc} \approx \omega_n$. For higher values of ζ, $\omega_{gc} < \omega_n$.
- In the interval $0 < \zeta < 1$, $\omega_{gc} \in (0.61BW, 0.75BW)$. By increasing the gain crossover frequency ω_{gc}, the closed-loop bandwidth also increases, and the closed-loop settling time decreases.
- Regarding the slope of the cut of $L_2(s)$ with 0 dB, it should not be larger than -40 dB/decade in absolute value, as this should produce low stability margins, as commented before.

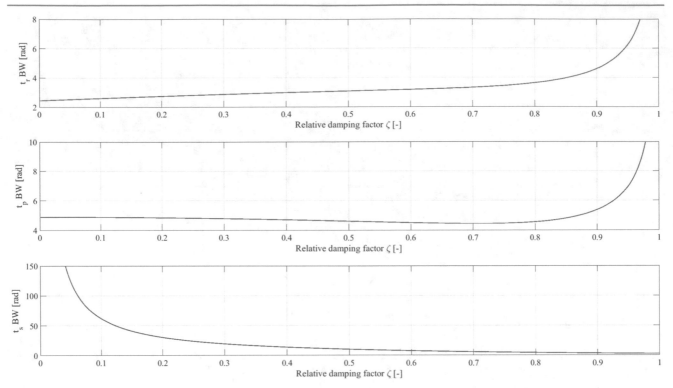

Fig. 8.5 Relations between closed-loop characteristic times with bandwidth, as a function of the relative damping factor ζ

Remark

All the previous statements have been made considering that the closed loop is underdamped second-order. If the closed-loop system can be modeled as a first-order system, the open-loop bandwidth of $L_1(s)$ is $|1/\tau| = |p|$, while the corresponding closed-loop bandwidth of $G_{cl_1}(s)$ is $|(1 + K)/\tau| = |p(1 + K)|$. Therefore, the closed-loop bandwidth is greater than the open-loop bandwidth and the closed-loop system is faster than the open-loop one, as $\tau_{cl} = \tau/(1 + K)$. This can easily be analyzed by the root locus method.

In the next section, more considerations are given about the relationship between open-loop and closed-loop frequency responses.

8.1.2   Sensitivity Functions, Disturbance Rejection, and Unmodeled Dynamics

Plotting gain curves for the *Gang of Four* transfer functions explained in the introduction to Chap. 7 is also a good way to obtain a quick overview of a feedback system and closed-loop specifications [1]. Figure 8.9 shows the typical profile of the magnitude plot of $L_2(j\omega)$ in (8.2) with $\zeta = 0.5$ and values of $\omega_n = 0.1$, 1 and 10 rad/s.

Figure 8.10 shows the corresponding magnitude of the characteristic closed-loop sensitivity functions given by (7.7)

- S: Sensitivity function, $S = \frac{1}{1+L}$.
- T: Complementary sensitivity function, $T = \frac{L}{1+L}$.
- GS: Load sensitivity function (also called input sensitivity function), $GS = \frac{G}{1+L}$.
- CS: Noise sensitivity function (also called output sensitivity function), $CS = \frac{C}{1+L}$.

Notice that, as a design principle (as already commented in the introduction), $|L(j\omega)|$ should be large at low frequencies for good load disturbance attenuation, small at high frequencies for reducing the effect of noise in the control loop, and having an adequate gain crossover frequency ω_{gc} to ensure good performance and robustness characteristics (around ω_{cg} tracking errors depend on the phase of $L(j\omega)$, as known from the analysis of relative stability margins). Moreover, the gain curve at crossover frequency must have a negative slope in the interval $[-20, -40]$ dB/decade.

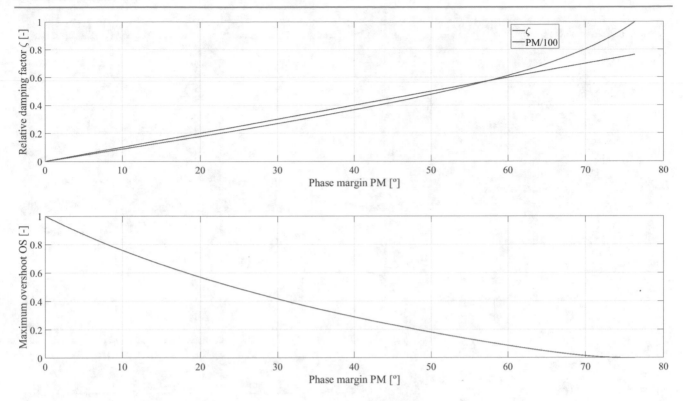

Fig. 8.6 Relations between PM and closed-loop relative damping factor and peak overshoot

The previous plots (or other frequency response plots) are a complement to time responses. Usually, the design is performed so that $T(\omega \approx 0) = 1$, and $T(\omega \to \infty)$ is small. The sensitivity function cannot be small for all frequencies. The sensitivity function S is zero at the poles of L and one at the zeros of L. $|S(j\omega)| \leq 1$ can be achieved for all frequencies only if the Nyquist curve of $L(j\omega)$ is in the first and third quadrants [1]. Moreover, if $L(s)$ has relative degree one or higher: $\lim_{s \to j\infty} S(s) = 1$. As pointed out by Gunter Stein [4], "sensitivity improvements in one frequency range must be paid for with sensitivity deteriorations in another frequency range, and the price is higher if the plant is open-loop unstable. This applies to every controller, no matter how it was designed". This comment relates to the "water-bed effect" of the frequency response (if $S(j\omega)$ is decreased at one frequency it must increase in another), and indicates that control design is a compromise. This observation is a consequence of Bode's sensitivity integral of which, only basic ideas are given here. Following [5], assuming that $L(s)$ goes to zero faster than $(1/s)$ as $s \to \infty$, then the sensitivity function satisfies the following integral:

$$\int_0^\infty \ln |S(j\omega)| d\omega = \pi \sum p_i - \frac{\pi}{2} \lim_{s \to \infty} sL(s), \tag{8.14}$$

where $p_i \in \mathbb{C}^+$ are the unstable poles of $L(s)$. According to the integral, fast poles (and slow zeros) in the RHP are bad for design purposes [1]. Equation (8.14) indicates that there is a redistribution of disturbance attenuation over different frequencies. The derivation of this formula is provided by [5], Sect. 11.8, where it is also stated that for MP systems, the phase is uniquely given by the shape of the gain curve, and vice versa, the phase curve being a weighted average of the derivative of the gain curve. If the gain curve has a constant slope $20n$ dB/decade in a Bode diagram, the phase curve has a constant value $n\pi/2$ rad. Moreover, this implies that the slope of the gain curve cannot be too steep (the interested reader can analyze the derivation in [5], Sect. 11.4). So, as an approximation of the limitations imposed by Bode's sensitivity integral, for MP systems the approximations in Table 8.2 are met.

This is the main reason why the slope of the magnitude curve around ω_{gc} must be between -20 and -40 dB/decade (preferably between -20 and -30 dB/decade), as a general rule.

If the case for disturbance attenuation (with $R = 0$) is considered, following the block diagram in Fig. 7.2, the measured output of the process without control (open loop) is $Y_{ol}(s) = N(s) + G(s)D(s)$, while the closed-loop output is given by

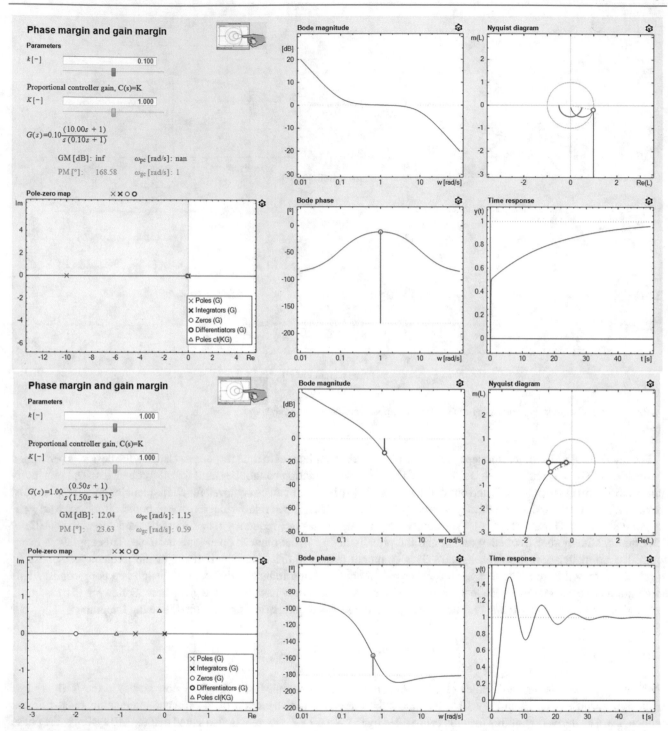

Fig. 8.7 Examples where the relationship between gain crossover frequency ω_{gc} and closed-loop rise time does not work

$$Y_{cl}(s) = \frac{G(s)}{1 + C(s)G(s)}D(s) + \frac{1}{1 + C(s)G(s)}N(s) = S(s)Y_{ol}(s). \tag{8.15}$$

Then, disturbances with frequencies such that $|S(j\omega)| < 1$ are attenuated by feedback, while those such that $|S(j\omega)| > 1$ are amplified by feedback [1]. The interpretation in the Nyquist diagram is that the sensitivity is less than 1 for all points outside the circle with radius 1 and center at the critical point $-1 + j0$. Disturbances at these frequencies are attenuated through feedback. Regarding the complementary sensitivity function, the *roll-off rate* is the slope of $|T(j\omega)|$ at the BW frequency.

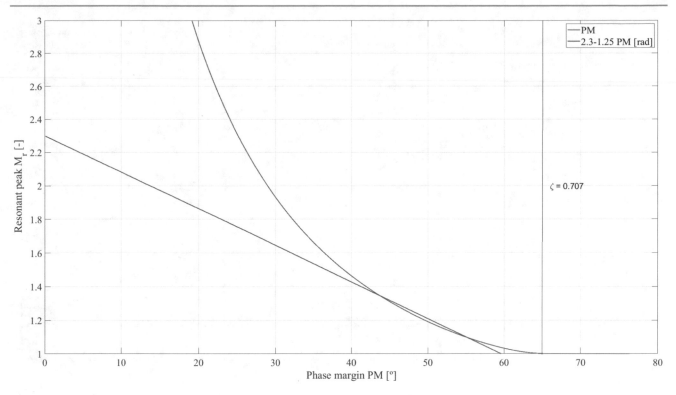

Fig. 8.8 Relationship between PM and resonance peak M_r

Table 8.2 Values of the slope of $|L(j\omega_{gc})|$ imposed by Bode's sensitivity integral (considering a constant slope)

PM [°]	Slope of $|L(j\omega_{gc})|$ [dB/decade]
30	−33.33
45	−30.00
60	−26.66

This and BW together specify the rejection of exogenous disturbances in transient terms. As pointed out by [6], the steeper the magnitude diagram at the BW frequency (higher roll-off rate), the better the disturbance rejection. In other words, the system better distinguishes the main signal (which is transmitted in the BW) from the disturbance (whose frequency is outside the BW and is filtered out). However, in general this results in poor stability margins, as the corresponding loop transfer function would also have a steeper magnitude diagram in ω_{gc}. So, closed-loop control design is always a compromise.

Regarding noise, it is easy to see from Fig. 8.10 that at high frequencies (characteristic of noise signals), $|S(j\omega)| \approx 1$, so that the requirements of both setpoint tracking and disturbance rejection contradict noise reduction. $|S(j\omega)| < 1$ in the frequencies of interest for reference tracking and disturbance rejection will produce the exact transmission of noise to the output. As pointed out by [6], *the feedback can be at most as good as the measurement is*, so it is important to invest in high-quality sensors and place them in the right positions.

The lowest frequency for which the sensitivity function has magnitude 1 is called the *sensitivity crossover frequency* ω_{sc}. The maximum sensitivity, which indicates the worst case in terms of disturbance amplification, is

$$M_s = \max_\omega |S(j\omega)| = \max_\omega \left| \frac{1}{1 + L(j\omega)} \right|, \tag{8.16}$$

where M_s is called *maximum sensitivity* and ω_{ms} is the *maximum sensitivity frequency*. In general, $s_m = 1/M_s$ is called the *stability margin*, as provides the worst-case distance to the critical point (thus being an alternative to classical GM and PM stability margins). As pointed out before, the sensitivity has a constant magnitude in circles with the center at the critical point. By looking at Fig. 8.11, it can be seen by trigonometry that M_s relates to gain and phase margins as [7]

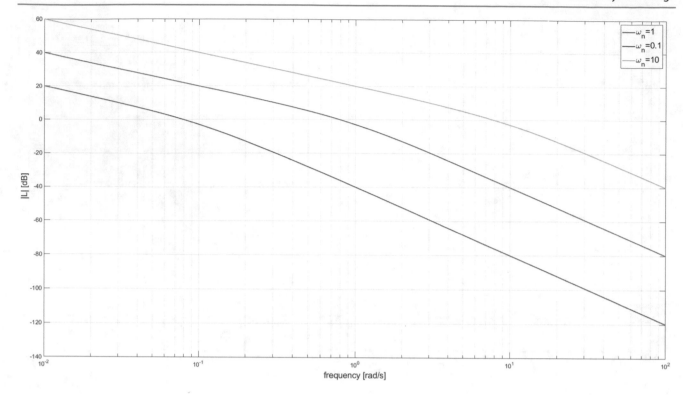

Fig. 8.9 Typical profile of loop transfer functions

Table 8.3 Relationships between M_s and PM

M_s	PM
2	$\geq 30°$
1.6	$\geq 36°$
1.4	$\geq 41°$
1	$\geq 60°$

$$GM \geq \frac{M_s}{M_s - 1}, \quad PM \geq 2\arcsin\left(\frac{1}{2M_s}\right) \approx 60\frac{1}{M_s}, \tag{8.17}$$

some representative values being those in Table 8.3.

A look at how small process changes ΔG affect closed-loop response provides some insights about robustness [7]:

$$G_{cl} = \frac{FCG}{1 + CG} \rightarrow \frac{\Delta G_{cl}}{\Delta G} = \frac{FC}{1 + CG} - \frac{FC^2 G}{(1 + CG)^2} = \frac{C}{(1 + CG)^2} = \frac{1}{1 + CG}\frac{CF}{1 + CG} \rightarrow$$

$$\rightarrow \frac{\Delta G_{cl}}{G_{cl}} = S\frac{\Delta G}{G}. \tag{8.18}$$

Relative variations in the process transfer function relate to relative variations in the closed-loop transfer function through the sensitivity function. If the process transfer function changes from G to $G + \Delta G$ (ΔG being stable), from Fig. 8.11, it can be seen that for ensuring stability ($L(j\omega)$ not reaching the critical point) $|C\Delta G| < |1 + L|$, so that $|\Delta G| < |(1 + L)/C|$ for all frequencies. So, the following inequality must be satisfied:

$$|\Delta G(j\omega)| < \frac{|G(j\omega)|}{|G_{cl}(j\omega)|}, \tag{8.19}$$

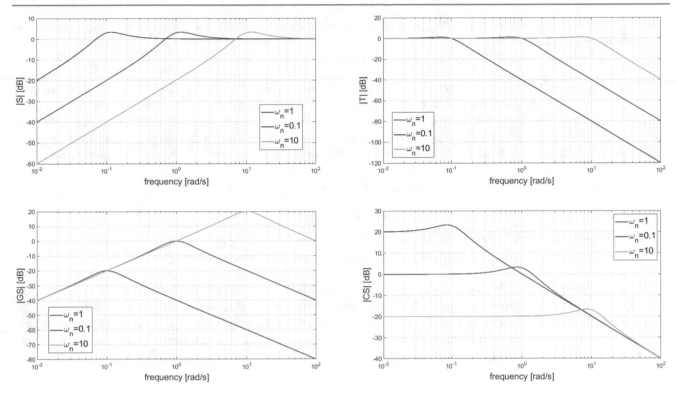

Fig. 8.10 Typical profile of sensitivity functions

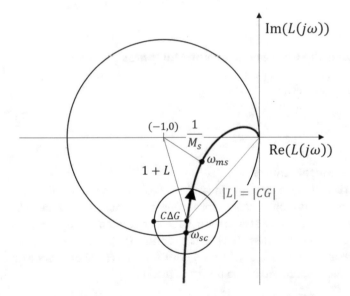

Fig. 8.11 Interpretation of sensitivity functions in the Nyquist diagram

which indicates that large plant variations are allowed where G_{cl} is small [7]. High model precision is required for frequencies where the gain of the closed-loop system is larger than the gain of the open-loop system.

All the previous considerations are related to control design specifications commented on at the beginning of this introduction.

When controlling a system, there are also fundamental limitations in the performance that can be achieved, which goes beyond the objective of this book. These can be summarized as [1]:

1. Large signal behavior: Constraints on actuators (saturation, slew rate, etc.) and process output.

2. Small signal behavior: Sensor noise, resolution of analog-to-digital and digital-to-analog converters, and friction (and other nonlinearities).

3. Dynamics: NMP dynamics, RHP zeros and poles, and time delays.

The first two are not dealt with in this book (only some aspects of noise are taken into account in this chapter), while the third case is considered in several chapters of this text. Time delays and zeros in the RHP limit the achievable bandwidth, and poles in the RHP require high bandwidth (strong control signals). Systems with poles and zeros in the RHP can be very difficult or even impossible to control robustly [1], being necessary to add sensors and actuators or redesign the process. In [5] (p. 399), a deep analysis of the dynamic limitations imposed by RHP poles and zeros is presented, the main conclusions being [1]:

- A RHP zero z limits the achievable gain crossover frequency $\omega_{gc} \leq z\sqrt{\frac{M_s-1}{M_s+1}}$.
- A RHP pole p requires a high gain crossover frequency $\omega_{gc} \geq p\sqrt{\frac{M_s+1}{M_s-1}}$.
- A RHP pole–zero pair gives a lower bound to the maximum sensitivity $M_s \geq \left|\frac{p+z}{p-z}\right|$.
- A RHP pole p and a time delay t_d gives a lower bound to the maximum sensitivity $M_s \geq e^{pt_d}$.

These concepts are useful when dealing with loop shaping design. Section 8.11 covers basic aspects of these control techniques.

In this book, disturbance compensation is dealt with by feedback, but there are feedforward control schemes that, using the actual value of measurable disturbances, are able to anticipate the effect that such disturbance produces in the process output. There are many books dealing with feedforward controllers, and the basics can be learnt from [7] and the accompanying interactive tools of the ILM project (https://arm.ual.es/ilm/, [8, 9]).

8.2   Steady-State Errors in Unit Feedback Control Systems

8.2.1   Interactive Tool: steady_state

8.2.1.1 Concepts Analyzed in the Card and Learning Outcomes
- Unit step, ramp, and parabola reference inputs.
- Steady-state or steady-state regime error.
- System type.
- Steady-state error constants.

8.2.1.2 Summary of Fundamental Theory
This card deals with concepts related to the closed-loop systems' design with unit feedback. In particular, *steady-state errors* produced by the inability of the closed-loop system to track certain test input signals, such as *step*, *ramp*, and *parabola*, are studied following a classical approach. Steady-state error in a feedback system means the error between the *reference* and the *output* when the system reaches a state that does not change over time [10].

Whether a unit feedback system presents a steady-state error for a given input depends on the *type* of the loop transfer function. Consider a generic system described by the transfer function:

$$G(s) = \frac{k \prod_{\ell=1}^{q}(\beta_\ell s + 1) \prod_{\ell=1}^{r}\left(\left(\frac{s}{\omega_{n_{z_\ell}}}\right)^2 + \left(\frac{2\zeta_{z_\ell}}{\omega_{n_{z_\ell}}}\right)s + 1\right)}{(s)^{\aleph} \prod_{i=1}^{p}(\tau_i s + 1) \prod_{i=1}^{h}\left(\left(\frac{s}{\omega_{n_i}}\right)^2 + \left(\frac{2\zeta_i}{\omega_{n_i}}\right)s + 1\right)}. \tag{8.20}$$

This system includes in its denominator the term $s^{\aleph}$, which represents a pole of multiplicity $\aleph$ in the origin ($\aleph$ *integrators*). $\aleph$ is also often referred to as the *type number*. Note that the concept of *type* should not be confused with the concept of *order* of the system (the latter is given by the degree of the polynomial of the denominator or *characteristic polynomial* of the system).

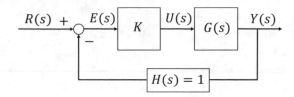

Fig. 8.12 Closed-loop system with unit feedback

As the type of a system increases, precision increases in a closed loop, but it exacerbates the stability problem (it is always necessary to find a compromise between precision and relative stability), as can be concluded from a simple root locus analysis.

Considering the closed-loop system with unit feedback represented in Fig. 8.12, the relationship between the *error* ($e(t)$, $E(s)$) and the *reference* ($r(t)$, $R(s)$) is given by

$$E(s) = \frac{1}{1 + KG(s)}R(s) \rightarrow e_{ss} = \lim_{t \to \infty} e(t) = \lim_{s \to 0} sE(s) = \lim_{s \to 0} \frac{sR(s)}{1 + KG(s)}, \quad (8.21)$$

where the expression of the *steady-state error* e_{ss} has been obtained from the application of *Laplace's transform final value theorem*. The case of non-unit feedback can be studied for example in the reference [10], Sect. 5.8, pp. 265–267. Notice that as pointed out by [6] (Sect. 4.2), this theorem is only valid for $sE(s)$ analytic in the closed RHP. If this condition is not satisfied, the method is not conclusive (for instance in the case of unstable closed-loop systems).

In the scope of steady-state error analysis of unit feedback systems, *position error* (e_{ss_p}) refers to the tracking error in steady state when the reference is a step, the term *velocity error* (e_{ss_v}) is used to express the steady-state error when the reference is a ramp, and *acceleration error* (e_{ss_a}) when it is a parabola. In cases where the closed-loop system is stable, so-called *steady-state error constants* or *static-error constants* can be defined as follows.

The *steady-state position error constant* (or static position error constant) K_p is defined as $K_p = \lim_{s \to 0} KG(s) = KG(0)$. For the system shown in Fig. 8.12, the steady-state error (e_{ss_p}) for a unit step input is given by

$$e_{ss_p} = \lim_{s \to 0} \frac{s}{1 + KG(s)} \frac{1}{s} = \frac{1}{1 + KG(0)} = \frac{1}{1 + K_p}. \quad (8.22)$$

Note that for a type-0 system, where the transfer function is represented in a standardized time constant form, $K_p = Kk$, and therefore $e_{ss_p} = 1/(1 + Kk)$; and for a type-1 (or higher) system, $K_p = \infty$ ($e_{ss_p} = 0$). This implies that the response of a unit feedback system to a step reference will have a steady-state error if there is no integrator in the open loop (in the controller or in the plant, whose product is represented here by the function $KG(s)$, because only a *proportional action* is being considered). As the static gain of the open loop increases, the steady-state error decreases, but it will be difficult in some cases to obtain reasonable relative stability margins, and the control signal will reach values which are not physically achievable due to actuator saturation. To achieve null steady-state error with a step reference, the system must be type-1 or higher.

The *steady-state velocity error constant* (or static velocity error constant) K_v is defined as $K_v = \lim_{s \to 0} sKG(s)$ for the system in Fig. 8.12. Following the same procedure for the position error constant, the steady-state error for a unit ramp input (slope equal one) $R(s) = 1/s^2$ is given by $e_{ss_v} = 1/K_v$. The term *velocity error* is used to express the steady-state error for a ramp input (position error due to a ramp input). For a type-0 system, $K_v = 0$ and therefore $e_{ss_v} = \infty$. For a type-1 system, $K_v = KG(0)$ and $e_{ss_v} = 1/Kk$. If the type is 2 or higher, $K_v = \infty$ and $e_{ss_v} = 0$.

The *steady-state acceleration error constant* (or static acceleration error constant) K_a is defined as $K_a = \lim_{s \to 0} s^2 KG(s)$. The same approach in previous paragraphs can be followed to reach the conclusions summarized in Table 8.4. As can be seen, type-2 systems have no steady-state error when the input is a step or ramp, presenting a constant error when the input is a parabola. Type-0 and type-1 systems are unable to track a parabolic input in a steady state.

The ∞ entries in the table correspond to the cases where $sE(s)$ is not analytic in the closed RHP [6], but they are correct, as computing $e(t)$ by applying the inverse Laplace transform to $E(s)$ in those cases produces unbounded error signals (functions of ramps, parabolas, and other unbounded signals). Also notice that to achieve bounded or zero steady-state error, the system

Table 8.4 Steady-state errors to typical reference inputs

	Unit step input $r(t) = 1; R(s) = \dfrac{1}{s}$	Unit ramp input $r(t) = t; R(s) = \dfrac{1}{s^2}$	Unit parabola input $r(t) = \dfrac{1}{2}t^2; R(s) = \dfrac{1}{s^3}$
Type-0 system	$\dfrac{1}{1+Kk}$	∞	∞
Type-1 system	0	$\dfrac{1}{Kk}$	∞
Type-2 system	0	0	$\dfrac{1}{Kk}$

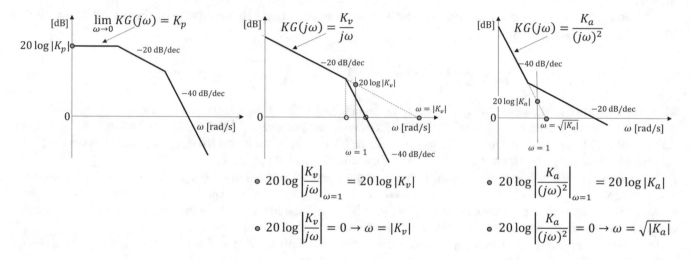

Fig. 8.13 Interpretation of steady-state error constants in Bode diagrams

type must be at least as large as the order of the input signal in the time domain (t^n). The higher the error constants $K_{p,v,a}$, the smaller the errors, but they also affect the transient response and stability.

Figure 8.13 provides an interpretation of the steady-state error constants in Bode diagrams. Position (K_p), velocity (K_v), and acceleration (K_a) error constants serve to describe the low-frequency behavior of type-0, type-1, and type-2 systems, respectively. Descriptive points of the frequency response from which the error constants can be obtained are also included in the figure.

A remark on load disturbance rejection

Consider that the reference to the loop is $R(s) = 0$, without loss of generality. From (7.4), the transfer function relating the load disturbance with the output is given by

$$\frac{Y(s)}{D(s)} = \frac{G(s)}{1 + KG(s)}.$$

If $D(s)$ is a unit step, the steady-state output is given by

$$Y(s) = \frac{G(s)}{1 + KG(s)} \frac{1}{s} \rightarrow \lim_{t \to \infty} y(t) = \lim_{s \to 0} sY(s) = \lim_{s \to 0} \frac{G(s)}{1 + KG(s)}.$$

If $G(s)$ is type-0, the output in the steady state is not zero (unless the system has a pure differentiator). This is also the case if $G(s)$ is type-1, meaning that the disturbance is not rejected. For instance, consider

$$G(s) = 1/s, \quad \lim_{s \to 0} \frac{G(s)}{1 + KG(s)} = \lim_{s \to 0} \frac{1/s}{1 + K/s} = \lim_{s \to 0} \frac{1}{s + K} = \frac{1}{K} \neq 0.$$

So, for a closed-loop system to reject step load disturbances, the *controller* must be at least type-1 (it is not enough that the plant is type-1). If in the previous examples the proportional controller is substituted by

$$C(s) = K/s \text{ (type-1)}, \quad \lim_{s \to 0} \frac{G(s)}{1 + C(s)G(s)} = \lim_{s \to 0} \frac{sG(0)}{s + KG(0)} = 0,$$

the same applies if $G(s)$ is type-1.

8.2.1.3 References Related to this Concept

- [3] Shahian, B., & Hassul, M. (1993). *Control system design using MATLAB®*. Prentice Hall, ISBN: 0-13-174061-X.
- [6] Bavafa-Toosi, Y. (2017). *Introduction to linear control systems*. Academic Press-Elsevier. ISBN: 978-0-12-812748-3. Chapter 4, Sect. 4.2, pp. 258–263.
- [10] Dorf, R. C., & Bishop, R. H. (2011). *Modern control systems* (12th ed.). Prentice Hall, ISBN: 978-0-13-602458-3. Chapter 5, Sect. 6, pp. 322–330.
- [11] Franklin, G. F., Powell, J. D., & Emani-Naeni, A. (2010). *Feedback control of dynamic systems* (6th ed.). Pearson. ISBN: 978-0-13-500150-9. Chapter 4, Sect. 2, pp. 196–200.
- [12] Golnaraghi, F., & Kuo, B. C. (2017). *Automatic control systems*. (10th ed.). McGraw Hill Education, ISBN: 978-1-25-964384-2. Chapter 7, Sect. 6.

Application   Interactive tool: steady_state

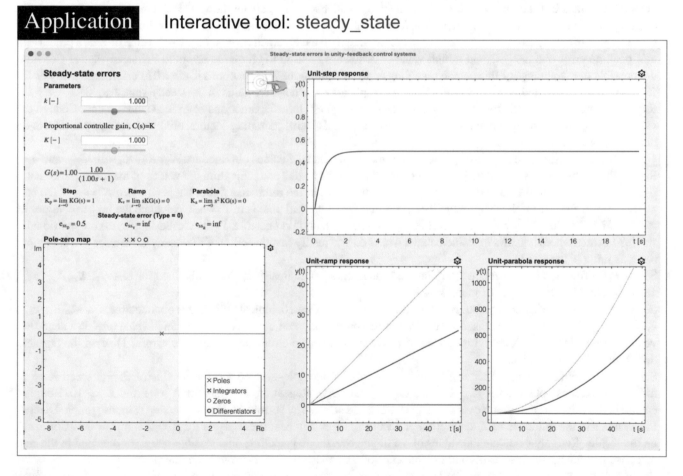

Steady State

The main objective of this card is to analyze steady-state errors of unit feedback control systems when different test inputs are used as a reference in the closed loop. It is therefore possible to visualize the errors in time response graphs and to analyze the value of the main steady-state error constants.

The upper-left area, as usual, is reserved for textual information, related in this case to open-loop system **Parameters** and error constants. From top to bottom, the first information presented is the static gain of the open-loop transfer function $G(s)$, which can be modified both through a slider or a textbox (where the sign can be changed). The same applies to the proportional gain of the controller K shown in Fig. 8.12. Below these values, the symbolic representation of the transfer function is placed in red color. The formulas and updated values of the constants K_p, K_v, and K_a are also displayed, as well as the system type, the corresponding input and the position (e_{ss_p}), velocity (e_{ss_v}), and acceleration (e_{ss_a}) errors.

In the **Pole–zero map**, all the dynamic elements can be dragged in and out (or double-clicking) from the repository, having special importance in this card the integrators, which define the type of the system. The Options menu incorporates the twelve transfer functions that are used as a basis for examples in various applications (see Table 1.2). These transfer functions simply provide different plant structures, which can be varied by dragging, adding, or subtracting dynamic elements on the **Pole–zero map**. It is also possible to enter the transfer functions in either (NUM,DEN) or ZPK formats.

The right part of the tool is devoted to graphically analyzing closed-loop responses to unit step, ramp, and parabola (which are represented by dotted lines in green color). When the mouse is positioned over the input and output curves, a label is visible providing the time values and the variable ($r(t)$, $y(t)$) associated with that point. All the graphs have the gearwheel icon whose settings help to change the scale when necessary.

8.2.1.4 Homework

1. Select $P_2(s)$ transfer function from the Options menu. Consider $K = 1$. Indicate the values obtained for K_p, K_v, and K_a, and e_{ss_p}, e_{ss_v}, and e_{ss_a}. Repeat the analysis for $K = 5$ and $K = 10$, commenting on the results. For a value of $K = 5$ and $k = 1$, repeat the analysis for $\tau = 5$ and $\tau = 0.1$ s. Comment on the obtained results. For which value of K is a practically null position error obtained? Could a high value of K always be used to eliminate errors in the steady state? Consider the possible disadvantages of this choice regarding actuator saturation. What happens if $K < 0$? Perform the analysis for $K = -0.1$, $K = -1$, and $K = -10$. Reason on the implications of negative gains in the steady-state errors.

2. Repeat the previous analysis by consecutively selecting all type-0 transfer functions from the Options menu. Change the values of the gain K and the dynamic elements (time constants, zeros, damping factors, etc.) and analyze the effects on the steady-state errors.

3. Using the transfer function $P_{11}(s)$ from the Options menu, indicate the values obtained for K_p, K_v, K_a, e_{ss_p}, e_{ss_v}, and e_{ss_a}. Repeat the analysis for $K = 5$ and $K = 10$, commenting on the results and indicating the values of maximum overshoot, rise time, and peak time obtained in each case for a step input in the reference. What happens when $K = 5$ if the real pole is moved away from the origin? Justify the results. When the real pole is in $s = -2$, calculate the range of values of K for which the closed-loop system is stable and those for which it is unstable. Is the closed-loop system conditionally stable? In cases where it shows unstable behavior, how do you justify that the value of the steady-state error to a step input remains finite in the tool?

4. Select $P_1(s)$ from the Options menu (by default it is an integrator). Indicate the values obtained for K_p, K_v, K_a, e_{ss_p}, e_{ss_v}, and e_{ss_a}. What is the influence of increasing K?

5. Select $P_{10}(s)$ from the Options menu. For $K = 0.1$, $K = 1$, and $K = 10$, indicate the values obtained for K_p, K_v, K_a, e_{ss_p}, e_{ss_v}, and e_{ss_a}. Discuss how the closed-loop transient response is affected in each case. For which value of K is a near-zero velocity error obtained? Could a high value of K always be used to eliminate steady-state errors? Discuss the possible drawbacks of such a choice.

6. Select $P_3(s)$ from the Options menu. Add an integrator using the pole–zero editor. For a value of $\beta = 1$ s and $K = 5$, indicate the values obtained for K_p, K_v, K_a, e_{ss_p}, e_{ss_v}, and e_{ss_a}. Repeat the analysis for $K = 1$ and $K = 10$. Roughly calculate the values of maximum overshoot, peak time, and steady-state error obtained in the step response graph. Discuss the results and the influence of K on them. For a value of $K = 5$, repeat the analysis for $\beta = 5$ and $\beta = 0.1$. Comment on the results. Roughly calculate the values of maximum overshoot, peak time, and steady-state error obtained in the step response graph. Discuss the results and the influence of β on them, and try to provide an interpretation.

7. Prove that if a system tracks a parabolic input with zero error, then its velocity response necessarily exhibits a peak overshoot (reference [6], Problem 4.6).

8. Given the following plants with the associated controllers ($L(s) = C(s)G(s)$) (reference [6], Problems 4.7–4.9), determine whether it is possible to have steady-state errors or maximum e_{ss} for the given inputs (where R_0 is the amplitude of the input, which can be freely selected by the user). If impossible, what is the minimum achievable error? Also determine some performance features and stability ranges using root locus analysis—root_locus (card 7.2).

$$G_1(s) = \frac{1}{(s+1)^2}, \quad C_1(s) = K\frac{s+2}{s+10}, \quad C_2(s) = K\frac{s+5}{s+30},$$
$$r_1(t) = R_0 u_s(t), \quad r_2(t) = R_0 t, \quad e_{ss} \leq 0.02 R_0,$$

$$G_2(s) = \frac{1}{s^2+1}, \quad C_1(s) = K\frac{s+1}{s+10}, \quad C_2(s) = K\frac{s+1}{s+20},$$
$$r_1(t) = R_0 u_s(t), \quad r_2(t) = R_0 t, \quad r_3(t) = 0.5 R_0 t^2, e_{ss} \leq 0.02 R_0,$$

$$G_3(s) = \frac{4}{s^2+5s+6}, \quad C_1(s) = \frac{K}{s}, \quad C_2(s) = \frac{K(s+1.5)}{s},$$
$$r_1(t) = R_0 t, \quad r_2(t) = 0.5 R_0 t^2, \quad e_{ss} \leq 0.05 R_0,$$

$$G_4(s) = \frac{1}{s^2(s+10)}, \quad C_1(s) = \frac{K(s+1)}{s+20},$$
$$r_1(t) = R_0 t, \quad r_2(t) = 0.5 R_0 t^2, \quad e_{ss} \leq 0.05 R_0.$$

9. Obtain the steady-state error and steady-state values of the following transfer functions with unit feedback. Find the unknown parameters such that the nonzero steady-state errors become 2% ($e_{ss} \leq 0.02 R_0$) (reference [6], Example 4.2, p. 261):

$$L_1(s) = \frac{K_1}{s+1}, \qquad r_1(t) = R_0 u_s(t),$$

$$L_2(s) = \frac{K_2(s+2)}{s(s+5)(s^2+2s+5)}, \quad r_2(t) = R_0 t,$$

$$L_3(s) = \frac{s+K_3}{s^2(s+10)^2}, \qquad r_3(t) = 0.5 R_0 t^2.$$

Check the results using the interactive tools steady_state and root_locus to verify that the values of the parameters fulfilling the steady-state specifications do not ensure closed-loop stability. Find the ranges of the parameters where stability is ensured. This is an example where it can be seen that two of the systems require compensation.

8.3 Proportional, Integral, and Derivative (PID) Controllers

8.3.1 Interactive Tool: PID_concept

8.3.1.1 Concepts Analyzed in the Card and Learning Outcomes
- Characteristics of proportional (P) control.
- Characteristics of integral (I) control.
- Characteristics of derivative (D) control.
- PID control.

8.3.1.2 Summary of Fundamental Theory
An automatic feedback controller compares the system output value with the reference, determines the error, and produces a control action that will attempt to reduce the error to zero, or to a very small value. Systems have three basic characteristics that must be considered when controlling them:

- Changes in the controlled variable due to changes in the reference, disturbances, or noise.
- Time required for the system output to reach a new steady state when a change in reference or disturbance occurs.
- Closed-loop system stability.

Depending on how these three aspects are considered, a particular control structure has to be used. *Proportional–Integral–Derivative* (PID) controllers are the most widely used in industry. According to an estimate by Profs. Åström and Hägglund [7], 95% of the control loops in the industry are of the PID type, and mainly PI. The *tuning* process is especially relevant, by means of which the values of the characteristic parameters of the PID controller are obtained, starting from certain performance specifications. It is therefore fundamental to understand the effect that its three main components have in a feedback loop, as summarized in what follows. Readers are encouraged to read the book [7], which from the authors' point of view is the best reference in the scope of PID control.

Proportional (P) control: A proportional controller provides a control signal proportional to the error. It acts as a gain amplifier K, as already shown in Chap. 7. The control action is represented by

$$u_c(t) = Ke(t), \tag{8.23}$$

where K is the *proportional gain*, which is the amount by which the error signal $e(t)$ is multiplied to obtain the control signal $u_c(t)$. Considering zero initial conditions, the transfer function $C(s)$ of the proportional controller can be obtained, which as can be seen is type-0:

$$U_c(s) = KE(s) \rightarrow C(s) = \frac{U_c(s)}{E(s)} = K. \tag{8.24}$$

Generally, if the reference of a proportional control system changes, the error cannot be set to zero (unless the plant is type-1 or higher), but a steady-state error arises, called *output offset*, consisting of the permanent deviation of the controlled variable from its reference value when the steady state is reached. This is one of the main drawbacks of proportional controllers. The steady-state error can be reduced by increasing the proportional gain K (see Sect. 8.2). In addition, an increase of K provides faster responses but reduces stability margins, and it can also lead to high control signal values that are not implementable in practice, due to actuator saturation. Moreover, the proportional controller does not increase the order of the system. A variant of the controller to reduce or cancel the steady-state error is to include a feedforward term [5]:

$$u_c(t) = Ke(t) + u_P \tag{8.25}$$

where u_P is adjusted to provide the desired value of the control action in steady state (usually $u_P = \bar{u}$ defining the operating point). If the reference r is constant and $G(s)$ is type-0, a typical election is based on steady-state analysis as $u_P = r/G(0) = r/k$, which will allow $\bar{y} = r$, if there are no disturbances. This choice requires an "exact" knowledge of the dynamics of the process (at least at low frequencies). In practice, the parameter u_P, called *reset* or *control offset*, is usually adjusted manually [5].

Integral (I) control: In a controller that uses an *integral action*, the control signal is modified at a speed proportional to the error signal, i.e. if the error signal is large, the control signal increases very quickly, if it is small, the control signal increases slowly. If zero initial conditions are considered, this controller can be represented mathematically by

$$\frac{du_c(t)}{dt} = K_i e(t) \quad \text{namely} \quad u_c(t) = K_i \int_0^t e(\xi)d\xi, \tag{8.26}$$

from which the transfer function of the integral controller can be obtained, which as can be seen is type-1:

$$U_c(s) = K_i\frac{E(s)}{s} \rightarrow C(s) = \frac{U_c(s)}{E(s)} = \frac{K_i}{s}, \tag{8.27}$$

K_i being the *integral gain*. By the definition of integral, this term provides the system with "memory", as the output u_c at an instant t will depend on the previous behavior from 0 to t. If a steady-state situation is considered, with $u_c = u_{ss}$ and $e = e_{ss}$, it should be fulfilled that

$$u_{ss} = K_i e_{ss} t,$$

which is only possible if $K_i = 0$ or $e_{ss} = 0$ [5]. Therefore, it can be deduced that with integral action the error will be zero if a steady-state situation is reached unless the closed loop is unstable. In the frequency domain, an integral controller has infinite gain at zero frequency. The closed-loop transfer function has unit static gain and there is no steady-state error when the reference is a step (in absence of disturbances). To some extent, the integral action can be interpreted as a method for generating the term u_P in a proportional controller automatically [5].

The basic features of an integral controller are as follows:

- When the controller output is constant, the steady-state error of the closed-loop system will be zero if the system is stable in the closed loop.

- When the controller output varies at a constant speed, the steady-state error must have a constant value, or in other words, when the error is constant, the integral grows with time at a constant speed and therefore the controller output also grows at a constant speed trying to reduce such error. Only when the error is zero, the value of the integral and, therefore, the control signal are constant ($u_c(t) = u_P$).

As can be seen, whenever an error occurs, the integral action changes the controller output to correct it, so this controller tends to cancel the errors in a steady state (note that this controller provides an integrator to the loop transfer function loop, increasing the system order and type by one, as is discussed in Sect. 8.2). This is the main advantage of this kind of control action, because it also has drawbacks, as its use tends to produce oscillatory responses. In practice, this term is combined with proportional and derivative actions, as described below.

Proportional–Integral (PI) control: As discussed in the previous paragraphs, the proportional control action generates a steady-state error. On the other hand, although the integral action overrides the error by itself, it may cause oscillatory behavior or very slow responses (if oscillatory behavior is not desirable). For these reasons, the two actions are combined to obtain a PI controller, which presents the advantages of each of them. Considering zero initial conditions, it can be represented by

$$u_c(t) = K\left(e(t) + \frac{1}{T_i}\int_0^t e(\xi)d\xi\right) = Ke(t) + K_i\int_0^t e(\xi)d\xi, \tag{8.28}$$

where T_i is called *integral time*, which represents in this case the time needed for the integral action to provide a signal equal to that given by the proportional action. The transfer function of a PI controller is given by

$$C(s) = \frac{U_c(s)}{E(s)} = K\left(1 + \frac{1}{T_i s}\right) = K\frac{(T_i s + 1)}{T_i s} = K + \frac{K_i}{s}. \tag{8.29}$$

With a PI controller, the steady-state error is avoided, but at a higher speed than with only integral action. The higher the PI controller's K gain, the faster the system will be, but more oscillations may occur. The PI controller introduces an integrator and a zero into the open loop, thus increasing the order and type of the system and its complexity.

Derivative (D) control: In derivative control, the controller output is proportional to the rate of change of the error signal with time:

$$u_c(t) = K_d\frac{de(t)}{dt}. \tag{8.30}$$

Parameter K_d is called *derivative gain*. The transfer function of a derivative controller is given by $C(s) = U_c(s)/E(s) = K_d s$ (differentiator).

This type of controller does not react to steady-state error signals, since its derivative is zero. For this reason, it must be combined with other control actions. Furthermore, it is not a causal element and, in practice, it is implemented as a pseudo-derivative, as discussed below.

The effect of derivative control action is to anticipate error changes and to provide a faster response to those changes. The fast response inherent in derivative control allows the system to be stabilized in a short period of time, especially when the error is constantly changing.

Proportional–Derivative (PD) control: When a proportional controller is combined with a derivative controller, the control action and the transfer function are described by

$$u_c(t) = K\left(e(t) + T_d\frac{de(t)}{dt}\right) = Ke(t) + K_d\frac{de(t)}{dt} \rightarrow$$

$$\rightarrow C(s) = \frac{U_c(s)}{E(s)} = K(1 + T_d s) = K + K_d s. \tag{8.31}$$

The output of the controller may vary when there is a changing error. When there is a change in the reference signal to the control loop, there is a rapid initial change in the controller output due to the derivative action, followed by a gradual change due to the proportional action. The *derivative time* T_d represents the time the derivative action anticipates the effect of the proportional action on the final control element. In theory, the derivative action has a positive stabilizing effect on the dynamics of the control system, causing oscillations to be damped, allowing to increase the proportional gain of the controller, and thus the speed of response, without increasing the oscillations.

The PD controller can be represented as

$$u_c(t) = K\left(e(t) + T_d\frac{de(t)}{dt}\right) \approx K\hat{e}(t + T_d),$$

where $\hat{e}(t + T_d)$ can be interpreted as the prediction of the error at the instant $t + T_d$ using a Taylor series expansion $\hat{e}(t + T_d) \approx \left(e(t) + T_d\frac{de(t)}{dt}\right)$ [5].

It should be noted that this is a type-0 controller and therefore is not able to eliminate steady-state errors (the steady-state error considerations previously made for P and D controllers apply here as well).

The major drawback of D and PD controllers is that derivative action can amplify the noise introduced into the system through the feedback loop, causing high-frequency oscillatory behavior in the control signal and hence in the process output. Furthermore, it is not a causal element and is usually implemented through a pseudo-derivative, so the ideal derivative is filtered through a first-order system with a time constant equal to T_d/N_d. The implementation is usually done in the following way:

$$C(s) = K(1 + T_d s) \approx K\left(1 + \frac{T_d s}{\frac{T_d}{N_d}s + 1}\right), \tag{8.32}$$

if $N_d = 1$, a low-frequency approximation of the derivative is achieved (for $|s| < 1/T_d$) [5].

In this card, as the objective is to introduce the fundamental concepts associated with PID controllers, this filter has not been implemented. Alternative options are to include a filter $F(s)$ in the reference (typically first-order) or to implement the derivative term acting on the process output, rather than on the tracking error (the error derivative and the output derivative only differ at the time of change of the reference).

Proportional–Integral–Derivative (PID) control: If the three control actions described in the previous paragraphs are combined, the PID controller is obtained, which helps to eliminate steady-state errors and reduce the trend of the system to oscillate in a closed loop, provided that its characteristic parameters are properly tuned. It is represented by the following equation (which is the ideal or non-interactive structure, as commented in the following paragraphs):

$$u_c(t) = K\left(e(t) + \frac{1}{T_i}\int_0^t e(\xi)d\xi + T_d\frac{de(t)}{dt}\right), \tag{8.33}$$

and therefore its transfer function is given by

$$C(s) = \frac{U_c(s)}{E(s)} = K\left(1 + \frac{1}{T_i s} + T_d s\right). \tag{8.34}$$

This is the most general and probably the most used type of controller, since it allows an optimal exploitation of the characteristics of the three control actions. It can be considered as a proportional controller, which has an integral control to eliminate steady-state error and a derivative control to improve stability and increase the speed of response. In this combination,

- Proportional control shapes the response curve of the controlled variable. It produces more output the higher the error.
- Integral control reduces the error. It produces more output the longer the error lasts.
- Derivative control decreases the time during which the error changes by predicting its value in advance.

The theoretical PID controller consists of two zeros and an integrator (pole at the origin). The practical implementation is usually done by using the filter in the derivative action to make the controller causal, as discussed above. The tuning of the PID parameters to achieve an adequate closed-loop response can be done theoretically or experimentally, recording in the latter case the dynamic response when reference changes or transient disturbances occur. If the controller is not properly tuned, the closed loop may become unstable, i.e. it may produce unbounded outputs when the inputs are limited, therefore not achieving the control objectives. There are multiple methods for tuning PID controllers; the most simple ones will be discussed in Sects. 8.5 (PI control using the pole-placement method), 8.6 (PI control using the pole-cancellation method), and 8.7 (PID control using the Ziegler–Nichols tuning rules) as main examples.

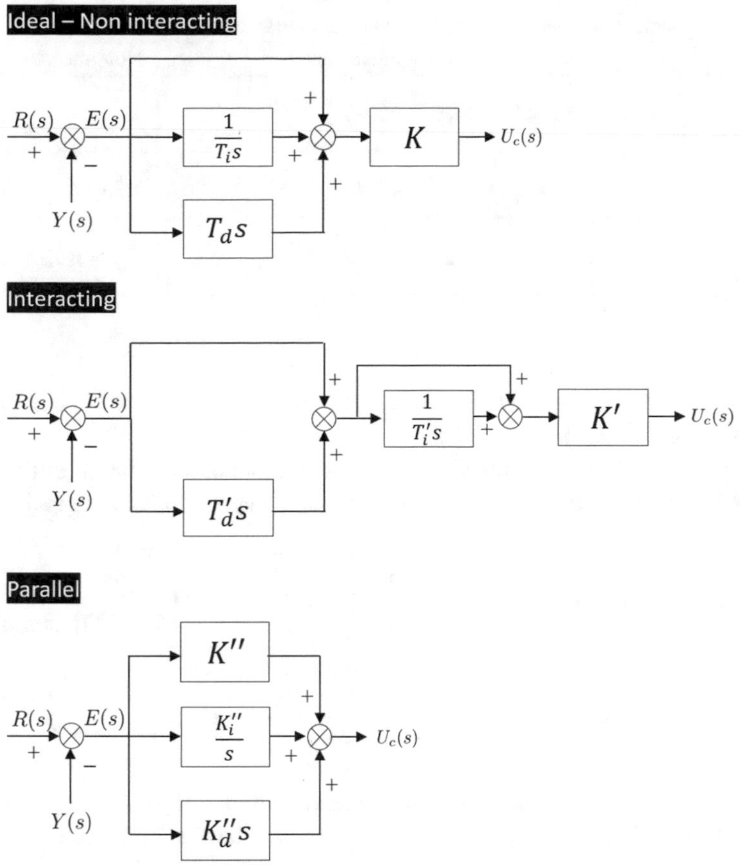

Fig. 8.14 PID control parameterizations: Non-interacting (ideal), interacting, and parallel

Notice that there exist different parameterizations of the PID controller in equation (8.34) that are useful depending on the design method used [5, 7]:

$$\text{Ideal or non interacting: } U_c(s) = K\left(1 + \frac{1}{T_i s} + T_d s\right) E(s), \tag{8.35}$$

$$\text{Interacting: } U_c(s) = K'\left(1 + \frac{1}{T_i' s}\right)\left(1 + T_d' s\right) E(s), \tag{8.36}$$

$$\text{Parallel: } U_c(s) = \left(K'' + \frac{K_i''}{s} + K_d'' s\right) E(s). \tag{8.37}$$

These implementations are summarized in Fig. 8.14.

The non-interacting algorithm or ideal PID control algorithm is the one considered as standard by the Instrument Society of America (ISA). It is the most cited in the literature and is currently the most widely used. It is characterized because the integral and derivative actions are independent (hence its name), and the proportional gain affects all three actions.

The interacting algorithm (also called a serial or classical algorithm) is the most widely used in traditional analog implementations (due to the simplicity of performing an analog PID with two amplifiers, as opposed to the non-interacting one that required three). It is characterized by the fact that any modification of the time constants, T_i and T_d, affects all three actions. It is also very useful for the use of analytical methods of the pole-placement type, since it allows dealing with some designs where the non-interacting scheme presents limitations.

The parallel algorithm is the only parametrization that allows the three actions to be modified separately. It is very useful in the frequency domain and analytical designs, as it allows to clearly study the contributions made by each of the controller

Table 8.5 Relationship between non-interacting (ideal), interacting, and parallel PID implementations

Interacting → Non-interacting	Non-interacting → Interacting	Non-interacting → Parallel
$K = K' \frac{T_i' + T_d'}{T_i'}$	$K' = \frac{K}{2}(1 + \sqrt{1 - 4T_d/T_i})$	$K'' = K$
$T_i = T_i' + T_d',\ T_i \geq 4T_d$	$T_i' = \frac{T_i}{2}(1 + \sqrt{1 - 4T_d/T_i})$	$K_i'' = \frac{K}{T_i}$
$T_d = \frac{T_i' T_d'}{T_i' + T_d'}$	$T_d' = \frac{T_i}{2}(1 - \sqrt{1 - 4T_d/T_i})$	$K_d'' = K T_d$

actions. It is undoubtedly the most flexible algorithm from a computational point of view, but it has limitations, as, for instance, the physical interpretation of the effect of the parameters.

From the transfer functions of the different controllers, it is possible to observe that all of them have a pole at the origin and two zeros that can be real or complex conjugates (non-interacting and parallel) or only real:

$$C(s) = K\left(1 + \frac{1}{T_i s} + T_d s\right) \rightarrow s = \frac{-1 \pm \sqrt{1 - \frac{4T_d}{T_i}}}{2T_d} \rightarrow \begin{cases} T_i > 4T_d \text{ real,} \\ T_i \leq 4T_d \text{ complex,} \end{cases}$$

$$C(s) = \left(K'' + \frac{K_i''}{s} + K_d'' s\right) \rightarrow s = K'' \frac{-1 \pm \sqrt{1 - \frac{4K_i'' K_d''}{K''^2}}}{2K_d''} \rightarrow \begin{cases} K''^2 > 4K_i'' K_d'' \text{ real,} \\ K''^2 \leq 4K_i'' K_d'' \text{ complex.} \end{cases}$$

Therefore, non-interacting and parallel algorithms are more general from a design point of view, as they can be tuned for processes with oscillatory modes (complex poles) due to the fact that they have complex zeros.

On the other hand, in the case of real zeros, one of them can be used to cancel the dominant pole of the process and the other to improve its stability. This is relatively straightforward for the non-interacting and interacting algorithms, where the proportional gain is available to place the closed-loop poles at the desired location. However, in the case of the parallel algorithm, the solution is not so simple, because the choice of zeros imposes constraints on all three parameters. All algorithms are equivalent to each other (with some restrictions); see Table 8.5.

Notice that the following section describes the interactive tool for understanding the basis of PID control. All interactive tools implementing PID control use the non-interacting (ideal) implementation. As commented in the introduction, for those readers who want to go deeper into the subject, the authors recommend reading the excellent book by K. J. Åström and T. Hägglund [7] (to understand for example different filters that can be used in the control loop, the integrator windup phenomenon, and how to compensate it with anti-windup mechanisms, different PID and feedforward control design methods, etc.). There are interactive tools that treat some of these aspects and can be found in https://arm.ual.es/ilm/, [8, 9]. Some examples from this book are included in the Homework section and also in Sect. 8.11 devoted to loop shaping design.

8.3.1.3 References Related to this Concept

- [3] Shahian, B., & Hassul, M. (1993). *Control system design using MATLAB®*. Prentice Hall, ISBN: 0-13-174061-X.
- [5] Åström, K. J., & Murray, R. M. (2014). *Feedback systems: An introduction for scientists and engineers* (2nd ed.). Princeton University Press, ISBN: 9780691193984. Chapter 10, pp. 10–1 to 10–14.
- [7] Åström, K. J., & Hägglund, T. (2006). *Advanced PID control*. ISA - The Instrumentation, Systems and Automation Society, ISBN: 978-15-561-7942-6.
- [10] Dorf, R. C., & Bishop, R. H. (2011). *Modern control systems* (12th ed.). Prentice Hall, ISBN: 978-0-13-602458-3, Example 4.4, pp. 259–267; example CP4.7, pp. 301–302.
- [13] Barrientos, A., Sanz, R., Matía, F., & Gambao, E. (1996). *Control de sistemas continuos. Problemas resueltos (Control of continuous systems. Problems solved)*. McGraw-Hill, ISBN: 84-481-0605-9. Chapter 9, Sect. 1, pp. 332–334.
- [14] Bolzern, P., Scattolini, R., & Schiavoni, N. (2009). *Fundamentos de control automático (Fundamentals of automatic control)*. McGraw-Hill, ISBN: 978-84-481-6640-3. Chapter 14, exercise 14.1, p. 391.

Application   Interactive tool: PID_concept

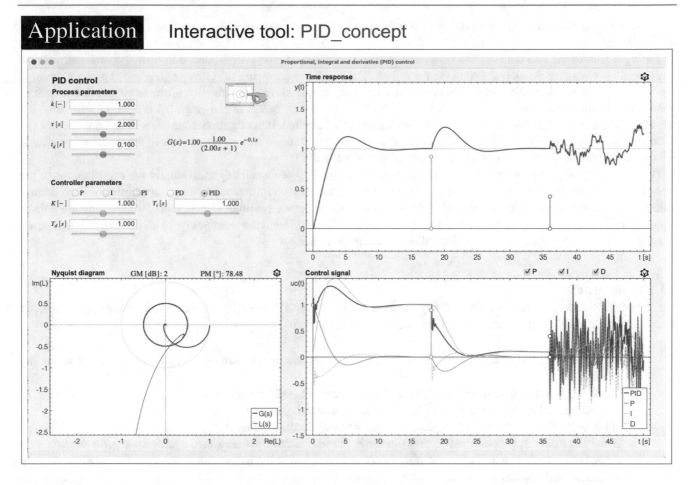

PID Concept

This card is devoted to analyzing the effect of the three control actions which a PID controller implements, both in the time and frequency domains. In the time domain, it is possible to visualize the contribution of each of the terms in the control signal. Moreover, the response to step changes in the reference and the load disturbance is possible to analyze, as well as the influence of noise. The representation selected to analyze the frequency response is the Nyquist diagram, which also provides GM and PM for the selected parameters. Different ideal PID structures can be selected: P, I, PI, PD, and PID.

The upper-left area of the tool shows the **Process parameters**, whose structure can be selected through the Options menu. As the objective of this card is to assimilate the main concepts of PID control, only three dynamic model structures widely used in industry are available: a multiple integrator with delay (Example 1, $P_{1r}(s)$ function of Table 1.2), a first-order system with time delay (Example 2, $P_{2r}(s)$), which is the default option when starting the tool, and a second-order system without time delay (Example 3, corresponding to $P_{11b}(s)$ without the real pole). In all cases, when selecting the system structure, the textboxes and sliders are displayed to facilitate the introduction or modification of the parameters. In the case of the multiple integrator, in addition to the gain and time delay, its order n can be selected. In the other cases, the characteristic parameters of first-order systems with time delay are shown (k, τ, t_d) and those of second-order systems (k, ζ and ω_n). A symbolic representation of the system's transfer function is always displayed.

Below this area, the selected controller structure (variations of the non-interacting PID) and **Controller parameters** are displayed. First, a series of radio buttons to select the structure of the controller are displayed: P, I, PI, PD, and PID. Once selected, the corresponding parameters can be modified through textboxes or sliders: K (P, PI, PD, PID), T_i (I, PI, PID), and T_d (PD, PID). The values of T_i and T_d must be positive. The lower-left area presents in this case a **Nyquist diagram** of the open-loop transfer function $G(s)$ (red) and of the loop transfer function $L(s) = C(s)G(s)$ (blue). It has been considered convenient to include this diagram so that the effect of PID-type controllers on closed-loop behavior in both the time and frequency domains can be simultaneously analyzed. The values of the gain margin (GM) and phase margin (PM) of the compensated system are also displayed.

The right area of the tool contains the **Time response** and **Control signal** plots. As explained in Sect. 7.5 in Chap. 7, these graphics allow changing the step amplitude in the reference at $t = 0$ (by accessing the circle or the dashed green horizontal line ○ · · ·). Moreover, there are two circles on the abscissa axis that, if moved horizontally, allow to change the instant in which the load disturbance or noise signal is introduced. If the one in the center of the graph (○) is dragged vertically, the amplitude of the step in the load disturbance at the system input is modified. If the rightmost one (○) is dragged vertically, the noise variance is changed. When the mouse pointer is placed over any of these circles, the associated information (amplitude and activation instant) is displayed in the lower-left corner of the tool.

The **Control signal** plot presents a remarkable feature, the possibility of visualizing individually the contributions to the control signal of the different components P, I, and D of the selected controller (depending on the application). Each control action uses a color that is detailed in a legend on the graph. In both the output and input signal graphs, when the mouse pointer is placed over them, labels are activated containing the time coordinates (t) and the variables (y, u) associated with that point.

All graphs have gearwheels whose settings allow to modify the scale.

8.3.1.4 Homework

1. Select from the Options menu the first example, corresponding to $P_{1r}(s) = \frac{k}{s^n}e^{-t_d s}$ with $k = 1$, $n = 1$, and $t_d = 0$ s. Set the amplitude of the load disturbance and the variance of noise to zero.

   a. For a P controller:
      i. According to the Nyquist diagram of the loop transfer function, is it possible to get an unstable response for any value of K?
      ii. Could the gain be increased infinitely to achieve a rise time $t_r \to 0$? Justify the answer.
      iii. Indicate the value of the gain providing a rise time of approximately 6 s.
      iv. What is the PM and the GM of the compensated system? Do their values depend on K?
      v. If the controller is proportional (type-0), what is the reason for obtaining zero steady-state error when a step reference is used? If a unit step load disturbance is introduced at the input of the system, is the steady-state zero error achieved? Justify the answer.
      vi. Enter a noise of variance 0.05. What is the amplitude of the output variance?
      vii. What happens if the sign of K is changed?
      viii. For K calculated in exercise 1.a.iii), select $n = 2$ ($G(s) = 1/s^2$, double integrator). Is it possible to obtain a non-oscillatory response for any value of K? Justify the answer also by using root locus analysis (root_locus, card 7.2). What happens if $n > 2$?
      ix. For $n = 1$ and the value of K obtained in exercise 1.a.3), now enter a value of $t_d = 0.5$ s. How much is the time response affected, and what are the new PM and GM? Answer the same questions with $t_d = 1$ s. Determine the value of t_d that causes the system to achieve a steady oscillation or limit cycle for a unit step reference (value of t_d for which the Nyquist diagram of the compensated system crosses the critical point (-1.0)). What is the value of the period of the oscillation? For $t_d = 1$, what is the value of K that causes the closed-loop system to enter a steady oscillation?

   b. For a PD controller (without load disturbance and noise),
      i. Select $G(s) = \frac{1}{s}$ (without time delay), $K = 1$, and $T_d = 0.5$ s. Note that when the PD controller is selected, the default parameter values seem to make the closed-loop system unstable, whereas a simple root locus analysis provides a stable closed-loop system. The reason for this unexpected behavior is the error in the numerical integration of the equations which, as discussed in Sect. 1.7.2, uses a sampling period of 0.05 s, so that with high values of T_d producing a very fast response, numerical problems arise. These problems disappear when selecting appropriate values of T_d, in this case below 0.8.

      Starting from the value $T_d = 0.5$ s, gradually increase its value to $T_d = 0.8$ and analyze what happens both to the control signals and time responses. Set again the value to $T_d = 0.5$ and gradually decrease its value to $T_d = 0.1$ s. In which case is the "best" response in closed loop achieved? How does reducing T_d improve the speed of response? Justify this result using the interactive tool (a root locus analysis is also recommended, card 7.2).

      ii. Repeat the previous exercise selecting $n = 2$ (double integrator). Does a decrease in the value of T_d improve the response in this case? Justify the answer using the interactive tool.

iii. For $G(s) = \frac{e^{-t_d s}}{s}$, $K = 1$, and $T_d = 0.2$, start to increase the time delay and indicate the value for which a steady oscillation is obtained at the output. Relate it to the Nyquist diagram of the compensated system. Now change the values of K and T_d to achieve a closed-loop behavior that, after the time delay has elapsed, is similar to that achieved with the initial parameter setting and $t_d = 0$ s.

iv. Analyze the effect of a unit load disturbance on steady-state error and of an output noise of 0.05 variance on the control signal.

c. For PI and PID controllers:

i. Justify if it is possible to improve the closed-loop behavior for this system ($k = 1, n = 1, t_d = 0$) with a PI or PID controller, in terms of transient oscillatory behavior, load disturbance rejection, and noise amplification.

ii. For $G(s) = \frac{e^{-t_d s}}{s}$ and a PI controller with $K = 1$, obtain the value of T_i for which the same peak overshoot is obtained when a unit step is entered in the reference and a unit step in the load disturbance. Justify if it is possible and why. Now consider $t_d = 0.5$ s. Modify the values of K and T_i to obtain response profiles similar to those achieved for the system without time delay. Is this possible? And with a PID controller? Analyze the particular contributions of the proportional, integral, and derivative action on the control signal profile.

2. Select from the Options menu the second example, corresponding to $P_{2r}(s) = \frac{k}{(\tau s+1)} e^{-t_d s}$ with $k = 0.5$, $\tau = 5$ s, and $t_d = 3$ s. Set to 1 the step load disturbance amplitude and to 0.05 the noise variance. Perform the following activities:

a. Using the tool, justify whether it is possible to obtain an overdamped closed-loop response with zero steady-state error using a proportional controller. And with a PI, PD, or PID controller? Find a set of PID controller parameters (there may be multiple combinations) which provide a settling time of less than 15 s after a unit step change at both the reference and the disturbance. Reason on the result also using the Nyquist diagram.

b. Enter the parameter values $k = 0.5$, $\tau = 5$ s, and $t_d = 3$ s. Select a PD controller. By trial and error, try to obtain a closed-loop behavior that does not exhibit overshoot and such that the settling time, after the time delay, is less than 3 s. Indicate the values of the corresponding PM and GM. Introduce a unit step load disturbance. Does the closed-loop system present an error in the steady state? If so, indicate its value. Reset the load disturbance (set its value to zero) and enter a noise with a variance of amplitude 0.05. What is the maximum deviation of the output from its previous steady-state value? What can be the reason for this undesired behavior?

3. Select from the Options menu the third example, corresponding to

$$P_{11b}(s) = \frac{k\omega_n^2}{s^2 + 2\zeta\omega_n s + \omega_n^2}.$$

a. Setting $k = 4$, $\zeta = 1$, and $\omega_n = 2$ rad/s, consider a system that controls average blood pressure during anesthesia described in reference [10], Example 4.4 (slightly modified). It is assumed that the level of blood pressure maintains a relationship with the amount of anesthesia during a surgical operation. The following figure represents a block diagram of the system, where the action of the surgery is represented through the disturbance $D(s)$. Determine an appropriate value of T_i that provides a satisfactory trade-off solution when at $t = 0$ s there is a change in the desired blood pressure as a unit step, at $t = 20$ s there is a unit step surgical disturbance, and at $t = 40$ s the sensor that measures the blood pressure introduces into the control loop a noise with a variance of approximately 0.01.

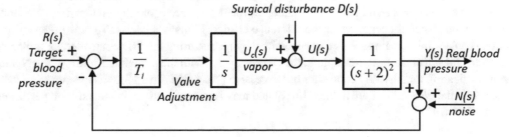

b. With the same settings of the load disturbance and noise as in the previous exercise, set $k = 1$, $\zeta = 0.5$, and $\omega_n = 1$ rad/s. Using the tool, try to obtain by trial and error a controller that provides an overdamped closed-loop system with no steady-state error when a step change in the reference is applied and with a PM of at least 90°. Justify the choice of controller structure using the root locus method. For the obtained parameters, indicate whether the system could be recovered if a unit step load disturbance is introduced at the system input. Also, comment on whether the controller would function well in the presence of output noise.

c. Consider the torsional mechanical system in the following figure, which is described in the reference [10], example CP4.7. The torque due to the twisting of the shaft is $-k_m\theta$, the damping torque due to the braking device is $-b_m\dot{\theta}$, and those of the disturbance and the input reference are $d(t)$ and $r(t)$, respectively. The moment of inertia of the mechanical system is J_m.

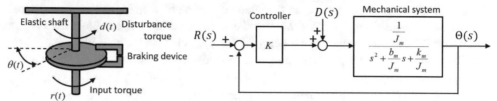

The transfer function of the open-loop system is

$$G(s) = \frac{\dfrac{1}{J_m}}{s^2 + \dfrac{b_m}{J_m}s + \dfrac{k_m}{J_m}}.$$

A proportional control loop is also shown in the figure. For the following set of parameters $k_m = 5$, $b_m = 0.9$, and $J_m = 1$ and assume that the desired angle is $\theta_d = 0°$.

  i. Enter the values of the parameters.
  ii. Determine the closed-loop response with a controller gain $K = 50$, when a disturbance in the form of a unit step is introduced into $d(t)$.
  iii. What improvements are expected when compared to the open-loop response?

d. For the system given by the transfer function:

$$G(s) = \frac{1}{(s+1)(s+2)},$$

design a P controller such that the maximum overshoot of the response to a unit step in the reference is 5%, the rise time about 2 s, and the steady-state error less than 50%. Comment on the effect on the output of a unit step load disturbance (reference [13], pp. 332–334).

4. Select the second example in the Options menu (without time delay). Following [15], Sect. 18.6.2, select the values of $K = 2.64$, $T_i = 3.4$, and $T_d = 0.3$ (modified from the reference). The process output response is satisfactory, but a look at the control signal shows a remarkably different interpretation of the apparently benign output response, because a sharp spike on the control signal occurs when the reference changes. This is a *derivative kick* effect, which in practice could represent a problem for actuators. This can be avoided by restructuring the PID controller, as explained in [15].

5. Reset the tool and select the third example in the Options menu. With the predefined settings, the Nyquist diagram of $L(s)$ has a loop, indicating that the PID controller has excessive phase lead. This loop disappears by reducing the value of T_d. This is the so-called *derivative cliff* defined by [7] and may have implications in closed-loop performance. More insights are given in Sect. 8.11.

Table 8.6 Different representations of phase-lead and phase-lag controllers

Representation	Zero–pole-gain	Time-constants
Standard	$C(s) = K_c \dfrac{s - z_c}{s - p_c}$	$C(s) = K \dfrac{\beta s + 1}{\tau s + 1}$
With attenuation factor	$C(s) = K_c \dfrac{s + \frac{1}{\beta}}{s + \frac{1}{\alpha\beta}}$	$C(s) = K_c \alpha \dfrac{\beta s + 1}{\alpha\beta s + 1}$

8.4 Phase-Lag and Phase-Lead Compensators

8.4.1 Interactive Tool: lead_lag_concept

8.4.1.1 Concepts Analyzed in the Card and Learning Outcomes
- Understanding the basic concepts of phase-lead and phase-lag compensator (or networks) in frequency-domain design.
- Maximum and minimum phase lag provided by phase-lead and phase-lag controllers.

8.4.1.2 Summary of Fundamental Theory
This card analyzes the characteristics and effects on the frequency response of simple and common controllers formed by a zero and a pole in the LHP, and so-called phase-lead and phase-lag controllers (compensators, networks, or compensation networks). Notice that these kinds of compensators were indirectly treated in the cards devoted to time and frequency response of a first-order system with a zero (Sects. 3.4 and 4.5, respectively). Their effect when used as controllers $C(s)$ in feedback loops are studied in other sections of this chapter, in the field of control system design methods in the frequency domain, therefore being advisable to previously assimilate the concepts introduced in this card.

The transfer function that describes this class of controllers is, in its different representations, the one shown in Table 8.6.

If the zero is located closer to the origin of the s-plane than the pole ($p_c < z_c < 0$, $0 < \tau < \beta$), the controller is named *phase-lead compensator* or *phase-lead network*. If the pole is located closer to the origin than the zero ($z_c < p_c < 0$, $0 < \beta < \tau$), it is named *phase-lag compensator* or *phase-lag network*. In phase-lead controllers, α is called *attenuation factor* (since its value is in the interval $0 < \alpha < 1$), and in phase-lag controllers $\alpha > 1$. It is an intuitive factor (geometric distance) because it establishes, depending on its interval, the structure in the form of a phase-lead or phase-lag network. This text uses the normalized structure in time constants (with $\beta > 0$ and $\tau > 0$), shaded in Table 8.6, existing a direct relationship between its parameters and those of the other representations: $K_c = K\beta/\tau$, $p_c = -1/\tau$, $z_c = -1/\beta$, and $\alpha = \tau/\beta = z_c/p_c$.

Figure 8.15 shows the polar diagrams and Bode diagrams of phase-lead and phase-lag controllers, where the pole and the zero are separated one decade in this example (for easier analysis), together with the formulas indicating the maximum or minimum phase contribution (ϕ_m) they provide, as well as the corresponding frequency (ω_m).

The decision to use a phase-lead or phase-lag network depends on the design specifications. When using a phase-lead or phase-lag compensator, the system order is increased by one (unless a cancellation occurs between the zero of the compensator and a pole of the open-loop system transfer function or vice versa). Unlike PID controllers, they do not increase the system type.

As analyzed in the introduction to this chapter, all controllers affect the stability, steady-state error, and bandwidth of the closed-loop system. Phase-lead compensators increase the gain crossover frequency ω_{gc}. This has the effect of decreasing the settling time in the step response (increases the system's damping). Although this is almost always desirable, the increase of ω_{gc} leads to an increase in the bandwidth of the closed-loop system, which can result in the amplification of high-frequency noise. The effect of phase-lag compensators is the opposite. They usually reduce the closed-loop system bandwidth, while helping to decrease steady-state errors. Intuitively, if $\tau \approx 0$ and so $\tau \ll \beta$, the compensator is almost similar to a PD controller (in fact, it could be considered a PD controller with a high-frequency filter, so that, as an approximation, it can be considered that a phase-lead network has similar characteristics than a PD controller. If $\tau \gg 1$ and $\tau \gg \beta$, the pole is quite close to the origin and the compensator resembles a PI controller, although in this case without increasing the system type.

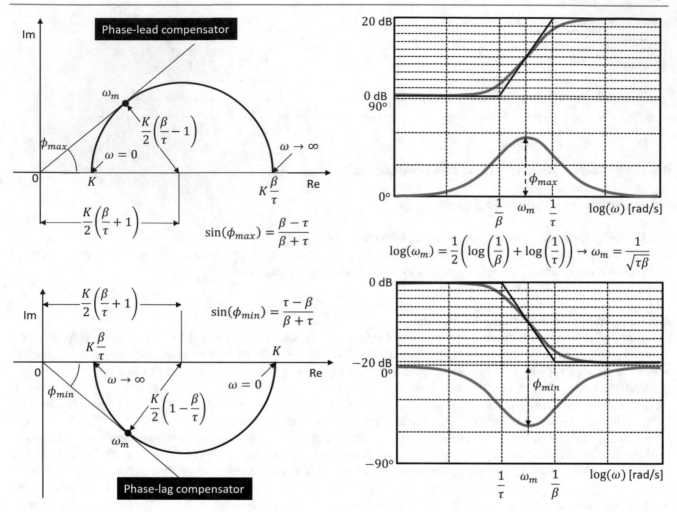

Fig. 8.15 Polar and Bode diagrams of phase-lead and phase-lag compensators

8.4.1.3 References Related to this Concept

- [3] Shahian, B., & Hassul, M. (1993). *Control system design using MATLAB*®. Prentice Hall, ISBN: 0-13-174061-X. Chapter 7, Sect. 4, paragraph 4, pp. 185–194.
- [16] Ogata, K. (2010). *Modern control engineering* (5th ed.). Prentice Hall, ISBN: 978-0-13-615673-4. Chapter 7, Sects. 10–13, pp. 491–511.

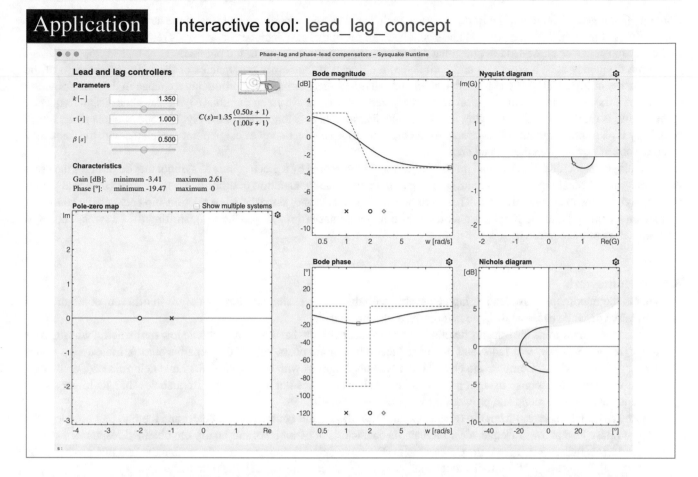

Phase-Lag and Phase-Lead Compensators

This card is devoted to analyzing the effect of phase-lead and phase-lag compensators in the frequency response. The influence of the location and relative position of the pole and the zero of the compensator can be easily analyzed in the Bode, Nyquist, and Nichols diagrams, and thus, their effect when used as controllers in frequency design methods using PM and GM specifications.

The **Parameters** area displays the parameters of the compensator in time constant format (K, τ, and β) and the symbolic representation of its transfer function. The parameters have been limited to the MP case (they are all positive). The **Characteristics** of minimum and maximum gain and phase of the network for the selected parameter configuration are also shown.

The **Pole–zero map** includes the pole and zero of the analyzed compensator to facilitate analysis based on their relative positions. If the Show multiple systems option is activated, new zero–pole pairs located at different positions on the negative real axis are included in the diagram, facilitating comparative analysis. These pairs are represented by different colors, corresponding to the colors of the frequency and time response graphs drawn in the right area of the tool. When an element (pole or zero) on the graph is selected, it automatically acquires the property of an active system and is represented using a black thick line. The representative values are simultaneously updated in the **Parameters** area. By placing the mouse over the interactive elements (poles or zeros), their position is indicated in the lower-left corner of the tool. The simultaneous inclusion of multiple systems can also be selected from the Options menu: a single Lag controller, Some lag controllers, a Lead controller, Some lead controllers, or several Lag controllers versus lead.

The right area of the tool is reserved for the various frequency response diagrams of the compensators. By default, the **Complete view** is shown (**Bode magnitude**, **Bode phase**, **Nyquist diagram**, and **Nichols diagram**), although through the Options menu it is possible to replace the Nichols diagram with a **Time response** plot showing the step response of the compensator in open loop (where it is possible to verify that being both the numerator and the denominator of degree one, it responds instantaneously to the unit step introduced in the instant $t = 0$, as already analyzed in Sect. 3.4). The scales can be modified using the settings available in the gearwheel icon. The Bode diagrams include the asymptotic representations of the active system, which have been discussed in Sect. 4.3. The Nyquist and Nichols diagrams in this case include shaded areas representing the geometric loci of the existence of phase-lead or phase-lag controllers for positive intervals of the parameters.

On the frequency axis, a cross (**x**) and a circle (**o**) are drawn at the corner frequencies of the pole and the zero of the compensation network. These two interactive elements can be dragged horizontally, producing a change in the corresponding parameter of the controller and the position of its pole or zero. An interactive green rhombus (◇) indicates the design frequency (plot scaling is required). When dragging it in the Bode diagram, its position varies in the Nyquist and Nichols plots, where frequency is an implicit variable. This interactive element permits the selection of a design point associated with the frequency it represents, at any point in the shaded area.

Bode diagrams include an additional interactive element represented by a green square □ symbol that determines the point in the diagram of maximum (or minimum) phase ϕ_m and amplitude (attenuation or amplification), whose numerical values are displayed in the **Characteristics** area. These symbols are interactive and, since there is a relationship between the controller parameters (τ and β) and the points of maximum or minimum phase (through ϕ_m and ω_m), shifting the □ symbol changes the parameters of the controller.

8.4.1.4 Homework

1. What is the maximum phase lead or lag achievable with this compensator structure? Analyze in different configurations the relative distance between the pole and the zero to achieve that.
2. Use the tool to determine the transfer function of a phase-lag network that does not attenuate low frequencies, whose lower corner frequency is $\omega_{cf_1} = 1$ rad/s, and that introduces a maximum phase lag of 30° (negative phase). Indicate the values of K, τ, β, ϕ_m, and ω_m. Compare the Nyquist and Nichols diagrams with those in Fig. 8.15 and determine ω_m on them.
3. Repeat the previous exercise using a phase-lead controller that does not amplify low frequencies, has its lower corner frequency at $\omega_{cf_1} = 1$ rad/s, and provides a maximum phase of 40°.
4. Choose Lag controllers vs lead from the Options menu. From the default compensators, determine the one that produces the greatest phase lead at lower frequencies. Indicate its characteristic parameters and justify the obtained response.
5. Choose Lag controllers vs lead from the Options menu. From the default compensators, determine the one that produces the greatest magnitude attenuation. Indicate its characteristic parameters and justify the obtained response.

8.5   PI Control of First-Order Systems Without Time Delay by Pole Placement

8.5.1   Interactive Tool: PI_pole_placement

8.5.1.1 Concepts Analyzed in the Card and Learning Outcomes
- Specifications in the pole-placement design of PI controllers.
- Design of PI controllers using the pole-placement method for first-order systems without time delay.
- 2-DoF structure: reference filtering.

8.5.1.2 Summary of Fundamental Theory
This section deals with the simplest case of application of the general pole-placement method, based on transforming the performance specifications to a desired position of the closed-loop poles.

In the simple case discussed in this section, a first-order system without time delay is considered to be controlled with an ideal PI controller, to obtain a second-order closed-loop system with LHP poles that produce a desired transient response in terms of peak overshoot (relative damping factor) and a characteristic time (rise time, peak time, or settling time). This is equivalent to the characteristic polynomial of the closed-loop system given by $J(s) = s^2 + 2\zeta_{cl}\omega_{n_{cl}}s + \omega_{n_{cl}}^2$, having its roots $p_1 = -\zeta_{cl}\omega_{n_{cl}} + j\omega_{n_{cl}}\sqrt{1 - \zeta_{cl}^2}$ and $p_1^* = -\zeta_{cl}\omega_{n_{cl}} - j\omega_{n_{cl}}\sqrt{1 - \zeta_{cl}^2}$. Notice that the subscript cl has been used to make explicit that the relative damping factor and the natural frequency are those of the closed-loop, but it should not be necessary as in this card the transfer function of the open-loop system is of the first order. Logically, as the system is type-0 and uses a PI controller, the closed-loop system will have zero steady-state error to a step reference. The parameters of the desired characteristic polynomial are related to the specifications by the following formulas:

Percentage overshoot [%]	Peak time	Rise time	Settling time [2%]

$$OS\,[\%] = 100\exp\left(-\frac{\zeta_{cl}\pi}{\sqrt{1-\zeta_{cl}^2}}\right),\quad t_p = \frac{\pi}{\omega_{n_{cl}}\sqrt{1-\zeta_{cl}^2}},\quad t_r = \frac{\pi-\phi}{\omega_{n_{cl}}\sqrt{1-\zeta_{cl}^2}},\quad t_s = \frac{4}{\zeta_{cl}\omega_{n_{cl}}},$$

$$\cos\phi = \zeta_{cl}$$

The loop transfer function with unit feedback $L(s) = C(s)G(s)$ is given by

$$L(s) = \frac{K(T_i s + 1)}{T_i s}\frac{k}{\tau s + 1}.$$

The closed-loop transfer function is therefore given by

$$G_{cl}(s) = \frac{\dfrac{Kk}{T_i\tau}(T_i s + 1)}{s^2 + \left(\dfrac{1+Kk}{\tau}\right)s + \dfrac{Kk}{T_i\tau}}.$$

By analyzing this transfer function, a number of conclusions can be drawn:

- The static gain[2] of the closed loop is 1. This is to be expected because by using a PI controller (type-1) the steady-state error of the closed-loop system when the reference is a step must be zero.
- The denominator of the closed-loop system transfer function is of second order, so to meet the specifications it may be equated to the standard denominator of a second-order transfer function, where ζ_{cl} and $\omega_{n_{cl}}$ are obtained from the specifications of transient behavior by applying the above formulas that relate them to peak overshoot and characteristic times. Therefore

$$s^2 + \left(\frac{1+Kk}{\tau}\right)s + \frac{Kk}{T_i\tau} = s^2 + 2\zeta_{cl}\omega_{n_{cl}}s + \omega_{n_{cl}}^2.$$

The following relationships are obtained:

$$2\zeta_{cl}\omega_{n_{cl}} = \frac{1+Kk}{\tau} \rightarrow K = \frac{2\tau\zeta_{cl}\omega_{n_{cl}} - 1}{k}, \tag{8.38}$$

$$\omega_{n_{cl}}^2 = \frac{Kk}{T_i\tau} \rightarrow T_i = \frac{Kk}{\tau\omega_{n_{cl}}^2} = \frac{2\tau\zeta_{cl}\omega_{n_{cl}} - 1}{\tau\omega_{n_{cl}}^2}.$$

The parameters of the PI controller are obtained according to the closed-loop behavior specifications (desired location of the poles in the s plane). If no logical values are obtained for the parameters K and T_i (for example a negative integral time), this would indicate that the imposed specifications cannot be met with this controller structure.

- The closed-loop transfer function has a zero at $s = -1/T_i$. The effect of this zero can be negligible if it is far enough from the closed-loop poles, but if it is not the case, the closed-loop response will be different from that expected from the specifications. In general, the closed-loop time response will exhibit a shorter peak time and higher overshoot (recall card 3.5 on the influence of a zero on the time response of a second-order system). In this case, the design should be repeated with less demanding closed-loop specifications in terms of response times or by performing an analysis using the root locus technique. Another simple option is to use a control system with two degrees of freedom $C(s)$ and $F(s)$, where $F(s) = \frac{1}{(T_f s + 1)}$ is a first-order filter acting on the reference signal, with unit static gain and a pole at $s = -1/T_f$. When $T_f = T_i$, it cancels the zero of the closed-loop system.

8.5.1.3 References Related to this Concept

- [3] Shahian, B., & Hassul, M. (1993). *Control system design using MATLAB®*. Prentice Hall, ISBN: 0-13-174061-X. Chapter 7, Sect. 3, paragraph 2, pp. 173–177.
- [5] Åström, K. J., & Murray, R. M. (2014). *Feedback systems: An introduction for scientists and engineers* (2nd ed.). Princeton University Press, ISBN: 9780691193984. Example 8.7, pp. 8–20; example 11.1, pp. 11–4.
- [7] Åström, K. J., & Hägglund, T. (2006). *Advanced PID control*. ISA - The Instrumentation, Systems and Automation Society, ISBN: 978-15-561-7942-6. Chapter 6, Sect. 4, pp. 174–186.

---

[2] Remember that it can be obtained directly by making $s = 0$ in the transfer function as a result of applying the final value theorem of the Laplace transform.

- [13] Barrientos, A., Sanz, R., Matía, F., & Gambao, E. (1996). *Control de sistemas continuos. Problemas resueltos (Control of continuous systems. Problems solved)*. McGraw-Hill, ISBN: 84-481-0605-9. Chapter 9, Sect. 2, pp. 335–339.

Application   Interactive tool: PI_pole_placement

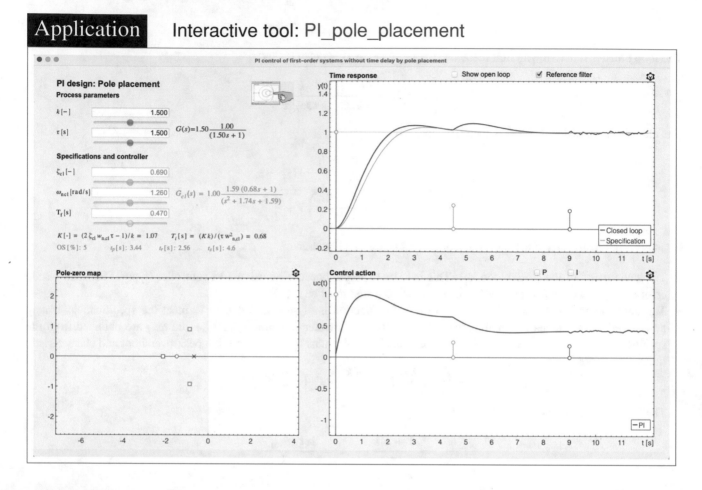

PI Design by Pole Placement

This card is devoted to studying the simplest pole-placement design, that is, a PI controller for a first-order system without time delay. This helps the reader to easily understand the concept of pole placement and also the role of reference filters in the closed-loop time response.

In the upper-left area, the **Process parameters** are shown: the static gain k and the time constant τ of a first-order transfer function without time delay. Next to the sliders and textboxes available for modifying their values, the symbolic representation of the transfer function of the selected system $G(s)$ is drawn in red color.

The **Specifications and controller** are displayed in the form of desired closed-loop relative damping factor ζ_{cl} and undamped natural frequency $\omega_{n_{cl}}$ (providing the desired poles of the closed-loop system), together with a symbolic representation of the resulting closed-loop transfer function in green color ($G_{cl}(s) = \frac{C(s)G(s)}{1+C(s)G(s)}$). It should be noted that in this case, this symbolic representation includes the closed-loop zero resulting from the use of a PI controller, but not the reference filter $F(s)$. Below this representation, the formulas for the proportional gain K of the controller and the integral time T_i are displayed (8.38), which are updated when the closed-loop specifications change.

This area also includes indices of the theoretical behavior of the closed-loop system, given by the maximum percentage overshoot (OS), the peak time (t_p), the rise time (t_r), and the settling time (t_s) of the system for those specifications, without taking into account the effect of the resulting zero in the numerator. These values are those corresponding to the green **Time response** plot, as discussed below.

In the **Pole–zero map**, the pole of the open-loop system is drawn in red color ($\times$), the zero of the controller (and thus of the closed-loop transfer function) is drawn in blue color ($\circ$), the poles of the closed-loop system matching the specifications in green color ($\square$), and the pole of the reference filter in gray color ($\square$). All elements except the zero are interactive objects, so when they are dragged, they produce changes in the **Process parameters** and the **Specifications and controller** areas. The zero of the closed-loop system is not interactive because its location is a result of imposing the closed-loop specifications (desired closed-loop poles). Moving the desired poles of the closed-loop system affects the controller specifications, this being an interactive and graphical way of enforcing specifications. Placing the mouse over the poles displays their position in the lower-left corner of the tool.

The right side of the tool contains the **Time response** plots. The upper part represents the closed-loop system output (blue solid line) and that corresponding to the ideal specifications without the zero (green solid line). The open-loop response to a step input can be also shown (red dashed line), activating the selection box available on the graph ($\square$ Show open loop). In all three cases, a legend with the corresponding lines is included in the lower-right corner of the graph. If the mouse is placed over any point of the response, the time (t) and output (y) values are displayed over the plot, being a useful tool for analyzing the closed-loop behavior. By clicking on the ($\square$ Reference filter) activation button, a reference filter with unit gain and time constant T_f is applied (by default with a value of $T_f = T_i$, although any other positive value can be selected to analyze its effect on the closed-loop performance). The way of introducing changes in reference, disturbances, and noise in this plot has been explained in card 8.4, dragging the circles vertically (amplitude) and horizontally (time of activation). These circles are also included in the lower graph corresponding to the **Control action**, which represents the input to the system (controller output, blue solid line). Through the selection boxes enabled on the graphic, the individual components of the proportional action (P, magenta solid line) and the integral action (I, cyan solid line) can be included. This graph includes a legend associated with the plots of P, I, and PI actions, and if the mouse is placed over the graph of the PI control signal, the associated values of time (t) and value of the signal (u) are displayed.

In all the graphs, the change of scaling is done using the settings available in the gearwheel positioned in the upper right corner of the chart, as explained in Chap. 1.

8.5.1.4 Homework

1. Using the interactive tool root_locus (card 7.2), represent a system with arbitrary parameters with a loop transfer function given by

$$L(s) = C(s)G(s) = \frac{K(T_i s + 1)}{T_i s} \frac{k}{(\tau s + 1)},$$

which represents a first-order system controlled by a PI controller. Analyze the root locus obtained according to the position of the zero of the system, useful when interpreting the following exercises.

2. Using the interactive tool PI_pole_placement, select a process with static gain $k = 0.5$ and time constant $\tau = 10$ s. Impose closed-loop specifications with 5% percentage peak overshoot and rise time of 1 s. Analyze the controller parameters obtained and the control signal. Do you consider it realistic to impose these specifications? Justify the answer. Analyze the effect of activating the reference filter, by modifying its time constant T_f.

3. Select a process with static gain $k = 1.5$ and time constant $\tau = 1.5$ s. Set specifications for a closed-loop response with a percentage overshoot of 5% and a rise time of 1.4 s. Indicate the associated values of ζ_{cl} and $\omega_{n_{cl}}$. Also calculate the theoretical peak time and settling time values and compare them to those provided by the tool. Does the actual system response meet the specifications? Justify the answer. Provide the actual overshoot and peak time values obtained. Using the tool, try to find the parameters leading to the fastest possible controller that will achieve a closed-loop percentage overshoot of 5%. Indicate the followed procedure and which parameter is affected. What is the peak time obtained in this case?

4. Select a process with static gain $k = 1.5$ and time constant $\tau = 1.5$ s. Set specifications to get a closed-loop system with two equal real poles and $\omega_{n_{cl}} = 2.5$ rad/s. Why is an overdamped response not achieved? Analyze the response. Indicate the maximum value of $\omega_{n_{cl}}$ from which an overdamped response is obtained. Justify the answer.

5. Select a process with static gain $k = 1.5$ and time constant $\tau = 2.0$ s. For $\omega_{n_{cl}} = 0.9$ rad/s, find the value of ζ_{cl} with which the theoretical response of the method (without zero) and the real response (with zero) are equal. What is the value of the obtained peak overshoot? And that of the peak time? Are the logical values of the controller parameters obtained? Justify the answer.

6. For the system used in the exercises of Sect. 8.3, given by the transfer function:

$$G(s) = \frac{1}{(s + 1)(s + 2)},$$

design a PI controller such that the peak overshoot of the closed-loop response to a unit step in the reference is 5%, the rise time is approximately 2 s, and the steady-state error is zero. Discuss the effect on the output of a unit step load disturbance (reference [13], pp. 335–339).

7. For the default setting when starting the tool, include a step load disturbance of amplitude 1. Activate the reference filter and check if the disturbance rejection can be improved by using more demanding specifications without strongly affecting reference tracking. Change the different parameters to analyze the trade-offs that can be achieved using a 2-DoF control structure.

8.6   PI Control of First-Order Systems by Pole Cancellation

8.6.1   Interactive Tool: PI_lambda

8.6.1.1 Concepts Analyzed in the Card and Learning Outcomes
- Application of the pole-cancellation method to a first-order system with no time delay using a PI controller.
  - Design specifications.
  - Control law.
  - Influence of modeling errors on the behavior obtained with the pole-cancellation method.
- Limitations in the application of the method.
- Extension of the method to first-order systems with time delay: Lambda method for PI and PID controllers.

8.6.1.2 Summary of Fundamental Theory
As the name suggests, the pole-cancellation technique is based on positioning the zeros of the controller in the LHP at the same locations where the most dominant poles of the open-loop system are located, using the remaining degrees of freedom so that the closed-loop poles are in locations that meet certain specifications.

Pole–zero cancellation techniques are used in automatic control, usually to control stable, low-order systems. A disadvantage of this technique is that even if one pole of the loop transfer function is canceled, it may be present in other transfer functions of the control loop (e.g. in the transfer function relating the load disturbance to the output). This is why it is not used with unstable systems, nor when the system has a very slow pole. However, in the case of stable systems, significant improvements in the evolution of the closed-loop system can be achieved with this intuitive control technique.

This section explains the simplest case, corresponding to the design of a PI controller to compensate a first-order system without delay. This case represents the situation where the specifications are posed in terms of zero steady-state error to a step reference and increasing the speed of response of the closed-loop system relative to the uncontrolled system. Moreover, the same case with time delay is also treated, providing the well-known Lambda tuning method [7].

Assume a system with a dominant pole which can be approximated by a first-order transfer function without delay. As the system is type-0, to avoid steady-state error to step reference, a controller with integral action (e.g. PI) must be introduced to increase the type of the loop transfer function. The use of the PI controller also makes it possible to increase the speed of the closed-loop response compared to that of the open loop.

The transfer function of the loop transfer function is given by

$$L(s) = C(s)G(s) = \frac{K(T_i s + 1)}{T_i s} \frac{k}{(\tau s + 1)},$$

K being the proportional gain of the PI controller and T_i the integral time.

Pole-cancellation compensation consists of canceling the system pole with the controller zero (place the controller zero at the same location as that of the system pole at the LHP), or making $T_i = \tau$. By doing this, the loop transfer function becomes

$$L(s) = \frac{Kk}{sT_i} \text{ and closing the loop: } G_{cl}(s) = \frac{L(s)}{1 + L(s)} = \frac{1}{(\frac{T_i}{Kk}s + 1)}.$$

The closed loop is represented by a first-order transfer function with equivalent time constant $\tau_{cl} = T_i/(Kk) = \tau/(Kk)$ and static gain equal to one (when using a PI controller the steady-state error of the closed-loop system to a step reference is zero and, therefore, it has unit static gain). It is easy and intuitive to see that by increasing K, the time constant of the closed-loop system can be made smaller than that of the open-loop system. It is to be expected that the form of the closed-loop response corresponds to that of a first-order system (and therefore without overshoot), but in practice, as cancellation is not perfect, the behavior of the closed-loop system may differ from that expected.

If a desired time constant τ_{cl} is imposed as a closed-loop specification, the PI controller parameters are given by $T_i = \tau$, $K = \tau/(k\tau_{cl})$.

For higher-order systems, a PID controller can be used whose zeros cancel out one or two dominant poles in the plant.

As commented before, this technique has to be used with caution in systems where the output may be affected by other signals (e.g. disturbances), as the characteristic polynomial of the transfer function relating another exogenous signal with the output may contain the pole that has been canceled in the reference-output relationship. A very intuitive example can be found in the speed control of a vehicle in [5], pp. 8–20 and 11–4. It can also be used in combination with the pole-placement technique studied in Sect. 8.6.

Considering a process with transfer function $G(s) = k/(\tau s + 1)$ with a PI controller which cancels the pole ($T_i = \tau$), the expression of the system output in the presence of reference changes and load disturbance is given by

$$Y(s) = \frac{C(s)G(s)}{1 + C(s)G(s)} R(s) + \frac{G(s)}{1 + C(s)G(s)} D(s) = \underbrace{\frac{1}{\frac{\tau}{Kk}s + 1}}_{G_{cl}(s)} R(s) + \underbrace{\frac{\frac{\tau}{K}s}{\left(\frac{\tau}{Kk}s + 1\right)(\tau s + 1)}}_{G_{yd}(s)} D(s). \qquad (8.39)$$

It can be seen how in this case, the factor $(\tau s + 1)$ does not appear in the closed-loop transfer function, but it appears in the transfer function that relates the disturbance to the output, so that if $\tau < 0$ (unstable pole), a small disturbance could destabilize the system. It is also observed that if the time constant of the open-loop system is very large, even if the pole-cancellation method requires a very fast closed-loop response to a step change in the reference, the system will compensate a step load disturbance very slowly (with a speed determined by τ). In fact, Eq. (8.39) is very useful to analyze the compromise that can be obtained between the reference tracking problem and the disturbance rejection problem with PI controllers. If the controller is designed to track a reference with a desired closed-loop time constant τ_{cl}, the controller gain is given by $K = \tau/(k\tau_{cl})$. However, depending on the value of τ, that gain may not be adequate to reject a load disturbance. Furthermore, in the numerator of the transfer function G_{yd} there is a pure derivative multiplied by the factor τ/K, which will have a significant influence on the transient regime after the disturbance affects the system.

Notice that an unstable pole (including integrators and those lying on the imaginary axis) can never be canceled with a controller zero, because the closed-loop system will not be internally stable, in the sense that there will be other closed-loop transfer functions that would be unstable. That is the reason why, although perfect cancellation would be possible (absence of modeling errors), it cannot be performed. This aspect is treated in [17] in a section called "Canceling unstable poles: A tempting, bad idea", where the author gives the following three reasons (already commented in the previous paragraph) of why canceling an unstable pole will not work:

1. Input disturbances blow up. Even though the transfer function between the reference and the output is stable, the transfer function between the system input and the output is not.
2. Initial conditions blow up. Any nonzero value of $y(0)$ will grow exponentially.
3. Perfect cancellation is unlikely. In practice, the controller zero never matches exactly the unstable system pole and as the zero position only approximates the pole one, control becomes increasingly difficult, and finally impossible.

The analysis can also be performed using the sensitivity functions introduced at the beginning of Chap. 7.[3] Following [5], a process $G(s) = 1/(s - a)$ controlled by a PI controller with pole cancellation $C(s) = K(s - a)/s$, provides a loop transfer function given by $L = k/s$, so that

$$S = \frac{1}{1 + L} = \frac{s}{s + k}, \quad T = \frac{L}{1 + L} = \frac{k}{s + k}.$$

---

[3] Many examples can be found. For instance, if a PI controller is used to control a plant with a pure derivative in the numerator, the complementary sensitivity function will have finite gain (different from zero) so that step references can be tracked (with a certain error), but if the control action is analyzed, it grows indefinitely.

Although the factor $(s - a)$ does not appear in $L(s)$, the sensitivity function $S(s)$, and the complementary sensitivity function $T(s)$, it cannot be canceled if $a >= 0$, as if the load and noise sensitivity functions are computed, it is obtained as

$$GS = \frac{G}{1 + CG} = \frac{s}{(s - a)(s + k)}, \quad CS = \frac{C}{1 + CG} = \frac{k(s - a)}{s + k},$$

notice that GS is unstable in that case, so that a small disturbance at the process input can lead to an unbounded output. So, although most classical control methods rely on the loop transfer function $L(s)$ for analysis and design purposes, sometimes this function gives only limited insight, being necessary to analyze all sensitivity functions, although this book focuses only on the basic concepts.

The method can be applied to first-order systems with time delay (FOTD), following the well-known *Lambda Tuning* method, a special case of pole placement that is commonly used in the process industry [7]. It is based on approximating the time delay (exponential function) by a Taylor series expansion that provides a quotient of polynomials and therefore a rational function. For a FOTD,

$$G(s) = \frac{k}{\tau s + 1} e^{-t_d s} \approx \frac{k(1 - t_d s)}{\tau s + 1}, \quad e^{-t_d s} \approx (1 - t_d s), \tag{8.40}$$

$$L(s) = C(s)G(s) = \frac{K(T_i s + 1)}{T_i s} \frac{k(1 - t_d s)}{\tau s + 1}, \quad \text{with} \quad T_i = \tau \rightarrow L(s) = \frac{Kk(1 - t_d s)}{\tau s},$$

$$G_{cl}(s) = \frac{L(s)}{1 + L(s)} = \frac{(1 - t_d s)}{\underbrace{\frac{\tau - Kk t_d}{Kk}}_{\lambda} s + 1} \rightarrow K = \frac{\tau}{k(t_d + \lambda)}. \tag{8.41}$$

Notice that λ is the closed-loop time constant τ_{cl} (in the original work [18] it was denoted as λ, providing the name to the method), and thus it is a control specification. As commented in [7], this is a special case of pole placement that provides good results if the design parameter λ is chosen properly. A common rule of thumb is to choose $\lambda = 3\tau$ for a robust controller and $\lambda = \tau$ for aggressive tuning when the process parameters are well determined. Both choices lead to controllers with zero gain and zero integral time for pure time delay systems. For delay-dominated processes (those in which $t_d > \tau$), it is recommended to select $\lambda = \max(\tau, 3t_d)$. For lag-dominated processes (those in which $t_d < \tau$), it is recommended not to cancel the process pole. By doing that (obtaining a second-order characteristic equation), it is reasonable to choose λ proportional to t_d, as explained in [7].

For PID design, a similar approach can be followed:

$$G(s) = \frac{k}{\tau s + 1} e^{-t_d s} \approx \frac{k(1 - s t_d/2)}{(\tau s + 1)(1 + s t_d/2)}, \quad e^{-t_d s} \approx \frac{(1 - s t_d/2)}{(1 + s t_d/2)}, \tag{8.42}$$

where a Padé approximation [19] to the time delay has been applied. Using the interacting version of a PID controller, the following expressions are obtained:

$$C(s) = K' \left(1 + \frac{1}{T_i' s}\right) \left(1 + T_d' s\right), \quad \text{with } T_i' = \tau, \ T_d' = t_d/2, \tag{8.43}$$

$$L(s) = K' \frac{(T_i' s + 1)(T_d' s + 1)}{T_i' s} \frac{k(1 - s t_d/2)}{(\tau s + 1)(1 + s t_d/s)} = \frac{K' k(1 - s t_d/s)}{\tau s}, \tag{8.44}$$

$$G_{cl}(s) = \frac{L(s)}{1 + L(s)} = \frac{(1 - s t_d/s)}{\underbrace{\frac{\tau - K' k t_d/2}{K' k}}_{\lambda} s + 1} \rightarrow K' = \frac{\tau}{k(\lambda + t_d/2)}. \tag{8.45}$$

If the controller has non-interacting structure, from Table 8.5, the values of the controller parameters are given as a function of the parameters of the open-loop transfer function and the specification $\tau_{cl} = \lambda$ as

$$U_c(s) = K\left(1 + \frac{1}{T_i s} + T_d s\right) \text{ with } K = \frac{\tau + t_d/2}{k(\tau_{cl} + t_d/2)}, \ T_i = \tau + t_d/2, \ T_d = \frac{\tau t_d}{t_d + 2\tau}. \tag{8.46}$$

It is important to remark that the method treated in this section is not limited to first-order systems, as it can be also applied to higher-order systems, sometimes in combination with pole-placement methods. For instance, for second-order systems, PID control can be used to cancel two open-loop stable poles trying to obtain closed-loop systems with a dominant pole τ_{cl} and unit static gain to achieve zero steady-state error to a step reference input:

$$G(s) = \frac{k\omega_n^2}{s^2 + 2\zeta\omega_n s + \omega_n^2}, C(s) = K\left(1 + \frac{1}{T_i s} + T_d s\right) \begin{cases} K = \frac{1}{\tau_{cl} T_d k \omega_n^2}, \\ T_i = \frac{2\zeta}{\omega_n}, \\ T_d = \frac{1}{2\zeta\omega_n}, \end{cases}$$

$$G(s) = \frac{k}{(\tau_1 s + 1)(\tau_2 s + 1)}, C(s) = K'\left(1 + \frac{1}{T_i' s}\right)\left(1 + T_d' s\right) \begin{cases} K' = \frac{\tau_1}{k\tau_{cl}}, \\ T_i' = \tau_1, \\ T_d' = \tau_2. \end{cases}$$

8.6.1.3 References Related to this Concept

- [3] Shahian, B., & Hassul, M. (1993). *Control system design using MATLAB*®. Prentice Hall, ISBN: 0-13-174061-X. Chapter 7, Sect. 2, pp. 170–171.
- [5] Åström, K. J., & Murray, R. M. (2014). *Feedback systems: An introduction for scientists and engineers* (2nd ed.). Princeton University Press, ISBN: 9780691193984. pp. 8–20 and 11–4.
- [7] Åström, K. J., & Hägglund, T. (2006). *Advanced PID control*. ISA - The Instrumentation, Systems and Automation Society, ISBN: 978-15-561-7942-6. Chapter 6, Sect. 5, pp. 186–189.

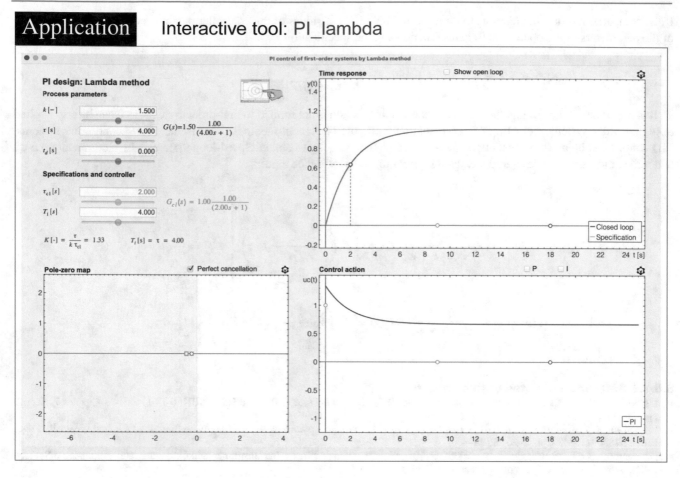

PI Design by Pole Cancellation and Lambda Tuning Method

This card is devoted to understanding the pole–zero cancellation design, using a PI for controlling a first-order system without and with time delay. The advantages and drawbacks of the technique are analyzed using illustrative examples and the corresponding interactive tool.

The upper-left area of the tool contains the **Process parameters** of the first-order system (whose values can be modified through textboxes and sliders) and a symbolic representation of its transfer function. Below is the area of **Specifications and controller**, in which the user can specify (again via a textbox and a slider) the desired value of the closed-loop system time constant (τ_{cl}). That of the integral time T_i is determined by the value of τ if the Perfect cancellation option is active, or can be freely set otherwise. The symbolic representation of the first-order closed-loop system $G_{cl}(s)$ is also displayed. Under the specifications, the expressions of the control law and the values associated with the parameters of the PI controller (K and T_i) are shown.

The **Pole–zero map** contains the pole of the open-loop system ($\times$), the zero of the controller ($\circ$, which by default is placed over the open-loop pole), and the closed-loop system pole ($\square$). Positioning the mouse over any of these elements, its coordinates are shown in the lower-left corner of the tool. Disabling the Perfect cancellation option above the diagram enables the analysis of cases where cancellation is not perfect. In this case, there is the possibility of dragging the zero of the controller to the right or the left of the system pole. It is also possible to change its value via a textbox or a slider in the **Specifications and controller** area.

On the right side of the tool, the closed-loop **Time response** and associated **Control action** are displayed. The circles over them allow changing the amplitude and the instant at which the reference, disturbance, and noise act on the system (see card 8.4 for a complete description). The **Time response** plot includes a selection box to activate the representation of the time evolution of the open-loop system. Within the graph, there is an explanatory label for the variables whose dynamic evolution is represented. In the **Control action** plot, the graphic representations of the proportional and integral actions can be selected through selection boxes independently (both contribute to the overall shape of the control signal curve). The scale can be modified using the settings available in the gearwheel icon.

8.6.1.4 Homework

1. Analysis without time delay ($t_d = 0$):
    a. For an open-loop system with static gain $k = 0.5$ and time constant $\tau = 1$ s, design a PI controller so that the closed-loop system has a time constant of 0.5 s. Enter a unit step load disturbance and indicate how long it takes to recover the reference value. Enter a noise with variance 0.02. Analyze the effect of changing the specification on τ_{cl} in the control signal in the presence of noise. Comment on the results.
    b. For the same case of the previous exercise, deactivate the option Perfect cancellation. Study the effect of shifting the controller zero to the right and left of the pole on reference tracking, load disturbance rejection, and noise immunity.
    c. With the default settings when opening the tool, adjust the time constant of the closed-loop system so that the control signal does not exceed amplitude 2. Enter a unit step load disturbance at $t = 8$ s. What is the settling time after the disturbance is entered? Deactivate the Perfect cancellation option. Move the zero of the controller so that the settling time after the disturbance is less than 5 s. Is it possible to achieve this result without the response to a step in the reference surpassing the value of such reference?
    d. For an open-loop system with static gain $k = 1$ and time constant $\tau = 10$ s, design a PI controller so that the closed-loop system has a time constant of 2 s. Enter a unit step load disturbance and indicate how long it takes to recover the reference value. Using the PID_concept interactive tool (card 8.3), select a first-order system without delay with the same parameters ($k = 1$, $\tau = 10$ s). Using a PI controller by pole–zero cancellation (selecting $T_i = \tau = 10$ s), analyze the effect of increasing the gain of the controller K on the tracking of a unit step reference and on the rejection of a unit step load disturbance. Analyze the value reached by the control signal $u_c(t)$ in each case and comment on the result.
    e. With the Perfect cancellation active, set $k = 1$ and $\tau = -5$. It can be verified that the tool provides a stable closed-loop response. Analyze the effect of not applying perfect cancellation (root locus analysis using card 7.2 is recommended). What happens if a step load disturbance is applied? Relate the results with the comments in the summary of the theory.
2. Repeat the previous exercises for three values of the time delay: $t_d = 0.3\tau$, $t_d = \tau$, and $t_d = 3\tau$.

8.7 PID Control Based on the Open-Loop Ziegler–Nichols Tuning Rules

8.7.1 Interactive Tool: PID_Ziegler_Nichols

8.7.1.1 Concepts Analyzed in the Card and Learning Outcomes
- Obtaining controller parameters from the reaction curve method.
- Heuristic rule-based design.
- PID controller design for open-loop overdamped systems with time delay.
- Manual fine-tuning from values obtained using the Ziegler–Nichols rules.

8.7.1.2 Summary of Fundamental Theory
Empirical or experimental methods of controller tuning are widely used in industry, as obtaining models based on first principles is often a complex task and requires knowledge on several disciplines. Heuristic methods usually follow three steps:

1. Estimation of certain characteristics of the process dynamics (open-loop or closed-loop).
2. Calculation of controller parameters from certain rules that relate these parameters to the system characteristics obtained in the previous step.
3. The behavior of the closed-loop system is analyzed and fine-tuning of the parameters is carried out, usually online.

The Ziegler–Nichols method, developed in 1942–43 for Taylor Instruments controllers, is a heuristic methodology for obtaining the parameters of PID controllers mainly for disturbance rejection purposes. It is used when the step response of the open-loop system has an overdamped or sigmoidal shape, that is, it does not present oscillations and the system also has time delay. The original method is based on the measurement of a part of the open-loop unit step response, characterized by parameters a and $t_1 = t_d$ in Fig. 8.16 (left side), which are respectively the intersections of the line tangent to the point with the greatest slope of the response curve with the ordinate and abscissa axes. Parameter t_1 is an approximation of the system time delay.

With this method, there is no need to wait for the system to reach a steady state, but it is subject to errors in the estimation of the tangent. The parameters of the PID controller are obtained from Table 8.7. In the original method, the parameters were obtained through extensive simulations using a set of representative processes whose controllers were tuned manually, trying to find a correlation between the controller parameters and the a and t_d parameters. The criterion they used was a damping ratio of 1/4 (the second overshoot of the closed-loop response has an amplitude less than 1/4 of the first one). This criterion prevents large deviations in the first peak of the system response when there are changes in the load or other disturbances on the system, but it produces a percentage overshoot of up to 50% when there are step changes in the reference.

One variant of the method, not applicable to underdamped systems or systems with integrators, consists in approximating the step response to that of a first-order system whose transfer function has three parameters: k (static gain), τ (time constant), and t_d (time delay):

$$G(s) = \frac{k}{(\tau s + 1)} e^{-t_d s}. \tag{8.47}$$

This is the so-called reaction curve method, which has been explained in Sect. 3.8.

If the system under control does not clearly present a pure time delay, t_d is considered equal to the time it takes the system to reach 5% of its steady state. With the values of the descriptive parameters of a first-order system with delay (see right part of Fig. 8.16), the characteristic parameters of a PID controller can be obtained through the expressions in Table 8.7. Note that in this case, the model parameters must be calculated taking into account the steady state of the system and therefore $a = k t_d / \tau$. Furthermore, if a unit step is entered in $t = 0$, the system takes 28% of its final value in $t_{28} = t_d + \tau/3$ and 63% in $t_{63} = t_d + \tau$. Therefore, in addition to what is indicated on the right side of Fig. 8.16, it is possible to obtain analytically $\tau = \frac{3}{2}(t_{63} - t_{28})$ and $t_d = t_{63} - \tau$.

The method provides acceptable results in the range $0.1 < t_d/\tau < 1$. In cases where this condition is not met, the results obtained from the application of the rules will not usually provide acceptable results, requiring manual tuning of the controller and even the use of other more advanced control schemes (e.g. those based on predictors [20]).

As can be seen, the application of the rule provides the values of the characteristic constants of the ideal PID controller: K, T_i, and T_d.

As previously stated, the Ziegler–Nichols tuning rules were developed to obtain closed-loop control systems with good attenuation of load disturbances (regulatory control problem). The cases considered were those in which the main factor limiting behavior was the dynamics of the process. The method placed little emphasis on compensation for measurement noise, sensitivity to process variations, and reference changes. These shortcomings are addressed and improved by the AMIGO method [7]. When controllers designed with Ziegler–Nichols rules are used for step reference tracking, they often cause high

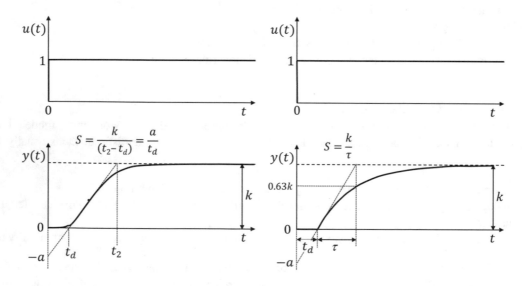

Fig. 8.16 Characterization of the unit step response in the open-loop Ziegler–Nichols method

Table 8.7 Ziegler–Nichols open-loop method rules

Controller	K	T_i	T_d
P	$\dfrac{1}{a} \approx \dfrac{\tau}{kt_d}$	–	–
PI	$\dfrac{0.9}{a} \approx \dfrac{0.9\tau}{kt_d}$	$3t_d$	–
PID	$\dfrac{1.2}{a} \approx \dfrac{1.2\tau}{kt_d}$	$2t_d$	$0.5t_d$

overshoot and low relative stability margins. In this case, the method is useful to provide an initial set of controller parameters from which, through manual tuning (with a considerable reduction of controller gain), the closed-loop response can be improved. From initial closed-loop experiments, the controller parameters are fine-tuned according to rules of the following type:

- Increasing the proportional gain decreases stability.
- The error decays faster if the integral time is decreased.
- Reducing integral time decreases stability.
- Increasing derivative time improves stability.

Another possibility is to use a 2-DoF control structure, where by including a filter in the reference $F(s)$, the tracking and disturbance rejection problems can be decoupled. As the output is given by

$$Y(s) = \frac{F(s)C(s)G(s)}{1 + C(s)G(s)} R(s) + \frac{G(s)}{1 + C(s)G(s)} D(s), \tag{8.48}$$

$C(s)$ can be designed using the Ziegler–Nichols rules for achieving a fast disturbance rejection, while the overshoot of the response to step references is highly attenuated by $F(s)$. In theory, once $C(s)$ has been tuned, $F(s)$ could be easily obtained through

$$G_{cl} = \frac{C(s)G(s)}{1 + C(s)G(s)} \rightarrow F(s) = \frac{1}{G_{cl}(s)} \frac{1}{(\tau_{cl}s + 1)(\frac{\tau_{cl}}{N_f}s + 1)^{n_f}},$$

where τ_{cl} is the closed-loop dominant time constant, $N_f \geq 1$ is a parameter allowing to place other closed-loop less dominant poles, and n_f is such that $F(s)$ is causal (depends on the relative degree of the closed-loop transfer function). In this development, it is supposed that G_{cl} has unit static gain. If this is not the case, the numerator of $F(s)$ has to include the closed-loop transfer function static gain, as $F(s)$ filter must have unit static gain.

Nevertheless, as done in this tool, very often it is enough using a first-order filter for PI control $F(s) = \frac{1}{(T_f s + 1)}$, as analyzed in Sect. 8.5, or a second-order filter in PID control, with $F(s) = \frac{1}{(T_f s + 1)^2}$, which are the implementations used in this tool.

8.7.1.3 References Related to this Concept

- [5] Åström, K. J., & Murray, R. M. (2014). *Feedback systems: An introduction for scientists and engineers* (2nd ed.). Princeton University Press. ISBN: 9780691193984. Chapter 10, Sect. 3, pp. 10–10 to 10–13.
- [7] Åström, K. J., & Hägglund, T. (2006). *Advanced PID control*. ISA - The Instrumentation, Systems and Automation Society, ISBN: 978-15-561-7942-6. Chapter 6, Sect. 2, pp. 159–169.
- [10] Dorf, R. C., & Bishop, R. H. (2011). *Modern control systems* (12th ed.). Prentice Hall. ISBN: 978-0-13-602458-3. Chapter 7, Sect. 6, pp. 488–492.
- [11] Franklin, G. F., Powell, J. D., & Emani-Naeni, A. (2010). *Feedback control of dynamic systems* (6th ed.). Pearson. ISBN: 978-0-13-500150-9. Chapter 4, Sect. 3, paragraph 4, pp. 210–212, exercise 4.33, p. 236.
- [14] Bolzern, P., Scattolini, R., & Schiavoni, N. (2009). *Fundamentos de control automático (Fundamentals of automatic control, in Spanish)*. McGraw-Hill, ISBN: 978-84-481-6640-3. Chapter 14, Sect. 4, paragraph 2, pp. 387–388.
- [21] Goodwin, G. C., Graebe, S. F., & Salgado, M. E. (2001). *Control system design*. Prentice Hall. ISBN: 0-13-958653-9. Chapter 6, Sect. 5, pp. 166–168.

Application   Interactive tool: PID_Ziegler_Nichols

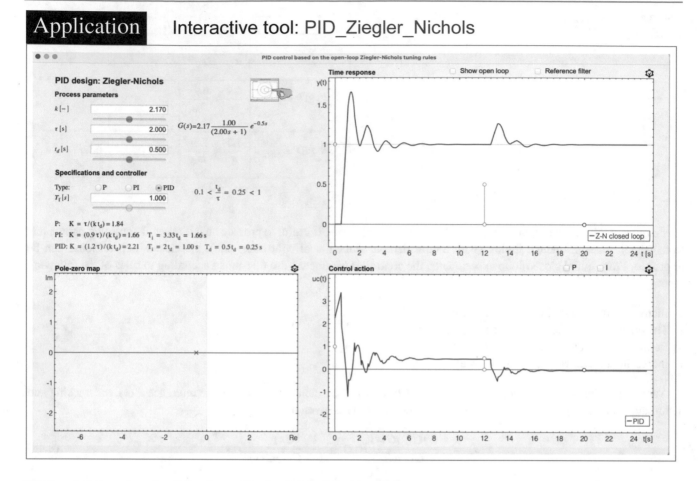

PID Control Based on the Open-Loop Ziegler–Nichols Tuning Rules

This card is devoted to studying the Ziegler–Nichols PID tuning rules based on the reaction curve method. This is a heuristic method through which the first approximation of PID control parameters can be obtained for overdamped systems with time delay. The interactive tool is very useful to analyze the closed-loop response achieved using the PID parameters provided by the rules, in terms of setpoint tracking, load disturbances rejection, and noise amplification. Although the tool incorporates an option to Activate free design, so that modifications to the set of controller parameters given by the rules can be performed, if the obtained parameters are introduced in the interactive tool PID_concept explained in card 8.3, this also helps to analyze the implications in the frequency response. The PID_Ziegler_Nichols tool also implements setpoint filtering (2-DoF control) as explained in the card, helping to decouple tracking and disturbance rejection problems, so that the design of the PID controller can be oriented to the disturbance rejection problem, while avoiding noise amplification in the loop.

The tool's interface follows the same distribution of elements as the previous ones. The **Process parameters** area contains the static gain k, time constant τ, and time delay t_d of the FOTD transfer function (whose symbolic representation is displayed), with corresponding textboxes and sliders to modify them.

The **Specifications and controller** area allows selecting the controller structure selecting the appropriate radio button P, PI, or PID, as well as the time constant of the reference filter T_f. An inequality is shown that indicates whether or not the relation $0.1 < t_d/\tau < 1$ is fulfilled and, therefore, provides information about the suitability or not of applying the method. If the inequality is met, it is represented in blue color. If it is not fulfilled it is drawn in green color. Below this line, the rules being applied according to the type of controller selected (the active rule is indicated in blue) are displayed, as well as the numerical values of the parameters. Notice that the Options menu includes the possibility of Activate free design, so that the controller parameters can be modified from the settings given by the Ziegler–Nichols rules.

The **Pole–zero map** by default shows the location of the pole of the FOTD transfer function, which can be modified interactively (note that this type of representation does not allow the inclusion of t_d and k, which must be changed in the **Process parameters** area).

The **Time response** shows the system output in the upper graph and the input in the lower part (**Control action**), where the individual contributions (in case the user marks the corresponding checkbox placed on the graph) of the proportional (P), integral (I), and derivative (D) actions in the final control actions can be displayed. As in the previous tools, the green circle (○, located on the ordinate axis of both graphs) allows modifying the reference step amplitude, while the two circles placed on the abscissa axis facilitate the inclusion of a step load disturbance (the central circle) or noise at the output (the right circle), when dragged vertically. In this case, other circles remain on the time axis that allow the instant in which the disturbance or noise is introduced to be modified. In the upper graph, the representation of the open-loop time response and the reference filter can be activated. When the mouse pointer is placed over any of the time response curves, the associated time t and signal values (input u and output y) are shown. Both graphs include the gearwheels whose settings are used to change their scales.

8.7.1.4 Homework

1. Select a process with $k = 1$, $\tau = 5$, and $t_d = 3$. Select a unit step reference in $t = 0$ s and a change in the load disturbance of amplitude 0.5 in $t = 30$ s (change the scale using the settings of the gearwheel icon). Analyze the results obtained by direct application of the Ziegler–Nichols rules with a PI controller without reference filter:
   a. Is the contribution of the proportional part or that of the integral part greater? Comment on the answer. What happens during the first t_d seconds to both control actions and what is the explanation?
   b. Using the Activate free design option or the interactive tool PID_concept explained in card 8.3, enter the same settings and the values of K and T_i provided by the Ziegler–Nichols rules and obtain new parameters for the PI controller so that reference tracking is improved (less overshoot and shorter peak time) and also does disturbance rejection (less overshoot and shorter settling time). Is this improvement possible? Discuss the steps taken to achieve it (if possible). Could it be achieved with a PID controller? Comment on your answer. Test now the reference filter, analyzing the effect of changing the time constant T_f on the response. Does the reference filter affect disturbance rejection? And noise attenuation? Justify the answer.
   c. Using the Activate free design option or the interactive tool PID_concept, set the value of K to that provided by the Ziegler–Nichols rules, and modify the value of the integral time T_i so that the maximum overshoot when the reference is a step is 30%. With which value of the integral time is this achieved? What is now the peak overshoot and settling time associated with the response to the disturbance?
   d. Using the Activate free design option or the interactive tool PID_concept, set the value of T_i to that provided by the Ziegler–Nichols rules, and modify the value of K gain so that there is no overshoot after a step change in the reference. With which value of K is this achieved? Which is now the overshoot and settling time associated with the response to the disturbance?
2. Select a system with static gain unit ($k = 1$) and time constant equal to 1 s ($\tau = 1$). Also select a load disturbance with an amplitude of 0.5 (the default one). For values of the time delay t_d in the interval [0.1, 10] in the interactive tool PID_Ziegler_Nichols,
   a. By selecting a P controller, indicate for which value of t_d the best results are obtained from your point of view. Comment on the results obtained in the whole t_d modification interval and justify the results from the point of view of the transient response and the steady-state behavior.
   b. Repeat the previous exercise for PI and PID controllers.
   c. For the case where $t_d = 0.5$ s, indicate if the response of the controller to a step load disturbance is better or worse than that to a reference change. Is the 1/4 ratio between the first and second overshoot met? Using the Activate free design option or the interactive tool PID_concept, try to modify the controller gain to achieve an overdamped response to a step reference. What happens in this case with the disturbance rejection characteristics?
   d. In the case where $t_d = 1$ s and PI control, Activate free design and modify the two controller parameters (gain K and integral time T_i) to obtain the best possible response (from your point of view) that represents a compromise solution to reference tracking, disturbance rejection, and steady-state error. Annotate the value of the parameters obtained. Then select a PID controller and analyze if it is possible to improve the results obtained by changing the value of T_d. Comment on the answer. With the last configuration of the PID controller selected, choose a noise value with an approximate variance of 0.01. Discuss the differences you find between the PID controller and the PI controller.

3. Consider a system described in [5] with a transfer function $G(s) = e^{-s}/s$. Approximate the integrator by a first-order system with a very large time constant in PID_Ziegler_Nichols and determine the parameters of P, PI, and PID controllers using the Ziegler–Nichols open-loop rules. Discuss the results obtained and analyze the effect of including setpoint filtering.

4. A papermaking machine uses a transfer function (reference [11], p. 236):

$$G(s) = \frac{e^{-2s}}{3s + 1},$$

which relates the incoming matter flow to the outgoing thickness. Apply the Ziegler–Nichols open-loop tuning method to this problem. Analyze the response to a unit step change in the reference and load disturbance. Also analyze the behavior of the closed-loop system when there is noise at the output. Using the PID_concept interactive tool with the obtained settings, calculate the critical proportional gain K from which the system becomes unstable.

5. An example of traffic congestion control on a link in TCP transmissions can be found in [5]. It uses a linearized model that describes the dynamics that relate the length of the queue q to the packet drop p:

$$G_{qp}(s) = \frac{b}{(s + a_1)(s + a_2)} e^{-t_d s}.$$

The parameters are given by $a_1 = 2N_f^2/(ct_d^2)$, $a_2 = 1/t_d$ and $b = c^2/(2N_f)$. Parameter c is the capacity of the bottleneck, N_f is the number of sources feeding that link, and t_d is the round-trip delay. Use the values of the parameters $N_f = 75$ sources, $c = 1250$ packets/s, and $t_d = 0.15$ s and find the parameters of a PI controller using the Ziegler–Nichols rules by approximating the response provided by $G_{qp}(s)$ by a FOTD transfer function (for instance the interactive tool t_generic (card 3.6) to approximate the model by a first-order model with time delay). Simulate the closed-loop system responses obtained with a PI controller.

6. For different values of k, τ, and t_d, different controller structures (P, PI, and PID), using a unit step reference, a load step disturbance of amplitude 0.5, and output noise with variance 0.01, check the results achieved with the Ziegler–Nichols rules and the effect of reference filtering.

8.8   Classical Design of Phase-Lag Controllers in the Frequency Domain

8.8.1   Interactive Tool: f_design_lag

8.8.1.1 Concepts Analyzed in the Card and Learning Outcomes
- Minimum set of specifications in the frequency domain: steady-state error and phase margin.
- Design of phase-lag compensators in the frequency domain by classical methods.

8.8.1.2 Summary of Fundamental Theory
As discussed in Sect. 8.4, phase-lag controllers reduce system gain and add lag, factors that do not seem very positive in a control loop. However, these controllers are useful in those cases where it is desired to reduce the steady-state error by maintaining certain relative stability specifications (usually linked to the PM) and with a consequent decrease in the gain crossover frequency ω_{gc}, and therefore in the bandwidth of the closed-loop system. Notice that in classical frequency design, only a relative stability specification is used (typically the PM). In Sect. 8.10, another specification related to the desired gain crossover frequency will be added to the design.

The design is typically performed into two phases:

1. The controller gain is computed to fulfill steady-state specifications (considering $L(s) = KG(s)$).
2. It is checked if with the obtained proportional controller, the desired PM (specification, PM_{des}) is met. If that is not the case, the pole and zero of the phase-lag compensator are placed at adequate frequencies to modify $L(s)$ to obtain the desired PM.

The following controller structure is used with $\tau > \beta$:

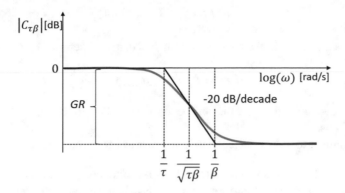

Fig. 8.17 Calculation of gain reduction GR

$$C(s) = K\frac{\beta s + 1}{\tau s + 1} = KC_{\tau\beta}(s). \tag{8.49}$$

Specifying in more detail the steps that are usually followed in classical design:

1. Obtain the proportional constant of the controller (K) directly from the steady-state error specification. For instance, if $G(s)$ is type-0, the specification is posed in the form of a position error constant K_p to be fulfilled by the compensated system. Therefore $K_p = \lim_{s\to 0} C(s)G(s) = Kk \to K = K_p/k$. In the same way, if $G(s)$ is type-1, $K = K_v/k$ and so on.

2. Check if $KG(j\omega)$ meets the PM specification. If it does, it will indicate that with a proportional controller it is possible to meet both steady-state and relative stability specifications.

3. If the gain increase provided by K causes the phase margin specification not to be met, a gain reduction must be performed that brings the gain crossover frequency ω_{gc} of $L(s)$ to a value where the desired PM is fulfilled. The compensator is introduced following additional steps:

   • Look for the frequency ω_{gc1} at which

   $$\lfloor KG\left(j\omega_{gc1}\right) = -180° + PM_{\text{des}} + SM,$$

   where SM is the so-called *security margin*, typically between 5° and 12°, which is used to compensate for the phase shift that could be introduced by the phase-lag controller at ω_{gc1}.

   • Determine the gain of $KG(s)$ in that frequency, such that the gain reduction or attenuation in dB to be applied by the compensator in ω_{gc1} is given by (Fig. 8.17)

   $$GR = -20\log|KG(j\omega_{gc1})|.$$

   • Calculation of the compensator parameters (β, τ).

   – Figure 8.17 shows compensator gain as a function of frequency, so that taking into account the slope of -20 dB/decade:

   $$GR = -20\left(\log\left(\frac{1}{\beta}\right) - \log\left(\frac{1}{\tau}\right)\right) = -20\log\left(\frac{\tau}{\beta}\right) = -20\log(\alpha) \to \alpha = 10^{-\left(\frac{GR}{20}\right)},$$

   where GR enters the formula with its negative sign and $\alpha = \frac{\tau}{\beta}$ determines the geometric distance between the pole and the zero of the compensator.

   – The value of β is then set so that the frequency of interest, ω_{gc1}, is sufficiently far to the right of $1/\beta$, so that the phase lag introduced by the compensator at that frequency is small enough (that is the reason of using SM). A typical choice is

   $$\frac{1}{\beta} = \frac{\omega_{gc1}}{10} \to \beta = \frac{10}{\omega_{gc1}}.$$

   – Finally $\tau = \beta\alpha$.

   • It is necessary to check if $\omega_{gc} \approx \omega_{gc1}$, so that the design is successful (Fig. 8.18). By the definition,

$$|C\left(j\omega_{gc}\right)G\left(j\omega_{gc}\right)| = 1,$$

and this will be the frequency at which the phase margin will actually be defined and given by

$$PM = 180° + \lfloor C\left(j\omega_{gc}\right)G\left(j\omega_{gc}\right).$$

When $GM > GM_{\text{spec}}$, the specifications are met and the design will be correct. Otherwise, it will be necessary to look for other design mechanisms.

- Finally, it is always convenient to perform a simulation in order to verify the time response of the closed-loop system.

As can be seen in Fig. 8.17, a phase-lag compensator has a low-pass filter component. This type of compensation allows a high gain at low frequencies (and therefore an improvement in steady-state error) and reduces the gain in the frequency range in which the phase margin must be improved, thus reducing the bandwidth of the closed-loop system (as a consequence of the decrease in ω_{gc}), and causing a slower transient response. In this compensator, the attenuation that occurs in the magnitude curve at high frequencies is mainly exploited, instead of the phase lag (which does not serve to achieve the control objectives).

8.8.1.3 References Related to this Concept

- [3] Shahian, B., & Hassul, M. (1993). *Control system design using MATLAB®*. Prentice Hall, ISBN: 0-13-174061-X. Chapter 7, Sect. 4, paragraph 4, pp. 185–194.
- [10] Dorf, R. C., & Bishop, R. H. (2011). *Modern control systems* (12th ed.). Prentice Hall, ISBN: 978-0-13-602458-3. Chapter 10, Sect. 8, pp. 772–776.
- [12] Golnaraghi, F., & Kuo, B. C. (2017). *Automatic control systems* (10th ed.). McGraw Hill Education, ISBN: 978-1-25-964384-2. Chapter 11, Sect. 5.
- [16] Ogata, K. (2010). *Modern control engineering* (5th ed.). Prentice Hall, ISBN: 978-0-13-615673-4. Chapter 7, Sect. 12, pp. 505–509.
- [22] D'Azzo, J. J., Houpis, C. H., & Sheldon, S. (2003). *Linear control system analysis and design with MATLAB®* (5th ed.). Marcel Dekker Inc., ISBN: 0-8247-4038-6. Chapter 11, Sect. 3, pp. 444–447.

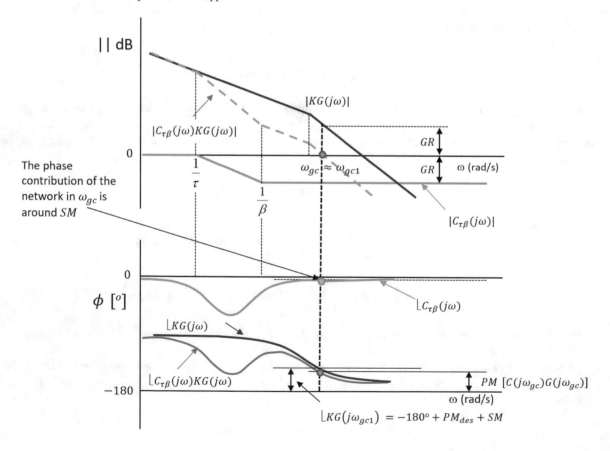

Fig. 8.18 Phase shift included by the phase-lag controller

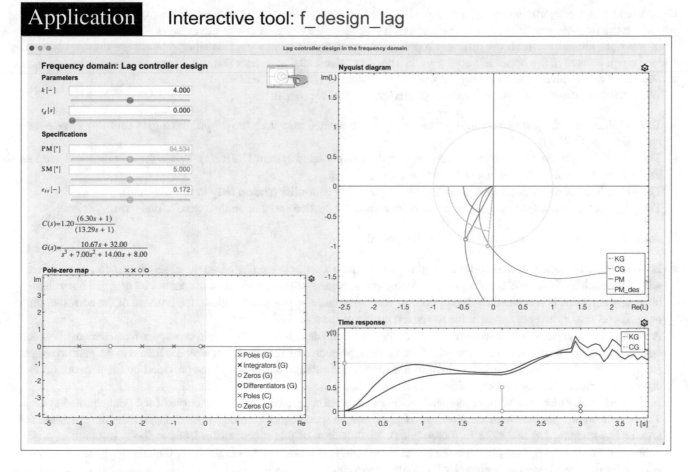

Frequency Design: Phase Lag

This card deals with the classical design of phase-lag compensators in the frequency domain based only on two specifications: steady-state error and phase margin. The underlying concepts and approximations are graphically treated, so that the user can verify all the steps explained in control textbooks and interactively analyze if the specifications can be met.

The **Parameters** area shows the static gain k and time delay t_d of the system, with associated textboxes and sliders. Setting poles and zeros of the transfer function is done directly using the repository of dynamic elements in the **Pole–zero map**.

In the **Specifications** section, the user can set the values of the design specifications of the phase-lag compensator treated in this card: Phase margin (PM), Safety margin (SM), and steady-state error (e_{ss}), this last linked to the type of $G(s)$. If $G(s)$ is type-0, e_{ss} refers to the steady-state error to unit step reference, if it is type-1, e_{ss} is the error when the reference is a unit slope ramp, and if it is type-2, the error to reference parabolic input. Under the textboxes and sliders that facilitate setting the specifications, the transfer function of the system to be controlled is represented in red and that of the resulting phase-lag controller in blue.

In this tool, the **Pole–zero map**, besides facilitating the inclusion of the dynamic elements of the system (poles ×, zeros ○, integrators ×, and differentiators ○), represents in blue color the zero ○ and the pole × of the phase-lag network, whose position is determined by the algorithm running on the tool (the position is automatically calculated according to the problem specifications), which is the one stated in the summary of fundamental theory. The legend displays the symbols representing these elements. A (G) in parentheses indicates that these elements belong to the plant and a (C) that they belong to the controller. A very interesting aspect of this representation is that as the dynamic configuration of the plant is changed by clicking and dragging dynamic elements from/to the repository (× ×○○), (removing can be also done by double-clicking on the elements), automatically the phase-lag network is computed to the actual configuration, taking into account the existence of a solution for the given system and specifications. If with a simple proportional gain the specifications can be reached, the $C(s)$ controller is represented as a proportional controller. This also applies to cases where the specifications cannot be met (which can be verified in the frequency response graph). Otherwise, it will adopt the complete phase-lag network structure.

In the right area of the tool, the different representations of the system in the frequency domain are drawn. From the Options menu located in the upper-left corner of the tool, the representation in Bode diagram, Nyquist diagram, Nichols diagram, and Complete view (which simultaneously displays all the diagrams) can be chosen.

When the **Bode diagram** is selected, a series of elements are drawn over it:

- The red line is the frequency response of the system compensated only with the proportional gain fulfilling steady-state specifications ($KG(s)$).
- The blue line represents the frequency response of the compensated system $C(s)G(s)$. When the specifications are met using a proportional controller, the red and blue curves overlap.
- The green line shows the frequency response of the phase-lag controller (shown only in this diagram).
- The gray dashed line on the magnitude curve represents the 0 dB line and that on the phase curve $-180°$.

In addition, other interactive elements appear in the graphics:

- In both the **Bode magnitude** and **Bode phase** plots, a blue cross and circle are placed on the abscissa axis ($\times$, o) associated with the position of the pole and zero of the phase-lag compensator. These two elements cannot be dragged horizontally, because their value is computed by the algorithm. When the mouse is placed over them, the position of the corresponding pole p_c and zero z_c is represented in the lower-left corner of the tool.
- A vertical black solid line is drawn on the Bode diagram at the design frequency (gain crossover frequency ω_{gc}), which cuts the frequency response of the controller C at a point represented by the symbol ◇, which is linked to the gain crossover frequency and, indirectly, to the safety margin SM used in the calculations of the phase provided by the network (at that crossover frequency the network provides a small phase shift).
- In the **Bode phase** plot, two vertical segments ending in a circle are represented on the curve of the compensated system. The one shown in light blue indicates the desired phase margin specification (which can be changed from the **Specifications** area or by dragging left and right in this graphic). Modification of this specification leads to the calculation of a new controller. The green segment indicates the actual phase margin $KG(s)$, which is logically not interactive because its value is fixed by the calculations performed by the control algorithm.

In the **Nyquist diagram** and the **Nichols diagram**, the same color code is used for the function $KG(s)$ and $C(s)G(s)$, and the segments finished in circles are also drawn, delimiting the desired and actual phase margins, as well as the ◇ symbol linked to the gain crossover frequency and the α factor of the network.

The **Time response** of the closed-loop system is drawn both for the compensated system with loop transfer function $C(s)G(s)$ and for $KG(s)$. The circles over them allow changing the amplitude and the instant at which the reference, disturbance and noise act on the system (see card 8.4 for a complete description). The scale can be modified using the settings available in the gearwheel icon.

The Options menu includes a set of practical examples, as well as the possibility of introducing generic transfer functions in format (NUM,DEN) or ZPK. Example 1 states the control of a first-order system with time delay. Example 2 introduces a double integrator, as an example of a system that cannot be controlled with this kind of network and requires adding zeros/poles to the controller. Example 3 deals with the control of a second-order system with an integrator. In all these examples, the user can interactively modify the specifications and analyze the obtained results.

8.8.1.4 Homework

1. Consider a system described by

$$G(s) = \frac{1}{s(s+1)(0.5s+1)}.$$

The closed-loop specifications for this system, described in [16] (p. 505), are

a. The static velocity error constant must be $K_v = 5 \text{ s}^{-1}$.

b. The PM must be at least 40°.

Design a controller to meet the specifications and check the results using the tool.

2. Select Example 1 from the Options menu of the tool, corresponding to a system with a time delay t_d greater than the system time constant τ:

$$G(s) = \frac{2}{s+1} e^{-2s}.$$

Check how the system compensated only with the proportional gain K is unstable, while the compensator manages to stabilize the closed-loop system through a reduction of the system's gain crossover frequency ω_{gc}. Find the phase-lag network with which a gain crossover frequency $\omega_{gc} = 0.75$ rad/s is achieved. What is the value of PM?

3. Select Example 2 from the Options menu, corresponding to a double integrator $G(s) = 0.1/s^2$ (type-2 system). Can the method analyzed in this card be applied to this case? Justify the answer (a root locus analysis may be helpful). Add a zero to the transfer function and find the range of values of the time constant of the zero for which there is a phase-lag network fulfilling the default specifications.

4. Select Example 3 from the Options menu, corresponding to a process with transfer function:

$$G(s) = \frac{0.1}{s(s+1)}.$$

Justify why the pole p_c and the zero z_c of the controller are so close to the imaginary axis. Why is the time response worse than that achieved by compensating the system with only K? What phase margin should be set as a specification so that the time responses are similar? From which value of PM is it sufficient to use a proportional controller K?

5. Design a phase-lag compensator for the following system:

$$G(s) = \frac{1}{s(s+2)},$$

so that $K_v = 20$ and the phase margin is 45° (reference [10], pp. 772–774). Interpret the results.

8.9 Classical Design of Phase-Lead Controllers in the Frequency Domain

8.9.1 Interactive Tool: f_design_lead

8.9.1.1 Concepts Analyzed in the Card and Learning Outcomes
- Basic frequency-domain specifications of phase-lead controllers (steady-state error and phase margin).
- Design of phase-lead compensators in the frequency domain.

8.9.1.2 Summary of Fundamental Theory
This card is devoted to designing phase-lead controllers (see card 8.4) in the frequency domain using only two specifications: steady-state error and PM. The transfer function that represents this compensator is

$$C(s) = K\frac{\beta s + 1}{\tau s + 1} = KC_{\tau\beta}(s) = K\frac{\beta s + 1}{\alpha\beta s + 1}, \quad \alpha = \frac{\tau}{\beta}. \tag{8.50}$$

As in the case of the design of phase-lag controllers using the Bode diagram, a number of steps are followed in the design of the controller:

1. First, the gain K that satisfies the steady-state error specifications is computed (in terms of the error constants K_p, K_v, or K_a given by Kk).

2. Check whether the system compensated with that gain ($KG(s)$) meets the PM specification in the new gain crossover frequency ω_{gc1} ($|KG(j\omega_{gc1})| = 1$). If not, additional phase must be provided by compensator $C_{\tau\beta}(s)$ trying that its maximum phase contribution $\phi_{max} = \phi_m$ is introduced in the new gain crossover frequency. Following the definition of phase margin:

$$\phi_m = PM_{\text{des}} + SM - 180° - \lfloor KG(j\omega_{gc1}),$$

where PM_{des} is the phase margin given by the specifications. Since $C_{\tau\beta}(s)$ also contributes to the magnitude plot, when applying the compensator the new gain crossover frequency will be slightly higher than ω_{gc1}, and for that reason a few degrees are added to the phase margin specification in the design (SM, known as the safety margin, usually between 5° and 12°). The phase introduced by the phase-lead network is

$$\lfloor C_{\tau\beta}(j\omega) = \arctan(\beta\omega) - \arctan(\tau\omega).$$

The frequency at which the controller contributes with the maximum phase can be found by deriving the above equation with respect to frequency or by analyzing the Bode plots in Figs. 8.15 and 8.19. It is obtained as

$$\omega_m = 1/\sqrt{\tau\beta}. \tag{8.51}$$

The maximum phase that can be provided by the controller is $\lfloor C_{\tau\beta}(j\omega_m) = \phi_m$. By applying trigonometry (see Fig. 8.15), the following relations can easily be obtained:

$$\sin(\phi_m) = \frac{\beta - \tau}{\beta + \tau} = \frac{1 - \alpha}{1 + \alpha} \rightarrow \alpha = \frac{1 - \sin(\phi_m)}{1 + \sin(\phi_m)} = \frac{\tau}{\beta}. \tag{8.52}$$

Once the value of α is determined from ϕ_m, the gain amplification GA in dB introduced by the compensator in $\omega = \omega_m$ (taking into account the positive slope of 20 dB/decade from $\omega = 1/\beta$; see Fig. 8.19) is

$$GA[\text{dB}] = |C_{\tau\beta}(j\omega_m)|_{dB} = 20 \cdot \left(\log\left(\frac{1}{\sqrt{\tau\beta}}\right) - \log\left(\frac{1}{\beta}\right)\right) = 10\log\left(\frac{\beta}{\tau}\right) = 10\log\left(\frac{1}{\alpha}\right). \tag{8.53}$$

The phase-lead network will not provide its maximum phase ϕ_m in ω_{gc1}, as the gain crossover frequency is being slightly changed due to the above-mentioned gain amplification GA, which causes the actual gain crossover frequency ω_{gc} to be slightly higher than ω_{gc1}. Therefore, the compensator is applied at a frequency where the system has an inverse gain to the one provided by GA:

$$20\log|KG(j\omega_{gc})| = -GA = -10\log\left(\frac{1}{\alpha}\right). \tag{8.54}$$

As α is already known from (8.52), Eq. (8.54) provides $\omega_{gc} = \omega_m$ that will be the gain crossover frequency of the compensated system.

3. As $\alpha = \tau/\beta$, from (8.51), (8.52), and (8.54), the values of τ and β can be obtained as

$$\tau = \frac{\sqrt{\alpha}}{\omega_{gc}}; \quad \beta = \frac{1}{\sqrt{\alpha}\omega_{gc}}. \tag{8.55}$$

4. Draw the Bode diagram of $C(j\omega)G(j\omega)$ to confirm the design. If the specifications are not fulfilled, these may be too demanding and other design methods should have to be used.

Phase-lead controllers are used to improve stability margins. They also increase the gain crossover frequency of the loop transfer function, increasing the closed-loop bandwidth and speed of response, but at the cost of augmenting the high-frequency noise sensitivity.

8.9.1.3 References Related to this Concept

- [3] Shahian, B., & Hassul, M. (1993). *Control system design using MATLAB®*. Prentice Hall, ISBN: 0-13-174061-X. Chapter 7, Sect. 4, paragraph 4, pp. 185–194.
- [10] Dorf, R. C., & Bishop, R. H. (2011). *Modern control systems* (12th ed.). Prentice Hall, ISBN: 978-0-13-602458-3. Chapter 10, Sect. 4, pp. 751–757; problem CP10.4, p. 831.
- [13] Barrientos, A., Sanz, R., Matía, F., & Gambao, E. (1996). *Control de sistemas continuos. Problemas resueltos (Control of continuous systems. Problems solved)*. McGraw-Hill, ISBN: 84-481-0605-9. Chapter 9, Sect. 9, pp. 362–369.
- [16] Ogata, K. (2010). *Modern control engineering* (5th ed.). Prentice Hall, ISBN: 978-0-13-615673-4. Chapter 7, Sect. 11, pp. 493–502.

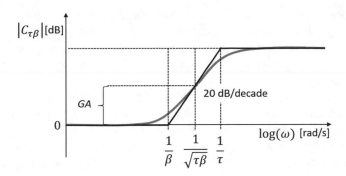

Fig. 8.19   Calculation of the gain amplification GA

Application   Interactive tool: f_design_lead

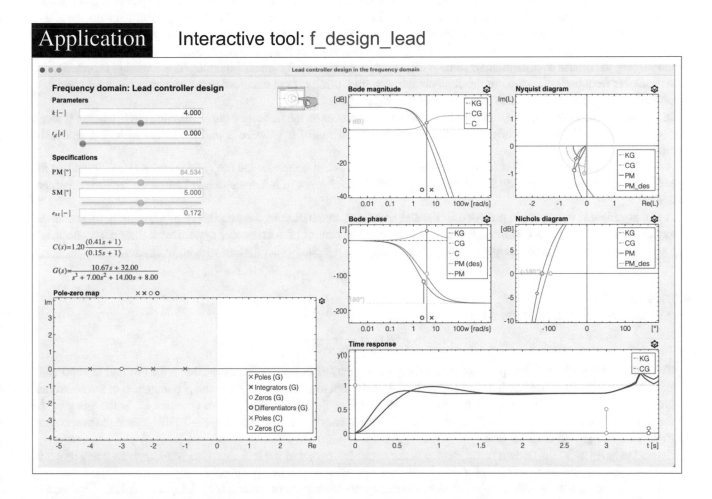

Frequency Design: Phase Lead

This card deals with the classical design of phase-lead compensators in the frequency domain, using only two specifications: steady-state error and PM. The underlying concepts and approximations are graphically treated, in such a way that the user can verify all the steps explained in the summary of fundamental theory and interactively analyze if the specifications can be met. This interactive tool is very similar to the f_design_lag in terms of its distribution and interactive elements, although the design seeks different objectives.

The upper-left area includes two **Parameters** of the transfer functions $G(s)$ which are not configurable from other parts of the tool, such as the static gain k and the time delay t_d, whose values can be changed via the red textboxes and sliders. The **Specifications** section contains a steady-state error specification e_{ss}, the desired phase margin (PM), and the safety margin (SM) which is added to compensate for the increase in the gain crossover frequency that produces the network. Under the textboxes and sliders associated with the specifications, the symbolic representation of the transfer function $G(s)$ is drawn in red and that of the phase-lead controller $C(s)$ in blue. It is important to note that the steady-state error specification e_{ss} is linked to the system type. If $G(s)$ is type-0, e_{ss} is the steady-state error specification to a unit step reference. Similarly, if $G(s)$ is type-1, it refers to a unit ramp input, and so on.

The **Pole–zero map** plot includes the dynamic elements of $G(s)$, represented in red color (poles, integrators, zeros, and differentiators), and the pole and zero of the phase-lead compensator in blue. The legend at the bottom right of the graph makes it easy to determine each of these elements. Next to the title of the graph, the repository of dynamic elements ($\times \times \circ \circ$) is located, each of which can be dragged in/out the plot (removing can also be done by double-clicking on the elements).

The top right area of the tool is devoted to displaying the **Bode diagram**, the **Nyquist diagram**, the **Nichols diagram**, or all of them at once. The choice of the representation is made through the Options menu, and includes the frequency response of $KG(s)$ (red), $C(s)G(s)$ (blue), and $C(s)$ (green, only in the **Bode diagram**). These plots also include two segments (or arcs in the case of the **Nyquist diagram**) which have a circle at their end and represent the phase margin of the system $KG(s)$ (green) and the desired phase margin (light blue). The latter is interactive, so the specification of PM can be modified from the graph. In the **Bode diagram**, the 0 dB and $-180°$ lines are represented in dashed gray line. On the frequency axis, a non-interactive cross ($\times$) and a circle ($\circ$) are represented, determining the position of the pole p_c and zero z_c of the compensator, respectively. When placing the mouse over them, their value is displayed in the lower-left corner of the tool. The $\diamond$ symbol is linked to the gain crossover frequency $\omega_{gc} = \omega_m$ of the compensated system $C(s)G(s)$ (and therefore to the maximum phase ϕ_m and gain amplification GA, directly related to the α parameter of the network).

In the **Nyquist diagram**, it can be observed how the symbol $\diamond$ moves on the frequency curve from the circle passing through the critical point $(-1, 0)$ to the origin (which corresponds to values of α between 1 and 0). In the **Nichols diagram**, the interpretation is similar.

As in the previous card, the **Time response** of the closed-loop system is represented ($C(s)G(s)$ and $KG(s)$), including the already explained circles to modify the instant and amplitude of changes in reference, load disturbance, and measurement noise.

A set of practical examples is provided in the Options menu, in addition to the possibility of entering generic transfer functions in (NUM,DEN) or ZPK formats. Example 1 deals with the control of a first-order system with time delay. Example 2 introduces a double integrator. Example 3 deals with the control of a second-order system with an integrator.

8.9.1.4 Homework

1. Assume a system with a transfer function:

$$G(s) = \frac{4}{s(s+2)}.$$

   The objective is to design a compensator so that the static velocity error constant is $K_v = 20 \text{ s}^{-1}$, the PM is at least 50°, and the GM is at least 10 dB (example from [16], p. 496). Use the tool to calculate the K, τ, and β parameters of the controller. Note that for type-1 systems such as this exercise, the value of the static velocity error constant K_v is the value of the frequency at the intersection of the extension of the low-frequency initial line with slope -20 dB/decade (corresponding to the integrator) with the 0 dB line.

2. Select Example 1 from the Options menu located in the upper-left corner of the tool. Analyze what happens when trying to impose a specification with a PM higher than the default one. How do you justify the behavior you observe in closed loop?

3. Select Example 1 from the Options menu, corresponding to a double integrator with no time delay. Justify if that system can be stabilized using only a proportional controller K. Which is its PM? Can it be stabilized using a phase-lead network? It is recommended to perform an analysis both with the interactive tool described in this section and using the root_locus interactive tool (card 7.2) to answer these questions. Discuss the structure and parameters of the phase-lead network obtained by applying this method. What other known controller does it resemble? What is the value of the system delay that would cause the closed-loop system to become unstable for the default specifications? Change the specifications and repeat the exercise.

4. Use Example 3 from the Options menu. Indicate the controller parameters to achieve $PM = 75°$.

5. Design a phase-lead network for the system given by

$$G(s) = \frac{(s+2)}{(s+0.1)(s^2 + 10s + 29)},$$

to have a steady-state error when the reference is a step of 1% and $PM = 60°$ (reference [13], pp. 362–369).

6. Given a process with representative transfer function $G(s) = 1/s(s+2)$, design a phase-lead compensator so that the steady-state error when the reference is a unit slope ramp is less than 5% and $PM \geq 45°$ (reference [10], pp. 592–595).

8.10   Design of Phase-Lead or Phase-Lag Controllers in the Frequency Domain

8.10.1   Interactive Tool: f_design_lead_lag

8.10.1.1   Concepts Analyzed in the Card and Learning Outcomes
- Design of phase-lead or phase-lag controllers in the frequency domain using three specifications: steady-state error, PM, and gain crossover frequency.

8.10.1.2   Summary of Fundamental Theory
Classical frequency methods use as specifications the compliance with a certain steady-state error e_{ss} and a certain phase margin PM. There is the possibility of adding a specification on ω_{gc}, as a phase-lead/-lag controller has three degrees of freedom (K, β, and τ), being possible to develop a mathematical method that allows to reach three specifications, in case a solution exists. By adding a specification on the gain crossover frequency, the bandwidth of the closed-loop system can be modified and thus the speed of response.

It is assumed in the following that the proportional gain of the controller K has been previously calculated based on steady-state specifications, as in the previous cards. By applying the concept of PM using the polar form,

$$|L(j\omega_{gc})| = 1; \; PM = 180° + \lfloor L(j\omega_{gc}) \rightarrow L(j\omega_{gc}) = 1e^{j(-180°+PM)}.$$

Taking into account the controller structure,

$$C(j\omega_{gc})G(j\omega_{gc}) = K \frac{(j\omega_{gc}\beta + 1)}{(j\omega_{gc}\tau + 1)} \rho_G e^{j\phi_G} = 1e^{j(-180°+PM)},$$

$$\frac{j\omega_{gc}\beta + 1}{j\omega_{gc}\tau + 1} = \frac{1}{K\rho_G} e^{j(PM-180°-\phi_G)} = \frac{1}{K\rho_G}\left(-\cos(PM - \phi_G) - j\,\sin(PM - \phi_G)\right), \tag{8.56}$$

where ρ_G and ϕ_G are the magnitude (not in dB) and phase of $G(j\omega_{gc})$. Splitting Eq. (8.56) into its real and imaginary parts, a system of two equations with two unknowns is obtained:

$$\beta = \frac{1 + K\rho_G \cos(PM - \phi_G)}{-\omega_{gc}K\rho_G \sin(PM - \phi_G)}, \quad \tau = \frac{\cos(PM - \phi_G) + K\rho_G}{\omega_{gc}\sin(PM - \phi_G)}. \tag{8.57}$$

As commented before, to use these equations it is necessary to first determine K from the steady-state specifications. ρ_G and ϕ_G can be determined analytically or by analyzing the Bode diagram of $KG(j\omega)$. The frequency ω_{gc} is a specification related to the speed of response of the closed loop, as analyzed in Eqs. (8.11) and (8.11) in the introduction to this chapter. It is important to remember that K and ρ_G are magnitudes not expressed in dB. If the obtained values of τ and β are negative, it means that the specifications are very demanding and cannot be met with such controller structure, so they have to be relaxed or another control strategy has to be used.

Remark
Although this card only deals with phase-lead and phase-lag controllers, the extension of the design method analyzed hereto PID controllers is straightforward, as there are three specifications (e_{ss}, PM, and ω_{gc}) and the controller has three degrees of freedom. The PID parallel configuration is used in this case:

$$C(s) = K'' + \frac{K_i''}{s} + K_d''s. \tag{8.58}$$

From the steady-state specification e_{ss}, which depends on the type of the system, $K_i^{''}$ is first obtained. From the transient performance and relative stability specifications, the system of equations is obtained by applying the definition of PM:

$$\left(K^{''} + \frac{K_i^{''}}{j\omega_{gc}} + j\omega_{gc}K_d^{''}\right)G(j\omega_{gc}) = 1e^{j\phi(\omega_{gc})}, \tag{8.59}$$

$$\phi(\omega_{gc}) = -180° + PM. \tag{8.60}$$

Solving the real and imaginary parts of the two equations, the two unknowns $K^{''}$ and $K_d^{''}$ are obtained.

8.10.1.3 Reference Related to this concept

- [3] Shahian, B., & Hassul, M. (1993). *Control system design using MATLAB®*. Prentice Hall, ISBN: 0-13-174061-X. Chapter 7, Sect. 4, paragraph 5, pp. 189–192.

Application   Interactive tool: f_design_lead_lag

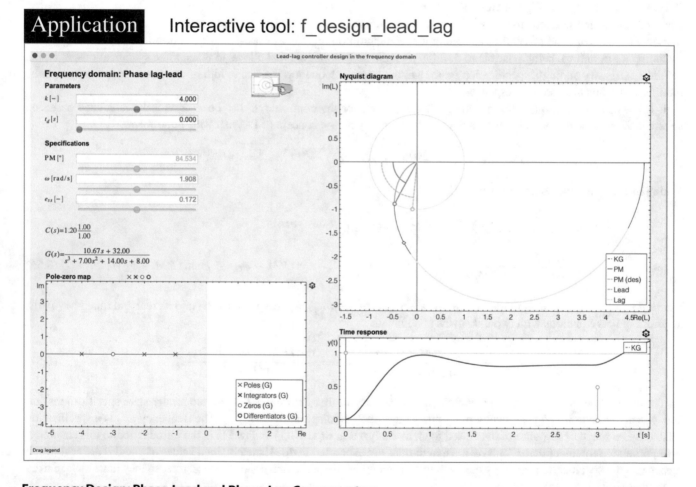

Frequency Design: Phase-Lead and Phase-Lag Compensators

This card deals with the design of phase-lead or phase-lag compensators imposing three specifications, one related to steady state and two to the relative stability and speed of response. This provides analytical formulas to directly design the controller, which limits the possibilities of interactive tools to graphically show the design and obtained results. It can be a starting point for loop shaping design and moreover the tool provides interesting information about the conditions for which to fulfill the specifications either with a phase-lead or a phase-lag compensator.

Parameters: The static gain k and time delay t_d values are set through textboxes and sliders.

Specifications: The three specifications are selected here, PM, gain crossover frequency ω_{gc}, and steady-state error e_{ss} referred to

- Position, with a unit step reference, if $G(s)$ is type-0.
- Velocity, with a unit slope ramp reference, if $G(s)$ is type-1.
- Acceleration, when the reference is a unit parabola, if $G(s)$ is type-2.

The system's and controller's symbolic transfer functions are also displayed.

The dynamic elements of $G(s)$ (poles, integrators, zeros, and differentiator: ×, ✗, ○, ●) are included and modified through the poles and zeros editor, associated with the **Pole–zero map**, by clicking and dragging items from/back to the repository next to the figure title (removing can be also done by double-clicking on the elements).

The upper-right area of the tool includes the different frequency representations of $L(j\omega) = C(j\omega)G(j\omega)$, each of which can be selected as a single view or displayed simultaneously, depending on the choice made in the Options menu in the upper-left corner of the tool.

In all cases the red color is used to draw the frequency response of $KG(s)$, blue for $C(s)G(s)$, and green for $C(s)$ (this is only in the Bode diagram). As a new feature in this tool, yellow lines are superimposed on the curves of the $KG(s)$ system representations to indicate the range of design frequencies where a phase-lag network is a controller that meets the specifications. Similarly, the segments of the time response of $KG(s)$ at frequencies where the solution to the problem is a phase-lead network are represented in brown color. The lines remaining in red indicate that the solution to the problem with the given specifications is a proportional controller with K gain, or that the specifications cannot be met with this controller structure. The graphs include explanatory legends for the color range used.

The PM of $KG(s)$ is represented by a green segment, ending in a circle of the same color on the frequency response plot. The desired phase margin, PM (des), is represented by a light blue segment, also ending in a circle of the same color that can be moved to change this specification (automatically modifying the rest of the interactive tool elements that are affected by this value).

The frequency diagrams also show a ◇ symbol representing the design frequency (gain crossover frequency). In the **Bode diagram**, that element is placed in the cut of a black interactive vertical line that allows modifying the gain crossover frequency ω_{gc} of the compensated system with the frequency curves of the compensator. In the **Nyquist diagram** and **Nichols diagram**, the symbol ◇ is placed on the $KG(j\omega)$ graph and facilitates visualizing the change of the design frequency ω_{gc}, which in these diagrams is an implicit variable.

The **Bode diagram** contains in the frequency axis a cross (×) and a circle (○) associated with the pole p_c and zero z_c of the network, respectively. Their position is automatically computed by the algorithm and thus they are not interactive elements, and their values are shown in the lower-left corner of the tool.

As in the previous cards, the **Time response** plot represents $KG(s)$ and $C(s)G(s)$ and allows introducing step changes in the reference and load disturbance and to inject output noise using the circles in the abscissa axis, as explained in other cards.

The Options include a set of practical exercises, besides the possibility of introducing transfer functions in generic formats (NUM,DEN) or ZPK. Example 1 deals with the control of a first-order system with time delay. Example 2 introduces a double integrator. Example 3 deals with the control of a second-order system plus an integrator.

8.10.1.4 Homework

1. When starting or resetting the tool, a third-order system with a zero and without time delay is displayed by default, with specifications of $PM = 84.5°$, $\omega_{gc} = 1.9$ rad/s, and $e_{ss} = 0.172$. As the system is type-0, this error refers to a step change in the reference. Determine the gain crossover frequency ω_{gc} ranges where the solution to the problem is a proportional controller, a phase-lag network, or a phase-lead compensator.
2. Repeat the exercise considering a specification of $e_{ss} = 0.1$ and static gain $k = 1$ in the system. Enter a unit load disturbance in $t = 3$ s and a noise of variance 0.01 in $t = 6$ s, scaling the time response graph accordingly. Modify the other specifications (PM and ω_{gc}) indicating, from your point of view, which of the controllers meeting the specifications represents the best compromise in terms of reference tracking, rejection of the disturbance, and low sensitivity to noise.
3. Repeat the two previous exercises considering a time delay $t_d = 0.2$ s.
4. Consider the system described by the transfer function (reference [3], p. 191):

$$G(s) = \frac{400}{s(s^2 + 30s + 200)}.$$

Given the following design specifications:

a. Steady-state error when the reference is a unit slope ramp less than 10%,
b. Gain crossover frequency equal to $\omega_{gc} = 14$ rad/s, and
c. $PM = 45°$,

Calculate a phase-lead or phase-lag compensator to meet these specifications.

8.11 Loop Shaping Design

The methods analyzed in cards 8.8, 8.9, and 8.11 belong to a large family of classical control system design methods, based on the open-loop transfer function $L(s) = C(s)G(s)$, within a unit feedback configuration. The design targets for LTI feedback systems may be phrased in terms of frequency response design goals and loop shaping. Loop shaping consists of designing the controller $C(s)$ to meet specifications on the open-loop transfer function $L(s)$, within the desired frequency spectrum, by manipulating the poles, zeros, and gain of $C(s)$. The basic idea behind loop shaping is to convert the desired specifications for the closed-loop system into constraints on $L(s)$, as analyzed in the introduction to this chapter. The controller $C(s)$ is then designed so that $L(s)$ satisfies these constraints. As the controller $C(s)$ influences the open-loop transfer function $L(s)$, for instance, to make an unstable system stable, all there is to do is to bend $L(s)$ away from the critical point. The relationship between open- and closed-loop specifications introduced in the introductory section of this chapter plays a major role.[4]

In previous sections, classical frequency-domain design methods that implicitly use loop shaping by lag compensation and lead compensation have been studied. This means that the user must consider how to change the controller parameters (pole, zero, and gain), in such a way that $L(s)$ fulfills the specifications. For this reason, the use of interactive tools such as ILM[5] loop shaping [8, 9] and LCSD[6] [24], which are the ones dealt with in this section, make it easy to design the controller. In these tools, primitives are available that allow the poles and zeros of the controller $C(s)$ to be placed interactively over the open-loop diagram (Bode, Nyquist, or Nichols) to shape the desired $L(s)$. Thus, the scheme is an indirect design technique where the closed-loop transfer function is synthesized by the open-loop transfer function as an intermediate step. In this conceptual framework of open-loop shaping, robustness can be easily analyzed using the concept of stability margins and sensitivity functions.

This section, therefore, summarizes many of the concepts treated in the book, and the two interactive tools used to visualize the concept of loop shaping are more complex than those developed to explain and understand basic concepts. Moreover, as they have originally been developed out of the scope of this book, their nomenclature does not exactly follow the one used, so the corresponding equivalences are provided. The next sections summarize the main concepts and provide illustrative examples.

ILM_PIDloopshaping allows designing PID controllers using specifications based on stability margins. This tool was designed in collaboration with professors Karl J. Åström and Tore Hägglund, from the University of Lund in Sweden, and is a complement of their excellent book on PID control [7]. Yves Piguet, the CEO of Calerga Sárl and developer of Sysquake, was also coauthor of the tool.

LCSD includes many controller structures and loop shaping can be done based on time-domain and frequency-domain specifications in different diagrams (root locus, and Bode, Nyquist, and Nichols diagrams). Many of the exercises in the book can be solved using LCSD. This tool was developed in collaboration with José Manuel Díaz and Rocío Muñoz, from UNED.

---

[4] Loop shaping can also be applied directly in closed loop [23]. Performance measures are easy to calculate by examining closed-loop transfer functions. For example, the sensitivity function $S(s)$ is directly related to the closed-loop performance (steady-state error) in the low-frequency range, while in the high-frequency range it can be related to noise attenuation. Furthermore, robustness can be analyzed in terms of the maximum value of $|S|$. An analogous analysis can be carried out using the complementary sensitivity function $|T|$. An important disadvantage of the closed-loop shaping transfer functions is that the relationship between the controller and the closed-loop transfer functions is not affine and therefore it is not intuitive how the poles and zeros of the controller should be modified to provide proper shape to the closed-loop transfer functions. This problem can be bypassed using the Youla parameterization, which transforms the closed-loop transfer function into the series connection of the plant $G(s)$ and a design filter. This scenario is thus analogous to that used in open-loop shaping, so it can be used to directly shape closed-loop transfer functions.

[5] Interactive learning modules.

[6] Linear control system design.

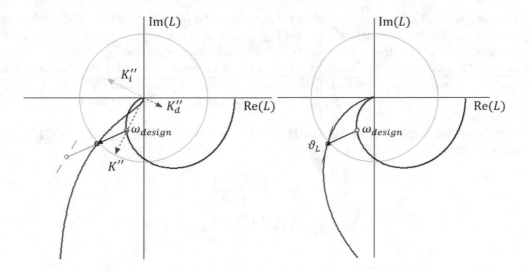

Fig. 8.20 Effect of controller gains and slope of $L(j\omega)$ in the polar plot

8.11.1 Interactive Tool: ILM_PIDloopshaping

8.11.1.1 Concepts Analyzed in the Card and Learning Outcomes
- Explain basic concepts of PID loop shaping without considering implementation aspects.
- Open-loop loop shaping design of PID controllers with parallel structure in the polar plot.
- Specifications in the frequency domain and related frequencies: GM, PM, maximum sensitivity, and maximum complementary sensitivity.
- Relationships between open-loop and closed-loop frequency domains: geometric locus of constant sensitivity and complementary sensitivity functions.

8.11.1.2 Summary of Fundamental Theory
This section summarizes the main concepts related to PID loop shaping in the polar plot [7–9]. The best structure to design PID controllers in the polar plot is the parallel one (see Eq. (8.37) and Fig. 8.14):

$$C(s) = K'' + \frac{K_i''}{s} + K_d'' s$$

which yields the loop transfer function

$$L(s) = C(s)G(s) = K'' G(s) + \left(\frac{K_i''}{s} + K_d'' s \right) G(s).$$

The point on the Nyquist curve of the loop transfer function corresponding to the frequency ω is given by

$$L(j\omega) = K'' G(j\omega) + j \left(-\frac{K_i''}{\omega} + K_d'' \omega \right) G(j\omega), \tag{8.61}$$

where ω specifies the design point, which can correspond to the unit circle, the sensitivity circles, or the real axis, so that specifications on gain and phase margins or sensitivities can be posed.

The focus of the design is placed on how the loop transfer function changes when controller parameters are modified, which reveals the parameter values required to obtain a given shape of the loop transfer function. For PI and PD control, the mapping is uniquely given by one point. For PID control, it is also possible to obtain an arbitrary slope ϑ_L of the loop transfer function at the target point.

From (8.61), the proportional gain changes $L(j\omega)$ in the direction of $G(j\omega)$, the integral gain K_i'' changes $L(j\omega)$ in the direction of $-jG(j\omega)$, and the derivative gain K_d'' changes $L(j\omega)$ in the direction of $jG(j\omega)$. This is visualized in Fig. 8.20, where the Nyquist plot of the process transfer function $G(s)$ is shown in black color, and $L(s)$ in red color. In the left figure (corresponding to a Free design), the length of the arrows corresponds to the values of the controller gains and the compensated point at the frequency ω is calculated as the sum of three vectors, namely the proportional, the integral, and the derivative vectors. In the right figure (Constrained PID design), the slope of the loop transfer function can be set at a design point, as commented in what follows.

The first possibility of loop shaping design is a PI design (named Constrained PI). Dividing (8.61) by $G(j\omega)$ and separating the real and imaginary parts:

$$K'' = \text{Re}\left(\frac{L(j\omega)}{G(j\omega)}\right), \tag{8.62}$$

$$-\frac{K_i''}{\omega} + K_d''\omega = \text{Im}\left(\frac{L(j\omega)}{G(j\omega)}\right) = A(\omega). \tag{8.63}$$

With $K_d'' = 0$, Eqs. (8.62) and (8.63) yield the two parameters of the PI controller.

An additional condition is required for the design of PID controllers (Constrained PID), based on the slope of $L(j\omega)$ (ϑ_L) at the design frequency:

$$\frac{dL(s)}{ds} = \frac{dC(s)}{ds}G(s) + C(s)\frac{dG(s)}{ds} = \frac{dC(s)}{ds}G(s) + \frac{L(s)\frac{dG(s)}{ds}}{G(s)} = \left(-\frac{K_i''}{s^2} + K_d''\right)G(s) + \frac{L(s)\frac{dG(s)}{ds}}{G(s)}. \tag{8.64}$$

The slope of the Nyquist curve is then given by

$$\frac{dL(j\omega)}{d\omega} = jL'(j\omega) = j\left(\frac{K_i''}{\omega^2} + K_d''\right)G(j\omega) + jC(j\omega)G'(j\omega). \tag{8.65}$$

The complex number represented by (8.65) has the phase angle ϑ_L if

$$\text{Im}\left(jL'(j\omega)e^{-j\vartheta_L}\right) = 0. \tag{8.66}$$

Results (8.64)–(8.66) imply that

$$\frac{K_i''}{\omega^2} + K_d'' = \frac{\text{Re}\left(L(j\omega)\dfrac{G'(j\omega)}{G(j\omega)}e^{-j\vartheta_L}\right)}{\text{Re}\left(G(j\omega)e^{-j\vartheta_L}\right)} = B(\omega). \tag{8.67}$$

Combining (8.67) with (8.62)–(8.63) provides the controller parameters

$$K_i'' = -\omega A(\omega) + \omega^2 B(\omega), \tag{8.68}$$

$$K_d'' = \frac{A(\omega)}{\omega} + B(\omega), \tag{8.69}$$

where $A(\omega)$ and $B(\omega)$ are given by (8.63) and (8.67), respectively.

Regarding specifications, they can be posed in terms of classical stability margins or sensitivities. It has been seen in the introduction to this chapter (Fig. 8.11) that the maximum sensitivity, which indicates the worst case in terms of disturbance amplification is

$$M_s = \max_\omega |S(j\omega)| = \max_\omega \left|\frac{1}{1 + L(j\omega)}\right|, \tag{8.70}$$

where M_s is called *maximum sensitivity*, ω_{ms} is the *maximum sensitivity frequency*, and $s_m = 1/M_s$ is called the *stability margin*, as it provides the distance to the critical point $(-1, 0)$. Sensitivity has a constant magnitude in circles with a center at that critical point. The condition that the largest sensitivity is less than M_s is equivalent to the condition that the Nyquist plot of the loop transfer function is outside a circle with center at $(-1, 0)$ and radius $1/M_s$ (Fig. 8.21).

There is a similar interpretation for the complementary sensitivity function. If $L(j\omega) = x + jy$:

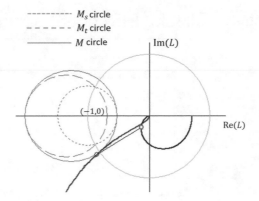

Fig. 8.21 Sensitivity circles

Table 8.8 Sensitivity circles: M_s, constant complementary sensitivity, M_t, constant mixed sensitivity, and equal sensitivities $M = M_s = M_t$

Contour	Center	Radius
M_s-circle	-1	$1/M_s$
M_t-circle	$-\dfrac{M_t^2}{M_t^2 - 1}$	$\dfrac{M_t}{M_t^2 - 1}$
M-circle	$-\dfrac{x_1 + x_2}{2}$	$\dfrac{x_1 - x_2}{2}$

where $x_1 = \max\left(\frac{M_s+1}{M_s}, \frac{M_t}{M_t-1}\right)$ and $x_2 = \max\left(\frac{M_s+1}{M_s}, \frac{M_t}{M_t-1}\right)$

$$|T| = \frac{\sqrt{x^2 + y^2}}{\sqrt{(1+x)^2 + y^2}},$$

so that the magnitude of the complementary sensitivity function M_t is constant if [7]:

$$x^2 + y^2 = M_t^2((1+x)^2 + y^2) \rightarrow \left(x + \frac{M_t^2}{M_t^2 - 1}\right)^2 + y^2 - \frac{M_t^2}{(M_t^2 - 1)^2} = 0,$$

which is a circle with center and radius as indicated in Table 8.8.

The requirements that the maximum sensitivity be less than M_s and that the complementary sensitivity be less than M_t imply that the Nyquist plot should be outside the corresponding circles. It is possible to find a slightly more conservative condition by determining a circle enclosing both circles (M circle) as illustrated in Fig. 8.21. A particular simple criterion is obtained by requiring that $M_s = M_t$ [7].

8.11.1.3 References Related to this Concept

- [7] Åström, K. J., & Hägglund, T. (2006). *Advanced PID control*. ISA - The Instrumentation, Systems and Automation Society, ISBN: 978-15-561-7942-6. Chapter 6, Sect. 8, pp. 206–221.
- [8] Guzmán, J. L., Åström, K. J., Dormido, S., Hägglund, T., Berenguel, M., & Piguet, Y. (2008). Interactive learning modules for PID control. *IEEE Control Systems Magazine, 28*(5), 118–134.
- [9] Guzmán, J. L., Åström, K. J., Dormido, S., Hägglund, T., & Piguet, Y. (2006). Interactive learning modules for PID control. *IFAC Proceedings Volumes, 39*(6), 7–12.

Application Interactive tool: ILM_PIDLoopShaping

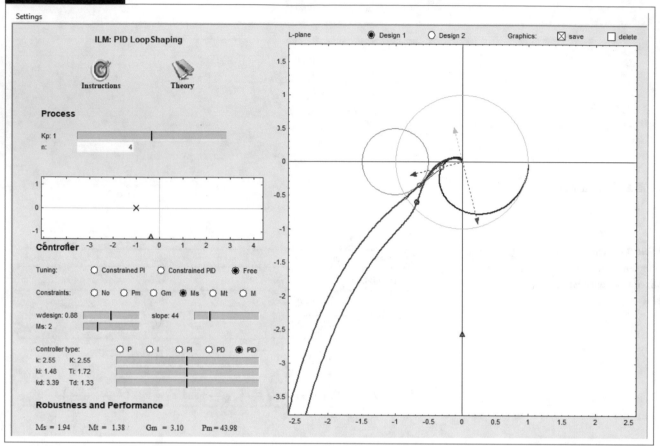

PID Loop Shaping Design

This tool focuses on using interactive graphics to learn PID control and develop intuition. The tool can be downloaded from https://arm.ual.es/ilm/pidloopshaping.php. The readers are encouraged to visit the website to experience the interactive features. The correspondence with the nomenclature used in this book is the following:

Element	This book	ILM_loopshaping
Process transfer function	$G(s)$	$P(s), G_p(s)$
Process static gain	k	K_p
Gain margin	GM	Gm
Phase margin	PM	Pm
Gain crossover frequency	ω_{gc}	Wpc
Phase crossover frequency	ω_{pc}	Wgc
Proportional gain of parallel PID	K''	k
Integral gain of parallel PID	K_i''	k_i
Derivative gain of parallel PID	K_d''	k_d
Controller output	u_c	u
Plant input	u	v
Plant output	y_p	x

The **Process** transfer function can be selected and modified depending on the option selected from the Settings menu. Different process models are available (when selected, their representative parameters can be modified using textboxes and sliders), including the possibility of including a symbolic representation (StringTF) and a free transfer function (InteractiveTF), where poles and zeros can be defined graphically in a pole–zero map.[7]

PID loop shaping provides three methods for tuning the parameters in order to move the process transfer function from the design point to the target point. These methods are listed in the Tuning zone as Free, Constrained PI, and Constrained PID. Free tuning allows an unconstrained loop to be shaped by dragging on the control parameters. Constrained PI and Constrained PID permit the calculation of the controller parameters based on some constraints on the target point, using the formulas obtained in the theory summary.

When the Free tuning option is selected, sliders are used to modify the controller gains k, k_i, and k_d. The controller gains can be changed by dragging arrows, the proportional gain changes $L(j\omega)$ in the direction of $G(j\omega) = P(j\omega)$ (blue arrow), the integral gain $K_i^{''} = k_i$ changes $L(j\omega)$ in the direction of $-jP(j\omega)$ (cyan arrow), and the derivative gain $K_d^{''} = k_d$ changes $L(j\omega)$ in the direction of $jP(j\omega)$ (magenta arrow).

For the Constrained PI and Constrained PID tuning options, the target point can be limited to move on the unit circle, the sensitivity circles, or the real axis.

The design frequency ω can be chosen using the slider wdesign or graphically by dragging the green circle on the process Nyquist curve (black curve). The target point on the Nyquist plot and its slope can be dragged graphically. The slope can also be changed using the slider slope. Furthermore, it is possible to constrain the target point using the Constraints radio buttons to the unit circle (Pm), the negative real axis (Gm), circles representing constant sensitivity (Ms), constant complementary sensitivity (Mt), or constant sensitivity combinations (M). When sensitivity constraints are active, the associated circles are drawn in the L-plane plot, and sliders can be used to modify their values. The circles are defined in Table 8.8.

The figure showing the main screen of the interactive tool above illustrates designs for two PID controllers and a given sensitivity. The target point is moved to the sensitivity circle, and the slope is adjusted so that the Nyquist curve is outside the sensitivity circle. The red design shows a PID controller using Free tuning, while the blue design shows a Constrained PID tuning. Specifications that cannot be reached are indicated in the tool by giving the integral or derivative gain negative values in these cases.

Robustness and Performance parameters are displayed on the screen below the controller parameters, and these parameters characterize robustness and performance. The values are maximal sensitivity (Ms), sensitivity crossover frequency (Ws), maximal complementary sensitivity (Mt), complementary sensitivity crossover frequency (Wt), gain margin (Gm), gain crossover frequency (Wgc), phase margin (Pm), and phase crossover frequency (Wpc).

The L-plane graphic is given on the right-hand side of the PID Loop Shaping menu. This graphic contains the Nyquist plots of the process transfer function $G(s) = P(s)$ in black and the loop transfer functions $L(s) = P(s)C(s)$ in red. Three different views can be shown depending on the tuning options. The design and target points can be modified interactively on this graphic. The design point is shown in green on the Nyquist curve of the process. The target point is represented in light green in the case of Free tuning, and in black for constrained tuning. The slope of the target point can also be changed interactively. For Free tuning, the controller gains are shown as arrows in the Nyquist plot. The controller gains can be modified interactively by dragging the ends of the arrows. The scale of the graphic can be changed using the red triangle located at the bottom of the vertical axis. Options save and delete can be found above the L-plane graphic, making it possible to save designs to perform comparisons. Once the save option is active, two pictures appear, one of which shows the current design in red while the other shows the current design in blue. Modifications of the controller parameters affect the current (active) design, which can be changed using the options Design 1 and Design 2, which appear on the top of the L-plane graphic. Once a design is chosen, the associated curve is switched to red, and the controller zone is modified based on that design. The controller gain values can be seen by moving the cursor on the curves.

The Settings menu, which is available in the main menu of PID Loop Shaping, is divided into four groups. The first entry, called Process Transfer Function, is used to choose between several predefined transfer functions or to include a user-specified transfer function through two options. The String TF option allows a transfer function to be entered symbolically. For instance, $P(s) = 1/\cosh\sqrt{s}$ can be represented as `P='1/cosh(sqrt(s))'`. Results can be stored and recalled using the Load/Save menu. The option Save Report can be used to save all essential data in text format, which is useful for documenting results. Specific values for control parameters can be entered with Parameters menu option. The last menu option (Examples Advanced PID Book) allows loading examples from [7].

---

[7] Note that this tool was developed in an older version of Sysquake and instead of having a repository of dynamical elements, they are added, removed, and moved by selecting the appropriates radio buttons.

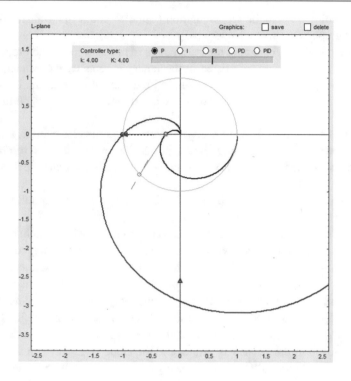

Fig. 8.22 Critical gain computation

8.11.1.4 Examples

Example 1: PID loop shaping in the Nyquist diagram and derivative cliff. Consider a plant described by

$$G(s) = \frac{1}{(s + 1)^4}.$$

1. Calculate the gain for a proportional controller for which the closed-loop system changes from stable to unstable. This simple exercise can be solved analytically using the definition of PM:

$$\lfloor L(j\omega_{gc}) = \lfloor C(j\omega_{gc})G(j\omega_{gc}) = -180°,$$

$$\lfloor \frac{K''}{(j\omega_{gc} + 1)^4} = -180°, \rightarrow \omega_{gc} = 1 \text{ rad/s}.$$

$$|L(j\omega_{gc})| = \left| K'' \frac{1}{(j\omega_{gc} + 1)^4} \right| = 1 \rightarrow K'' = 4.$$

By using the tool, the solution is easily achieved, as shown in Fig. 8.22.

2. Design a PID controller to achieve $M_s \leq 1.4$, trying to maximize the integral gain K_i''. For load disturbances with low-frequency content, the integral gain K_i'' is a measure of load disturbance attenuation [7, 8], but large values of K_i'' imply large peaks of the sensitivity function. Therefore, a trade-off becomes necessary between load-disturbance rejection and robustness.

A PID controller that satisfies the specifications is obtained when $K'' = 0.925$, $K_i'' = 0.9$, and $K_d'' = 2.86$; the Nyquist plot of the loop transfer function is shown in Fig. 8.23. It can be seen that the Nyquist curve has a loop, called a *derivative cliff*. As explained in [7], this feature, which is due to excessive controller phase lead, results from having a PID controller with complex poles, which occurs when $T_i < 4T_d$. In this example, the relation is $T_i = 0.33T_d$. The closed-loop time response can be seen in Example 3 of the next section (LCSD tool). If $T_i = 4T_d$ (as suggested by Ziegler–Nichols tuning rules), with $K'' = 1.1$, $K_i'' = 0.36$ and $K_d'' = 0.9$, the derivative cliff disappears.

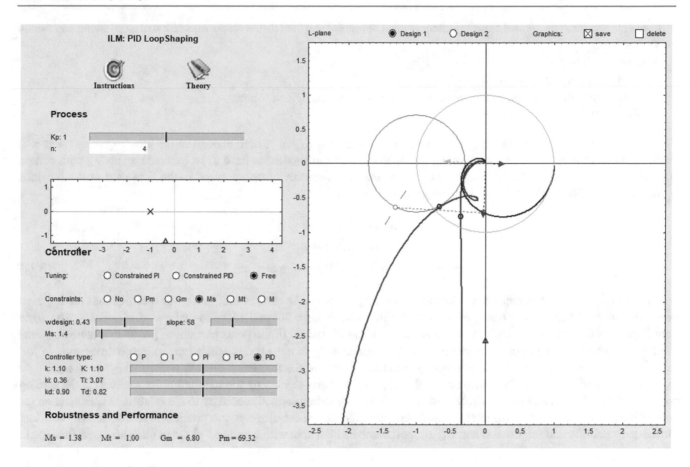

Fig. 8.23 Control design and derivative cliff

8.11.2 Interactive Tool: LCSD

8.11.2.1 Concepts Analyzed in the Card and Learning Outcomes
- Fundamentals of linear control system design (LCSD).
- Classical loop shaping as a design procedure that explicitly involves the shaping of the open-loop transfer function $L(s)$, within a desired frequency spectrum by manipulating the poles, zeros, and gain of the controller $C(s)$.
- Extended specifications in the frequency domain.
- Relationships between time-domain and frequency-domain specifications.

8.11.2.2 Summary of Fundamental Theory
Most of the required knowledge to perform loop shaping has been explained in the introduction to this chapter and different sections of the book. Classical loop shaping design is typically an iterative procedure where the designer shapes and reshapes $|L(j\omega)|$ to meet all specifications. With interactive tools, the procedure can be highly simplified, as the results of such shaping are immediately shown in different diagrams and also in the closed-loop response. The graphic image is as if the loop gain $|L(j\omega)|$ were an elastic band that it is possible to bend in the desired frequency range to meet the specifications given [24]. As in the previous cards, the basic block diagram is that given in Fig. 7.2. As derived from the analysis of closed-loop sensitivity functions done in the introduction of the chapter, the control system design is a compromise, as the complementary sensitivity function cannot be one for all frequencies neither the load sensitivity and sensitivity can be zero (8.14), as $S(j\omega) + T(j\omega) = 1$ for all frequencies. For that reason, it is normal to make certain assumptions on the frequency content of the inputs to the control loop: reference R, load disturbance D, and noise N. In most common applications, references and disturbances are constant (or step-shaped) or slow-varying signals, while noise is a rapidly varying signal. Therefore, R and D have their main frequency content in the low-frequency range, while N takes small values at low frequencies and becomes relevant at high frequencies. Considering this scenario, it is possible to formulate the control objectives:

- Low-frequency range (*control band*), references must be tracked and load disturbances rejected, so the following conditions have to be met:
  - Complementary sensitivity $T(j\omega) = \frac{L(j\omega)}{1+L(j\omega)} \approx 1$.
  - Load sensitivity $GS(j\omega) = \frac{G(j\omega)}{1+L(j\omega)} \approx 0$.
- High-frequency range (*cutoff band*), noise would not be amplified, which means:
  - Sensitivity $S(j\omega) = \frac{1}{1+L(j\omega)} \approx 1$.

These conditions require selecting a controller that guarantees that $L(j\omega)$ is high enough in the low-frequency range and small enough in the high-frequency range. Closed-loop stability and robustness have to be ensured complying with certain stability margins, GM and PM in classical control, or maximum sensitivity (as analyzed in the ILM card in the previous section). The previous statements can be summarized as follows:

$$|L(j\omega)| \geq l(\omega), \quad 0 \leq \omega \leq \omega_B,$$
$$|L(j\omega)| \leq u(\omega), \quad \omega \geq \omega_C,$$
$$-150° \leq \lfloor L(j\omega) \leq -30°, \quad \omega_B \leq \omega \leq \omega_C \tag{8.71}$$

where interval $[0, \omega_B]$ defines the control band and interval $[\omega_C, \infty]$ is the cutoff band. Crossover specifications are imposed between both bands (*crossover band*) in (8.71) to fulfill a PM between 30° and 150°, so that $\omega_B < \omega_{gc} < \omega_C$. The same could apply to GM and ω_{pc}. Figure 8.24 visualizes these specifications. In the polar plot, those points that have the same module are over the same circle centered in the origin; accordingly to meet the specifications in the control band, $L(j\omega)$ must be out of a big circle of radius defined by the allowed error in the frequency operation range, $0 \leq \omega \leq \omega_B$, while in the cutoff band $L(j\omega)$ must be inside a small circle with a radius defined by the required noise attenuation. Regarding robustness, $L(j\omega)$ must be out of a circle centered in $(-1, 0)$ of radius defined according to the robustness specification (as seen in the previous card). As the polar plot combines phase and gain in one diagram, it is possible to simultaneously consider closed-loop stability and robustness, and $L(j\omega)$ must combine high gain with a reduced phase to guarantee performance in

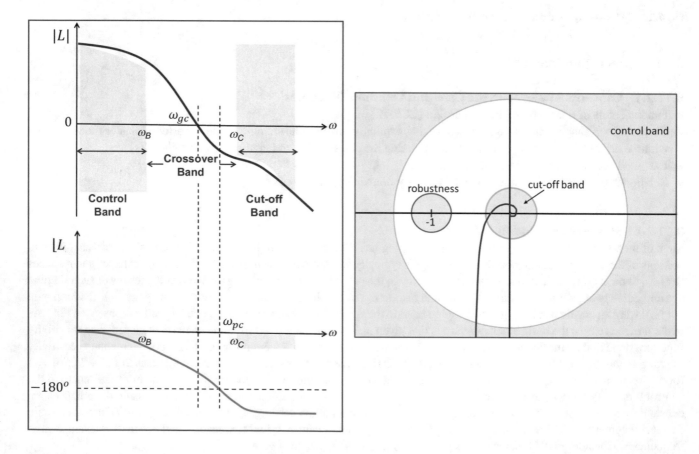

Fig. 8.24 Loop shaping specifications in Bode (left) and Nyquist (right) diagrams

a robust manner. Moreover, it is possible to compute those points that would correspond to the same value of $|S(j\omega)|$. This directly connects the open-loop system with the closed-loop one, and it is a very interesting hint during the loop shaping procedure. One drawback of using the Nyquist plot for loop shaping is that there is no straightforward connection between the movements of the poles and zeros of the controller and the changes in the curve at a given point. Additionally, it needs to display values of $L(j\omega)$ with big modules, in the control band, and others with small modules, in the cutoff band. This is, in general, difficult to be done because of the linear nature of the Nyquist plot [24].

Loop shaping can be performed in a natural way in the Nichols diagram; the interested reader is encouraged to read [24]. In manual loop shaping, the steps to design a controller are as follows:

1. Select a frequency response diagram (Bode, Nyquist, or Nichols) in which the design is carried out.
2. Draw the frequency specifications in the selected diagram.
3. Select a controller structure (lag compensator, lead compensator, PID controller, etc.).
4. Adjust the controller elements (gain, poles, and zeros) in a pole–zero map or a root locus, or their corner frequencies in the selected frequency response diagram, to shape the open-loop transfer function $L(s)$, and meet the specifications.
5. If the specifications cannot be met with the controller structure selected in step 3, the designer has to try another control structure.

The manual loop shaping design is an iterative procedure. Obviously, the performance of this design technique requires the use of software tools. Several examples are introduced in this card.

8.11.2.3 References Related to this Concept

- [5] Åström, K. J., & Murray, R. M. (2014). *Feedback systems: An introduction for scientists and engineers* (2nd ed.). Princeton University Press, ISBN: 9780691193984.
- [24] Díaz, J. M., Costa-Castelló, R., Muñoz, R., & Dormido, S. (2017). An interactive and comprehensive software tool to promote active learning in the loop shaping control system design. *IEEE Access*, 5, 10533–10546.
- [25] Skogestad, S. & Postlethwaite, I. (2005). *Multivariable feedback control, analysis and design* (2nd ed.). Wiley-Interscience, ISBN: 978-0470011683.
- [26] Vargas, H., Marín, L., de la Torre, L., Heradio, R., Díaz, J. M., & Dormido, S. (2020). Evidence-based control engineering education: evaluating the LCSD simulation tool. *IEEE Access*, 8, 170183–170194.

Application   Interactive tool: LCSD

Loop Shaping Design Through LCSD

This card is devoted to studying the basics of loop shaping design. The tool can be downloaded from https://www2.uned.es/itfe/LCSD/LCSD.html. The correspondence with the nomenclature used in this book is the following:

Element	This book	LCSD
Process transfer function	$G(s)$	$P(s)$
Process static gain	k	K
PID controller gain	K	K_p
Gain margin	GM	g_m
Phase margin	PM	φ_m
Relative damping	ζ	δ
Maximum overshoot	OS	P_O
Time constant	τ	T
Time delay	t_d	d
System type	$\aleph$	h
Steady-state error	e_{ss}	e_s
Controller output	u_c	u
Plant input	u	v
Plant output	y_p	x

The tool allows for multiple control designs; only those related to the concepts treated in this book are introduced. The web page includes a complete user guide, and the users are encouraged to read it. Here, only a short summary is included.

The LCSD main window is organized into six zones, represented in the previous figure:

- **Block diagram:** It allows implementing a 2-DoF structure. It has three blocks, corresponding with the prefilter F, controller C, and plant P, three inputs (reference r, load disturbance d, and output noise n), measured plant output (y), and four intermediate signals (tracking error e, controller output u, plant input v, and plant output x). In this area, users can perform the following actions:
  - Select the block or the input whose structure and parameters have to be configured in the rest of the zones of the tool. The block or input selected is represented in light green color. LCSD has implemented several predefined structures of transfer functions, controllers, and prefilters (same options that for the controllers). LCSD has implemented six types of inputs: pulse, step, ramp, parabola, sinusoid, and white noise (see Fig. 8.25).
  - Select the signals that are represented in the time response zone. The selected signals are represented in red color.
  - Enable feedback. If the feedback is enabled, the associated line of the block diagram is plotted in a solid line, otherwise the line is plotted in a dotted line.
  - Enable disturbances d and/or n. The arrow associated with an enabled disturbance is plotted in a solid line. If the disturbance is disabled, then it is plotted in a dotted line.
- The parameters setting zone contains several sliders and textboxes to configure the parameters of the block or input selected in the block diagram zone. Besides, the symbolic transfer function of the selected block or the mathematical expression of the selected input is shown. It helps the user to remember the meaning of the configurable parameters. Once the plant has been selected from the block diagram, double-clicking on the elements in this area opens new windows where to configure the parameters.
- The performance/specifications area contains two buttons (Frequency specifications and Time specifications). When there are no specifications enabled in a certain domain, the associated button is represented in yellow color. If the enabled specifications are all met the button is represented in green color, otherwise the button is represented in red color. All the relationships between the time and frequency domains (open and closed loop) analyzed in the introduction of this chapter can be applied here. The main frequency-domain specifications used by the tool are shown in Table 8.9 and the time-domain ones in Table 8.10.
  The low- and high-frequency disturbance attenuation allows interactively to define the control and cutoff bands by modifying the shape of the associated rectangle. Dragging first the upper-left corner allows entering trapezoidal shapes (see the manual of the tool for more details).

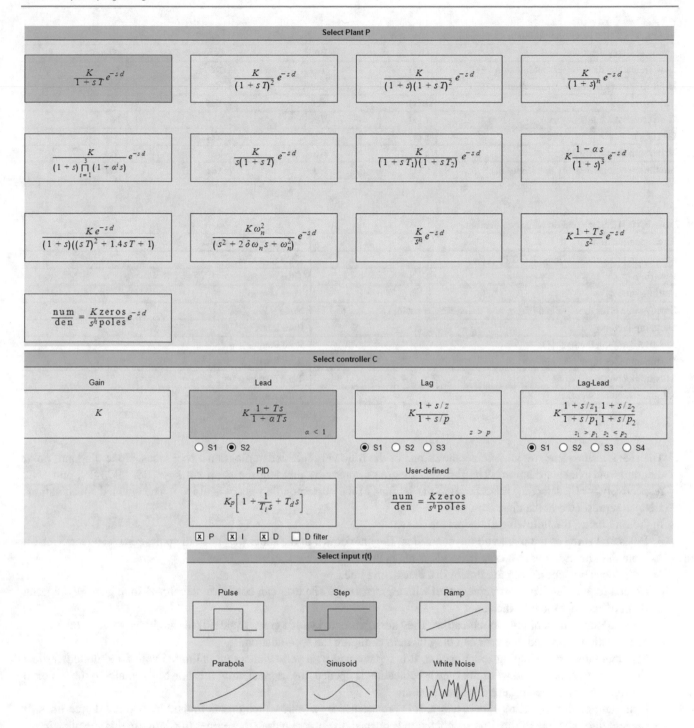

Fig. 8.25 Blocks implemented in LCSD for the plant, the controllers, and input selection

If there is some enabled specification, then this table is replaced with the enabled specifications and the system values for these specifications. Each specification has associated a small circle whose color indicates whether (green color) or not (red color) the specification is being met for the system.

- The **Pole–zero map** shows the poles ($\times$) and zeros ($\circ$) of prefilter (light blue color), controller (blue color), and plant (red color). For the selected block in the block diagram zone, the user can set the position of a pole or a zero by dragging the associated interactive element. Moreover, a repository of dynamic elements is placed over the plot, slightly modified from the one used in the book. If the mouse is placed over any of the elements, a label indicates which kind of element is (real pole, complex conjugate poles, integrators, etc.).

Table 8.9 LCSD: Frequency-domain specifications

Frequency specification	Of	Bode	Nyquist	Nichols
Low-frequency disturbances attenuation LF	$L(s)$	Yes	No	No
High-frequency disturbances attenuation HF	$L(s)$	Yes	No	No
PM (φ_m)	$L(s)$	Yes	Yes	Yes
Gain crossover frequency ω_{gc}	$L(s)$	Yes	No	No
GM (g_m)	$L(s)$	Yes	Yes	Yes
Maximum sensitivity M_s	$S(s)$	No	Yes	Yes
Maximum complementary sensitivity M_t	$T(s)$	No	Yes	Yes

Table 8.10 LCSD: Time-domain specifications

Time specification	Zone where it is drawn
Maximum percentage overshoot OS [%](P_O)	Time response
Rise time t_r	Time response
Settling time t_s	Time response
Steady-state error to step input [%] e_s and static-error constant (K_p, K_v, K_a)	None
Damping factor σ	Root locus
Relative damping factor ζ (δ)	Root locus
Dominant poles d_p	Root locus
Maximum control signal u_h	Time response
Minimum control signal u_l	Time response

- The root locus/frequency response zone shows one of the following interactive diagrams: root locus, Bode diagram, polar diagram, and Nichols diagram. The different diagrams appear by pressing the short-cut key "R", "B", "P", and "N", respectively, or by selecting them through the settings of the corresponding gearwheel icon. The enabled specifications are also represented in the diagrams.

  In the root locus, the following elements are drawn:

  – Poles and zeros of the plant and the controller. They are drawn in the same way as in the pole–zero map (see Sect. 6). Note that the user can also drag these elements in this zone.
  – Root locus branches. They are drawn in a black solid line.
  – Closed-loop poles. They are represented with a green cross. The user can configure the closed-loop gain by dragging them on the root locus branches.
  – Closed-loop dominant pole specification. They are represented with a green square. If the auxiliary window of the time specification is opened, the user can drag them to configure the specification.
  – Minimum absolute damping specification. It is represented by a yellow area that is limited with a red dotted vertical line. If the auxiliary window of the time specification is opened, the user can drag this line horizontally to the left or to the right in order to configure this specification.
  – Minimum relative damping specification. It is represented by a yellow area that is limited by two red dotted lines. If the auxiliary window of the time specification is opened, the user can drag these lines to configure this specification.

  To configure the scale of the root locus zone, the user has to click on the gearwheel icon located in the right upper corner of the zone.

  In the frequency response diagrams, the following elements are available:

  – Magnitude, phase, polar, and Nichols curves for the prefilter F (blue), the controller C (blue), the plant P (red), the open-loop transfer function L (black), the closed-loop transfer function $T = L/(1 + L)$ (dark green), the transfer function FT (black), and the sensitivity transfer function $S = 1/1 + L$ (gray). The user can choose the curves that are plotted by the selector located in the left upper square. These curves are plotted in solid lines. Note that the available functions in the selector depend on the selected block in the block diagram zone. For example, when the block C is selected, then the available functions in the selector are P, C, L, T, and S.

- Gain selector K. It is located in the central upper part of the zone. To enable or disable the selector, the user has to click on it. When the gain selector is enabled, it is enclosed within a light green square, and the user can configure the gain of the frequency response. To increase (or decrease) the gain, the user has to drag vertically upward (or downward) the Bode magnitude curve, the polar curve, or the Nichols curve. Once the user has finished configuring the gain, he/she has to click on the gain selector to disable it.
- Corner frequencies are represented by the same elements shown in the repository of poles and zeros. The user can drag them to the appropriate locations, and removing by double-clicking on them.
- Damping ratio of a pair of complex poles and zeros. The user has to locate the mouse pointer on the interactive element, and drag vertically upward or downward.
- Frequency specifications. Already described.

Besides, LCSD allows the user to add and remove poles and zeros in the pole–zero map zone, and in the root locus/frequency response zone.

- The time response zone contains one or two subplots with the time response of the signals that are selected in the block diagram zone. The enabled time specifications are also represented in the diagrams.
- The upper menu bar includes the following options:
  - Session menu: It allows the loading and saving of sessions. LCSD also allows saving a report file with the following data: P, C, F, time and frequency specifications, and system performance. The reset option is included in this menu.
  - Control menu: LCSD allows the saving the current controller C and prefilter F in a text file.
  - Options menu: allows the setting of a frequency range, the activating of the internal derivative filter, and setting of the time simulation parameters (see the manual).

The general design procedure using this tool is the following:

1. Select the plant $P(s)$.
2. Set the specifications.
3. Select from block $C(s)$ the type of compensator.
4. In the frequency diagram, drag the gain or the frequency corners of the compensator poles and zeros to modify the shape of the open-loop function $L(s)$. The user can see immediately what specifications are met because LCSD validates the specifications each time the user modifies $L(s)$. In the main window, there are circular indicators associated with each specification. If an indicator is in green color that means its associated specification is met. Otherwise, the indicator is in red color.

8.11.2.4 Examples

Example 1: Loop shaping using the Bode diagram. The objective is to design a compensator for the plant described by

$$G(s) = \frac{2}{s},$$

using the following specifications:
1. The velocity error constant K_v must be 10.
2. $PM \approx 45°$.
3. Sinusoidal inputs in the reference of up to 0.1 rad/s should be reproduced with less than 2% error.
4. Output noise with frequency greater than 100 rad/s should be attenuated at the controller output u_c least by 20%.

Although the system is just an integrator and from root locus analysis it seems the system is easy to control, this is an example of trade-offs to be achieved in different frequency bands, as a consequence of constraints imposed by Bode's sensitivity integral. The first two specifications are classical ones and should be met easily.

As $K_v = kK$ and the static gain $k = 2$, the controller gain must be at least $K = 5$ to fulfill the steady-state specification.

The relationship between the different inputs and outputs of the control loop were analyzed in equation (7.4) in Chap. 7.

Specification 3 means that

$$\left|\frac{E(s)}{R(s)}\right| = \left|\frac{1}{1 + L(s)}\right| \leq 0.02 \rightarrow |L(s)| > 49 \rightarrow |L(s)|_{dB} \geq 33.8 \text{ dB}, \forall \omega \leq 0.1 \text{ rad/s}.$$

Specification 4 is given by

$$\left|\frac{U_c(s)}{N(s)}\right| = \left|\frac{C(s)}{1 + L(s)}\right| = \left|\frac{L(s)}{1 + L(s)}\frac{1}{G(s)}\right| \leq 0.8,$$

and as $|L(s)|$ is much less than 1 in that frequency range, the previous inequality can be approximated by

$$|L(s)| \leq 0.8|G(s)|, \forall \omega \geq 100 \text{ rad/s},$$

meaning that $|L(j\omega)|_{dB}$ must be -2 dB below $|G(j\omega)|_{dB}$ in the cutoff band. Notice that as $|S(j\omega)|$ must be close to one at high frequencies, noise cannot be highly attenuated if also tracking and disturbance rejection specifications have to be fulfilled.

From the analysis of the specifications and the structure of the transfer function, at high frequency $L(s)$ must have at least -20 dB/decade slope. The slope around ω_{gc} must be between -20 and -40 dB/decade, as analyzed in this chapter. At low frequencies, the shape must fulfill the control band specification. The design is therefore challenging.

Selecting the specifications in the tool, inserting the transfer function, and using a phase-lag compensator, the result in Fig. 8.26 can be obtained. The figure also includes the response to sinusoidal references (signals r, y, and e) and output noise (signals n and $u = u_c$). The settings of these inputs are also shown in the figure. It can be checked that specifications 3 and 4 are met. For activating these inputs, the time specifications have to be disabled.

Example 2: Loop shaping using the Nyquist diagram. This example shows a design in the Nyquist diagram already treated in the case of the ILM loop shaping tool for showing the idea of *derivative cliff*. The design is done for the following transfer function:

$$G(s) = \frac{1}{(s + 1)^4}.$$

To design a PID controller to achieve $M_s \leq 1.4$ trying to maximize the integral gain, the ideal PID controller parameters are $K = 0.93$, $T_i = 1.03$ s, and $T_d = 3.09$ s. It can be seen in Fig. 8.27 that the response has oscillations. Figure 8.28 shows a modification without a derivative cliff (upper plot) and a comparison with another solution with a derivative cliff (lower plot).

Example 3: LCSD with root locus. The objective in this example is to control a plant whose transfer function is a double integrator:

$$G(s) = \frac{1}{s^2},$$

using the following specifications:
1. Settling time $t_s \leq 4$ s.
2. Relative damping factor $\zeta \geq 0.36$.
3. Maximum percentage overshoot for a step reference $OS \leq 35\%$.
4. Location of closed-loop dominant poles: $p_1, p_1^* = -1 \pm j1.5$.

Notice that those specifications are redundant, so that the most restrictive ones should be applied (that is done by the tool):

$$OS [\%] = 100 \exp\left(-\frac{\zeta\pi}{\sqrt{1 - \zeta^2}}\right) \leq 35 \rightarrow \zeta \geq 0.32.$$

So, the relative damping factor to be selected is $\zeta \geq 0.36$, which is the one provided by the specifications:

$$t_s = \frac{4}{\zeta\omega_n} = 4 \rightarrow \zeta\omega_n = 1 \rightarrow \omega_n = 2.79 \text{ rad/s}.$$

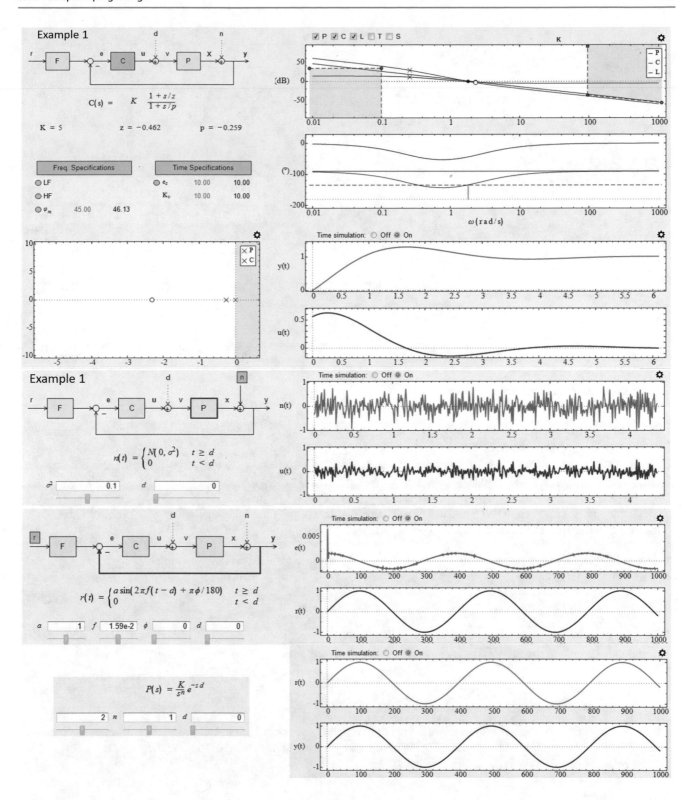

Fig. 8.26 Configuration of LCSD tool for example 1

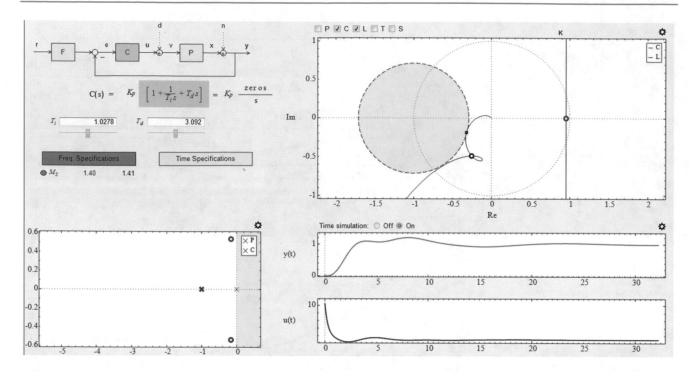

Fig. 8.27 Configuration of LCSD tool for example 2

The location of the dominant closed-loop roots is $p_1, p_1^* = -\zeta\omega_n \pm \omega_n\sqrt{1-\zeta^2} = -1 \pm j2.59$, so that the selected ones are those given by the specifications.

From root locus analysis, it is easy to see that a phase-lead network is a simple way to achieve the specifications, although the user is encouraged to test other controllers. Figure 8.29 shows a possible solution to this example obtained just by placing the zero of the compensator so that the shape of the locus traverses the closed-loop poles given by the specifications. Then the controller gain K is set. Notice that one of the time specifications is not completely met, so that the user can try to find a better solution to this problem.

Example 4: DC motor control using LCSD. In Chap. 5, the transfer function relating the input voltage to the angular position of the DC motor was obtained (5.18) and given by

$$G(s) = \frac{\Theta(s)}{E_a(s)} = \frac{K_t}{s(L_a J_l s^2 + (L_a b_l + R_a J_l)s + R_a b_l + K_t K_b)}.$$

Introducing appropriate values of the parameters [26], the following transfer function is obtained:

$$G(s) = \frac{10}{s(s^2 + 14s + 41)}.$$

In reference [26], this transfer function is used to study proportional control using LCSD. Here, a different design is performed, using the specifications studied in the introductory section of this chapter. The specifications are as follows:
1. Steady-state error to ramp input less than 20%.
2. Relative damping factor $\zeta \geq 0.69$.
3. Rise time $t_r \leq 0.5$ s.
   These are demanding specifications, and the user is encouraged to "play" with the tool to fulfill the specifications:

- The steady-state specification implies $K_v = 5$, and therefore a gain of the controller $K = 20.5$, as the static gain of the plant is $k = 10/41 = 0.2439$.
- The specification of the relative damping factor implies $OS\,[\%] \leq 5\%$ (3.26) and $PM \approx 100\zeta = 69°$ (8.10). From the rise time specification, as PM is close to $70°$, this provides $t_r \approx \frac{\pi}{\omega_{gc}}$ (8.12) $\rightarrow \omega_{gc} = 6.28$ rad/s.

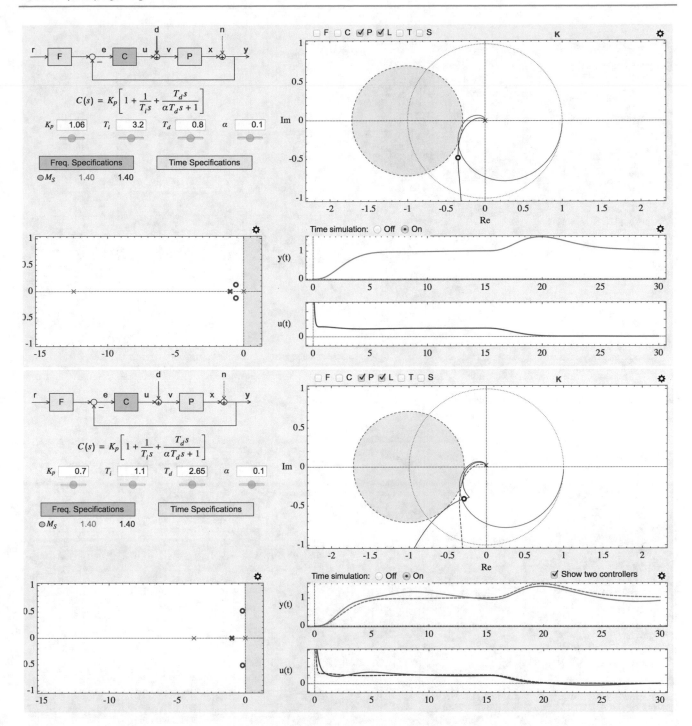

Fig. 8.28 Solution of example 2 without and with derivative cliff

The plant can be introduced using the $\frac{num}{den}$ option. The selection of the controller structure can be done by root locus analysis or by testing with different structures. Here, the design has been done using a PD controller. The gain of the controller is selected to fulfill the steady-state error specification, so that the remaining degree of freedom is T_d, which can be dragged over the Bode or root locus diagrams to try to fulfill the specifications, although with this control structure not all of them can be met. Figure 8.30 shows in a single picture the main configured elements and achieved results. The user has to improve this design using the tool.

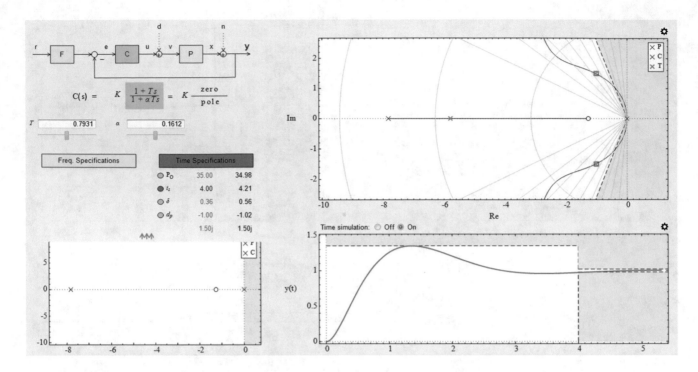

Fig. 8.29 Configuration of LCSD tool for example 3

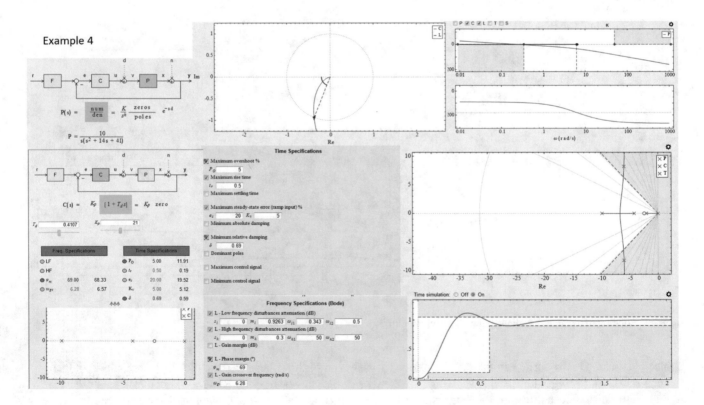

Fig. 8.30 Configuration of LCSD tool for example 4

References

1. Åström, K. J. (2004). *Introduction to control*. Department of Automatic Control, Lund Institute of Technology, Lund University.
2. Nise, N. S. (2015). *Control systems engineering* (7th ed.). Wiley.
3. Shahian, B., & Hassul, M. (1993). *Control system design using MATLAB®*. Prentice Hall.
4. Stein, G. (2003). Respect the unstable. *IEEE Control Systems Magazine, 23*(4), 12–25.
5. Åström, K. J., & Murray, R. M. (2014). *Feedback systems: An introduction for scientists and engineers* (2nd ed.). Princeton University Press.
6. Bavafa-Toosi, Y. (2017). *Introduction to linear control systems*. Academic Press-Elsevier.
7. Åström, K. J., & Hägglund, T. (2006). *Advanced PID control*. ISA - The Instrumentation, Systems and Automation Society.
8. Guzmán, J. L., Åström, K. J., Dormido, S., Hägglund, T., Berenguel, M., & Piguet, Y. (2008). Interactive learning modules for PID control. *IEEE Control Systems Magazine, 28*(5), 118–134.
9. Guzmán, J. L., Åström, K. J., Dormido, S., Hägglund, T., & Piguet, Y. (2006). Interactive learning modules for PID control. *IFAC Proceedings Volumes, 39*(6), 7–12.
10. Dorf, R. C., & Bishop, R. H. (2011). *Modern control systems* (12th ed.). Prentice Hall.
11. Franklin, G. F., Powell, J. D., & Emani-Naeni, A. (2010). *Feedback control of dynamic systems* (6th ed.). Pearson.
12. Golnaraghi, F., & Kuo, B. C. (2017). *Automatic control systems* (10th ed.). McGraw Hill Education.
13. Barrientos, A., Sanz, R., Matía, F., & Gambao, E. (1996). *Control de sistemas continuos. Problemas resueltos (Control of continuous systems. Problems solved)*. McGraw-Hill.
14. Bolzern, P., Scattolini R., & Schiavoni, N. (2009). *Fundamentos de control automático (Fundamentals of automatic control)*. McGraw-Hill.
15. Wilkie, J., Johnson, M., & Katebi, R. (2002). *Control engineering. An introductory course*. Palgrave Macmillan.
16. Ogata, K. (2010). *Modern control engineering* (5th ed.). Prentice Hall.
17. Bechhoefer, J. (2021). *Control theory for physicists*. Cambridge University Press.
18. Dahlin, E. B. (1968). Designing and tuning digital controllers. *Instruments and Control Systems, 41*(6), 77–84.
19. Baker, G. A., & Graves-Morris, P. (1996). *Padé approximants*. Cambridge University Press.
20. Normey-Rico, J. E., Guzmán, J. L., Dormido, S., Berenguel, M., & Camacho, E. F. (2009). An unified approach for DTC design using interactive tools. *Control Engineering Practice, 17*, 1234–1244.
21. Goodwin, G. C., Graebe, S. F., & Salgado, M. E. (2001). *Control system design*. Prentice Hall.
22. D'Azzo, J. J., Houpis, C. H., & Sheldon, S. N. (2003). *Linear control system analysis and design with MATLAB®* (5th ed.). Marcel Dekker Inc.
23. Díaz, J. M., Costa-Castelló, R., & Dormido, S. (2019). Closed loop shaping linear control system design: An interactive teaching/learning approach. *IEEE Control Systems, 39*(5), 58–74.
24. Díaz, J. M., Costa-Castelló, R., Muñoz, R., & Dormido, S. (2017). An interactive and comprehensive software tool to promote active learning in the loop shaping control system design. *IEEE Access, 5*, 10533–10546.
25. Skogestad, S., & Postlethwaite, I. (2005). *Multivariable feedback control, analysis and design* (2nd ed.). Wiley-Interscience.
26. Vargas, H., Marín, L., de la Torre, L., Heradio, R., Díaz, J. M., & Dormido, S. (2020). Evidence-based control engineering education: Evaluating the LCSD simulation tool. *IEEE Access, 8*, 170183–170194.

9.1 Introduction

In the previous chapters, different control design and tuning methods have been explained. In all of them, emphasis was placed on the need to simulate the behavior of the closed-loop control system to verify whether or not the design specifications are met. This chapter makes practical use of the knowledge acquired up to now to implement closed-loop control on the chosen examples from Chaps. 2 and 5. The entire process of modeling and system linearization has been studied in Chap. 2. In Chap. 5, the relationships between the physical parameters of the process and the inputs that define the operating point with the parameters of the transfer function that models the behavior of the process around that operating point were established.

As usual in industrial practice, a linear system model, valid around an operating point, is used to design a controller that meets certain specifications. The linearized models of the systems introduced in Chap. 5 in the form of transfer functions are used in this chapter to design feedback control algorithms. Depending on the model complexity and the specifications, different control approaches are implemented, using either 1-DoF or 2-DoF structures and several implementations of PID-type controllers. All the PID-type controllers in this chapter are implemented using an ideal (non-interacting) structure, although the design can be performed using interacting or parallel algorithms (see Sect. 8.3) and the obtained PID parameters can be transformed into the ideal ones using Table 8.5.

All the interactive tools used in this chapter include the Session menu with the Reset option, which forces the tool to return to the default settings, as well as the possibility of saving (Save session) and restoring Load session sessions. There is also the option to Pause simulation.

9.2 The Tank Level Control

9.2.1 Interactive Tool: tank_level_control

9.2.1.1 Concepts Analyzed in the Card and Learning Outcomes
- PI controller tuning based on a first-order transfer function valid for the operation around an operating point.
- Control of a nonlinear dynamical system about the nominal operating point (that from which the linearized model has been obtained).
- Effect of changing the operating point in the closed-loop response.
- Simulation of the process in closed loop using the nonlinear model as simulation model and the linearized one for control design purposes. Analysis of the control signal.

9.2.1.2 Summary of Fundamental Theory
This card allows the user to analyze a practical example of automatic control of the level of fluid inside a tank, which model has been developed in Sect. 2.2 (interactive tool tank_level_lin) and transfer function in Sect. 5.2 (interactive tool tank_level_tf). All the required concepts have been studied in previous chapters, so that this card will focus on the practical aspects of analysis, synthesis and simulation using interactive tools.

The transfer function for the tank level system from a linearization process was obtained in equation (5.3). As it is a first-order transfer function, PI design based on pole cancellation is a control design method suitable for this kind of description,

© Springer Nature Switzerland AG 2023

J. L. Guzmán et al., *Automatic Control with Interactive Tools*,

https://doi.org/10.1007/978-3-031-09920-5_9

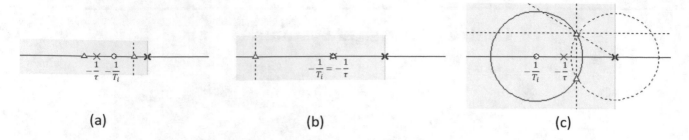

Fig. 9.1 Root locus depending on the relative position of the zero of the controller: **a** $T_i > \tau$, **b** $T_i = \tau$ (pole cancellation), **c** $T_i < \tau$

so that application of the control law obtained in Sect. 8.6 provides a suitable set of control parameters which can be obtained by imposing a desired closed-loop time constant. Notice that the model can also be obtained by analyzing the open-loop step response of the system (reaction curve method). Summarizing the procedure, as the process can be represented by a first-order transfer function, selecting the integral time of the controller equal to the time constant of the system $T_i = \tau$, a pole-zero cancellation is carried out, so that the closed loop has unit static gain (zero error to reference step input) and the time constant of the closed loop is selected to be $\tau_{cl} = \varsigma\tau$, typically with $0 < \varsigma < 1$, a parameter to modulate the speed of response of the closed loop in relation to the open loop one. Notice that the closed-loop time constant cannot be arbitrarily small in practice, as control signals can saturate the actuators:

$$L(s) = C(s)G(s) = \frac{K}{T_i}\frac{(T_i s + 1)}{s}\frac{k}{(\tau s + 1)} = \frac{Kk}{T_i s},$$

$$G_{cl}(s) = \frac{1}{\frac{\tau}{Kk}s + 1} \rightarrow \tau_{cl} = \frac{\tau}{Kk} \rightarrow K = \frac{\tau}{\tau_{cl}k} = \frac{1}{\varsigma k}. \tag{9.1}$$

The obtained controller is valid for the closed-loop operation around the particular operating point. The user can experiment with the tool explained in the next paragraphs, so that analysis of the modification of the closed-loop response when the operating point is changed due to the modification of the setpoint can be performed, as well as repeating the design when the operating point changes.

Also, other possible PI designs can be carried out and implemented in the tool, for instance using root locus analysis (interactive tool root_locus in card 7.2), as shown in Fig. 9.1, or frequency-design methods, as those treated in the Homework section.

Table 2.1 contains references on how this process is controlled in different textbooks as further reading. As an example, Problem 1.16 (pp. 64) of [1] explains the elements of a liquid tank system and main control objectives.

9.2.1.3 References Related to this Concept

- [2] Åström, K. J., & Murray, R. M. (2014). *Feedback systems: An introduction for scientists and engineers* (2nd ed.). Princeton University Press, ISBN: 9780691193984. Exercise 4.2, page 4–33.
- [3] Golnaraghi, F., & Kuo, B. C. (2017). *Automatic control systems.* (10th ed.). McGraw Hill Education, ISBN: 978-1-25-964384-2.
- [4] Johansson, K. H. (2000). The quadruple-tank process: A multivariable laboratory process with an adjustable zero. *IEEE Transactions on Control Systems Technology, 8*(3), 456–465.
- [5] Ogata, K. (2010). *Modern control engineering* (5th ed.). Prentice Hall, ISBN: 978-0-13-615673-4. Chapter 2, Sect. 7, pages 43–45; Chap. 4, Sect. 2, pages 101–106.

Application   Interactive tool: tank_level_control

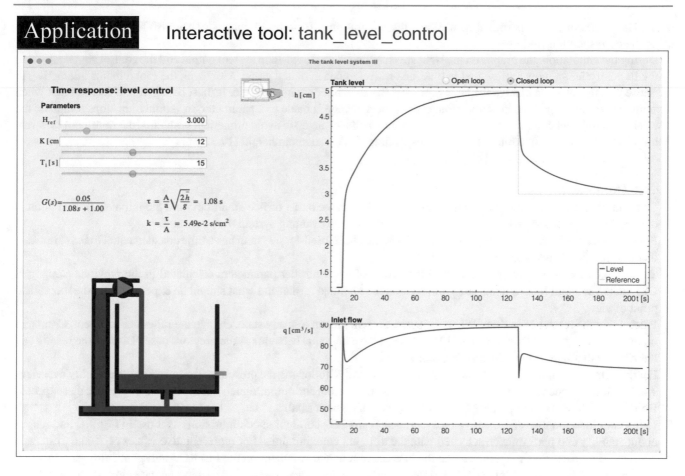

Tank Level System: Closed-Loop Control

This card deals with PI control of tank level. In this case, the closed-loop response of the controlled system is shown trying to keep interactivity features as much as possible.

The default window is the same as the one discussed in Sect. 5.2 (interactive tool tank_level_tf), being recommended that the reader analyze this card beforehand. The only new feature is that the upper-right side has two radio buttons that allow the user to activate the Open loop (default) analysis or Closed loop simulation.

The open-loop configuration window allows the user to define the system structure (initial equilibrium point $(\overline{q}, \overline{h})$, tank section A, and discharge hole area a) and to obtain the linearized model in the form of a first-order transfer function (whose static gain and time constant are related to the defined operating point and the physical parameters of the system). A thick green line appears in the tank wall, indicating the selected operating point, although in Open loop it is not an interactive element.

The graphics on the right side of the tool represent the time evolution of the **Tank level** in open loop (both of the nonlinear and the linearized model) and the **Inlet flow**. Placing the mouse pointer over a certain point of curves, a label that shows the properties associated with that point (time t and output y or input u) is displayed. In the Open-loop mode, the lower-right graphic of the tool represents a step signal, configurable through two horizontal lines and one vertical dashed black line, as in card 5.2.

Once the system is configured, if the Closed loop option is selected, a dynamic simulation of the closed-loop process (the nonlinear differential equation is used for simulation purposes) controlled with a PI is activated. The initial parameters of the PI controller have been obtained for an arbitrary configuration. The user can replace them with adequate ones obtained, for instance, by the pole-cancellation method (Sect. 8.6). In the **Parameters** area, the reference value H_{ref} can be modified (via a textbox or a slider) or by vertically dragging the horizontal green segment on the process diagram indicating the operating point. The values of the proportional gain K and the integral time T_i of the PI controller can also be interactively changed using the corresponding sliders.

During the simulation, the graphs on the right are changing. The closed-loop system output and the control signal provided by the PI controller are now represented in blue, and the reference in dotted green. Moreover, the tool remains interactive, in the sense that the user can change the tank geometry, the controller parameters, etc. It must be taken into account that when running the closed-loop dynamic simulation, every time a change is made to a parameter, the simulation stops to update the new value entered and continue the simulation according to the change. To avoid numerical problems, the minimum flow rate is limited to 0.01 and the maximum to that corresponding to the maximum height (17.32 cm).

9.2.1.4 Homework

1. Select an initial configuration where the tank area is $18\,\text{cm}^2$, the outlet orifice area $1\,\text{cm}^2$ and the initial operating point is defined by $\overline{h} = 8$ cm. Analyze the value of $\overline{q}$, the gain and the resulting system time constant.
2. For the above configuration, design a PI controller so that the closed-loop system has a time constant equal to 0.8τ. Indicate its proportional gain K and integral time T_i.
3. Select the Closed loop option and introduce the values of the controller parameters calculated in the previous paragraph. Change the setpoint from 8 to 8.5 cm. Comment on the obtained results and what should be expected from the theoretical point of view.
4. Place again the reference in 8 cm. When the system reaches again a steady state, change the reference to 12 cm. Comment on the obtained results. Does the closed-loop system have the same behavior as the previous case? Discuss the results and the reasons that may lead to differences, if any.
5. For the default configuration, determine the values of T_i and K following the procedure summarized in the theory overview. Analyze the influence of ς in the closed-loop response (both input and output signals). Is the obtained response that expected from the theoretical development? Justify the answer using both small and large changes in the setpoint.
6. Using root locus analysis (interactive tool root_locus, card 7.2), build three configurations as those in Fig. 9.1, and reason about stability and performance achieved with the different configurations. For representative values of K and T_i in each configuration, perform simulations using the interactive tool and comment on the results. Analyze stability as a function of K using the Nyquist stability criterion (interactive tool Nyquist_criterion, card 7.3). Compute the gain and phase margins of the closed-loop system (interactive tool stability_margins, card 7.4) and try to relate them to the root locus and observed time response.
7. Repeat the previous exercise by selecting a controller structure with a pole and a zero (phase-lead or phase-lag controller). Using the interactive tool f_design_lead (card 8.9) or f_design_lag (card 8.8), try to design a controller to obtain the same closed-loop responses as those in the previous exercises.
8. Design a PI controller using the interactive tool LCSD (card 8.11.2) selecting reasonable closed-loop specifications. Test the obtained parameters in the tool and analyze the results of the simulations.

9.3   Variable Section Tank Level Control

9.3.1   Interactive Tool: spherical_tank_level_control

9.3.1.1 Concepts Analyzed in the Card and Learning Outcomes
- PI controllers tuning based on the step response around an operating point.
- Control of a nonlinear dynamical system around an operating point.
- Effect of changing the operating point in the closed-loop response.
- Closed-loop control and simulation of the spherical section tank level.

9.3.1.2 Summary of Fundamental Theory
This case is very similar to the one analyzed in the previous card. The main difference is that the nonlinearity of the process is more accentuated, what can be observed when performing simulations. The same control approach can be used (PI by pole cancellation, not repeated here again), although in this case, the derivative action is also available. Notice that this system is used in [6] as an example to design gain-scheduling controllers, which are controllers which parameters are adapted in open

loop when it is known how the dynamics of a process change with the operating conditions, as in this example. It is then possible to change the parameters of the controller by monitoring the operating conditions. As in this case τ and k depend on the level $\bar{h}$, if this working level is known, then the parameters of the model and thus the parameters of the controller can be changed accordingly. As pointed out in [6], gain scheduling is a nonlinear feedback of special type, it has a linear controller whose parameters are changed as a function of operating conditions in a preprogrammed way. This kind of control approach is out of the scope of this book.

Many control algorithms can be used to design a PID controller for this process. Notice that a PID controller has two zeros and one integrator. As the linearized system is first-order, the closed loop will not be strictly causal, unless a filter of the derivative action is used (8.32). Anyway, at the design level, the ideal transfer function can be considered to design by pole placement, pole cancellation, root locus, or other techniques. As an example to extend the case treated in the previous card, let's consider an interacting PID as treated in Sect. 8.3 in series with the plant:

$$L(s) = C(s)G(s) = \frac{K'}{T_i'} \frac{(T_d's + 1)(T_i's + 1)}{s} \frac{k}{\tau s + 1}. \tag{9.2}$$

If $T_i' = \tau$, from root locus analysis, the closed-loop system will have a pole and a zero, both in the real axis. The closed-loop pole is to the right of the zero and is therefore more dominant. A reference filter could also be used (not implemented in the interactive tool explained in this section). The closed-loop transfer function is therefore:

$$G_{cl}(s) = \frac{T_d's + 1}{\frac{\tau + K'kT_d'}{K'k}s + 1} \rightarrow \tau_{cl} = \frac{\tau + K'kT_d'}{K'k}. \tag{9.3}$$

As pointed out in Sect. 8.3, to transform the obtained parameters of the PID controller to the non-interacting structure, it is required that $T_i \geq 4T_d$. If it is selected $\tau_{cl} = \varsigma_1\tau$ and $T_d' = \varsigma_2\tau$, calling $\varsigma = \varsigma_1 - \varsigma_2$ such that $1 > \varsigma_1 > \varsigma_2 > 0$, the same control law as such obtained in the previous section is achieved:

$$K' = \frac{\tau}{k(\tau_{cl} - T_d')} = \frac{1}{\varsigma k}. \tag{9.4}$$

So, in this case, the closed-loop time constant is modulated again through ς, which represents the distance between the closed-loop pole and zero. The conversion of the obtained control parameters (K', T_i', T_d') to the non-interacting (ideal) ones is done using the formulas in Table 8.5:

$$K = K'\frac{T_i' + T_d'}{T_i'}, \quad T_i = T_i' + T_d', T_i \geq 4T_d, \quad T_d = \frac{T_i'T_d'}{T_i' + T_d'}. \tag{9.5}$$

As in the previous card, other possible PID designs can be carried out and implemented in the tool, for instance using root locus analysis (interactive tool root_locus, card 7.2, Fig. 9.1).

9.3.1.3 Reference Related to this Concept

- [7] Tavakolpour-Saleh, A., Setoodeh, A., & Ansari, E. (2016). Iterative learning control of two coupled nonlinear spherical tanks, *International Journal of Mechanical and Mechatronics Engineering, World Academy of Science, Engineering and Technology, 10*(11), 1862–1869.

Application — Interactive tool: spherical_tank_leavel_control

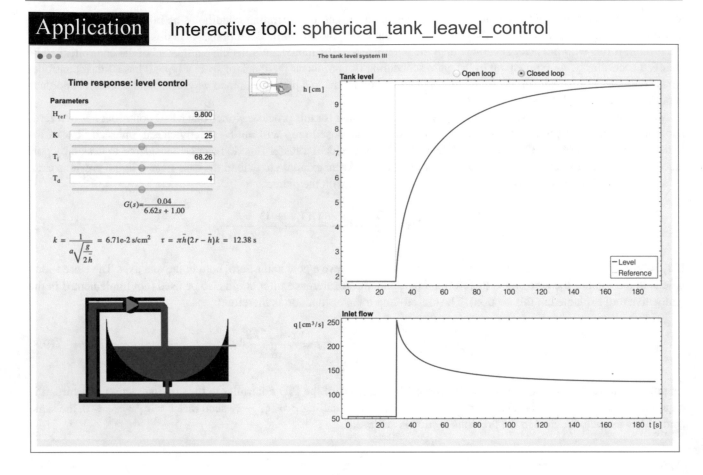

Variable Section Tank Level System: Closed-Loop Control

> In the application described in the interactive tool it is possible to simulate the closed-loop control of the spherical tank level system under PID control.

The layout of this tool is similar to the previous one. By default, the Open loop radio button is active, so that the parameters of the process (sphere radius and outlet hole section) can be fixed through sliders or textboxes, while also the operating point $(\overline{q}, \overline{h})$ can be selected (see Sect. 5.3). The plots on the right-hand side of the tool help to visualize the differences between the nonlinear and linear models, which depend on the amplitude of the selected step in the inlet flow, being in this case more appreciable than in the previous case.

When the Closed loop radio button is selected (once the open-loop configuration has been defined), the closed-loop system using as level reference $H_{ref} = 8$ is simulated using a PID controller which default parameters are $K = 25$, $T_i = 68.26$ and $T_d = 4$ (in adequate units). The user can freely modify these parameters by acting on the associated sliders or textboxes. Moreover, the reference value can be modified (via a textbox or a slider) in the **Parameters** area, or by vertically sliding the horizontal green segment on the process diagram indicating the operating point on the tank.

During the simulation (which is done using the nonlinear model), the graphs on the right are changing and the tool remains interactive (see more details in the previous card).

9.3.1.4 Homework

1. Select an initial configuration where the sphere radius is 18 cm, the outlet hole section is 1.0 cm² and the initial operating point is defined by $\overline{h} = 8$ cm. Analyze the value of $\overline{q}$, the gain and the resulting system time constant.

2. For the above configuration, design a PID controller following the approach summarized in the theory overview. Analyze the influence of ς in the closed-loop response (both input and output signals). Is the obtained response that expected from the theoretical development? Justify the answer using both small and large changes in the setpoint. Analyze the stability using root locus analysis (interactive tool root_locus, card 7.2) and Nyquist stability criterion (interactive tool Nyquist_criterion, card 7.3). Compute the gain and phase margins of the closed-loop system (interactive tool stability_margins, card 7.4) and try to relate them to the root locus and observed time response.
3. Design PID controller using the interactive tool LCSD (card 8.11.2) selecting reasonable closed-loop specifications. Test the obtained parameters using the structure of an ideal PID controller in the tool.

9.4 Ball & Beam Control

9.4.1 Interactive Tool: ball_and_beam_control

9.4.1.1 Concepts Analyzed in the Card and Learning Outcomes
- System type and steady-state errors.
- Stabilization using feedback.
- Setpoint tracking using a PD controller with and without reference filter.
- Disturbance rejection using a PD controller.
- Setpoint tracking and disturbance rejection using a 2-DoF control structure with PID control and reference filter.
- Stability analysis using root locus, Nyquist criterion and stability margins.
- Other control design approaches.

9.4.1.2 Summary of Fundamental Theory
The transfer function of this process has been obtained in equation (5.9) and it is given by a double integrator:

$$G(s) = \frac{X(s)}{\Theta(s)} = \frac{mg}{\left(\frac{I_b}{r^2} + m\right)} \frac{1}{s^2} = \frac{g}{1 + \frac{2}{5}\left(\frac{R}{r}\right)^2} \frac{1}{s^2} = \frac{k}{s^2}, \tag{9.6}$$

where $X(s)$ and $\Theta(s)$ represent the Laplace transform with zero initial conditions of the translational position x and the angle of the beam shaft θ (deviation variables), g is the acceleration of gravity at sea level, R is the radius of the ball and r is the rolling or effective radius of the ball (Fig. 2.7).

If the steady state is firstly analyzed, as the process is type-2 (double integrator), in theory there will be zero steady-state error when step changes in the reference are introduced (notice that ramp changes make no sense in this process). Therefore, a P or PD controller should be enough for controlling closed-loop reference tracking response of this process (but not for load disturbance rejection, as commented in the last remark of Sect. 8.2). From root locus analysis (Fig. 9.2), it is easy to see that using a P controller provides a critically stable closed-loop response (sinusoidal one), so that it is not an adequate controller election.

If an ideal PD controller is implemented, the closed-loop transfer function is given by:

$$G_{cl}(s) = \frac{X(s)}{X_r(s)} = \frac{K(T_d s + 1)\frac{k}{s^2}}{1 + K(T_d s + 1)\frac{k}{s^2}} = \frac{Kk(T_d s + 1)}{s^2 + KT_d s + Kk}. \tag{9.7}$$

A second-order closed-loop transfer function is obtained, so that pole-placement design can be applied based, for instance, in maximum overshoot and a characteristic transient time (rise, peak or settling time) providing the desired relative damping factor ζ and natural frequency ω_n, so that $K = \omega_n^2/k$ and $T_d = 2k\zeta/\omega_n$. Notice that the steady-state error to a step input in the reference is zero, as the static gain of the closed-loop transfer function is 1 (making $s = 0$ in $G_{cl}(s)$ as a consequence of the application of the final value theorem of the Laplace transform). The open-loop zero $(T_d s + 1)$ is also in the closed-loop transfer function, so that a reference filter given by:

$$F(s) = \frac{1}{T_d s + 1}, \tag{9.8}$$

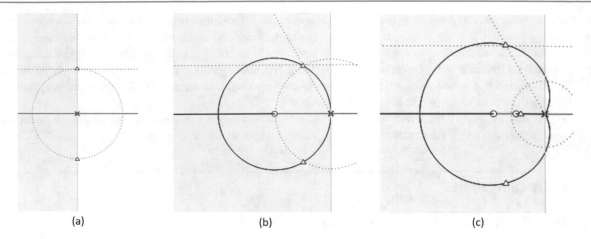

Fig. 9.2 Root locus: **a** P controller, **b** PD controller, **c** PID controller

has to be used to ensure that specifications imposed through the pole-placement technique are fulfilled.

From the load disturbance point of view (in this case the disturbance represents a step in the beam angle), the transfer function relating the disturbance with the output (remember the *Gang of Six* in Chap. 7) when a PD controller is used is given by:

$$G_d(s) = \frac{X(s)}{D(s)} = \frac{G(s)}{1 + C(s)G(s)} = \frac{k}{s^2 + KkT_ds + Kk}, \tag{9.9}$$

where it has been considered that the reference is zero. In this case the static gain is $1/K$, so that, with high values of the controller gain the steady-state error to a step change in the disturbance will be reduced as expected from theory (with the logical limitations imposed by the actuators saturation limits), but it will never be zero. Therefore, in order to reach zero steady-state error to step load disturbance, a PID controller should be used as explained in Chap. 8. Computing again $G_d(s)$ in equation (9.9) with a PID control structure, it results in:

$$G_d(s) = \frac{G(s)}{1 + C(s)G(s)} = \frac{\frac{k}{s^2}}{1 + K\left(1 + \frac{1}{T_is} + T_ds\right)\frac{k}{s^2}} = \frac{ks}{s^3 + KkT_ds^2 + Kks + \frac{Kk}{T_i}}. \tag{9.10}$$

In this case, if the final value theorem of the Laplace transform is applied considering a unit step load disturbance ($D(s) = 1/s$):

$$\lim_{t \to \infty} x(t) = \lim_{s \to 0} sX(s) = \lim_{s \to 0} sG_d(s)D(s) = \lim_{s \to 0} sG_d(s)\frac{1}{s} = 0,$$

as the numerator of $G_d(s)$ contains a differentiator (term s) that comes from the integrator of the PID controller. So, using a PID controller, after a step load disturbance the output of the process will reach again the initial value after a transient that will depend on the values of the parameters of the PID controller. Regarding reference tracking, the closed-loop transfer function is given by:

$$G_{cl}(s) = \frac{C(s)G(s)}{1 + C(s)G(s)} = \frac{\frac{Kk}{T_i}(T_iT_ds^2 + T_is + 1)}{s^3 + KkT_ds^2 + Kks + \frac{Kk}{T_i}}. \tag{9.11}$$

A third-order characteristic polynomial is obtained this way. In this case, a common design method is to impose closed-loop dynamics dictated by the characteristic polynomial $(\tau_{cl} + 1)^3 = (\lambda s + 1)^3 = s^3 + \frac{3}{\lambda}s^2 + \frac{3}{\lambda^2}s + \frac{1}{\lambda^3}$. Equating this polynomial to the characteristic polynomial of the closed loop (denominator of the equation (9.11)), the following expression is obtained:

$$K = \frac{3}{k\lambda^2}, \quad T_i = 3\lambda, \quad T_d = \lambda.$$

Notice that in this case, the reference filter is given by:

$$F(s) = \frac{1}{T_i T_d s^2 + T_i s + 1} \ . \tag{9.12}$$

The previous equations have been implemented in the tool shown in this card. From root locus analysis (Fig. 9.2), it can be seen that when using a PID controller, there is a minimum controller gain from which the closed-loop system can be stabilized.

Many other different design methods can be used to control this plant, which can be found in the references given in Table 2.3.

9.4.1.3 References Related to this Concept

- [8] Åström, K. J., & Hägglund, T. (2006). *Advanced PID control*. ISA - The Instrumentation, Systems and Automation Society, ISBN: 978-15-561-7942-6.
- [9] Hirsch, R. (1998). *EDUMECH - Mechatronic Instructional Systems - Ball on Beam System*. Shandor Motion Systems. https://cutt.ly/KUogfV1.
- [10] Shahian, B., & Hassul, M. (1993). *Control system design using MATLAB®*. Prentice Hall, ISBN: 0-13-174061-X.
- [11] Wellstead, P. E., Chrimes, V., Fletcher, P. R., Moody, R., & Robins, A. J. (1989). Ball and beam control experiment. *The International Journal of Electrical Engineering & Education, 16*, 21–39.

Application   Interactive tool: ball_and_beam_control

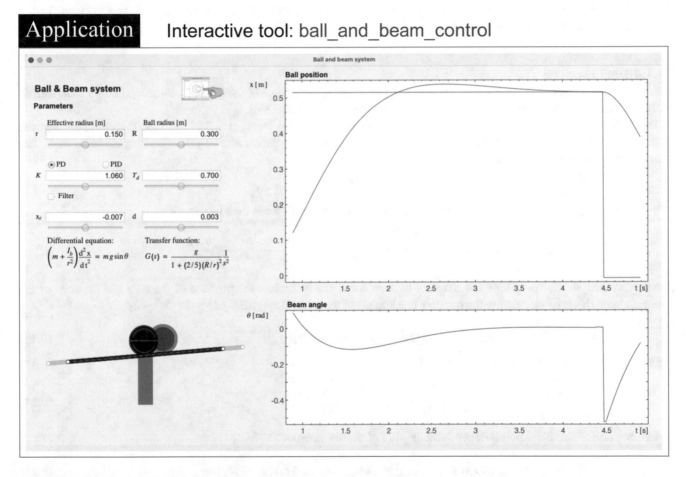

Ball & Beam System: Closed-Loop Control

In the application described in the interactive tool it is possible to simulate the closed-loop control of the ball & beam system. Step changes in both the ball reference position and a load disturbance can be introduced to analyze the closed-loop response, both in terms of transient response and steady-state error. 1-DoF and 2-DoF PD and PID control structures can be selected, without or with a reference filter to cancel the closed-loop zeros.

The **Parameters** area of the tool include the possibility of setting (through sliders and textboxes) the ball rolling r and radius R. From Fig. 2.7 it can be seen that if the channel width L is fixed, a relationship between R and r can be stated through the area of the isosceles triangle (this is taken into account in this tool to constrain the parameters into intervals to avoid losing physical sense). It is also possible to select ideal PD or PID control structures (and representative parameters K, T_i and T_d), to include the reference filter in equations (9.8) and (9.12) (using a checkbox) and to modify through textboxes the ball position reference (x_r) and load disturbance value d (step shape). Notice that the value of the disturbance is constrained in the range $[-\frac{\pi}{4}, \frac{\pi}{4}]$, as also is the beam angle (validity neighbourhood in the linearization). A symbolic representation of the nonlinear differential equation used to simulate the system and the transfer function representing the linearized dynamics is included above the process diagram, where its characteristic parameters can also be changed by interacting over it: The ball radius can be changed by clicking and dragging on the green circle representing its contour; the reference position can also be changed by selecting the red circle with the mouse and dragging it to the desired position on the beam; the beam angle can be disturbed by clicking over the beam, which length can also be changed by dragging the cyan segment. These changes can also be performed by pausing the simulation in the Options menu.

As usual, the right part of the tool represents the time evolution of the ball position x in black and the beam angle θ. The nonlinear differential equation is used for simulation purposes, so that it is integrated using a numerical approximation that uses an integration step of 0.01 to achieve a tradeoff between visualization and accuracy in response. The reference position x_r is shown in red color. The time axis is automatically updated to show a realistic evolution of the variables. Constraints on the variables have been implemented, so that when reaching such constraints the simulation will show horizontal lines representing the saturated signals.

9.4.1.4 Homework

Pre-analysis: This part of the homework helps to analyze some characteristics of the system with several of the tools treated in previous chapters and controllers different from those implemented in the tool.

This example is explained in [12]. In this case, the interactive tools root_locus (card 7.2) and LCSD (card 8.11.2) are used to analyze the closed-loop performance of the system as a function of the parameters of phase-lead/lag compensators. The controller is represented in ZPK format as:

$$C(s) = K_c \frac{s - z_c}{s - p_c},$$

with $p_c < 0$ and $z_c < 0$ as explained in Sect. 8.4. The loop transfer function can thus be written as:

$$L(s) = C(s)G(s) = K_c \frac{s - z_c}{s - p_c} \frac{k}{s^2} = K \frac{s - z_c}{(s - p_c)s^2} = K \frac{b(s)}{a(s)}.$$

One way to analytically study how the shape of the root locus changes when the position of the pole and the zero of the compensator varies is by calculating the possible breakpoints of the locus through $dK/ds = 0$: $(3s^2 - 2p_c s)(s - z_c) - s^2(s - p_c) = 0$, so that:

$$s(2s^2 - (p_c + 3z_c)s + 2p_c z_c. \tag{9.13}$$

The breakpoints are thus $s = 0$ and:

$$s = \frac{1}{4}(p_c + 3z_c) \pm \frac{1}{4}\sqrt{(p_c + 3z_c)^2 - 16p_c z_c}, \tag{9.14}$$

the number of possible breakpoints depends on the sign of the discriminant:

$$D = (p_c + 3z_c)^2 - 16p_c z_c = (p_c - z_c)(p_c - 9z_c). \tag{9.15}$$

Examining the sign of D, five cases can be identified. The case $p_c = z_c$ is not going to be considered as it produces a pole-zero cancellation in the compensator. Notice that from (9.13) the two roots corresponding to breakpoints are always negative, as the term inside the square root is always less than $(p_c + 3z_c)$.

Case 1: $p_c > z_c$. From (9.14) the discriminant $D > 0$ and the two roots are real and negative. It is easy to analyze that these two points do not belong to the root locus located on the real axis and, therefore, they are not valid breakpoints. The root locus on the real axis is between p_c and z_c ($z_c = rp_c$ with $r > 1$). From the coefficient of s^0 in (9.13), the product of the two roots is $p_c z_c = rp_c^2$. If one of these roots is to the right of p_c, then the other root will lie to the left of rp_c (to the left of z_c) and therefore both roots are outside of the segment belonging to the root locus. It follows then that it is adequate to show that one root of (9.13) does not belong to the segment of the root locus on the real axis because then the other root will not belong to the root locus either. If $z_c = rp_c$ is substituted into (9.14) with: $r = 1 + \delta, \delta > 0$ it is obtained:

$$s = p_c + \frac{3}{4}\delta p_c \pm \frac{1}{4}p_c\sqrt{9\delta^2 + 8\delta}. \tag{9.16}$$

The positive square root in (9.16) corresponds to a root that is to the right of p_c, so the other root will be to the left of z_c and both roots are not acceptable as breakpoints. In this case, the only valid break point is $s = 0$. The angles of the asymptotes are 90° and −90° and the centroid is $\eta = (p_c - z_c)/(3 - 1) = (r - 1)(p_c/2) = -\delta(p_c/2) > 0$, which is a point on the positive real axis. Figure 9.3a represents the root locus for this case. As can be seen, the closed-loop system is unstable for all values of the gain K since the two branches of the locus are in the RHP. This is to be expected since the plant, a double integrator, is marginally stable and the compensator is a phase-lag network ($p_c > z_c$) which has a destabilizing effect on the system.

Case 2: $9z_c < p_c < z_c$. The discriminant D in (9.15) is negative, so that the roots of (9.14) are complex conjugates and the only possible breakpoint is $s = 0$, as in the previous case. If $z_c = rp_c$ with $r < 1$, the centroid will be given by $\eta = (1 - r)(p_c/2) < 0$. This indicates that the asymptotes are in the LHP. The system is stable, which would be expected because $z_c > p_c$ corresponds to a phase-lead compensator. Figure 9.3b plots the root locus for this case.

Case 3: $p_c = 9z_c$. The discriminant D in (9.15) is zero, so that the roots of (9.14) are equal located at $s = 3z_c$, which is an acceptable breakpoint, as it is located in the segment of the root locus belonging to the real axis (between z_c and p_c). Moreover, if the breakpoint condition gives two identical roots, the characteristic equation must have 3 equal roots (at $s = 3z_c$). The centroid is $\eta = 4z_c$. Figure 9.3c plots the root locus for this case, which corresponds to a stable closed-loop system for all values of K, what is expected as it uses as phase-lead compensator.

Case 4: $p_c < 9z_c$. The discriminant D in (9.15) is positive ad the roots of (9.14) are also positive. Using a similar reasoning to case 1, it can be shown that both roots correspond to valid breakpoints since both roots fall between z_c and p_c, implying that there are 3 breakpoints. Figure 9.3d depicts the locus of the roots for this case, which corresponds to a stable system for all values of K.

This example shows that sometimes drawing a root locus without fully using all the information can lead to errors. In particular, and as it is necessary in this example, the centroid and the breakpoints must be obtained before drawing it with guarantee. Depending on the nature of this information, the root locus can vary significantly even when the transfer functions of the system are very similar.

The use of the interactive tools root_locus and LCSD with their high interactivity capability to do the drawing is very valuable, and shows that a conjunction of analytical analysis of the problem with an interactive visualization of the root locus is the right strategy to enhance the learning process. Another observation that can be made with this example is that a dominant zero in the LHP will tend to pull the root locus towards the LHP (stabilizing effect), and a dominant pole will tend to push it away (destabilizing effect).

Implemented in the tool

1. For the default configuration of the tool, select a proportional controller by fixing $T_d = 0$ and $K = 2$. Change the reference to $x_r = 0.1$. Check if a sinusoidal time response is obtained as expected from theory (notice that the nonlinear model is used for simulation purposes in a discrete-time implementation). By root locus analysis (interactive tool root_locus, card 7.2) or using the closed-loop characteristic polynomial, explain how increasing K influences the frequency and amplitude of oscillation. Taking into account (9.6), explain how an increase in R influences the amplitude and frequency of oscillations. Does the ball mass m influence the response? Explain the answer. What are the corresponding gain and phase margins? Check them with the interactive tool stability_margins (card 7.4). Analyze also the stability using the Nyquist criterion (interactive tool Nyquist_criterion, card 7.3).
2. For the configuration that appears by default when starting the interactive tool, design a PD controller following the procedure explained in the summary of fundamental theory, using as specifications a 5% percentage overshoot and peak

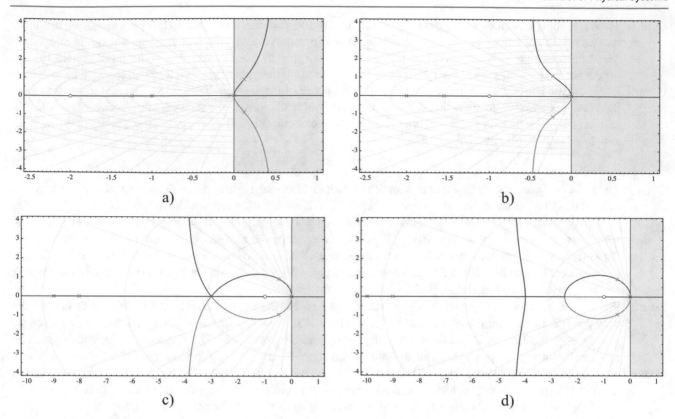

Fig. 9.3 Root locus for the ball & beam system with a phase-lead/lag compensator: **a** $p_c > z_c$, **b** $9z_c < p_c < z_c$, **c** $p_c = 9z_c$, **d** $p_c < 9z_c$

time of 2 s. Include the values of K and T_d in the tool and without including reference filtering (checkbox Filter not selected) perform a change of the reference position to $x_r = 0.1$. Analyze both the transient and steady-state response and the effect of changing the controller parameters K and T_d. Repeat the exercise using different values of the descriptive parameters. What are the corresponding gain and phase margins? Check them with the interactive tool stability_margins (card 7.4). Analyze also the stability using the Nyquist criterion (interactive tool Nyquist_criterion, card 7.3).

3. With the same PD controller of the previous exercise, instead of changing the reference position, introduce a disturbance in the beam angle $d = 0.1$. Does the closed-loop system has zero steady-state error? How this error can be reduced? Comment on the effect of increasing the controller gain K on the transient and steady-state responses. Reason on what should be the time response if the disturbance is included at the plant output instead of at the plant input (use the "Gang of Six" in Chap. 7).

4. Repeat the previous exercises including the reference filter. Comment on the obtained results.

5. Repeat the previous exercises using a PID controller instead of a PD controller. Analyze the main differences encountered both in terms of transient and steady-state responses and compare with what is expected from theory. Also analyze stability with the Nyquist criterion and stability margins.

6. For the configuration selected in question 2, design a lead compensator[1] following the procedure in Sect. 8.9 (interactive tool f_design_lead, card 8.9) so that $PM \approx 100\zeta$ (as analyzed in Chap. 8) and the closed-loop system has the same steady-state error to a parabola input as the PD designed in question 2. Using the root_locus interactive tool (card 7.2), compare this design with the PD controller obtained in question 2 and check if both designs provide the same closed-loop poles. Do both controllers have the same steady-state error to a reference step input? And to a step input disturbance? In the presence of noise, why does a lead compensator have fewer problems than a PD controller?

7. Design a PD and a PID controller using the interactive tool LCSD (card 8.11.2) selecting reasonable closed-loop specifications and implement the obtained parameters in the tool, commenting on the closed-loop simulation results. An example of loop shaping design taking into account the dynamics of the motor turning the shaft can be found in [13], homework E.18.

---

[1] The design of phase-lead compensators for this system can be analyzed in [10], page 470.

9.5   Inverted Pendulum on a Cart Control

9.5.1   Interactive Tool: inverted_pendulum_control

9.5.1.1 Concepts Analyzed in the Card and Learning Outcomes
- Stabilization of a nonlinear system using feedback.
- Different PID control structures: ideal, parallel and interactive.
- 2-DoF control strategies with reference filtering.
- Stability analysis using root locus, Nyquist criterion and stability margins.
- Other control design approaches.

9.5.1.2 Summary of Fundamental Theory
The equations of motion of the inverted pendulum on a cart system (Fig. 2.8) have been analyzed in Sect. 2.5. In Sect. 5.5, the transfer function relating the angle of rotation of the pendulum to the force applied to the cart (deviation variables) was obtained as:

$$\frac{\Theta(s)}{U(s)} = \frac{-1}{Mls^2 - (M+m)g} = \frac{-1}{Ml\left(s + \sqrt{\frac{M+m}{Ml}g}\right)\left(s - \sqrt{\frac{M+m}{Ml}g}\right)} = \frac{\frac{-1}{(M+m)g}}{\frac{Ml}{(M+m)g}s^2 - 1}, \tag{9.17}$$

having one pole on the negative real axis and another on the positive real axis:

$$\text{Stable pole} \quad p_s = -(\sqrt{g(M+m)}/\sqrt{Ml}),$$
$$\text{Unstable pole} \quad p_u = (\sqrt{g(M+m)}/\sqrt{Ml}). \tag{9.18}$$

thus being an open-loop unstable plant so that it must be stabilized through feedback control. In this section, due to the complexity of the system, the proposed control solution is to use of a 2-DoF control structure with an interacting PID as feedback controller and a second-order reference filter to cancel the zeros appearing in closed loop (see Sect. 8.3). If a PD is used, the system will have steady-state error. A load disturbance at the plant input is considered in the force applied to the cart.

Let's call:

$$k = -\frac{1}{(M+m)g}, \quad \tau^2 = \frac{Ml}{(M+m)g}.$$

This is typically a disturbance rejection problem, as the pendulum has to be stabilized around the vertical position. Let's first consider the case of PD control. The transfer function relating the load disturbance to the angle of the pendulum is given by:

$$G_d(s) = \frac{\Theta(s)}{D(s)} = \frac{G(s)}{1 + C(s)G(s)} = \frac{\frac{k}{\tau^2 s^2 - 1}}{1 + \frac{Kk(T_d s + 1)}{\tau^2 s^2 - 1}} = \frac{\frac{k}{\tau^2}}{s^2 + \frac{KkT_d}{\tau^2}s + \frac{(Kk-1)}{\tau^2}}. \tag{9.19}$$

Following a pole-placement approach, from the specifications of maximum overshoot and a characteristic time (rise, peak or settling times) to disturbance rejection, the desired closed-loop relative damping factor ζ_{cl} and natural frequency $\omega_{n_{cl}}$ can be obtained, so that the desired characteristic polynomial of the closed loop is given by $s^2 + 2\zeta_{cl}\omega_{n_{cl}}s + \omega_{n_{cl}}^2$. Equaling the denominator of equation (9.19) to this characteristic polynomial, the following expression is obtained:

$$K = \frac{1 + \tau^2\omega_{n_{cl}}^2}{k}, \quad T_d = \frac{2\zeta_{cl}\omega_{n_{cl}}\tau^2}{1 + \tau^2\omega_{n_{cl}}^2}. \tag{9.20}$$

The static gain of $G_d(s)$ is $k/(Kk - 1)$ and thus the process will have a steady-state error when a step input disturbance is introduced (it is interesting to analyze the physical meaning). Therefore, integral action has to be included in $C(s)$ to remove this steady-state error. As pointed out in [2], page 3–18, a PD control law stabilizes the pendulum, but it does not stabilize the motion of the cart. To do this it is necessary to introduce feedback from cart position and cart velocity, not considered in this simplified example.

To avoid steady-state error, as the plant is a second-order system with an unstable pole and that pole cannot be cancelled due to internal stability reasons, a third-order closed-loop system will be obtained when using PID control. So, as specification for control design, the following closed-loop transfer function is proposed, with static gain equal to one (to fulfill the steady-state specifications) and transient specifications given in terms of three real poles with a time constant $\tau_{cl} = \lambda$:

$$G_{ds}(s) = \frac{1}{(\lambda s + 1)^3} = \frac{\frac{1}{\lambda^3}}{s^3 + \frac{3}{\lambda}s^2 + \frac{3}{\lambda^2}s + \frac{1}{\lambda^3}}. \tag{9.21}$$

An interactive PID structure is used to fulfill the specifications:

$$C(s) = K'\frac{(T_i's + 1)(T_d's + 1)}{T_i's}, \tag{9.22}$$

$$G_d(s) = \frac{\Theta(s)}{D(s)} = \frac{G(s)}{1 + C(s)G(s)} = \frac{\frac{ks}{\tau^2}}{s^3 + \frac{K'kT_d'}{\tau^2}s^2 + \frac{K'k(T_i'+T_d')-T_i'}{T_i'\tau^2}s + \frac{K'k}{T_i'\tau^2}}. \tag{9.23}$$

Equaling the denominators of equations (9.21) and (9.23) a system of three equations is obtained:

$$K'kT_d'\lambda - 3\tau^2 = 0,$$
$$\lambda^2[(K'kT_d' + T_i'(K'k - 1)] - 3\tau^2T_i' = 0,$$
$$K'k\lambda^3 - T_i'\tau^2 = 0.$$

It can be seen that the system has no steady-state error when a step input disturbance is introduced:

$$\lim_{t \to \infty} \theta(t) = \lim_{s \to 0} s\Theta(s) = \lim_{s \to 0} sG_d(s)D(s) = \lim_{s \to 0} \not{s}G_d(s)\frac{1}{\not{s}} = 0,$$

so that in steady state the output of the system returns to the value previous to the introduction of the disturbance. Notice that the differentiator that appears in the numerator of $G_d(s)$ comes from the integrator of the PID controller, this being the reason for obtaining zero steady-state error to a step input disturbance. The values of K', T_i' and T_d' can be obtained and then translated to the ideal (non-interacting) structure as indicated in Table 8.5:

$$K = K'\frac{T_i' + T_d'}{T_i'}, \quad T_i = T_i' + T_d', T_i \geq 4T_d, \quad T_d = \frac{T_i'T_d'}{T_i' + T_d'}. \tag{9.24}$$

In this system, the reference is always $\theta_r = 0$, as the objective is to maintain a rod in the upright position, so it does not make sense to analyze the resulting closed loop for tracking a reference neither including reference filters.

An example of design using root locus can be found in [12] using the LCSD tool analyzed in Sect. 8.11.2, but for reference tracking. In that example, a configuration of the parameters describing the inverted pendulum is selected so that $k = -1$ and $\tau^2 = 1$, providing the transfer function $G(s) = -1/((s - 1)(s + 1))$. The specifications are percentage overshoot $OS \leq 5\%$ and settling time $t_s \leq 1.5$ s. From these specifications, $\zeta \geq 0.69$, $\varphi \leq 45°$, $\sigma \geq 2$, that can be translated into areas of possible closed-loop roots location. It is easy to see in Fig. 9.4 (where the specifications are also shown) that with a proportional controller it is not possible to fulfill the specifications. In this case, the characteristic polynomial of G_d and G_{cl} (for which root locus is usually designed) are the same, the differences in this case are that the numerator of G_d is -1 and that of $G_{cl} = -K$. Moreover, the closed-loop system will always have an unstable pole or poles on the j-axis. It is therefore necessary to change the asymptotes and corresponding centroid to the left. An strategy can be to move the asymptotes so that the centroid (and also the breaking point) is placed at $s = -2$, while maintaining $n - m = 2$. Thus, a phase-lead compensator can selected $C(s) = K\frac{\beta s + 1}{\tau s + 1} = K\frac{\beta s + 1}{\alpha\beta s + 1}$, $\alpha = \tau/\beta < 1$. If the controller zero cancels the stable pole of the plant ($\beta = 1$), the "centre of mass" of the poles moves leftwards and then α can be obtained from the formula of the centroid to be $\alpha = 0.2 = \tau$ (notice that here τ is the time constant of the compensator). K is then selected to place the closed-loop poles in the intersection of the specifications (around $-2 \pm j2$), so that $K = -2.6$, which is the result obtained when solving the problem analytically. Notice that the open-loop stable pole of the plant also appears as closed-loop pole, so that those imposed by the specifications

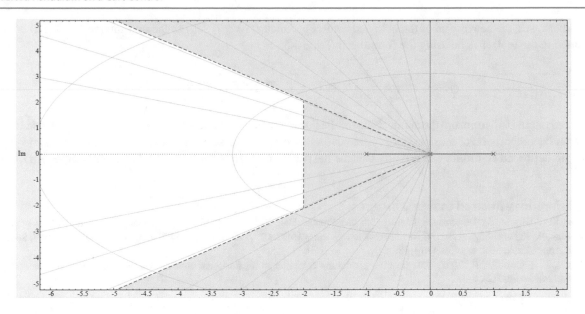

Fig. 9.4 Specifications and locus of the closed-loop poles with a proportional controller

are not the dominant ones and the settling time is higher than expected. Moreover, the system exhibits a large steady-state error (equal to $1/(1 + K)$), as expected because the system is type-0.

Analytical methods in the time domain have been applied here as an example, but also frequency-based design methods could be used. An example of loop shaping design can be found in [13], homework B.18.

Although it has not been implemented in the tool, the design of the feedback controller could be performed by considering disturbances at the plant output (for instance manually perturbing the pendulum position). In this case, by block algebra it can easily be seen that the transfer function relating the output disturbance to the plant output is given by:

$$G_{doy} = \frac{1}{1 + C(s)G(s)}.$$

To ensure the system completely rejects the disturbance at steady state, the controller must incorporate integral action, so that PI or PID controllers can be implemented. If an interacting PID structure is selected, it results in:

$$C(s) = K'\left(1 + \frac{1}{T_i's}\right)(1 + T_d's) = K'\left(\frac{s + \frac{1}{T_i'}}{s}\right)(1 + T_d's).$$

Taking into account the plant model (9.17) and the poles (9.18), the transfer function can be expressed in ZPK form as:

$$G(s) = \frac{\kappa}{(s - p_s)(s - p_u)}$$

with $\kappa = -1/(Ml)$. If $1/T_i' = -p_s$, then:

$$G_{doy} = \frac{1}{1 + \frac{K'\kappa(T_d's+1)}{s(s-p_u)}} = \frac{s(s - p_u)}{s^2 + (K'\kappa T_d')s + K'\kappa}$$

where the controller parameters K' (in this case negative) and T_d' can be obtained by imposing a closed-loop second-order polynomial obtained from specifications of, for instance, maximum overshoot and settling time (pole placement). Notice that the differentiator in the numerator ensures zero steady-state error to output step disturbance and the system will exhibit an inverse response as the open-loop unstable pole appears as an RHP zero (it is also interesting to interpret the physical meaning).

As in other cases, the conversion of the obtained control parameters (K', T_i', T_d') to the non-interacting (ideal) ones (those implemented in the tool) is done using the formulas in Table 8.5:

$$K = K' \frac{T_i' + T_d'}{T_i'}, \quad T_i = T_i' + T_d', T_i \geq 4T_d, \quad T_d = \frac{T_i' T_d'}{T_i' + T_d'}. \tag{9.25}$$

As pointed out in [14], to ensure the applicability of the linearized model, it is assumed that the nonlinear system is operated within a small neighborhood of the operating system. Nevertheless, in the case of the inverted pendulum on a cart system, the linearized model-based controller can actually balance the pendulum even when an initial deviation is large ($\pm\pi/3$).

9.5.1.3 References Related to this Concept

- [5] Ogata, K. (2010). *Modern control engineering* (5th ed.). Prentice Hall, ISBN: 978-0-13-615673-4. Chapter 3, Example 3–5, pages 68–72.
- [10] Shahian, B., & Hassul, M. (1993). *Control system design using MATLAB®*. Prentice Hall, ISBN: 0-13-174061-X. Chapter 7, Sect. 9, pages 217–219; Appendix A, Sect. 4, pages 476–488.
- [15] Aracil, J., & Gordillo, F. (2005). The inverted pendulum: A challenge for nonlinear control. *Revista Iberoamericana de Automática e Informática Industrial, 2*(2), 8–19.

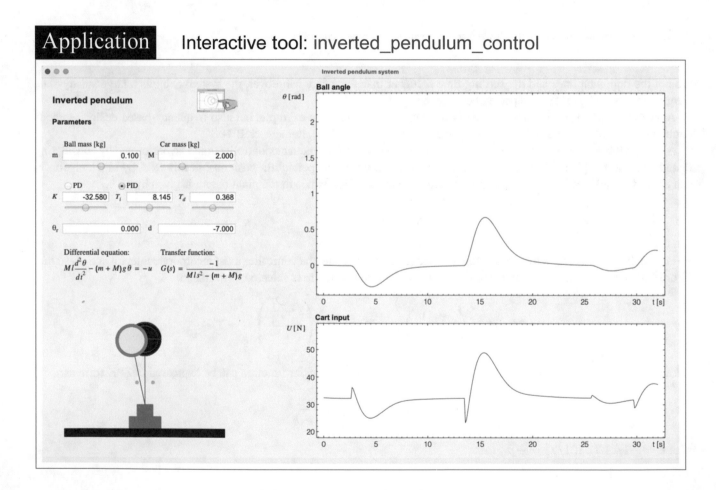

Inverted Pendulum on a Cart: Closed-Loop Control

In the application described in the interactive tool it is possible to simulate the closed-loop control of the inverted pendulum on a cart system. In this case, only the dynamics relating the force applied to the cart to the ball angle are included in the analysis. PID control is used to stabilize the system and achieve zero steady-state error to load disturbance.

The **Parameters** area of the tool includes the possibility of changing (through sliders and textboxes) the ball mass m and cart mass M, as well as selecting non-interacting (ideal) PD or PID control structures, setting their representative parameters K, T_i and T_d. Default setting of parameters produces a stable closed-loop response. When switching to PD or PID through the associated checkbox, these default settings are selected. The users can modify these values, being aware that the system is prone to instability. Notice that the pendulum length $l = 2$ m is fixed to avoid problems in the graphical representation and ensuring a realistic simulation. The acceleration of gravity is $g = 9.8$ m/s^2. In this tool, the reference position is fixed and cannot be changed (vertical position), because it is oriented to the disturbance rejection problem, not requiring the implementation of a reference filter, and a load disturbance in the cart input can be introduced through a textbox and a slider (limited to ± 20). A symbolic representation of the differential equation used to simulate the system and the transfer function representing the linearized dynamics is included above the process diagram, which in this case does not present interactive features. Notice that the cart movement has not been implemented due to space limits in the interface, so that only the movement of the pendulum is animated in this case.

The right area of the tool represents the time evolution of the ball angle θ in black and the cart input u. The time axis is automatically updated to show a realistic evolution of the variables. The nonlinear differential equation is used for simulation purposes, so that it is integrated using a numerical approximation that uses an integration step of 0.01 to achieve a tradeoff between visualization and accuracy in response.

9.5.1.4 Homework

Pre-analysis: This part of the homework helps to analyze different characteristics of the system with some of the tools treated in previous chapters. Although the problems is typically a disturbance rejection one, many textbooks also analyze the system from the point of view of the reference, thus using classical analysis techniques that apply to G_{cl}. In this section, some insights on the problem can be obtained from this approach:

1. Select $M = 1$ kg and $m = 0.64$ kg. Suppose that the system is to be controlled with a proportional controller with gain K:
   a. Using root locus analysis (interactive tool root_locus, card 7.2), analyze the stability and qualitative behaviour of the closed-loop system for positive and negative values of K. Reason on the obtained results and on the physical meaning.
   b. Repeat the previous exercise, but using in this case the Nyquist stability criterion (interactive tool Nyquist_criterion, card 7.3). An example of this exercise can be found in [16], example 3.13.
   c. For a value of $K = -32$, compute the gain and phase margins of the closed-loop system (interactive tool stability_margins, card 7.4). Reason whether the application of these relative stability margins makes sense in this system.
2. Taking into account the analysis in exercise 1.c, and for the same value of $K = -32$, explain whether it is possible to design a phase-lead or a phase-lag compensator providing a phase margin of $PM = 25°$. Design the compensator (interactive tool f_design_lead, card 8.9). Analyze the root locus of the resulting system (interactive tool root_locus, card 7.2) and comment on the obtained response in terms of transient response and steady state (and related physical meaning).
3. Design a controller to reject load disturbances (the reference is $\theta_r = 0$) in the form of a unit step at the plant input, so that the pendulum returns to the vertical position, with dynamics dictated by a relative damping coefficient of 0.6 and a settling time equal to 2.2 s (notice that less-dominant additional poles should have to be added to the given specifications). What would the closed-loop output be with such a controller if the disturbance enters the feedback loop at the plant output?

Implemented in the tool

1. For a configuration of the process (different values of the ball mass m and car mass M, the length is fixed to $l = 2$), obtain the parameters of the interacting PID explained in the theory section and convert them to the ideal (non-interacting) formulation, as this is the structure implemented in the tool. Introduce those values in the interactive tool and perform a unit step in the disturbance $d = 1$. Analyze both the transient and steady-state response and the effect of changing

the controller parameters K and T_d. Analyze the stability of the closed-loop system using the Nyquist stability criterion (interactive tool Nyquist_criterion, card 7.3).

2. Repeat the exercise using only a PD controller. Is the steady-state error zero in this case? Analyze the stability of the closed-loop system using the Nyquist stability criterion (interactive tool Nyquist_criterion, an example can be found in [2], Example 9.6).

9.6 DC Motor Control

9.6.1 Interactive Tool: DC_motor_control

9.6.1.1 Concepts Analyzed in the Card and Learning Outcomes
- System type and steady-state errors.
- Setpoint tracking using a PD controller with reference filtering.
- Disturbance rejection using a PD controller.
- Setpoint tracking and disturbance rejection using a 2-DoF control structure with PID control and reference filter.
- Stability analysis using root locus, Nyquist criterion and stability margins.
- Other control design approaches.

9.6.1.2 Summary of Fundamental Theory
The simplified transfer function of the DC motor was obtained in Sect. 5.6 as:

$$G(s) = \frac{\Theta(s)}{E_a(s)} = \frac{k}{s(\tau s + 1)} \, ,$$

with

$$k = \frac{K_t}{(R_a b_l + K_t K_b)} \quad \text{and} \quad \tau = \frac{R_a J_l}{(R_a b_l + K_t K_b)} \, .$$

It is a second-order transfer function with a real pole at $s = -1/\tau$ and a pole at the origin $s = 0$. Therefore, the system is type-1 and closed-loop steady-state error to a step reference input is zero, even if the controller does not incorporate integral effect. That is not the case if input disturbances are considered, in such case the controller should have to include integral effect to completely eliminate the effect of the disturbance in steady state.

If a PD controller is used, both the pole-cancellation and pole-placement methods can be used. If an ideal PD formulation is used:

$$G_{cl}(s) = \frac{\Theta(s)}{R(s)} = \frac{K(T_d s + 1)\frac{k}{(\tau s + 1)s}}{1 + K(T_d s + 1)\frac{k}{(\tau s + 1)s}} \, . \tag{9.26}$$

If it is selected $T_d = \tau$, then pole-cancellation method applies, if $T_d \neq \tau$ then pole placement can be performed. In pole cancellation:

$$G_{cl}(s) = \frac{1}{\frac{1}{Kk}s + 1} \, ,$$

so that the steady-state error to a step reference input is zero (as the consequence of the plant having an integrator, so that the static gain of the closed-loop system is one) and the closed-loop time constant is $\tau_{cl} = \lambda = 1/(Kk)$, so that increasing the gain of the controller, the response will be faster. The controller parameters are thus $T_d = \tau$ and $K = 1/(\lambda k)$, where λ is a specification.

In the case of pole placement:

$$G_{cl}(s) = \frac{\frac{Kk}{\tau}(T_d s + 1)}{s^2 + \frac{(1 + Kk)}{\tau}s + \frac{Kk}{\tau}} \, , \tag{9.27}$$

again the static gain of the closed-loop transfer function is one (zero error to a step reference) and the denominator can be equated to a characteristic polynomial obtained from the time-domain specifications: $s^2 + 2\zeta_{cl}\omega_{n_{cl}}s + \omega_{n_{cl}}^2$, where ζ_{cl} is obtained

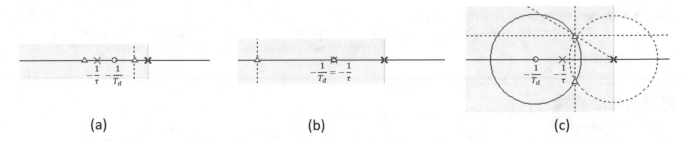

Fig. 9.5 Root locus depending on the relative position of the zero of the controller: **a** $T_d > \tau$, **b** $T_d = \tau$ (pole cancellation), **c** $T_d < \tau$

from the specification of peak overshoot and $\omega_{n_{cl}}$ from any of the characteristic times (rise, peak or settling time). After the design, it has to be checked that the closed-loop zero $(T_d s + 1)$ has not a dominant effect in the response (as if it is not the case the time response will not fulfill the specifications). At conceptual level, a filter in the reference $F(s) = 1/(T_d s + 1)$ could be included to fulfill the specifications, but the interactive tool included in this section does not include reference filtering capabilities.

The previous reasoning can be complemented with root locus analysis (using the root_locus interactive tool, card 7.2), as depending on the relative position of the pole $s = -1/\tau$ and the zero $s = -1/T_d$, different layouts for the position of the closed-loop poles can be obtained (Fig. 9.5).

In practice, PID control is used instead of PD control to face motor nonlinearities (dead-zone) or for disturbance rejection purposes. Notice that using a PD controller, if a step load disturbance is introduced at the motor input:

$$G_d(s) = \frac{\Theta(s)}{D(s)} = \frac{G(s)}{1 + C(s)G(s)} = \frac{\frac{k}{\tau}}{s^2 + \frac{(1 + KkT_d)}{\tau}s + \frac{Kk}{\tau}}, \tag{9.28}$$

then, the steady-state error is not zero (even the system is type-1), as can be analyzed using the final value theorem of the Laplace transform:

$$\lim_{t \to \infty} \theta(t) = \lim_{s \to 0} s\Theta(s) = \lim_{s \to 0} s G_d(s)D(s) = \lim_{s \to 0} s G_d(s)\frac{1}{s} = G_d(0) = \frac{1}{K} \neq 0,$$

so that, a ideal or non-interacting PID controller should have to be included for disturbance rejection purposes. In that case:

$$G_d(s) = \frac{\Theta(s)}{D(s)} = \frac{G(s)}{1 + C(s)G(s)} = \frac{\frac{ks}{\tau}}{s^3 + \frac{(1 + KkT_d)}{\tau}s^2 + \frac{Kk}{\tau}s + \frac{Kk}{T_i \tau}}, \tag{9.29}$$

and the steady-state error for a step input disturbance will be zero from the application of the final value theorem as the numerator of $G_d(s)$ has a term s that makes the limit equal zero. As in the previous case, the values of the controller parameters can be found by pole placement, using a characteristic polynomial from the peak overshoot a characteristic closed-loop time given by $s^2 + 2\zeta_{cl}\omega_{n_{cl}}s + \omega_{n_{cl}}^2$, multiplied by a non-dominant pole $(s - p)$, with $p < -\sigma_{cl} = -\zeta_{cl}\omega_{n_{cl}}$. A reference filter has to be added in this case as the closed-loop transfer function is given by:

$$G_{cl}(s) = \frac{C(s)G(s)}{1 + C(s)G(s)} = \frac{\frac{Kk}{T_i \tau}\left(T_i T_d s^2 + T_i s + 1\right)}{s^3 + \frac{(1 + KkT_d)}{\tau}s^2 + \frac{Kk}{\tau}s + \frac{Kk}{T_i \tau}}, \tag{9.30}$$

so that the filter has the transfer function is:

$$F(s) = \frac{1}{\left(T_i T_d s^2 + T_i s + 1\right)}. \tag{9.31}$$

Notice that the tool focuses on time-domain design, but also frequency-domain design methods can be used. These will be included in the Homework section.

9.6.1.3 References Related to this Concept

- [5] Ogata, K. (2010). *Modern control engineering* (5th ed.). Prentice Hall, ISBN: 978-0-13-615673-4. Chapter 3, Example A-3-9, pages 95–97.
- [17] Dorf, R. C., & Bishop, R. H. (2011). *Modern control systems* (12th ed.). Prentice Hall, ISBN: 978-0-13-602458-3. Chapter 2, Example 2.5, pages 70–74.

Application   Interactive tool: DC_motor_control

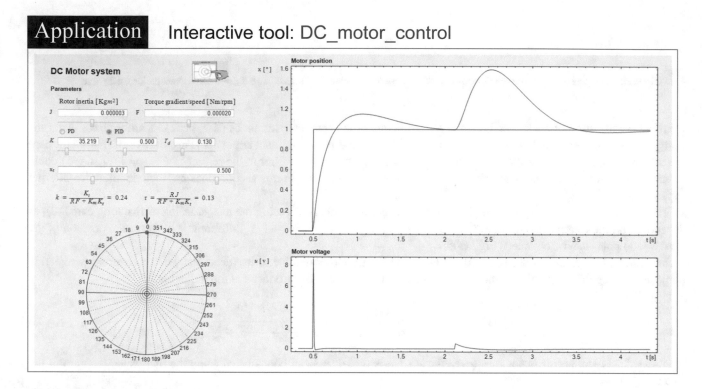

DC Motor: Closed-Loop Control

In the application described in the interactive tool it is possible to simulate the closed-loop control of the angular position of the DC motor. Ideal (non-interacting) PI and PID control are used in the face of changes in the angular reference and load disturbance (input voltage disturbance). Both transient and steady-state responses can be analyzed.

The **Parameters** area of this tool includes the possibility of changing the mechanical parameters of the DC motor (those with more influence on the fundamental time constant of the system): The moment of inertia J_l and the viscous-friction coefficient b_l. When these parameters are changed, the symbolic representation of k and τ also does, displaying the corresponding values. The user can select both ideal PD and PID controllers, which parameters can be included by textboxes and sliders and also the angular position reference θ_r and load disturbance values can be set in this area. Due to the difficulty of simulating a change in the angular position reference, this is fixed by a red arrow and the circular diagram rotates so that the green point has to be aligned to the red arrow. The same applies when a disturbance is introduced. Notice that the green point can be dragged with the mouse to produce those effects. The time evolution is shown in the right plots representing the motor angular position and the input voltage.

9.6.1.4 Homework

1. Consider proportional control of DC motor angular velocity. Indicate the expected steady-state errors to step, ramp and parabola inputs. Repeat the analysis considering the control of angular position. The interactive tool steady_state (card 8.2) can be used for that analysis, considering any value of k and τ in the model.
2. Consider proportional control of DC motor angular velocity. Analyze stability and performance using root locus analysis. Repeat the analysis considering the control of angular position. The interactive tool root_locus (card 7.2) can be used for that analysis, considering any value of k and τ in the model. Make an analysis on how the root locus (and thus stability and

performance) changes when adding the following controllers (selecting appropriate parameters): PD, PI, PID, phase-lead and phase-lag. Examples can be found in [18], Sect. 5.2.

3. Repeat the previous exercise, but use frequency-domain analysis based on stability_margins (card 7.4) and Nyquist_criterion (card 7.3) interactive tools. Examples can be found in [18], Sect. 6.7.

4. Exercise 3.18 from [19]. Consider the DC-motor described by its representative ODE $\tau\ddot{\theta}(t) + \dot{\theta}(t) = e_a(t)$ controlled in angular position by a proportional controller that acts with delay, so that $e_a(t) = K(r(t - t_d) - y(t - t_d))$, where τ and t_d are positive constants. K is slowly increased until the system oscillates with the angular frequency $\omega = 1$ rad/s. K is then set to 33% of this value. After a while, the system starts to oscillate again, now with the angular frequency $\omega = 0.5$ rad/s. This is due to the fact that the time delay t_d has changed to t_{d1}. Can the parameters τ, t_d and t_{d1} being estimated from these data? If so, determine them. This exercise can be solved analytically by using the definition of phase margin and the results can be checked using the interactive tool limitations_delay in Sect. 7.5.

5. Select the mechanical parameters of the DC motor (moment of inertia J_l and viscous-friction coefficient b_l) so that k and τ can be obtained from the tool. Select a PD controller and make $T_d = 0$, so that a proportional controller is obtained. Perform a step of 90° in θ_r and reason about transient response and steady-state error. Take into account possible saturations in the control signal. The use of the root_locus interactive tool is suggested for analyzing the transient, while the steady_state for the steady state. Obtain expressions for the undamped natural frequency and relative damping ratio. Give an expression for the controller gain in terms of the system parameters such that the closed-loop system is critically damped [20]. Now, perform a unit step in the load disturbance d. Is there steady-state error? Justify the answer. Calculate analytically and using the interactive tool the gain K of the feedback controller such that the phase margin is 70°. What is the value of the gain crossover frequency? And the gain margin?

6. Select the mechanical parameters of the DC motor (moment of inertia J_l and viscous-friction coefficient b_l) so that k and τ can be obtained from the tool. Notice that the design of the PD controller by pole cancellation and pole placement in the previous section is the same that if a PI controller is used with those methods for a first-order system (the integrator in the process to relate angular velocity and angular position can be considered as a part of the controller for reference tracking purposes, not for disturbance rejection):

$$C(s)G(s) = K(T_d s + 1)\frac{k}{s(\tau s + 1)} \text{ can be interpreted as } \frac{K(T_d s + 1)}{s}\frac{k}{\tau s + 1}.$$

Use the interactive tools PI_pole_cancellation and PI_pole_placement to obtain K and T_d to fulfill the specifications selected by the user. Analyze the results to a step reference input and a step disturbance input.

7. Add integral action in the previous design (selecting a PID controller with the same gain and derivative time) and modify T_i to achieve an adequate tradeoff between setpoint tracking and disturbance rejection.

8. For the default settings when starting the tool, design a PD or PID controller for the DC servo that satisfies as specifications a phase margin of $PM = 70°$ and gain crossover frequency $\omega_{gc} = 5$ rad/s. The interactive tools PID_concept (card 8.4) or those explained in Sect. 8.11 can be used. Test the obtained controller in the tool.

9. For the angular position control problem, design a lead or lag network so that the steady-state error to a unit ramp reference is less than 10% and the phase margin is $PM \geq 45°$. Use the interactive tools f_design_lag (card 8.8) or f_design_lead (card 8.9) to find the controller parameters. This design cannot be implemented in the tool as it is oriented to PD and PID design, but the obtained parameters can serve as a guide for solving the following exercise.

10. Using the interactive tool LCSD (card 8.11.2), design a PD or a PID controller to fulfill the specifications given in the previous exercise. Obtain the controller parameters in form of ideal PD or PID controller and implement them in the interactive tool analyzed in this section, comparing the results to those from time-domain design specifications. Relate time-domain to frequency-domain specifications.

11. The tool uses the simplified transfer function (5.19) obtained assuming that the armature inductance L_a is generally small and can be neglected. If this is not the case, the transfer function is given by a third-order system with an additional pole (Eq. (5.18)). Reason, using root locus, which is the main difference in the closed-loop performance and stability when considering the DC motor has a third-order transfer function.

12. If in the ball & beam system the dynamics of the motor are taken into account ([18], pp. 830) two transfer functions in series are obtained:

$$\frac{X(s)}{\Theta(s)} = \frac{k}{s^2}; \frac{\Theta(s)}{E_a(s)} = \frac{k'}{\tau s + 1}.$$

By root locus analysis using the interactive tool root_locus (card 7.2), reason on which control strategy is appropriate to control this system. Does it makes sense to control it using a multi-loop control approach where a controller (master) provides the desired beam angle θ to an inner loop where such angle is controlled by the input voltage e_a? Write the block diagram of this system. This is a *cascade control* approach extensively used in industry to linearize actuator dynamics and to reject load disturbances. This control approach has not been treated in this book, but the interested reader can find an excellent description in [8, 21].

References

1. Bavafa-Toosi, Y. (2017). *Introduction to linear control systems*. Academic Press-Elsevier.
2. Åström, K.J., & Murray, R.M. (2014). *Feedback systems: An introduction for scientists and engineers* (2nd ed.). Princeton University Press.
3. Golnaraghi, F., & Kuo, B. C. (2017). *Automatic control systems* (10th ed.). McGraw Hill Education.
4. Johansson, K. H. (2000). The quadruple-tank process: A multivariable laboratory process with an adjustable zero. *IEEE Transactions on Control Systems Technology, 8*(3), 456–465.
5. Ogata, K. (2010). *Modern control engineering* (5th ed.). Prentice Hall.
6. Åström, K. J., & Wittenmark, B. (1989). *Adaptive control*. Addison-Wesley Publishing Company.
7. Tavakolpour-Saleh, A., Setoodeh, A., & Ansari, E. (2016). Iterative learning control of two coupled nonlinear spherical tanks. *International Journal of Mechanical and Mechatronics Engineering, World Academy of Science, Engineering and Technology, 10*(11), 1862–1869.
8. Åström, K. J., & Hägglund, T. (2006). *Advanced PID control*. ISA—The Instrumentation Systems and Automation Society.
9. Hirsch, R. (2008). EDUMECH—Mechatronic Instructional Systems—Ball on Beam System. Shandor Motion Systems. Retrieved July 01, 2021, from https://cutt.ly/KUogfV1.
10. Shahian, B., & Hassul, M. (1993). *Control system design using MATLAB®*. Prentice Hall.
11. Wellstead, P. E., Chrimes, V., Fletcher, P. R., Moody, R., & Robins, A. J. (1989). Ball and beam control experiment. *The International Journal of Electrical Engineering & Education, 15*, 21–39.
12. Díaz, J. M., Costa-Castelló, R., & Dormido, S. (2021). An interactive approach to control systems analysis and design by the root locus technique. *Revista Iberoamericana de Automática e Informática Industrial, 18*(2), 172–188.
13. Beard, R. W., McLain, T. W., Peterson, C., & Killpack, M. (2017). *Introduction to feedback control using design studies*. Independently Published.
14. Qian, D., Yi, J., & Tong, S. (2013). Understanding neighborhood of linearization in undergraduate control education. *IEEE Control Systems Magazine, 33*(4), 54–60.
15. Aracil, J., & Gordillo, F. (2005). The inverted pendulum: A challenge for nonlinear control. *Revista Iberoamericana de Automática e Informática Industrial, 2*(2), 8–19.
16. Åström, K. J. (2004). *Introduction to control*. Department of Automatic Control, Lund Institute of Technology, Lund University.
17. Dorf, R. C., & Bishop, R. H. (2011). *Modern control systems* (12th ed.). Prentice Hall.
18. Hernández-Guzmán, V. M., & Silva-Ortigoza, R. (2019). *Automatic control with experiments*. Springer.
19. KTH—Royal Institute of Technology and Linköpings Universitet, Sweden. (2016). Reglerteknik ak med utvalda tentamenstal (automatic control exercises: Computer exercises, laboratory exercises). Retrieved July 01, 2021, from https://cutt.ly/jYkcFZV.
20. de Silva, C. W. (2009). *Modeling and control of engineering systems*. CRC Press, Taylor & Francis Group.
21. Seborg, D. E., Edgar, T. F., Mellichamp, D. A., & Doyle III, F. J. (2011). *Process dynamics and control* (3rd ed.). International Student Version. Wiley.

10.1   Problem 1. Stability Analysis

Select positive values of parameters β, τ_1, and τ_2 [s] in the following transfer functions:

$$L_1(s) = K\frac{(\beta s + 1)}{s(\tau_1 s + 1)(\tau_2 s + 1)} \; ; \quad L_2(s) = K\frac{(\beta s + 1)}{s^3} .$$

1. Analyze the root locus when the gain K varies between 0 and ∞. Indicate the values of the gain for which the closed-loop system is unstable. Confirm the expected results from theory using the interactive tool root_locus (card 7.2). For $L_2(s)$, propose a controller that stabilizes the closed-loop system.
2. Analyze the stability as a function of K making use of the Nyquist stability criterion. Confirm the results using the interactive tool Nyquist_criterion (card 7.3).
3. Using the interactive tool stability_margins (card 7.4) and for a value of $K = 1$, determine the PM and the GM of the feedback systems with loop transfer functions $L_1(s)$ and $L_2(s)$. Indicate the value of K for which these systems would become unstable and compare it with the results of exercises 1 and 2. For $L_2(s)$, propose a controller designed in the frequency domain that stabilizes the closed-loop system.
4. For a value of K providing a stable closed loop, analyze how much time delay can be added to $L_1(s)$ until the closed loop becomes unstable.
5. Supplementary material: Closed-loop stability using the Routh–Hurwitz criterion.

10.1.1   Solution

As an example, this exercise is solved using $\beta = 0.1429$, $\tau_1 = 0.125$, and $\tau_2 = 0.333$, providing

$$L_1(s) = K\frac{(0.1429\,ss + 1)}{s(0.125s + 1)(0.333\,s + 1)} = 3.435\,K\frac{(s+7)}{s(s+8)(s+3)},$$
$$L_2(s) = \frac{K(0.1429\,ss + 1)}{s^3} = 0.1429K\frac{(s+7)}{s^3} .$$

Analysis for $L_1(s)$

1. The system has a zero at $s = -7$ and three poles at $s = 0$, $s = -3$, and $s = -8$. Therefore, the locus on the real axis belongs to the intervals $(-8, -7) \cup (-3, 0)$, as when dealing with positive values of K the locus on the real axis leaves an odd number of real poles and zeros to its right. There are three poles ($n = 3$) and one zero ($m = 1$) that determine $n - m = 2$ asymptotes at $90°$ and $270°$. These asymptotes intersect at the centroid:

$$\eta = \frac{0 - 3 - 8 - (-7)}{3 - 1} = -2.$$

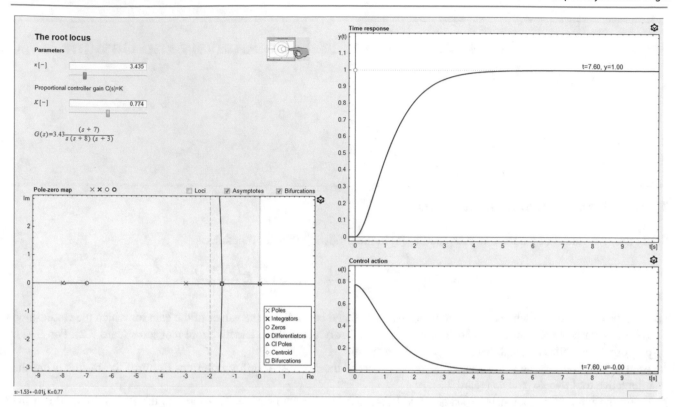

Fig. 10.1 Problem 1: Root locus of $L_1(s)$ using the interactive tool root_locus

The system clearly has an exit point from the real axis as there is a root locus between the poles at $s = 0$ and $s = -3$. Bifurcations are characterized by multiple roots of the characteristic equation (a multiple root cancels both the equation and its derivative). Therefore, it is calculated by the following procedure:

$$1 + KG(s) = 0 \rightarrow K = -\frac{1}{G(s)} = \frac{\tau_1 \tau_2 s^3 + (\tau_1 + \tau_2)s^2 + s}{\beta s + 1}; \quad \frac{dK}{ds} = 0 \rightarrow$$

$$\rightarrow -[(3\tau_1 \tau_2 s^2 + 2(\tau_1 + \tau_2)s + 1)(\beta s + 1) - \beta(\tau_1 \tau_2 s^3 + (\tau_1 + \tau_2)s^2 + s)] = 0.$$

By distributing and rearranging the last expression:

$$2\beta \tau_1 \tau_2 s^3 + [\beta(\tau_1 + \tau_2) + 3\tau_1 \tau_2]s^2 + 2(\tau_1 + \tau_2)s + 1 = 0.$$

The solutions to this equation are $s_1 = -1.532$ and $s_{2,3} = -7.238 \pm j1.567$, so that the exit point is $s = -1.532$. The value of the gain K at that point is obtained by substituting $s = -1.532$ in the characteristic equation, that is,

$$1 + KG(s = -1.532) = 0 \rightarrow K = 0.7745.$$

The value of the critical gain at which the closed-loop system becomes unstable corresponds to the cut with the imaginary axis. As the asymptotes are at $90°$ and $270°$ and meet at a negative value of s, the closed-loop system is stable for any value of K. However, at very large gains the system will exhibit many oscillations, since the poles of the closed loop will have a very large imaginary part (possibly saturating physical actuators in a practical implementation). This can be analyzed using the interactive tool root_locus (Sect. 7.2), where the value of $K = 0.7745$ has been selected to obtain closed-loop poles with $\zeta = 1$ and $\omega_n = 1.532$ rad/s (Fig. 10.1).

Notice that a different election of the parameters of $L_1(s)$ could provide a different root locus due to the relative position among poles and zeros. For instance, for $\beta = 0.111$, $\tau_1 = 1$, and $\tau_2 = 0.1429$ s (let's call this loop transfer function $L_1^\#(s)$), the centroid is in

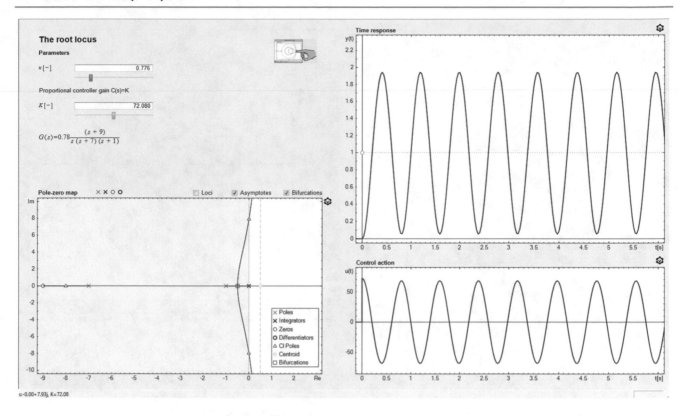

Fig. 10.2 Problem 1: Root locus of $L_1^{\#}(s)$ using the interactive tool root_locus

$$\eta = \frac{0 - 7 - 1 - (-9)}{3 - 1} = 0.5,$$

which indicates that the root locus is going to intersect with the imaginary axis. Substituting $s = j\omega$ in the characteristic equation and equaling to zero both the real part and the imaginary parts:

$$1 + KG(j\omega) = 0 \rightarrow 1 + K\frac{0.7775(j\omega + 9)}{j\omega(j\omega + 1)(j\omega + 7)} = 0,$$

$$-j\omega^3 - 8\omega^2 + (7\omega + 0.7775K)j\omega + 6.9972\,KK = 0,$$

$$\text{Re}: -8\omega^2 + 6.9972\,KK = 0; \quad \text{Im}: (7 + 0.7775K)\omega - \omega^3 = 0.$$

Solving the system of equations, it is obtained as $\omega = (0, -7.94, 7.94)$, so that the one used for computing the critical gain is 7.94 rad/s, providing $K = 72.08$. The results can also be verified using the interactive tool root_locus, as done in Fig. 10.2.

2. Before analyzing closed-loop stability for $L_1(s)$ using the Nyquist criterion, it is expected that $Z = P + N = 0$, as from the root locus it has been verified that the closed-loop system is stable for any value of K. Recall that N is the number of encirclements of Γ_L around the critical point $-1 + j0$ in the $L(s)$ plane and P the number of poles of the open loop in the RHP (P = 0 in this example). A contour in s must be selected that encloses the entire right-half s-plane, avoiding the passage through the integrator with a small semicircle of radius $\varepsilon \rightarrow 0$, as shown in the next figure (representing the s-plane). Substituting $s = j\omega$ in $L_1(s)$, the real and imaginary parts are obtained:

$$\text{Re} = \frac{K(\beta - \tau_1 - \tau 2 - \beta\tau_1\tau_2\omega^2)}{\tau_1^2\tau_2^2\omega^4 + (\tau_1^2 + \tau_2^3)\omega^2 + 1} \ ; \ \text{Im} = \frac{K(-1 + \omega^2(\tau_1\tau_2 - \beta(\tau_1 + \tau_2)))}{\tau_1^2\tau_2^2\omega^4 + (\tau_1^2 + \tau_2^3)\omega^2 + 1},$$

where points C and D on the imaginary axis are those corresponding to the usual Nyquist diagram. These points are mapped to the $L(s)$ plane:

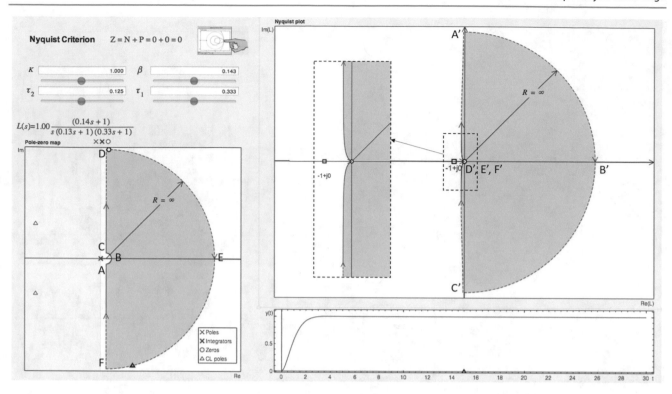

Fig. 10.3 Problem 1: Nyquist contour and stability analysis using the interactive tool Nyquist_criterion

$$C' : \omega = 0^+, \quad \text{Re} = -0.315K, \quad \text{Im} = -K/0 = -\infty,$$
$$D' : \omega \to \infty^+, \qquad \text{Re} = 0, \qquad \text{Im} = 0,$$
$$\text{Sample point } \omega = 1 \text{ rad/s}, \quad \text{Re} = -0.285K, \quad \text{Im} = -0.907K.$$

The points A and F are symmetric to C and D respectively and E is mapped to $(0, 0)$. Point B in the s-plane is placed at $s = \varepsilon e^{j0}$, therefore it is mapped to B': $\lim_{\varepsilon \to 0} L_1(s) = \lim_{\varepsilon \to 0} \frac{K}{\varepsilon} = \infty$.

This can be verified using the interactive tool Nyquist_criterion (Sect. 7.3), selecting $P_{16}(s)$ as a plant structure. It can be observed in Fig. 10.3 that $N = 0$ and $Z = 0$, so that the closed-loop system is stable for any value of K.

For the modified $L_1^{\#}(s)$ with critical gain $K = 72.08$, it can be observed using the same tool in Fig. 10.4, that for that value the Nyquist diagram crosses the critical point $-1 + j0$ (the scale for representing $R \to \infty$ makes it difficult to completely see the crossing, even when zooming the window). The reader can verify that by substituting $s = j\omega$ in the characteristic polynomial and making both the real and imaginary parts equal to zero, it is obtained that for $\omega = 7.94$ rad/s the system becomes critically stable, providing a value of $K = 72.08$ as expected. For $K > 72.08$, $Z = N + P = 2 + 0$ as the contour in the $L(s)$ plane turns clockwise around the critical point two times.

3. Figure 10.5 (interactive tool stability_margins, Sect. 7.4) shows the PM and GM for a value of $K = 1$. It can be seen that for a small gain, the system has a good phase margin.

In the Bode diagram, it is observed that as K increases, the gain crossover frequency increases, the phase approaches $-180°$, and the PM approaches $0°$. On the other hand, the phase never cuts the $-180°$ axis, so the GM is infinite. It can be verified, by increasing the gain, that the system never becomes unstable, although it does present a transient with too many oscillations, as expected from the root locus (in a real process the physical actuators would saturate). Analytically, applying the definition of PM:

$$|L(j\omega_{gc})| = \left| \frac{j0.1429\omega_{gc} + 1}{j\omega_{gc}(j0.125\omega_{gc} + 1)(j0.333\omega_{gc} + 1)} \right| = 1,$$

$$0.017\omega_{gc}^6 + 0.1267\omega_{gc}^4 + 0.9796\omega_{gc}^2 - 1 = 0 \to \omega_{gc} = 0.956 \text{ rad/s}.$$

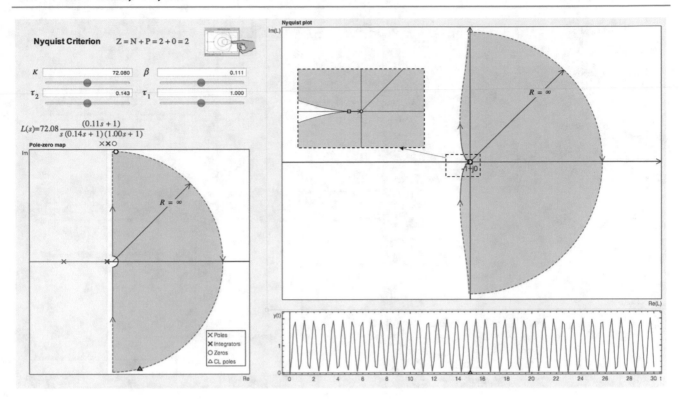

Fig. 10.4   Problem 1: Stability analysis of a conditionally stable transfer function $L_1^{\#}(s)$ for $K = 72.1$ using the interactive tool Nyquist_criterion

$$\lfloor L(j\omega_{gc} = \arctan(0.1429\omega_{gc}) - 90° - \arctan(0.125\omega_{gc}) - \arctan(0.333\omega_{gc}) =$$
$$= -106.709° \rightarrow PM = 73.29°.$$

For $L_1^{\#}(s)$, by applying the conditions of GM:

$$\lfloor L_1^{\#}(j\omega_{pc}) = \lfloor K \frac{0.782(j\omega_{pc} + 9)}{j\omega_{pc}(j\omega_{pc} + 1)(j\omega_{pc} + 7)} = -\pi,$$

and this provides $\omega_{pc} = 7.94$ rad/s and introducing in $|L_1^{\#}(j\omega_{pc})| = |L_1^{\#}(j7.94)| = 0.0139K$. For the closed-loop system to be unstable, $GM < 1$, so that $0.014 < 1 \rightarrow K = 72.08$, which is the same value obtained with other methods. This can be confirmed using the interactive tool stability_margins.

4. To analyze the influence of the time delay, as $L_1(s)$ is stable for any value of K, the case where $K = 1$ will be studied (it provides $PM = 73.3°$ and $GM = \infty$). The closed-loop system without delay is stable. It should be remembered that the delay causes a logarithmic spiral y the Nyquist diagram, surrounding the critical point. Regarding the Bode plots, the time delay $e^{-t_d s}$ does not affect the magnitude plot, but increases the phase lag as frequency does. Therefore, the minimum time delay to make the closed-loop system unstable is the one that at the gain crossover frequency adds a phase equal to the PM of the system without delay. From this critical time delay, the system will become unstable. In the particular case studied, the maximum delay that can be added is the one that adds a phase of $73.3°$ at the frequency 0.956 rad/s, given by $t_d = 73.3° \cdot (\pi/180°)/0.956 \approx 1.34$ s. The same analysis can be done for $L_1^{\#}(s)$.

5. As can be analyzed for instance in [1], the Routh–Hurwitz criterion is based on the closed-loop characteristic equation $J(s) = 1 + L_1(s) = 0$:

$$1 + K \frac{(\beta s + 1)}{s(\tau_1 s + 1)(\tau_2 s + 1)} = 0 \rightarrow (0.0417\, s s^3 + 0.4583\, s s^2 + (1 + 0.1429K)s + K = 0.$$

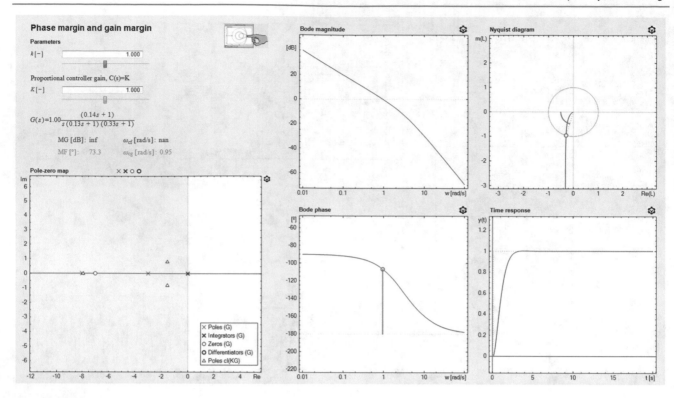

Fig. 10.5 Problem 1: Gain and phase margins for $L_1(s)$ using the interactive tool stability_margins

Table 10.1 Problem 1: Routh array for $L_1(s)$

s^3	a_3	a_1		s^3	0.0417	$1+0.1429K$
s^2	a_2	a_0	$\rightarrow$	s^2	0.4583	K
s^1	b_1	0		s^1	$1+0.0519K$	0
s^0	c_1	0		s^0	K	0

The Routh array is given in Table 10.1 where $b_1 = \frac{(a_2a_1-a_3a_0)}{a_2}$ and $c_1 = a_0$. As it has been assumed that the values of K will always be positive, all the coefficients in the first column of the table are positive, so there is no change of sign and the closed-loop system will always be stable, regardless of the value of K.

Analysis for $L_2(s)$

1. The system has a zero at $s = -7$ and three poles at $s = 0$ (type-3 system). The locus on the real axis is $[-7, 0]$ and as $n = 3$ and $m = 1$ the asymptotes will be 90° and 270°. These asymptotes intersect at the centroid:

$$\eta = \frac{0+0+0-(-7)}{3-1} = 3.5,$$

located at the RHP. This indicates that at least part of the root locus will lie on the RHP providing closed-loop unstable poles. To analyze the exit point from the real axis:

$$1 + KG(s) = 0 \rightarrow K = -\frac{1}{G(s)} = -\frac{s^3}{\beta s + 1},$$

$$\frac{dK}{ds} = 0 \rightarrow -3s^2(\beta s + 1) + \beta s^3 = 0 \rightarrow s_{1,2} = 0, s_3 = -10.497.$$

As s_3 does not belong to the locus, the exit point is $s = 0$ with $K = 0$. The closed-loop system will be unstable for any value of $K > 0$ as there will be two complex conjugate poles in the RHP, as can be seen in Fig. 10.6. Next, a controller

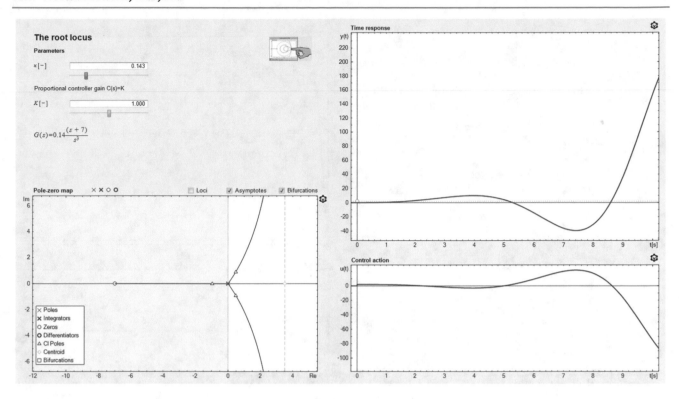

Fig. 10.6 Problem 1: Root locus of $L_2(s)$ using the interactive tool root_locus

must be proposed to stabilize the closed-loop system. This will be done by adding zeros in the loop transfer function in the interactive tool, so that the location of the roots of the closed-loop characteristic equation is modified. However, for open-loop unstable systems, aggressive control actions must be exercised to stabilize them in a closed loop, achieving stability above certain values of K.

For instance, by adding a zero between the integrators and the zero of the plant, a strictly causal system is still treated, but the closed loop becomes stable above a certain value of proportional gain. The closer the aggregated zero is to the imaginary axis, the smaller the gain beyond which the closed-loop system is stable. If another zero is added to the left of the plant zero, the system remains causal but unstable for low values of K in the closed loop. However, it is possible to make the system stable with a lower K value than if a single zero is added as in the previous case. The root locus resulting from the addition of one zero is shown in Fig. 10.7, where a zero has been added at $s = -1$ and $K = 20$. The only way to obtain a stable closed-loop system for all values of K should be to cancel an integrator with a pure derivative in the controller numerator, but this never can be done due to internal stability reasons (see Sect. 8.6).

2. Regarding the analysis of stability using the Nyquist criterion, $P = 0$ as the selected contour in the s-plane does not encircle any unstable pole of $L_2(s)$. Selecting the plant structure $P_{27}(s)$, the contours in the following figure are obtained, indicating that the system is unstable in the closed loop. By analyzing the frequency response in the Nyquist plane:

$$L_2(j\omega) = \frac{K(j\beta\omega + 1)}{(j\omega)^3} = -\frac{K\beta}{\omega^2} + j\frac{K}{\omega^3},$$

$$|L_2(j\omega)| = \sqrt{\frac{K^2\beta^2}{\omega^4} + \frac{K^2}{\omega^9}}; \quad \angle L_2(j\omega) = -270° + \arctan(\beta\omega).$$

The mapping of selected points in the s-plane contour to the $L(s)$ plane can also be seen in the figure. Points A, B, and C are located at $s = \varepsilon e^{j\phi}$, with $\phi = -90°, 0°$, and $90°$, respectively, so that the mapping to the $L_2(s)$ plane is given by

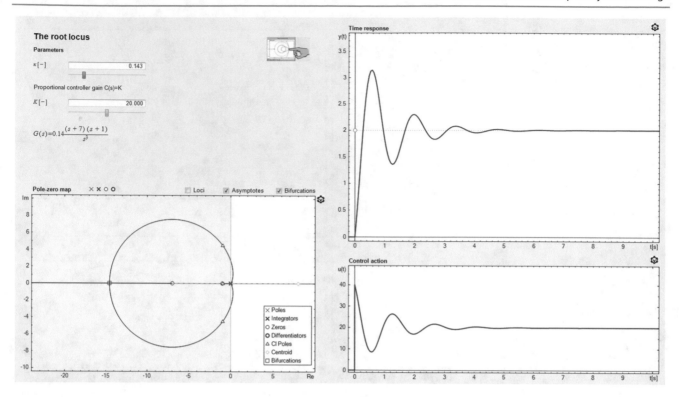

Fig. 10.7 Problem 1: Controller to stabilize $L_2(s)$ adding a zero using the interactive tool root_locus

$$\lim_{\varepsilon \to 0} \frac{K(\beta \varepsilon + 1)}{\varepsilon^3} = \lim_{\varepsilon \to 0} \frac{K}{\varepsilon^3} \cdot e^{-j3\phi},$$

$$C' : s = j0^+, \quad |L_2| = \infty, \ \lfloor L2 = -270°,$$

$$D' : s \to j\infty^+, \quad |L_2| = 0, \ \lfloor L2 = -270° + 90° = -180°.$$

By looking at Fig. 10.8, points A' and F' are symmetric to C' and D', respectively, that is, point F' is located at $L_2(s)$ plane at $(0, 0)$ and A' has a phase of 270°. In addition, point E' is located together with points D' and F'. Finally, point B' is mapped at $+\infty$. Therefore, the diagram makes two clockwise encirclements around the critical point $(-1, 0)$, leaving $Z = N + P = 2 + 0 = 2$ unstable poles in closed loop regardless of the value of K. The clockwise direction of rotation is determined by the phases of the points A', B', and C'. From A' to B', it goes from 270° to 0°, so the angle is decreasing. The same happens for the path from B' to C', whose angle goes from 0° to $-270°$.

3. Figure 10.9 shows the stability margins for $K = 1$. $L_2(s)$ has an infinite gain margin, since its phase never crosses the $-180°$ axis. However, it presents a negative phase margin for any value of K, since it starts with a phase of $-270°$ (three integrators) at low frequencies and ends at $-180°$ at high frequencies in the Bode diagram. From this diagram, it can also be concluded that the only way to make the system stable for some values of K is by adding zeros (differentiators cannot be added due to internal stability reasons).

If a zero is added there will be a frequency interval for which the phase is above $-180°$, providing a positive PM. It can be observed that the Nyquist diagram is always in the second quadrant, and therefore positive phase must be added to achieve a positive PM. In the specific case of $K = 1$, the phase margin is $PM = -81.84°$, with a gain crossover frequency $\omega_{gc} = 1$ rad/s. As the stabilizing controller, a phase-lead controller is used in this example, with the pole far from the origin of the s-plane to achieve a very large gain amplification.

It can be observed in Fig. 10.10 that this controller can stabilize the closed-loop system but at the cost of very high gains (the critical gain is approximately 49), which in practice would not work as the control signals will be much higher than the saturation limits of the actuator (see Fig. 10.11). The interactive tool shows the stability margins of the loop transfer function with the phase-lead network for high values of the gain. It is interesting to note that for a high gain, the system has a positive PM and a negative GM, being stable in a closed loop. This fact has already been commented on in Sect. 7.4. If, on the other hand, the relative stability margins are analyzed for the same loop transfer function (with the same phase-lead

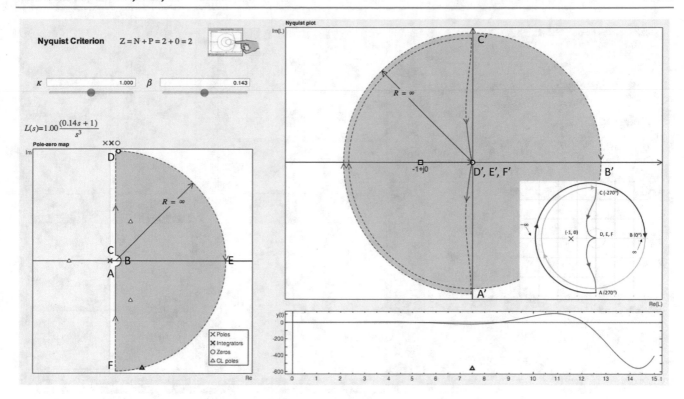

Fig. 10.8 Problem 1: Nyquist stability criterion with $L_2(s)$ using the interactive tool Nyquist_criterion

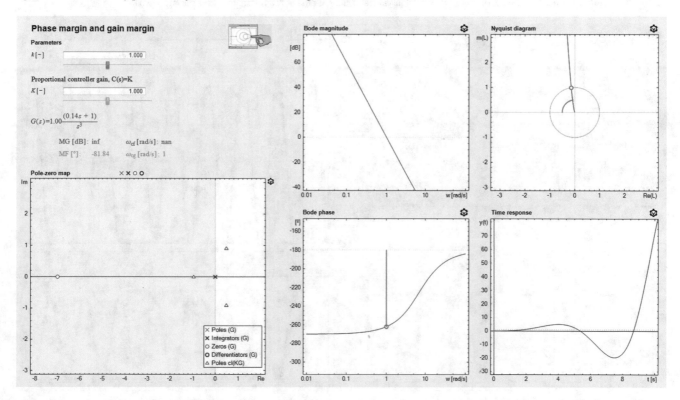

Fig. 10.9 Problem 1: Stability margins of $L_2(s)$ with $K = 1$ using the interactive tool stability_margins

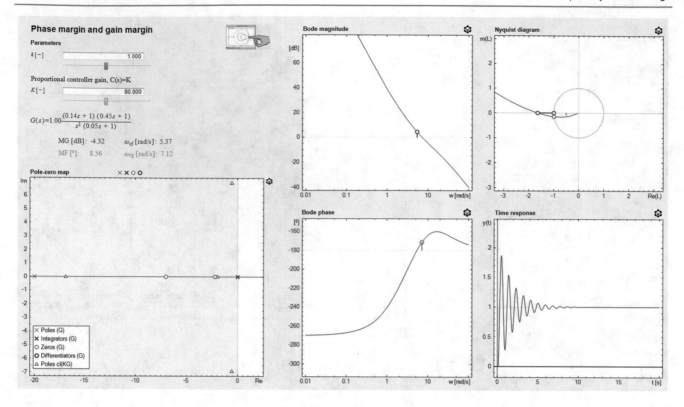

Fig. 10.10   Problem 1: Phase-lead control of $L_2(s)$ with high values of the gain using the interactive tool stability_margins

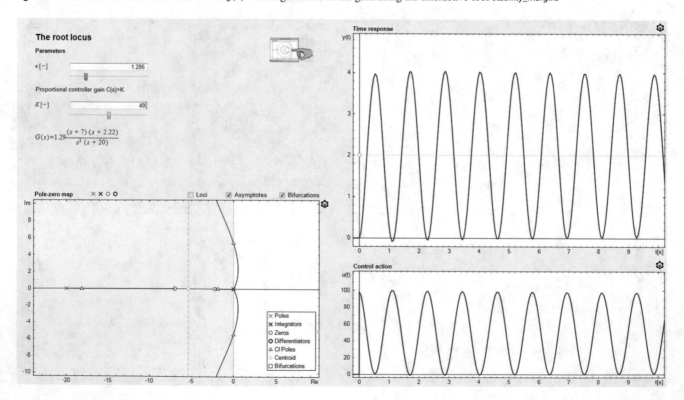

Fig. 10.11   Problem 1: Phase-lead control of $L_2(s)$ with $K = 49$ using the interactive tool root_locus

Table 10.2 Problem 1: Routh array for $L_2(s)$

s^3	1	$0.1429K$
s^2	ε	K
s^1	$-K/\varepsilon$	0
s^0	K	0

network), but with a gain lower than 49, it can be seen that the system has a positive GM but a negative PM and the closed-loop system is unstable.

4. The Routh–Hurwitz array for $L_2(s)$ is formed using the characteristic equation $1 + L_2(s) = s^3 + 0.1429Ks + K = 0$. In Table 10.2, the value associated with the coefficient of s^2 has been substituted by ε close to zero.

   Since the function is being studied for positive values of K, the table will have two sign changes in the first column, which means the existence of two closed-loop poles located in the RHP. This coincides with what was expected when studying the locus of the roots.

10.2 Problem 2. Design in Time Domain

A dynamical system is represented by the following transfer function:

$$G(s) = \frac{1}{(s+1)(s+3)} .$$

Design a PID controller such that the closed-loop system peak percentage overshoot is $OS = 3\%$ and its rise time $t_r = 1.111$ s (other different values can be selected by the reader). Indicate the associated relative damping factor and undamped natural frequency values. Compute the theoretical peak time and settling time values and compare them with those provided by the selected interactive tool. Provide the actual peak overshoot and peak time values obtained. Detail the different options evaluated to achieve a controller that meets the established specifications, justifying, if necessary, the need to use a structure with more degrees of freedom. Comment on the effect on the output of a load disturbance in the form of a unit step at the input of the system. Use appropriate interactive tools to verify the design results.

10.2.1 Solution

From the design specifications, first compute the associated relative damping factor ζ_{cl} and undamped natural frequency $\omega_{n_{cl}}$:

$$OS\,[\%] = 100\exp\left(-\frac{\zeta_{cl}\pi}{\sqrt{1-\zeta_{cl}^2}}\right) \rightarrow \zeta_{cl} = \sqrt{\frac{\ln(0.03)^2}{\pi^2 + \ln(0.03)^2}} = 0.7448,$$

$$\omega_{n_{cl}} = \frac{\pi - \arccos(\zeta_{cl})}{t_r\sqrt{1-\zeta_{cl}^2}} \rightarrow \omega_{n_{cl}} = 3.2519 \text{ rad/s}.$$

With these values, the expected peak time t_p and settling time t_s are

$$t_p = \frac{\pi}{\omega_{n_{cl}}\sqrt{1-\zeta_{cl}^2}} = 1.4478 \text{ s}, \quad t_s = 4/(\zeta_{cl}\omega_{n_{cl}}) = 1.6515 \text{ s}.$$

To design the controller, a combination of the pole-placement and pole-cancellation methods will be applied using an interacting PID controller. The integral time is selected so that it cancels the pole at $s = -1$ ($T_i' = 1$ s), since it is the dominant one, and the proportional and derivative degrees of freedom will be used to fulfill the transient specifications. If a PI is used and a plant pole is not canceled, the result is a third-order system, so the specifications of a second-order system are difficult

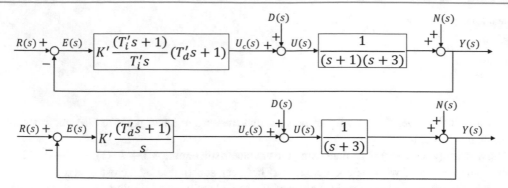

Fig. 10.12   Problem 2: Block diagram of the interacting PID control scheme

to fulfill unless a closed-loop pole is located far from the imaginary axis (non-dominant one). On the other hand, if a PI is used and the dominant pole is canceled, the controller has only one degree of freedom and the two closed-loop specifications cannot be met in general. The block diagram is represented in Fig. 10.12.

After pole–zero cancellation, the closed-loop transfer function is given by

$$G_{cl} = \frac{C(s)G(s)}{1 + C(s)G(s)} = \frac{\frac{K'(s+1)(T'_d s+1)}{s}\frac{1}{(s+1)(s+3)}}{1 + \frac{K'(s+1)(T'_d s+1)}{s}\frac{1}{(s+1)(s+3)}} = \frac{K'(T'_d s + 1)}{s^2 + (3 + K'T'_d)s + K'}.$$

By comparing with the normalized characteristic equation $s^2 + 2\zeta_{cl}\omega_{ncl}s + \omega_{ncl}^2$, the following expression is obtained:

$$K' = \omega_{ncl}^2 = 10.575; \quad 3 + K'T'_d = 2\zeta_{cl}\omega_{ncl} \rightarrow T'_d = 0.1744 \text{ s}.$$

The resulting controller is

$$C(s) = K'\frac{(T'_i s + 1)(T'_d s + 1)}{T'_i s} = 10.575\frac{(s+1)(0.1744\,ss + 1)}{s}.$$

The interactive tool PI_pole_placement (Sect. 8.5) can be used considering that the cancellation has been previously performed since the tool does not allow to introduce more poles. As the interactive tool uses an ideal PI controller, the correspondence with the obtained results is the following: $T_i = T'_d = 0.1744$, $K = K'T'_d \approx 1.8$. It can be seen in Fig. 10.13 that the specifications are not fulfilled in the blue response, having a smaller rise time and larger overshoot than that expected from the specifications (green response). This is due to the fact that the open-loop zero is also a zero in the closed loop, so the response is faster. In order to fulfill the specifications, a filter in the reference $F(s)$ can be added, with transfer function $F(s) = \frac{1}{T_i s + 1}$, by checking the corresponding checkbox in the tool.

The transfer function that relates the load disturbance to the output is given by

$$Y(s)/D(s) = \frac{G(s)}{1 + C(s)G(s)} = \frac{s}{(s+1)(s^2 + 1.8443s + 10.575)},$$

so that it will have the profile of an impulse response due to the existence of a pure derivative in the numerator, as can be analyzed in the figure.

10.3   Problem 3. Design in Frequency Domain

A dynamical system is represented by the following transfer function:

$$G(s) = \frac{6}{s(s + 1)}.$$

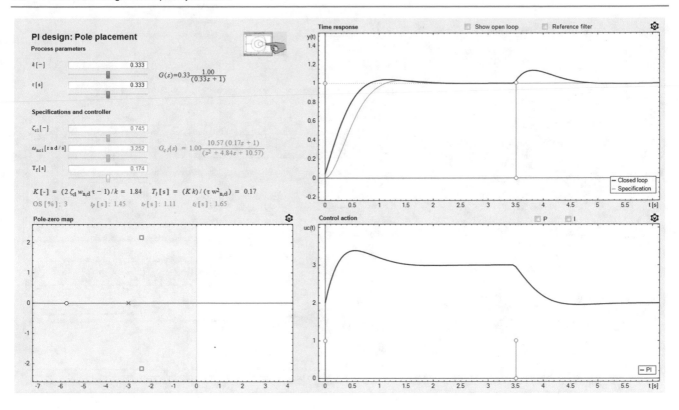

Fig. 10.13 Problem 2: Closed-loop response using the interactive tool PI_pole_placement

1. Design a controller so that the velocity error constant is $K_v = 1 \text{ s}^{-1}$ and $PM \geq 60°$. Analytically calculate the parameters K, τ, and β of a phase-lead or phase-lag compensator. Making use of the appropriate interactive tool, compare the obtained results. Check that for type-1 systems like the one in this exercise, the value of the velocity error constant K_v is the value of the frequency at the intersection between the -20 dB/decade slope asymptote of the integrator and the 0 dB line.

2. Assuming a time delay of $t_d = 0.3$ s is included in $G(s)$, add a new specification to those of the previous exercise so that the system has a gain crossover frequency $\omega_{gc} = 1.3$ rad/s. Obtain the new controller analytically and compare the results with those provided by the tool f_design_lead_lag (Sect. 8.10). There may be no solution, in which case the reason should be indicated.

10.3.1   Solution

1. First compute the proportional gain of the controller to fulfill the steady-state specification, so that $K_v = \lim_{s \to 0} KG(s) = kK \to K = 1/6 = 0.1667$. Now, analyze the PM of $L(s) = KG(s)$ to select the kind of compensator to use. By plotting the Bode diagram of $L(s)$, it can be seen that the system has a $PM = 51.8°$ at $\omega_{gc} = 0.786$ rad/s (this can be seen using the stability_margins interactive tool). A phase-lag controller seems to be an easy way of fulfilling the specifications (by looking at the phase plot) without considerably reducing the bandwidth of the closed-loop system. Using a security margin of 5°, the PM specification is fulfilled if $\lfloor KG(j\omega_{gc1}) = -180° + 60° + 5° = -115°$, so that $\omega_{gc1} \approx 0.47$ rad/s. This can also be obtained analytically:

$$\lfloor KG(j\omega_{gc1}) = -90° - \arctan\left(\frac{\omega_{gc1}}{1}\right) = -115° \to \arctan\left(\frac{\omega_{gc1}}{1}\right) = 25° \to \omega_{gc1} = 0.467 \text{ rad/s}.$$

The gain of $L(s)$ at that frequency provides the gain reduction:

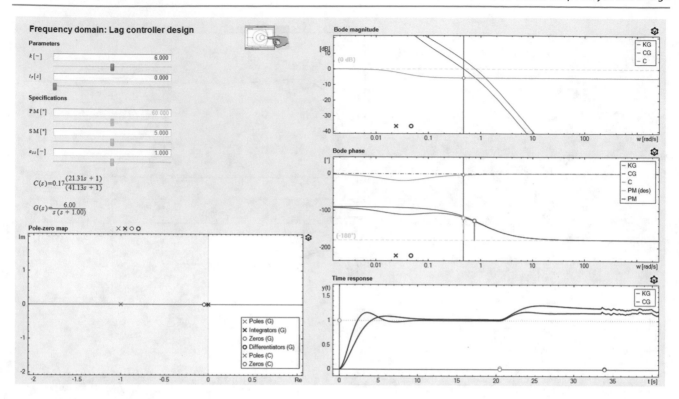

Fig. 10.14 Problem 3: Phase-lag design using the interactive tool f_design_lag

$$GR = -20 \log |L(j\omega_{gc1})| = -5.76 \text{ dB}.$$

Now compute the parameters of the phase-lag network:

$$GR = -20 \log(\alpha) \rightarrow \alpha = 10^{-GR/20} = 10^{-(-5.76/20)} = 1.9409,$$

$$\frac{1}{\beta} = \frac{\omega_{gc1}}{10} \rightarrow \beta = \frac{10}{\omega_{gc1}} = 21.41 \text{ s}.$$

$$\tau = \beta\alpha = 41.56 \text{ s}.$$

The controller is thus given by

$$C(s) = \frac{1}{6} \cdot \frac{21.41s + 1}{41.56s + 1}.$$

This can be analyzed with the interactive tool f_design_lag (Sect. 8.8), where very close results are obtained (not equal due to decimal rounding). It is important to check that $\omega_{gc1} \approx \omega_{gc}$. Figure 10.14 also shows the load disturbance and noise rejection characteristics of this controller. Although the plant is type-1, providing zero error to a step reference, as the controller does not have an integral effect, there is a steady-state error to a step load disturbance.

2. From the previous exercise, $K = 1/6$. By substituting the desired ω_{gc} in $L(j\omega_{gc})$ without taking into account the time delay:

$$L(j\omega_{gc}) = \frac{1}{j\omega_{gc}(j\omega_{gc} + 1)} = -\frac{1}{(\omega_{gc}^2 + 1)} - j\frac{1}{\omega_{gc}(\omega_{gc}^2 + 1)},$$

$$L(j1.3) = -0.3717 - j0.28,$$

$$|K(G(j1.3)| = K\rho = |L(j1.3)| = 0.469,$$

$$\underline{L(j1.3)} = \arctan\left(\frac{-0.286}{-0.3717}\right) = -2.486 \text{ rad } = -142.43°.$$

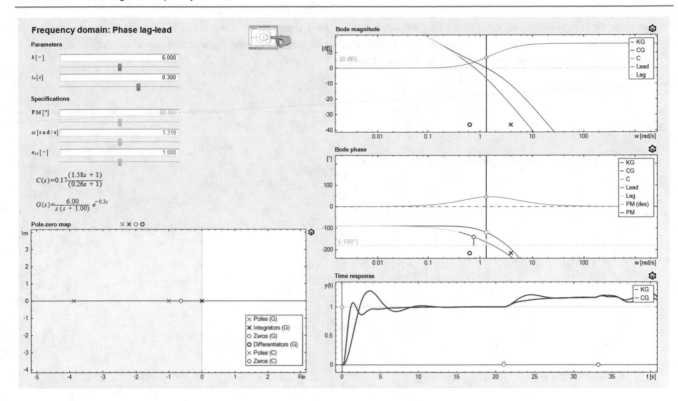

Fig. 10.15 Problem 3: Phase-lead design using the interactive tool f_design_lead_lag

The time delay does not affect the magnitude, but the phase lag by $-\omega_{gc}t_d = -1.3 \cdot 0.3 = -0.39$ rad $= -22.345°$, providing a total phase of $\phi = \lfloor L(j\omega_{gc}) = -142.42° - 22.35° = -164.78° = -1.876$ rad. Then, applying the phase margin definition:

$$C(j\omega_{gc})G(\omega_{gc}) = \frac{j\omega_{gc}\beta + 1}{j\omega_{gc}\tau + 1}K\rho e^{j\phi} = 1 \cdot e^{j(-180° + PM)}$$

where PM is that given by the specifications. Substituting the obtained parameters, a system of equations with two unknowns and two equations are obtained (also Eq. (8.58) can be directly applied):

$$\frac{j1.3\beta + 1}{j1.3\tau + 1} = 2.132 \cdot e^{j0.7816} = 2.132(\cos 0.7816 + j \sin 0.7816) = \underbrace{1.513}_{a} + j\underbrace{1.502}_{b}.$$

By multiplying the left-hand part of the equation by the denominator conjugate and operating:

$$1 + \omega_{gc}^2\tau\beta + j\omega_{gc}(\beta - \tau) = (a + jb)(\tau^2\omega_{gc}^2 + 1),$$

$$\beta = \frac{a(1 + \tau^2\omega_{gc}^2) - 1}{\tau\omega_{gc}^2}; \quad \tau = \frac{a - 1}{b\omega_{gc}},$$

so that $\beta = 1.55$ s and $\tau = 0.26$ s, providing the following controller:

$$C(s) = K\frac{\beta s + 1}{\tau s + 1} = 0.167\frac{1.55\, s + 1}{0.26\, s + 1}.$$

The analytical result can be checked using the interactive tool f_design_lead_lag (again, in Fig. 10.15, there are small differences in β due to the number of decimals used). Response to load disturbance and noise has also been included. The same conclusions as in the previous exercise can be drawn.

Reference

1. Ogata, K. (2010). *Modern control engineering* (5th ed.). Prentice Hall.

Index

© Springer Nature Switzerland AG 2023
J. L. Guzmán et al., *Automatic Control with Interactive Tools*,
https://doi.org/10.1007/978-3-031-09920-5

Printed in the United States
by Baker & Taylor Publisher Services